T0296772

Medical
MODELING

Woodhead Publishing Series in Biomaterials

Medical
MODELING

THE APPLICATION OF ADVANCED DESIGN AND ADDITIVE MANUFACTURING TECHNIQUES IN MEDICINE

THIRD EDITION

RICHARD BIBB

School of Art & Design, Nottingham Trent University, United Kingdom

DOMINIC EGGBEER

Surgical & Prosthetic Design, PDR, Cardiff Metropolitan University, Cardiff, United Kingdom

ABBY PATERSON

School of Design & Creative Arts, Loughborough University, Loughborough, United Kingdom

MAZHER IQBAL MOHAMMED

School of Design & Creative Arts, Loughborough University, Loughborough, United Kingdom

WOODHEAD PUBLISHING

An imprint of Elsevier

Woodhead Publishing is an imprint of Elsevier
50 Hampshire Street, 5th Floor, Cambridge, MA 02139, United States
125 London Wall, London EC2Y 5AS, United Kingdom

ISBN: 978-0-323-95733-5

For information on all Woodhead Publishing publications visit our website at https://www.elsevier.com/books-and-journals

Publisher: Matthew Deans
Acquisitions Editor: Sabrina Webber
Editorial Project Manager: Tom Mearns
Production Project Manager: Kamesh R.
Cover Designer: Vicky Pearson Esser

Typeset by TNQ Technologies

Working together
to grow libraries in
developing countries

www.elsevier.com • www.bookaid.org

Contents

PART 5.1   Implementation

PART 5.6 Research applications

About the authors

Richard Bibb, Associate Dean Research, Nottingham School of Art and Design, Nottingham Trent University, Nottingham, UK

Prof Richard Bibb is an Associate Dean Research and Professor of Medical Applications of Design at Nottingham Trent University, UK. He graduated from Brunel University, UK (1995), with a BSc (Hons) in Industrial Design. He then undertook doctoral research in Rapid Prototyping at the National Centre for Product Design and Development Research (PDR), Cardiff Metropolitan University, UK. This study involved the development of a computerized rapid prototyping selection system for designers in small companies.

After gaining his PhD in 1999, he established the Medical Applications Group at PDR to conduct collaborative applied research in medical applications of design technologies such as CAD and 3D Printing. He rose to the position of Director of Research for PDR before moving to Loughborough University in 2008. In 2014, he established the Digital Design and Fabrication Research Lab (DDF), which focused on advanced computer-aided design (CAD), 3D printing, and additive manufacturing technologies. During 15 years at Loughborough, he rose to the position of a Professor and served terms as Associate Dean Research, Acting Dean, and Director of Research. He moved to Nottingham Trent University in 2023 to take up the position of Associate Dean Research in Nottingham School of Art and Design. Professor Bibb's personal research focus is the application of advanced product design and development technologies in medicine, surgery, rehabilitation, and assistive technology.

Dominic Eggbeer, Professor of Healthcare Applications of Design, PDR, Cardiff Metropolitan University, Wales, UK

Prof. Dominic Eggbeer is a Professor of Healthcare Applications of Design at PDR, Cardiff Metropolitan University. His research focuses on the design and development of personalized medical devices, applying his knowledge to surgical implants, facial prosthetics, dental devices, and other areas of rehabilitative medicine.

In addition to his academic research, he has experience managing a small, ISO 13485 compliant commercial team in the design of patient specific implants and other devices. Eggbeer also has a leading role in collaboration, dissemination and in supporting broad uptake of novel design engineering approaches in healthcare.

Abby Paterson, Senior Lecturer, Digital Design and Fabrication, School of Design and Creative Arts, Loughborough University, Loughborough, UK

Dr. Abby Paterson is a Senior Lecturer in Product Design and Digital Design and Fabrication at Loughborough University, UK. She specializes in 3D scanning, CAD, and automated fabrication (specifically CNC milling and additive manufacture).

Abby graduated with a BSc in Product Design and Technology and a PhD in 3D Scanning, CAD and Additive Manufacture for Medical Applications from Loughborough University. After completing her PhD, she was appointed as a lecturer at the University of Manchester in the School of Materials; she continued her research in digital design and fabrication for medical devices and then returned to Loughborough University as a lecturer in 2014.

In addition to securing funding from the Royal Academy of Engineering, Abby has received funding from Versus Arthritis to develop specialized 3D CAD software for the design of customized 3D-printed wrist splints. Abby also engages with consultancy work through Loughborough University Enterprises Ltd.

Mazher Iqbal Mohammed, Reader of Product Design Engineering, School of Design and Creative Arts, Loughborough University, Loughborough, UK

Dr. Mazher Iqbal Mohammed is a Reader of Product Design Engineering within the School of Design and Creative Arts at Loughborough University. He graduated from Edinburgh University, UK, with an MPhys in General Physics. He then undertook an Engineering Doctorate (EngD) in Medical Device Bioengineering at the Centre for Doctoral Training within the Biomedical Engineering Department at Strathclyde University, UK.

Dr. Mohammed is currently part of the Digital Design and Fabrication Research Group at Loughborough University. He has worked in several prestigious academic institutes across both the United Kingdom and

Australia, including roles at Heriot Watt University, the Australian National University (ANU), and Deakin University. During this time, he has developed strong interdisciplinary skills working in product design, biomedical engineering, advanced manufacturing, material sciences, and biosensing.

Dr. Mohammed's research interests focus on design for health and well-being, working at the intersection of healthcare, sustainability, and equitable design to advance solutions, which positively impact human development and longevity. He is particularly interested in leveraging disruptive technologies such as additive manufacturing, nature-inspired design, and data-driven processes to develop innovative and practical solutions in healthcare, across diagnostic, prosthetics, surgical systems, and rehabilitation technologies.

Preface

The principal aim of this book is to provide a genuinely useful text that can help professionals from a broad range of disciplines to understand how advanced product design and development technologies, techniques, and methods can be employed in a variety of medical applications. The book describes the technologies, methods, and potential complexities of these activities as well as suggesting solutions to some of the commonly encountered problems and highlighting potential benefits. This book is based on the collective experience of the authors spanning more than 25 years of research and practice in medical applications. The majority of the research has been conducted through collaboration with clinical, academic, and industrial partners by the authors and their colleagues at Cardiff Metropolitan University, Loughborough University, Deakin University, and Nottingham Trent University.

The book is presented in two main sections. In the first section, the technical Chapter 1 through Chapter 4 introduce the various technologies involved covering data acquisition, data manipulation and preparation, and the manufacture of physical models, instrument, and products. The second section, Chapter 5, provides a wide variety of case studies that collectively cover the application of most, if not all, of the technologies introduced in the previous chapters. The case studies are collected into subsections addressing implementation, surgical applications, prosthetic applications, orthotic applications, and dental applications. A further subsection contains a variety of interesting research case studies in allied areas such as archaeology and science. To ensure that these case studies are relevant and appropriate, most have been drawn from work previously published in internationally peer-reviewed journals or conference proceedings with full acknowledgement, proper citation, and permission where appropriate. Where appropriate, these papers have been updated to reflect recent technologic advances. Other case studies have been invited from leading researchers.

This text also aims to encourage what is, by its very nature, a multidisciplinary and collaborative endeavor. The selected case studies reflect this by describing a broad range of techniques and applications. Although much work has been done in this area, there is a tendency for people to publish in the journals, language, and context of their own professional practice.

While this text does not purport to be the most comprehensive review of the work done to date, it is a conscious effort to overcome these professional interfaces and encourage multidisciplinary collaboration and indeed interdisciplinary learning by providing a single source of useful reference material accessible to readers from any relevant background.

Therefore, it is hoped that this book will appeal equally to medical and technical specialties including, for example, industrial/product designers, biomedical engineers, clinical engineers, rehabilitation engineers, medical physicists, radiologists, radiographers, surgeons, prosthetists, orthotists, orthodontists, anatomists, medical artists and anthropologists, and perhaps even veterinarians, archaeologists, and paleontologists.

The text provides an excellent resource for postgraduate students, doctoral candidates, and postdoctoral researchers working in this rapidly developing, important, and exciting area.

Richard Bibb
Nottingham
November 2023

Acknowledgements

This book would not have been possible without the help and support of many colleagues in our home institutions and among the many institutions we have been fortunate enough to work with over many years. It is appropriate therefore to offer our thanks to our colleagues at PDR, Cardiff Metropolitan University; Loughborough University; Deakin University; and Nottingham Trent University for their assistance and support.

As the central theme underpinning this book is multidisciplinary collaboration, it is important to recognize the input of all who have contributed to it. We thank all of our collaborators and co-authors without whom none of the work reported in this book would have been possible. Each case study is fully acknowledged, and we would also like to thank the various publishers for their kind permission to reproduce our previous papers and articles. We also thank the invited authors for their contributions and the use of images and those that have supplied technical guidance and models to support our research over the years. Each instance is acknowledged in the relevant place in the text.

We would also like to thank everyone at Woodhead/Elsevier for all their help and professional expertise in turning our manuscript into the book you see here.

**Richard Bibb, Dominic Eggbeer,
Mazher Iqbal Mohammed, Abby Paterson**
2023

CHAPTER 1

Introduction

1.1 Background

The purpose of this book is to describe some of the many possibilities, techniques, and challenges involved when utilizing advanced design and product development technologies in medicine. This ranges from the creation of models of anatomy to the design and manufacture of bespoke custom-fitting and personalized medical devices and prostheses.

The origin of this field lies in medical modeling, sometimes called biomodeling. **Medical modeling** is the term we use to describe the creation of highly accurate physical models of human anatomy directly from medical scan data. The process involves capturing three-dimensional human anatomy data, processing the data to isolate individual tissues, structures, or organs, optimizing the data for the manufacturing technology to be used, and finally, building the model using additive manufacturing (AM) techniques. Additive manufacturing is the general name coined to describe manufacturing processes of joining materials to make parts from 3D model data, usually layer upon layer, using computer-controlled machines that manufacture physical items directly from three-dimensional computer data. This is commonly referred to as 3D printing. Originally, these machines were developed to enable designers and engineers to build prototype models of objects they had designed using computer-aided design (CAD) software. These so-called "rapid" prototypes allowed them to ensure that what they had designed on-screen fitted together with the other components of the product being developed. Therefore, the machines were quickly developed to produce models of high accuracy as rapidly as possible.

In the 1990s, it was realized that rapid prototyping (RP) machines could utilize other types of three-dimensional computer data, such as that

Medical Modeling
ISBN 978-0-323-95733-5
https://doi.org/10.1016/B978-0-323-95733-5.00007-7

obtained from medical scanners. Software was developed to enable the medical scan data to interface with the machines, and medical modeling began. Since then, the field has developed to cover all kinds of applications ranging from archeology to reconstructive surgery. Early success and clear demonstration of benefits has led to widespread interest in the technologies from many medical specialties. However, with each development, more clinicians, surgeons, engineers, and researchers are realizing the potential benefits of AM techniques, which in turn places new challenges on those people whose job it is to produce these models.

This book aims to describe the processes required to produce high-quality medical models and offers an insight into the techniques and technologies that are commonly used. Chapters 2—4 follow the logical sequence of stages in the medical modeling process as shown in Table 1.1. Each chapter describes the technologies and processes used in each stage in general terms for those not familiar with them or new to the field, while the following case studies illustrate diverse applications that have been carried out in recent years. Where appropriate, the case studies include cross-references to related sections of Chapters 2—4 as a reminder, to eliminate repetition, or to enable the reader to begin by reading case studies and then find the relevant technical information easily without necessarily having to read the whole book in chapter order.

Table 1.1 The stages of the medical modeling process.

Step 1
Medical imaging
Select the optimal modality; set the appropriate protocols; scan patient.
Step 2
Export data format
Export the data from the scanner in an appropriate format and transfer data to AM laboratory
Step 3
Working with scan data
Isolate data relating to the tissues or organs to be modeled; design and or adapt the data as required; save and transfer the data in the correct format for the AM process
Step 4
Physical reproduction
Build the model; deliver the model to the clinician, cleaning, finishing, and/or sterilizing as required

By its very nature, medical modeling has brought together the fields of engineering and medicine. Consequently, this book aims to satisfy the needs of both fields as they work together on medical modeling. Therefore, although it is not possible to cover every medical or technical definition here, this chapter offers a brief introduction to some anatomic terminology for the benefit of those new to the field as well as an introduction to some specialist technical terms. Where a longer or more detailed description is required, an explanatory note may be found in Chapter 7, which also contains glossaries of technical and medical terms and abbreviations. There are also recommendations for further reading at the end of chapters to enable those with particular interests to develop their knowledge further.

Although this book is essentially technical in nature, it is important to consider that it also addresses genuine human needs, and consequently, there is consideration for patients and ethics. Therefore, throughout this book, illustrations and case studies have been made anonymous, and where necessary, permission has been granted.

1.2 The human form

The human body is the most significant physical form that we possess or encounter. Our physical form is inextricably bound up with our minds and behavior. It influences but also responds to our lifestyle choices and combined with our character defines us as individuals. Our physical form defines how we appear to others, and it affects our perception of ourselves. It displays our health and fitness and even our attractiveness to our loved ones. It enables us to recognize any one individual among the seven billion fellow humans with which we share the planet.

In addition to its undeniable importance, our physical form is one of the most complex shapes we encounter in life. Its importance to us makes us sensitive to the tiniest of details and the subtlest of contours. This complexity combined with our sensitivity to the human form has provided the preeminent challenge to artists in our history. Through drawing, painting, and sculpture, artists have strived to capture what it is that makes us human and how that is expressed through our physical form and appearance.

In terms of medicine, the human body is both subject and object. The study of the human form is the basis for all medicine as it strives to correct our malfunctions and degradation. It is from these noble aims that we are constantly trying to apply the latest in technology to improve our treatment of all kinds of illness. When such unfortunate things as disease or trauma damage our physical form, it not only physically debilitates us but also affects our psychological health. Therefore, the ability to capture and reproduce human anatomy to the infinite subtlety that we desire is a pursuit that is as important as it is challenging.

The age of computer technology has not necessarily eased this process. The reconstruction and rehabilitation of people can consist of any combination of dressing, rehabilitation, prosthesis, and surgery. Skills employed range from the artistry of the prosthetist to the engineering of artificial implants. Until recent times, reconstruction and rehabilitation has relied almost solely on the dexterity and artistry of a small but highly dedicated range of health professionals. However, in this modern age, the pressures on these people grow as survival rates increase and surgical interventions become ever more sophisticated. It is therefore not surprising that medicine looks toward advanced technologies to provide the effort, time, and cost savings that have been so successfully achieved in product design and engineering.

This text aims to describe some of the product design technologies that have been successfully used in the field of human modeling, reconstruction, and rehabilitation and to illustrate their application through case studies. As we will discover, there are many benefits to be found from applying modern technologies, yet they are not without their obstacles. The nature of the human form makes the transfer of techniques that are well suited to product design and engineering a particularly challenging yet rewarding field of work.

1.3 Basic anatomic terminology

Although this book does not intend to be used as a guide to human anatomy, the descriptions of techniques, medical conditions, and treatments require the use of accepted anatomic nomenclature. For those readers with medical training, this nomenclature will be well known. However, for those from a technical, design, or engineering background, some basic terminology will prove useful. This section will introduce some basic terms that will enable the reader to proceed with the rest of the text, but further reading on anatomy and physiology is strongly recommended. There are many excellent texts on anatomy, and a selection of titles is provided in the bibliography. Attending a short course in human anatomy and physiology would be highly recommended to any engineer or designer wishing to specialize in clinical or medical applications, and many universities offer such courses.

When referring to human anatomy, the relative positions of organs, limbs, and features are only useful if the body is in a known pose. Therefore, it is standard practice to assume the body is in the "anatomic position" when describing relative positions of anatomy. The anatomic position is with the body and limbs straight, feet together, head looking forward, arms at the sides of the torso with the palms facing forward and fingers straight. The principal axis of the human is through the center of the body running from head to feet; this is referred to as the long axis. This is shown in Fig. 1.1.

Once the anatomic position is known, perpendicular planes can divide the body. The plane through the body perpendicular to the long axis is known as the axial plane. The planes perpendicular to this are known as the coronal and sagittal. Directions and distances are described as they relate to the center of the body. Parts nearer to the center of the body are known as

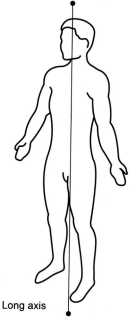

Long axis

Figure 1.1 The anatomic position and the long axis.

proximal, and those furthest away are distal. Parts that are nearer the midline of the body are known as medial, and those further from it are described as lateral. These terms are summarized in Table 1.2 and illustrated in Figs. 1.2 and 1.3.

Table 1.2 Anatomic directional terms.

Term	Definition
Superior	Toward the head (upper)
Inferior	Away from the head (lower)
Anterior	Toward the front of the body
Posterior	Toward the rear of the body
Medial	Nearer the midline of the body
Lateral	Further from the midline of the body
Contralateral	On the opposite side of the body
Ipsilateral	On the same side of the body
Proximal	Relating to limbs, nearer to the body
Distal	Relating to limbs, further from the body
Superficial	Toward the surface of the body
Deep	Into the body, away from the surface

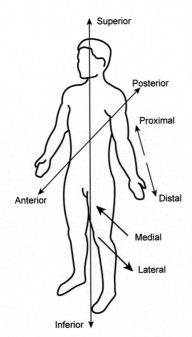

Figure 1.2 The direction terms used in human anatomy.

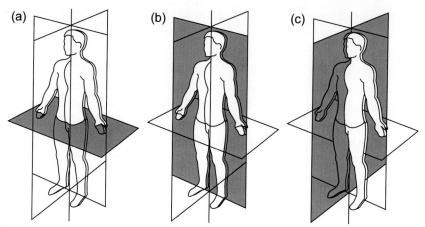

Figure 1.3 The major reference planes used in human anatomy: axial (left), coronal (middle), and sagittal (right).

1.4 Technical terminology

The technical terms used throughout this book are introduced here so that they can subsequently be abbreviated in later chapters. Much of the book is devoted to explaining AM technologies. AM is an umbrella term that evolved from the term "rapid prototyping," which was coined in the 1980s to refer to automated manufacturing technologies that produce physical items in an additive layer-by-layer manner from three-dimensional computer data. In the 1980s and 1990s, these technologies revolutionized the production of prototype parts during new product development that was increasingly being done using CAD. These technologies enabled the direct manufacture of prototype parts that previously were handmade by skilled technicians in a slow and laborious process. Hence, the technologies became known as "rapid" prototyping or RP. In the early 2000s, the layer AM principle was applied to a variety of materials and applications including the manufacture of end-use products, and consequently the phrase "rapid manufacturing" or RM was frequently used.

To reflect an increasing shift toward manufacturing as opposed to prototyping, more recently the umbrella term "additive manufacturing" has become widely accepted and defined in ISO/ASTM 52900:2021.

Additive manufacturing — general principles and terminology: AM is defined as technologies that build physical 3D geometries by successive addition of material (https://www.iso.org/obp/ui/#iso:std:iso-astm:52900: ed-2:v1:en). The mainstream media have adopted the term 3D printing, which is perhaps a more descriptive term for the lay reader. While the term 3D printing is now in common use, it is frequently associated with low-cost machines that may be purchased for as little as few hundred dollars and are increasingly popular in schools, colleges, and as a hobby. However, in this book, we will use the term AM to refer to layer additive technologies and use the terminology found in ISO/ASTM 52900, but case studies may use the terms rapid prototyping and or 3D printing.

The term CAD is now in common use and refers to a vast array of software applications that enable the design and definition of objects. In the simplest terms, CAD could refer to simply drawing on a computer screen. The technologies and processes described in this book rely on three-dimensional data, and consequently, the term 3D CAD will be used, and where the term CAD is used, three-dimensional is inferred. 3D CAD applications work by providing a manner of defining a three-dimensional surface using a computer program. In some versions, the surfaces are generated using mathematical geometry, and others allow a more arbitrary construction or manipulation of the data. Where these differences are significant, they will be explained in the following chapters.

CHAPTER 2

Medical imaging

2.1 Introduction to medical imaging

To manufacture a virtual or physical model of any human anatomy, it must first be captured in three dimensions in a manner that can be utilized by the computer processes to be used. Scanning modalities range from substantial hospital facilities normally found in radiology departments to small hand-held scanners that can be used in laboratories, clinics, or wards.

There are two main categories of scanning modality for human bodies, those that capture data from the whole body both internally and externally and those that capture only external data. Most hospital-based scanners capture data from the whole body both internally and externally. These are normally large, sophisticated medical imaging machines capable of scanning the complete human body. Examples include computed tomography (CT), magnetic resonance imaging (MRI), and positron emission tomography. Each modality uses a different physical effect to generate cross-sectional images through the human body. Typically, the patient is placed lying down on a table that is moved through the scanner while the images are acquired. The cross-sectional images are arranged in order so that the computer can construct a three-dimensional dataset of the patient. Software can then be used to isolate specific organs or tissues. These data can then be used to reconstruct an exact replica of the organ using computer-aided techniques. The different physical effects used by each type of scanner result in different tissues being imaged. These machines require trained and qualified operators to operate and require a large capital investment. The use of the most common modalities will be described in more detail later in this chapter.

The other type of scanning is used to capture only the external surface topography of a patient. A wide range of technologies can be used for capturing three-dimensional surface data. Three-dimensional surface

Medical Modeling
ISBN 978-0-323-95733-5
https://doi.org/10.1016/B978-0-323-95733-5.00012-0

scanning, sometimes referred to as digitizing, has been used in engineering and product design for many years as a method of integrating the surfaces of existing physical objects with computer-aided design (CAD) models. Consequently, this process is often referred to as "reverse engineering." There are many types of surface scanners or digitizers available to the engineer or designer. They can be separated into two main categories, "contact" or "touch probe digitizers" and noncontact scanners. Touch probe digitizers use a pressure-sensitive probe tip and calibrated motion to map out the surface of an object point by point. Digitizers can be operated by hand or using automated coordinate-measuring machines. Depending on the quality manufacture, digitizers can be extremely accurate. However, it is a slow and laborious process, sometimes taking hours to capture the surface of an object. While this is acceptable when scanning inanimate objects, it is clearly not appropriate for capturing the surface of human anatomy. Therefore, noncontact scanners are typically used when capturing surface data from people. Noncontact scanners utilize light and digital camera technologies to capture many thousands of data points on the surface of an object in a matter of seconds. The fast capture of data and the harmless light used make these types of scanners ideal for capturing human anatomy. These scanners are typically like very large cameras and may be tripod mounted or handheld. Despite the variety of surface scanners available on the market, the general principles of their operation and application are the same, and these principles are described later in this chapter.

This chapter is not intended to provide a definitive description of the technology and practice of each scanning modality but will establish criteria and guidelines that may be employed to optimize their use in the production of virtual or physical medical models. Many texts are available that describe each modality fully and some are listed in the recommended reading list at the end of this chapter.

2.2 Computed tomography

2.2.1 Background

The Greek word *tomos* means a slice, section, or cut. *Tomography* translates as the process of imaging a cross-section. Tomography is a topic of its own, which is applied in mathematics, science, engineering, and of course, medical imaging. It is a mathematical problem, where views captured from different angles around an object can be used to build a picture of that object. Increasing the number of views captured increases the ability to define the details of the object. The mathematics and fundamentals behind tomography are described in much more detail in radiography and medical imaging textbooks (for example Ref. [1,2]).

CT works by passing focused X-rays through the body and measuring the amount of the X-ray energy absorbed. The amount of X-ray energy absorbed by a known slice thickness is proportional to the density of the tissue. By taking many such measurements from many angles, the tissue densities can be composed as a cross-sectional image using a computer. The computer generates a grayscale image where the tissue density is indicated by shades of gray. The Hounsfield scale is a quantitative scale for describing radiodensity in medical CT and provides an accurate density for the type of tissue. On the Hounsfield scale, air is represented by a value of -1000 (black on the gray scale) and bone between $+700$ (cancellous bone) to $+3000$ (dense bone) (white on the gray scale). As bones are much denser than surrounding soft tissues, they show up very clearly in CT images, as can be seen in Fig. 2.1. This makes CT an important imaging modality when investigating skeletal anatomy. Similarly, the density difference

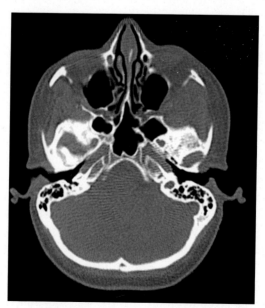

Figure 2.1 A CT image of the head.

between soft tissues and air is great, allowing, for example, the nasal airways to be clearly seen. Soft tissues and organs represent narrow Hounsfield value ranges and are therefore more difficult to differentiate between adjacent structures, such as between fat and muscle when viewing and segmenting CT data. Artificial contrast agents that absorb X-ray energy may be introduced into the body, which makes some structures stand out more strongly in CT images.

As CT uses ionizing radiation in the form of X-rays, exposure must be minimized, particularly to sensitive organs such the eyes, thyroid, and go-nads. The X-rays are generated and detected by a rotating circular array through which a moving table can travel. Typically, the patient lies on their back and is passed through the circular aperture in the scanner. The detector array acquires cross-sections perpendicular to the long axis of the patient. The images acquired are therefore usually termed axial or transverse images.

Most modern scanners perform a continuous spiral around the long axis of the patient. This innovation enables three-dimensional CT scanning to be performed much more rapidly, and consequently, three-dimensional CT scans may be referred to as helical CT. In addition, modern CT scanners employ multiple arrays to enhance the rate of data capture and improve three-dimensional volume acquisition.

CT images are generated as a grayscale pixel image, just like a bitmap computer image. If the distance between a series of axial images, called the slice thickness, is known, they can be interpolated from one image to the next to form cuboids, known as voxels. Therefore, a 3D CT scan generates a voxel representation of the human body. Software can be used to reslice these voxel data sets in axes perpendicular to the long axis, enabling different cross-sectional images to be generated from the original axial data. This is typically done in the sagittal and coronal planes; however, in theory, images may be generated in any plane.

The radiographers who conduct CT scans have specific parameters and settings for different types of scans. These are typically standardized and referred to as protocols. When embarking on using CT data for medical modeling, it is helpful to discuss it first with the radiographers, and they may well develop a protocol specifically for medical modeling.

CT scans are time consuming, expensive, and potentially harmful, so every care must be taken to ensure that the scan is conducted correctly the first and only time.

2.2.2 Partial pixel effect

When CT data are captured, the resulting images are divided up into a large number of pixels (typically a 512×512 matrix, but more modern scanners have a higher 1024×1024 matrix). Each pixel is a shade of gray that relates to the density of the tissue at that location. The resulting images are therefore an approximation of the original tissue shapes according to their density. The quality of that approximation is a function of the number and relative size of the pixels as well as other aspects of the CT scanner. The discrete size of the pixels means that edges between different anatomic structures are affected by this image quality.

The effect can be to effectively "blur the edges" due to the partial pixel effect. If the boundary between two different structures crosses a given pixel, that single pixel cannot represent both densities. Instead that pixel displays an intermediate density that is somewhere between the two. The effect can be illustrated by considering the shape in Fig. 2.2, which consists of two densities, low density being gray and high density being white. When the CT scan is performed, the cross-section is broken down into pixels, as shown in Fig. 2.3. In this view, it can be seen that some squares contain both high- and low-density areas. These pixels will therefore be shown as an intermediate gray depending on their relative proportions. This leads to the partial pixel effect which can be seen in the resulting tomographic image as shown in Fig. 2.4.

2.2.3 Anatomic coverage

The coverage of a CT volume is defined in two ways, by the number of axial images taken and the field of view (FOV) used for those images. In the long axis, it is defined by the table position where the first and last images

Figure 2.2 The original object shape.

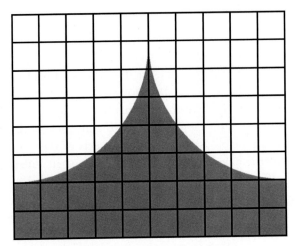

Figure 2.3 The effect of pixel size.

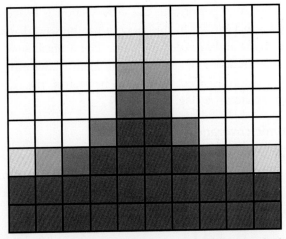

Figure 2.4 The resulting tomographic image (blurring of edges).

are acquired. The series of axial images must begin and end on either side of the anatomy of interest, but it is often important to begin and end the scan some distance on either side of the anatomy of interest. When conducting three-dimensional series scans, the data should be continuous. Noncontinuous sets of data may be combined in software later, but this introduces a risk of error as the patient may have shifted position slightly, and the separate series may not align perfectly.

The area covered by each axial image is referred to as the FOV and is typically square. The axial image consists of a fixed number of pixels, typically 512×512. The FOV is the physical distance over which the image is taken. Therefore, altering the FOV will alter the pixel size in the axial image. Usually, a FOV large enough to capture the whole cross-section of the anatomy is used. However, where specific small areas of anatomy are required, the FOV may be reduced to capture only that area. This results in a smaller pixel size, which increases the physical accuracy of the scan. For example, a typical FOV used to perform CT on the head would be 25 cm, resulting in a pixel size of 0.49 mm (assuming a 512×512 array), while a FOV of 13 cm may be used to capture data relating only to the mouth, which would result in a pixel size of 0.25 mm. While small pixel size is a desirable factor, it is more important that the images cover all the anatomy of interest plus some margin.

Although exposure to X-rays should be minimized, wherever possible, it is more important that the scan covers all the required anatomy, plus some additional margin. It is better to perform one extensive CT scan than a minimal one that is later found to be inadequate and necessitates subsequent scans. Basic mistakes in capturing the required coverage can be avoided through clear communication with the radiographer.

2.2.4 Slice thickness

This is the distance between the axial scans taken to form the three-dimensional scan series. In the case of helical CT scanning, the parameter applies to the distance between the images calculated during the scan (distance between cuts). To maximize the data acquired, this distance should be minimized. Some scanners can go as low as 0.5 mm, which gives excellent results, but this must be balanced against increased X-ray dose. Typically, distances of 1—1.5 mm produce acceptable results. A scan distance of 2 mm may be adequate for larger structures such as the long bones or pelvis. A scan distance greater than 2 mm will give poorer results as the scan distance increases.

Collimation is the term used to describe the thickness of the X-ray beam used to take the cross-sectional image. In combination with scan distance, consideration may be given to collimation and overlap. In most circumstances, the scan distance and collimation should be the same. However, using a slice distance that is smaller than the collimation gives an overlap. When scanning for very thin sections of bone that lie in the axial plane, such as the orbital floor or palate, an overlap may give improved results.

Even with a very small scan distance, some detail will be lost where thin sections of bone exist between the scan planes (although this is not true for volumetric acquisition). Typically, these are areas in the skull such as the palette and orbital floors. In addition, parts that are very small, or connected by thin sections, may not survive subsequent data or physical manufacturing processes.

2.2.5 Gantry tilt

For the purposes of virtual or physical modeling, gantry tilt should be avoided as it does not significantly improve the quality of the acquired data and provides an opportunity for error when reading the images. Large gantry tilt angles are clearly apparent on visual inspection and can be corrected. However, small angles may not be easy to check visually and may be compensated for incorrectly. Even the use of automatic import of the medical image standard DICOM is no guarantee, as although the size of gantry tilt angle is included in the format, the direction of tilt is not. Failure to compensate for the direction of the tilt correctly will lead to an inaccurate model, wasting time and money and potentially leading to errors in surgery or prosthesis manufacture. This effect is illustrated by the example shown in Fig. 2.5.

2.2.6 Orientation

Anterior—posterior and inferior—superior orientation is usually obvious on visual inspection, but lateral (left—right) orientation may be ambiguous. This is not a problem with automatically imported data, but when manually importing data, it is important that the correct lateral orientation can be ascertained. If there is an obvious lateral defect, then a note from the

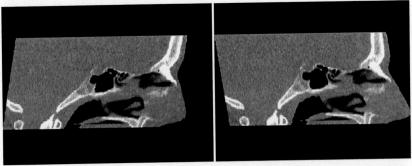

Figure 2.5 Incorrect gantry tilt compensation (left) and correct compensation (right).

clinician describing it is usually sufficient. Where the lateral orientation cannot be easily determined from the anatomy, extra care should be taken to verify the orientation before building a potentially expensive medical model.

2.2.7 Artifacts

This is the general term for signals within an image that do not correspond to anatomy. These may result from patient movement or X-ray scatter. Examples of medical modeling problems that have been encountered because of artifacts are discussed in the case studies in Chapter 5. Scattering is typically caused by metal objects such as dental crowns and fillings or surgical plates, screws, or even shrapnel.

2.2.7.1 Movement

A good quality CT scan depends on the patient remaining perfectly still throughout the acquisition. Movement during the acquisition will lead to distortions in the data (analogous to a blurred photograph). This has become less of a problem as acquisition times have decreased with the advent of helical multislice CT. However, it can still present a problem in some cases. For example, involuntary movement of the chest, neck, head, or mouth can occur through breathing or swallowing. Movement can be particularly difficult to control when scanning babies, small children, and claustrophobic patients, in which case a sedative or even general anesthetic may be required.

2.2.7.2 X-ray image scatter by metal implants

Dense objects such as amalgam or gold fillings, braces, bridges, screws, plates, and implants scatter X-rays, resulting in a streaked appearance in the scan image. The scatter results in significant image errors where false data appears with corresponding false missing data or shadows. Due to the nature of X-rays, little can be done to eliminate these effects, although manufacturers are now offering image-processing algorithms that can reduce its effect. Fig. 2.6 shows an axial CT image with significant artifact from scatter. Fig. 2.7 (left) shows how the scatter will be demonstrated on a three-dimensional reconstruction of the data, apparent as spikes radiating from the source of the scatter. These effects can be manually edited in software to produce a normal looking model (right). However, this does depend on the expertise of the operator, and consequently, the accuracy of the model in the affected areas cannot be guaranteed. In most cases, this

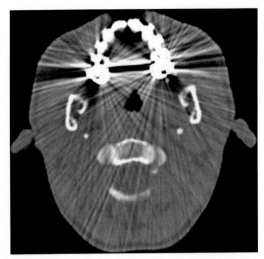

Figure 2.6 X-ray scatter artifact in a CT image (caused by dental crowns).

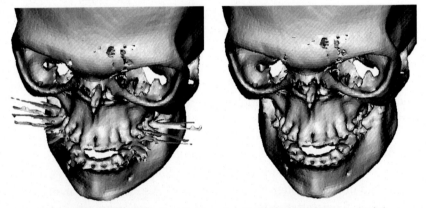

Figure 2.7 3D reconstruction (left) and edited 3D reconstruction (right).

does not usually affect the usefulness of the whole model. As cases showing artifacts usually occur in and around the teeth, a dental cast can be used in conjunction with the medical model and indeed may be combined with a physical model.

2.2.7.3 Noise
Noise is a fundamental component of a CT image and is especially prevalent in dense tissues. Although these images may be visually

acceptable, they are impractical for modeling. Good modeling from CT data depends on identifying a smooth boundary between bone and soft tissue. Noise reduces the boundary, which results in poor 3D re-constructions and consequently poor models. This commonly affects areas where the X-rays have to penetrate deep sections of tissue, for example, through the shoulders so vertebrae in the lower neck and upper back (C6–T4). A typical example is shown in Fig. 2.8. The effect becomes much more apparent when zooming into image data, as can be seen in the close-up view of the same data shown in Fig. 2.9. Three-dimensional reconstructions from these data will lead to a poor result, as shown in Fig. 2.10. Typically, such reconstructions will appear rough surfaced or porous.

If the effect only occurs in a few images, it may be possible to edit them out to produce a normal-looking model. However, this does mean that the accuracy of the model in these areas cannot be guaranteed. If the whole dataset is affected, this editing is unfeasible, and the resulting model may be too poor to be useful. Image processing may be used to "filter out" these effects, but this may affect the accuracy of the original data. Noise reduction can be achieved through spatial smoothing filters, which replace the intensity value of each voxel with weighted averages of the neighbors. There are two common methods: Gaussian filtering and bilateral filtering.

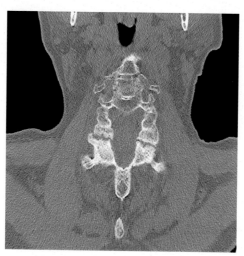

Figure 2.8 Noise in a CT image.

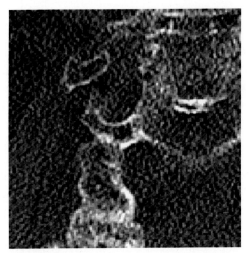

Figure 2.9 Close-up of noise in the same CT image.

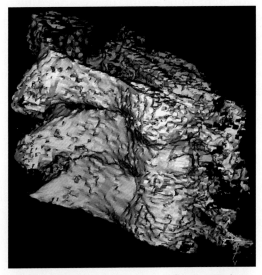

Figure 2.10 3D reconstruction of noisy CT data.

2.2.8 Kernels

Modern CT scanning software allows different kernels (digital filters) to be used. These modify the data to give better three-dimensional re-constructions and can help to reduce noise. Typically, the options will range from "sharp" to "smooth." Sharpening filters increase edge sharpness but at a cost of increasing image noise. Smoothing filters reduce noise content in images but also decrease edge sharpness. In general, when building medical models, smooth filters tend to give better results and are easier to work with. The effect of sharp versus smooth filtering is illustrated in Figs. 2.11 and 2.12, where the arrow represents a density profile, which is shown as the graph on the right.

Although the smooth image contrast appears poor on screen, taking a density profile shows that the actual contrast is good and that the smooth data allows a much lower threshold to be used.

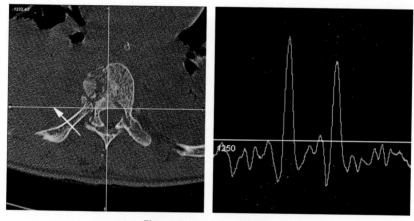

Figure 2.11 Sharp dataset.

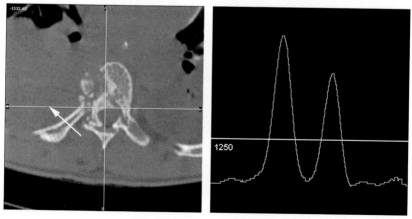

Figure 2.12 Smooth dataset.

2.3 Cone beam computed tomography

2.3.1 Background

Cone beam computed tomography (CBCT) is a more recent development in medical imaging that has become popular in orthodontic, dental, and maxillofacial applications, but it also has a foundation in interventional radiology, image-guided surgery, and radiation therapy. It utilizes the same basic principle as conventional CT scanning but in a unique way that results in far lower patient radiation doses. As well as fundamental technical differences in the way X-rays are delivered and detected and data is processed, there are more obvious differences in the appearance of dental/craniofacial CBCT scanners. The first obvious difference is that the patient is usually scanned in the seated position as opposed to lying down (see Fig. 2.13). Whereas conventional CT utilizes a narrow fan of X-rays and a narrow detector, the X-rays in CBCT are emitted in a cone shape and detected by

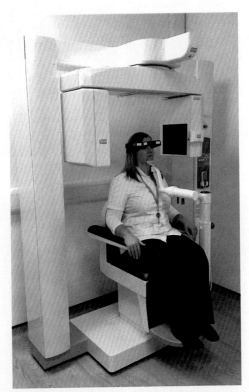

Figure 2.13 The Morita 3D Accuitomo 170 cone beam CT scanner based at Morriston Hospital, Swansea, UK.

a flat panel detector. Multiple images (usually hundreds) from different angles are captured through a single rotation of the emitter and detector around the head. Software algorithms are used to calculate volumetric data based on these multiple angles of capture around the fixed central point (termed isocenter).

2.3.2 Advantages

The primary advantages of CBCT over conventional CT are lower radiation doses, higher pixel/voxel resolution, ability to focus in great detail on a small area without irradiating surrounding tissue, lower operational cost, and the smaller footprint of the scanner.

2.3.2.1 Reduced radiation dose

CBCT offers significantly lower radiation dose than conventional CT when scanning the same area. However, the actual difference will depend on the machine and the parameters used; therefore care should be taken. Studies that have utilized dosimetry phantoms based on a craniofacial (upper and lower jaw) model cite reductions on the order of 15 times less than conventional CT for the equivalent scanned area.

2.3.2.2 Voxel size—Resolution

Whereas voxel size for conventional CT can be as small as 0.5 mm (at the compromise of increased radiation dose), CBCT produces isotropic (equal in the x-y-z dimensions) voxels as low as 0.076 mm (e.g., Kodak 9000 3D manufacturer specifications). This is a distinct advantage when evaluating bone quality for procedures such as implant planning.

2.3.2.3 Ability to focus on small areas

The FOV is dependent on the detector size and shape, beam projection geometry, and the ability to collimate the beam. Small FOVs correlate to increased resolution in a very local area. A small or focused FOV would be in the order of 5 cm or less, whereas 15 cm or more may be required for craniofacial applications. As with conventional CT, the reduction in FOV and associated radiation dose must be balanced against acquiring a sufficient region of interest for the intended application.

2.3.3 Limitations

The advantages of CBCT must be balanced against a number of limitations, including artifact, the relatively limited FOV (limited to the anterior portions of the head, usually focused on upper and lower jaws), grayscale values that change throughout the data volume (thereby causing an inconsistency in being able to measure Hounsfield units), and lower soft tissue contrast than conventional CT (further compounding difficulties in differentiating between tissue types across the dataset). These limitations are also highly dependent on the machine type, scanning, and reconstruction algorithms used, making it even more difficult to compare and measure datasets from different CBCT scans.

2.3.3.1 Artifact

Like conventional CT, CBCT also suffers from artifact such as those discussed in Sections 2.2.2 and 2.2.7. The way the data are affected is different, particularly in the presence of dense metallic objects (such as dental crowns and fillings). Due to the cone shape of the X-ray beam and large detector, scatter in CBCT can cover the whole dataset rather than being limited to the single slice of a conventional CT image. Like conventional CT, various proprietary software algorithms and scan strategies have been developed and adopted to minimize and overcome the problems associated with artifact.

2.3.3.2 Field of view

Whereas conventional, helical CT scanners can continuously scan large areas in one session, CBCT scanners are restricted in their FOV to less than 20 cm height depending on the emitter and detector size and scan protocols.

2.3.3.3 Inconsistency of grayscale values across the volume

A major limitation is the inconsistency of grayscale values across the data volume. This is a particular problem at the extremities of the FOV, where noise increases and contrast decreases. An example reconstructed axial image slice is shown in Fig. 2.14. Contrast between soft tissues is also relatively poor, making CBCT less useful for the evaluation of some diseases and pathologies. Both issues can make subsequent data segmentation in postprocessing software using automatic algorithms challenging.

2.3.4 Applications

Despite the limitations of CBCT scanning, it has found widespread application in implant dentistry, orthodontics, temporomandibular joint dysfunction, and increasingly, areas of maxillofacial and craniofacial surgery where the limited FOV and inconsistent grayscale values are less of a

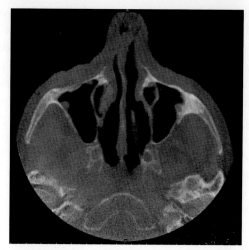

Figure 2.14 Axial slice from cone beam CT showing varied grayscale values.

problem. Like conventional CT, contrast between very dense tissues (such as cortical bone or enamel) contrasts well against air, making it suitable for acquiring volumes around the anterior midface. This makes CBCT a valuable tool in the process of planning intraoral implant placement, which is now one of the most common applications. From a practical perspective, the small machine footprint and comparatively lower purchase and operating costs also make it feasible for even small laboratories and clinics to operate CBCT scanners. CBCT scan data can also be complemented with color surface 3D topography scans, which can be visualized in addition to the underlying bony data [3]. This development is in response to the desire to better visualize and predict soft tissue movements in relation to orthodontic and surgical procedures, where there has been significant research in recent years.

There have also been attempts at using CBCT mounted on a C-arm arrangement for use in operating theaters [4]. This has led to applications in interventional radiology and image-guided surgery, where it can be necessary to rapidly evaluate spatial locations of anatomic structures or surgical instruments, without having to transport the patient.

2.4 Magnetic resonance

2.4.1 Background

Magnetic resonance (MR) imaging exploits the phenomenon that all atoms have a magnetic field that can be affected by radio waves. Atoms have a natural alignment, and MR works by using powerful radio waves to alter this alignment temporarily. When the radio waves are turned off, the atoms return to their natural alignment and release the energy they absorbed as radio waves. To construct an MR image, the strength of the radio waves emitted by the atoms is measured at precise locations. By collecting signals from many locations, a cross-sectional image can be created. As in CT scanning, the resulting cross-sectional image is a grayscale pixel image, the shade of gray being proportional to the strength of the signal.

As the human body is composed mostly of water, MR scanning targets the hydrogen nuclei present in water molecules. Therefore, locations that have a high water content show up in lighter shades of gray, and areas containing little or no water show up darker. For example, air shows up black as will the densest bone, while tissues highest in water content, such as fat, will show up white.

As the water content of different soft tissues differs, MR is an excellent modality for investigating the anatomy of soft tissue organs, as can be seen in a typical MR image of the abdomen shown in Fig. 2.15. However, unlike CT, MR is not good for visualizing bone. The boundary between air and soft tissue is also good, allowing models to be made of the skin surfaces of patients. Unlike CT, by altering the direction of the radio waves used, MR images can be acquired in cross-sectional slices at any angle. MR also differs from CT scanning in that there are more parameters that can be altered to improve the results for specific tissues. It is therefore important that the radiographer knows precisely which tissue type is being targeted before conducting the scans.

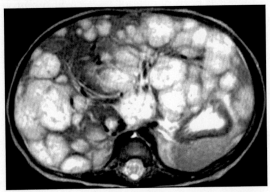

Figure 2.15 A typical MR image through the abdomen.

Due to the strong magnetic fields encountered during MR, the presence of metal may cause problems. Therefore, jewelry and watches must be removed, and patients' notes must be checked to ensure that they do not have attached or implanted devices that may be adversely affected such as heart pacemakers. As with CT scanning, patient movement will lead to distorted images, and babies, small children, and claustrophobic patients may require sedation or anesthesia. MR scanners also generate loud noises, which even with ear protection may not be pleasant for the patient.

Although MR does not utilize ionizing radiation, it may present a risk for certain patients. They are time consuming and expensive, so every care must be taken to ensure that the scan is conducted correctly the first and only time. It is also important to consider the dangers that any magnetic metal implants may have before conducting MR scans.

2.4.2 Anatomic coverage

As with any radiographic procedure, basic mistakes can be made through poor communication with the radiographer. Detail can be lost when the scans do not cover the whole anatomy of interest or do not include sufficient margins surrounding the anatomy of interest. Detail may also be lost by using a FOV that is too small. When conducting three-dimensional series scans, the data should be continuous. Noncontinuous sets of data may not be satisfactorily combined in software later, as the patient may have shifted position slightly, and the separate series may not align perfectly.

2.4.3 Missing data

Even with a very small scan distance, some detail may be lost where thin sections of tissue exist between the scan planes. In addition, parts that are very small, or connected only by thin sections, may not survive subsequent data processing or physical manufacture. Parts that are not connected will not be present unless they are artificially attached to surrounding anatomy.

Due to the time taken to acquire each image, flowing fluids will have moved between the excitation and emission stages of the scan. With multiple images being taken, this may result in the signal being reduced or reinforced. Therefore, blood vessels, for example, may appear too dark or too bright.

2.4.4 Scan distance

This is the distance between the scans taken to form the three-dimensional scan series (unlike CT, data capture is not limited to the axial plane). This may also be referred to as "pitch" or "distance between cuts." To maximize the data available to produce a smooth model, this distance should be minimized. Typically, distances of 1–1.5 mm produce good results. A scan distance greater than 2 mm will give increasingly poor results as the scan distance increases. However, taking thinner slices results in less signal strength per pixel being detected by the scanner. Therefore, more echoes are required to boost the signal strength, which results in significantly longer scan times.

Unlike most CT scanners, the number of pixels used in a cross-section is a variable parameter. Typically, the cross-section will be broken down into a relatively small number of larger pixels compared with CT. For example, a typical CT image may be 512 × 512 pixels at a pixel size of 0.5 mm, whereas an MR image may be 256 × 256 pixels at a pixel size of 1 mm. The main reason for this is to maintain signal strength. A larger number of smaller pixels results in less signal strength per pixel. Again, therefore, more echoes are required to boost the signal strength, increasing scan times.

For three-dimensional modeling, it may be necessary to alter the compromise between scan time and signal strength compared with the protocols normally used for diagnostic imaging. MR is often a preferred imaging methodology due to its inherent safety compared with CT. However, the application of MR for three-dimensional modeling should be carefully considered due to the increased scan times. Although MR is safe, the procedure may be uncomfortable and perhaps distressing for the patient, and the added costs and delays incurred should be considered.

2.4.5 Orientation

As with all medical imaging, anterior–posterior and inferior–superior orientation is usually visually obvious, but lateral orientation is ambiguous. Usually this is not a problem with automatically imported data, but when manually importing data, it is important that the correct orientation can be ascertained.

2.4.6 Image quality and protocol

MR image data is typically taken for diagnostic reasons, to investigate areas of specific illness or locate pathology, such as a tumor. Usually, the minimum number of images required to identify the problem is acquired; consequently, it is not common practice to undertake three-dimensional MR scans. Conducting a three-dimensional MR scan may take significantly longer than a normal session, and the compromise between scan time and the necessity of the three-dimensional data must be considered. As conducting MR scans is expensive and a critical resource in most hospitals, increasing the scanning time may cause problems and increases the inconvenience for the patient. Close collaboration with the radiographer is recommended to ensure that the data are of sufficient quality without creating problems.

The configuration of the MR machine may be enhanced by the addition of more coils, which has the effect of increasing the signal strength. Such configurations may be used by specific specialties such as neurosurgery.

Unlike CT images, altering the protocol of an MR scan can dramatically alter the nature of the image. By varying the sequence and timing of excitation and emission of the radio waves, different effects can be achieved. These may serve to improve the image quality for specific tissues or improve contrast between similar adjacent tissues.

2.4.7 Artifacts

This is the general term for corrupted or poor data in MR images. These may result from patient movement or magnetic effects.

2.4.7.1 Movement

A good quality MR scan depends on the patient remaining perfectly still throughout the acquisition. Movement during the acquisition will lead to distortions in the data (analogous to a blurred photograph). For example, involuntary movement of the chest, neck, head, or mouth can occur through breathing or swallowing. When taking multiple scans, it may be possible to synchronize the timing of the sequences used with breathing. Movement can be particularly difficult to control when scanning babies, small children, and claustrophobic patients, in which case a sedative or even general anesthetic may be required.

2.4.7.2 Shadowing by metal implants

All metal objects must be removed before MRI scanning, but some fixed or implanted items that cannot be removed can negatively affect the image quality. Dense metal objects such as amalgam or gold fillings, braces, bridges, screws, plates, and implants affect the magnetic field around them, leading to the appearance of artifacts in the scan image. The effect is normally apparent as a lack of data shown as dark patches or shadows surrounding the location of the metal object. The extent of this effect depends on the type of metal present.

2.4.7.3 Noise

Noise occurs in all MR scans and may reduce image quality, but taking multiple acquisitions can reduce this. Noise blurs the boundaries between different adjacent soft tissues. When observing the image, the human eye can account for this (we are very good at recognizing shapes we are familiar with), and the boundaries appear visible. However, for successful modeling from the data, the boundary must be clear and distinct to a much higher degree than when the images are only used visually.

Taking multiple acquisitions reinforces the signal, improving the quality of the image data. However, this may double or quadruple the time taken to complete a scan session. In practice, a compromise between noise reduction and scan time must be reached with your radiographer.

2.5 Noncontact surface scanning

2.5.1 Background

When attempting to capture human topography, that is the external shape or skin surface, it is frequently more practical and comfortable to use noncontact scanning systems, which typically rely on light-based data acquisition. Noncontact surface scanning uses light-based techniques to calculate the exact position in space of points on the surface of an object. Computer software is then used to create surfaces from these points. These surfaces can then be analyzed or integrated with CAD models.

Unlike CT and MR, these techniques capture only the exterior topography of the patient. This allows models to be made of the skin surfaces of patients. Although increasingly common, these techniques are not yet considered routine medical imaging modalities, and it is therefore possible that a nonmedical scanning facility will have to be used to capture the data. Many product development facilities have access to this kind of equipment, although the nature of the equipment can vary significantly.

2.5.2 Optical scanning technologies

There are several different technologies that use different forms of light and camera technology to capture three-dimensional surfaces including photogrammetry, structured light, and lasers. While they have much in common, the key differences are covered below before the rest of this section describes the common issues encountered with noncontact light-based 3D scanning. Most 3D scanners will produce data in the form of a massive number of 3D data points usually referred to as a point cloud. Point cloud data is in itself of little use, and it is normally used as the basis for a three-dimensional surface or solid model. This is described in more detail in 2.7 and Chapter 4: Using medical scan data.

2.5.2.1 Structured light scanners

Structured light scanners rely on high-contrast alternating "fringe" linear and or grid-like pattern images that are then projected onto an object at constant or varying frequencies (see Fig. 2.16). One or more cameras are then used to record the position of the projected images on the object, and supporting software then uses the deformed image projection data based on image intersections to form a virtual 3D model. Both white and blue light sources are used, with blue sources offering higher accuracy due to wavelengths reducing effects of reflection and image interference. Some scanners show highly visible projected patterns, while others are invisible to the naked eye. For example, the Artec Eva, Leo, and Spider scanners project an invisible infrared grid onto the object, while scanners such as Zeiss GOM scanners project visible blue fringe patterns onto the object (see Fig. 2.16).

2.5.2.2 Laser scanning

Laser scanners work in practice in a very similar way to structured light scanners; only, instead of projecting a fringe pattern, a laser is used to project a line of light onto the surface being captured. Cameras then capture

Figure 2.16 Scanning showing fringe pattern (courtesy of Zeiss).

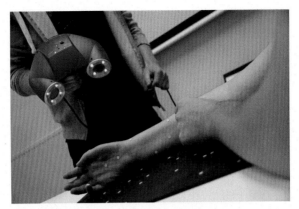

Figure 2.17 Laser scanning in progress.

the changing shape and distance of the projected line enabling the calculation of data points along that line. In practice, this can be fast enough to enable handheld scanners to capture data from patients accurately (see Fig. 2.17).

2.5.2.3 Light detection and ranging (LiDAR)

LiDAR scanners also work in practice in a similar way to other light-based scanners. LiDAR differs from laser scanning by sending out thousands of short pulses of laser light and measuring the time it takes for the reflected light to be detected (a bit like radar), enabling the calculation of many data points in rapid succession. To capture these data points in three-dimensional space, the position of the scanner needs to be also simultaneously acquired. For scanning geographic areas from aircraft, for example, a global positioning system data is used. For handheld scanners, integrated inertial measurement units are used. The technology is increasingly available on smart phones and tablets, enabling easy and convenient 3D data capture for very low cost.

2.5.2.4 Photogrammetry and stereophotogrammetry

Photogrammetry can be performed with any digital imagery of an object provided there are multiple images of the object and there is sufficient image count and overlap of the object in question shown in the images. Any digital image hardware can be used, including compact cameras, camera phones, and SLR cameras. The user must capture multiple images with varying positions and angles around the object, ensuring overlap to help assist in image alignment. High-contrast or reflective markers (typically reflective dots and/or black and white QR codes; see Fig. 2.18) may be used to enable accurate image alignment by providing known references in acquired images using simple geometric forms to assist the alignment algorithms to detect similarities in the images.

Photogrammetry typically involves using only one camera to capture multiple images around the object. Depending on the size of the object and the scan resolution required, this can be a time-consuming process to ensure sufficient image count and relies on user experience to ensure enough images are captured to successfully align the images together to form a suitable quality mesh. To overcome these weaknesses, stereo-photogrammetry involves positioning multiple cameras around an object to collect multiple images simultaneously. Cameras may be arranged in a linear, polar, or hemispherical array or an alternative array depending on the need of the user, but it is imperative that there is sufficient overlap so the collected images can be aligned with suitable alignment software. The ability to capture multiple images simultaneously is a significant benefit, offering time and cost benefits as well as reducing the need for specialist skills for image capture as expected with photogrammetry. The high speed also helps with minimizing involuntary movement between images being taken.

Figure 2.18 Using markers in photogrammetry. *(Reprinted with permission from Artec 3D (artec3d.com).)*

2.5.2.5 Color capture

Many forms of light-based 3D scanners can also capture color data, sometimes referred to as texture data. This enables photorealistic 3D models to be captured. The accuracy of the color data can vary, but some systems will enable sufficiently accurate data to be captured for clinical purposes.

2.5.3 Preparation and resources

Many 3D scanners benefit from additional resources to aid in data capture. Turntables are often used to capture the object from numerous angles to overcome line-of-sight restrictions by taking multiple scans while reducing/eliminating the need for manual handling, which can affect the objects' properties and positioning, particularly those that require prior preparation or are easily deformable. Individual scans are taken from the separate positions, which are then aligned in the scanning postprocessing software. Turntables may be automated, synchronized, and calibrated with the scanner equipment for faster, easier, and more accurate data capture. Manual turntables can also be used but require more care by the scanner operator as well as additional time and care in postprocessing the multiple scans to make a complete mesh.

Scanners may come with tripods or stands, while others can be handheld. The control of lighting can be important, and typical photographic studio setups are often utilized including moveable diffuse lights and neutral backgrounds.

2.5.4 Optical scanning safety considerations

Noncontact surface scanning can be time consuming and potentially expensive but is mostly completely safe. The noncontact nature of the scanning means there is less discomfort for the patient and no distortion of soft tissues caused by the pressure applied when taking casts or impressions (sometimes called a moulage). This advantage in combination with the ability to manipulate data makes the approach particularly well suited to applications in prosthetic reconstruction and rehabilitation. It is difficult, for example, to take a satisfactory impression of a breast; therefore noncontact scanning may be used in the creation of symmetrical prostheses for mastectomy patients. However, care should be taken when scanning the body surface to ensure it is in the position that relates to the intended use. For example, body parts that are weight bearing will distort according to the position and posture of the patient. Additional care must be taken if scanning the face as different scanning technologies may expose the patient to potentially unsafe or uncomfortable light sources; for example, certain laser-based scanners may be unsafe with direct vision, which may cause ophthalmologic damage, while other scanning technologies may present high-frequency flashing lights that may trigger photosensitive seizures for individuals with conditions such as epilepsy. Methods to circumvent such risks include wearing safety glasses or using alternative optical scanning equipment if available. Case studies illustrating some applications of noncontact scanning are described in Chapter 5.

2.5.5 Anatomic coverage

Basic mistakes can be made through poor communication between the clinician and operator acquiring the scans. Detail can be lost when the scans do not cover the whole region of interest or do not include sufficient margins around the anatomy of interest. Most scanners have a limited FOV, and several overlapping scans may be required.

In addition, these techniques rely on "line of sight." This means that areas that are obscured or at too great an angle to the line of sight will not appear in the scan data. Therefore, several scans may have to be taken from different viewpoints to ensure all the required details are captured. This can be achieved by moving either the patient or the scanner and repeating the process. Depending on the shape of the object or area being scanned, many scans may be necessary. When scanning faces, for example, a single scan will not acquire data where the nose casts a shadow, as shown in Fig. 2.19 (left).

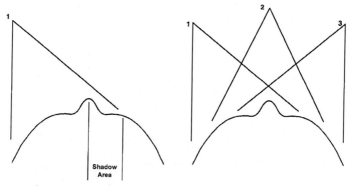

Figure 2.19 Line of sight issue (left) and multiple viewpoints for scanning (right).

This is overcome by taking several overlapping scans. For example, Fig. 2.19 shows three overlapping scans, one from the left, two from straight ahead, and three from the right to capture the whole face.

The data from each of these scans can then be aligned using software to give a single coherent dataset. When conducting a series of scans, it is therefore important that they overlap so that the individual scans can be put together accurately.

Despite using multiple scans, some areas may remain difficult to capture. For example, it is very difficult to capture data from behind the ear, at the nostrils, and between the digits of the hands and feet. Line of sight issues can be particularly challenging when attempting to capture an entire structure from 360 degrees, when for example capturing data of a whole limb. Capturing data of the hands is very challenging due to the complex shapes, the spaces between fingers, and the enormous variability of pose and position the fingers and hand can take. It is also very challenging for many patients to keep perfectly motionless (see 2.5.4 below) while the scanner is maneuvered around the whole limb.

To overcome line-of-sight limitations, unique specific scanning solutions have been developed for arms, hands, legs, and feet. For example, the Mano-X-2 upper extremity scanner, developed by Manometric, uses a polar array of cameras positioned with illumination to enable simultaneous image capture, which are then aligned in bespoke software to generate a 3D model of the forearm and hand (see Fig. 2.20).

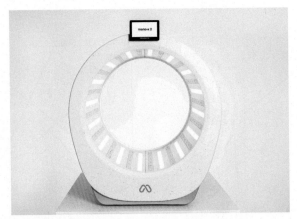

Figure 2.20 Manometric mano-x-2 hand scanner. *(Courtesy of Manometric.)*

2.5.6 Missing data

Because these techniques use light to calculate the points, transparent, dark, or highly reflective surfaces can cause problems. Usually, human skin performs very well in this respect, but steps may be required to dry particularly greasy or moist skin. These effects may also be overcome by applying a fine powder, such as talcum powder, which will give the object an opaque matt finish. The eyes may cause problems due to their shiny surface and watering. All optical scanners should be inherently safe; however, care should be taken when scanning the eyes to ensure that bright light or laser light does not directly enter the pupil.

Another inherent problem is caused by hair. Long, dense hair does not form a coherent surface and absorbs or randomly scatters the light from the scanner. However, the fine hair present on most body surfaces does not normally affect the captured data, although excessively thick body hair may reduce the quality of the data captured.

In most cases, the presence of hair will lead to gaps in the data. Fine or downy body hair often does not cause significant problems, but there may be little that can be done about a full head of hair, and it would not be normal to consider shaving the head for such a procedure unless the clinical benefits were overwhelming. In situations where the entire head must be scanned, a latex cover such as a swimming cap may be used to conceal hair, but one must bear in mind the effects of altering the topography because of constrained hair, particularly individuals with long, thick hair encased on the cap. When considering scanning the face, gaps are likely to arise from areas of significant facial hair, such as beards and moustaches, but may also be encountered around the eyebrows and lashes. These gaps can be clearly seen as black areas in the surface scan data shown in Fig. 2.21.

2.5.7 Movement

As with other scanning modalities, any patient movement during the scan will lead to the capture of poor data. The length of time required varies depending on the exact type and specification of the scanner being used but may range from a fraction of a second to as much as 10 minutes. Consequently, it is important that the subject be scanned in a comfortable and steady posture. A consequence of movement may be noise, false representations of anatomic topography, and subsequent difficulty if aligning multiple scans. This does not normally pose a problem when dealing with cooperative adults and older children, but small children, babies, the

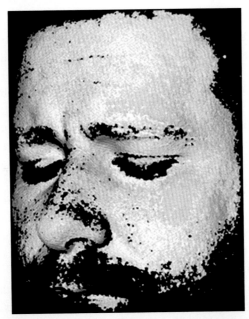

Figure 2.21 Scan data of the face of a male subject.

elderly, or patients with conditions such as Parkinson's disease may be difficult to keep still during the scanning. Unlike MR or CT scanning, it is highly unlikely that sedation would be justified for this type of scan. Capturing a specific posture may benefit from rigging structures, but it requires careful consideration to avoid soft tissue deformation, as shown by the flattened areas of the hand in Fig. 2.22.

With most scanning software, if there is sufficient overlap between adjacent scans, they can be aligned by suitable computer software. This means the patient does not have to remain perfectly still between successive scans. However, the same body posture or facial expression should be maintained throughout. For the best results, the patient may have to be braced in a comfortable position with suitable rigging apparatus during each scan. Depending on the scanner being used, it may be simpler to keep the subject still and move the scanner around them. Alternatively, a series of multiple scanners can be positioned around the subject to capture different views simultaneously or in rapid succession.

However, despite these steps, some movement is likely to be encountered such as breathing, swallowing, blinking, or involuntary tremors, so care should be taken when scanning the face, neck, chest, and extremities.

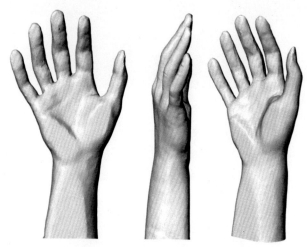

Figure 2.22 Soft tissue deformation as a result of rigging during scanning.

If the patient is able, a breath–hold may help if the scan time is only a matter of seconds.

2.5.8 Noise

Noncontact scanners capture many thousands of points at a time. The vast majority of these points will fall accurately on the surface of the object being scanned. However, due to tolerances and optical affects, some of these points will deviate from the object surface. If enough points deviate from the surface by a sufficient amount, it will affect the quality of the data. These errant points are usually referred to as "noise" in the data. The amount of noise present in captured data will depend on the type of scanner and the optical properties on the surface being scanned. Smooth, matt surfaces usually produce less noise than reflective or textured surfaces. Although it is usually necessary to take multiple overlapping scans to cover the whole surface of an object, large overlapping areas are likely to result in increased levels of noise.

Noise can be reduced by data processing usually called "filtering." Filtering selectively removes data points that deviate greatly from the vast majority of neighboring points. This is illustrated schematically in steps 1–5 in Fig. 2.23:

(1) the object surface to be scanned;

(2) the scan data points;

(3) using all of the data points creates a poor surface;

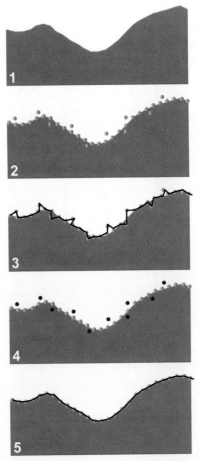

Figure 2.23 The effect of filtering to remove noise from optical scan data.

(4) points are selectively filtered according to their deviation from the majority of neighboring points;

(5) deleting filtered points leaves a closer fitting surface than the actual object surface.

The magnitude of the deviation can be defined by the user to vary the effect of the filtering. Filtering functions are typically included in the software that is used to operate the scanner.

The effect of noise can be seen in Fig. 2.24 (left) showing a three-dimensional polygon model created from optical scan data of a dental cast (in this example, the STL file format is used). Noise in the original scan

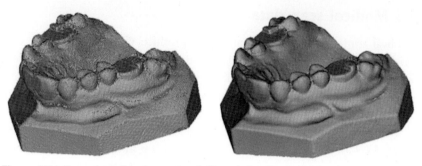

Figure 2.24 Scan data showing noise (left) and the same data with noise reduced (right).

data has resulted in a polygon surface that appears rough and pitted. Postprocessing software can be used to improve the quality of the surface model. These functions work by averaging out the angular differences across neighboring polygons within user-defined tolerances. The effect is to smooth out the surface, but without losing critical characteristics of the topography. When dealing with human anatomy, the smooth, curved surfaces typically encountered mean that this approach often leads to a more accurate model despite the additional data operations. Fig. 2.24 (right) shows the same three-dimensional model that has been "smoothed" to mitigate the effects of noise.

2.5.9 Low-cost and open-source methods for surface capture

In the past, organizations paid a premium for adequate resolution optical surface capture hardware and software. Now in the wake of the open sourcing era for hardware and software algorithms, the use of low-cost hardware such as LiDAR-equipped mobile phones and the Microsoft Xbox Kinect are increasingly used for digitizing artifacts and environments, providing a suitable supporting program/algorithm is in place. The low cost of hardware makes the approach more accessible to individuals or organizations on a restricted budget, and depending on the application, the resolution may be suitable for end use of the intended application or part if used to support additive manufacture. However, one must bear in mind that problems with optical data capture still exist, such as line of sight. These low-cost methods are enabling much more research to be done. However, quality assurance and certification of 3D scanners should also be considered when undertaking clinical work.

2.6 Medical scan data

Medical scanners such as CT and MR produce pixel-based images in a series of slices, while noncontact surface scanners produce three-dimensional point clouds. Therefore, the formats used to describe them are completely different, and separate software technologies are required to use them.

Most radiology departments can make image data available over hospital networks. Good third-party radiography software should be able to import the commonly used proprietary formats. The industry standard is called DICOM, and it should be used whenever possible, as all software should support it. An example of the practical implications of transferring data from a hospital to a service provider can be found in case study 5.1.2.

2.6.1 DICOM

DICOM stands for Digital Imaging and Communications in Medicine and is an internationally agreed standard for all medical imaging modalities. The standard was initiated in response to the development of computer-aided imaging in the 1970s by a joint committee from the American College of Radiology (ACR) and the National Electrical Manufacturers Association (NEMA). They first published an ACR-NEMA standard in 1985 and updated it in 1988. Version 3 saw the name changed to DICOM and was published in 1993. The standard now covers all kinds of medical images but also includes other data such as patient name, reference number, study number, dates, and reports. Since then, most manufacturers have adhered to the standard, and data transfer problems are much less likely to occur than was previously the case.

The DICOM standard (ISO 12052) enables the transfer of medical images to and from software and scanners from different manufacturers and aided the development of picture archiving and communication systems (PACS), which can be incorporated with larger medical information or records systems.

More information on DICOM can be found at http://medical.nema. org/

2.6.2 Automatic import

When using medical data manipulations software, there is usually an automatic import facility. The facility will automatically recognize many manufacturers' formats and will certainly recognize data written in the international image standard DICOM. The software will automatically read in the data and convert it into its own format ready for manipulation.

However, depending on the specification of the automatic import software being used, there may be some instances where the user will require some knowledge of the scan parameters to complete the import. Usually, these factors are not fully described in the DICOM standard. Gantry tilt is one such example. Although DICOM supports the magnitude of gantry tilt, it does not provide the direction. The software will automatically apply a correction, but the user will be required to either confirm or reverse the direction to ensure that the data are correct. The second example is the anatomic orientation of the dataset. DICOM will provide left, right, and anterior—posterior orientation but may not provide inferior—superior information (because a patient may enter the scanner head or feet first). The user may be required to select this orientation during import. The inferior—superior orientation is usually anatomically obvious, so this rarely creates a problem.

2.6.3 Compression

Image data can be large, and compression is sometimes used, especially for archiving. Data compression may incur a loss of information, which is called "lossy compression," or retain all data but write it in a more efficient manner, called "lossless compression." For modeling purposes, compression should be lossless or preferably avoided completely. Any loss of information may reduce the accuracy of models made from the dataset. In addition, some proprietary compressed formats may be unreadable by third-party software, rendering the data useless. Compression for medical modeling often is not necessary.

2.6.4 Manual import

When the data format is not in a DICOM compatible format or from a manufacturer that is not directly supported by the software application being used, manual import must be used. Error-free manual import of medical scan data will require access to all the parameters of the data. Table 2.1 lists the parameters that must be known to import data successfully.

Table 2.1 Information required for manual data import.

Data	Units
Number of images	Integer
File header size	In bytes
Interimage file header size	In bytes
Image size	In pixels e.g., 512 by 512
Slice distance	In millimeters
FOV or pixel size	In millimeters
Gantry tilt	In degrees
Scan orientation	Left—right or right—left

2.7 Point cloud data

The data captured by a noncontact surface scanner is merely a "point cloud." This is a collection of point coordinates in three-dimensional space and may range from hundreds to many millions of points. It is typical to convert this into a more useful topology (e.g., polygon meshes) and indeed a more useful file format before applying the data to analysis, manipulation, or visualization. There are a number of formats used for point cloud data (e.g., .xyz or .obj), and the format will depend on the type of scanner being used. Furthermore, point cloud data can also store RGB color data; this is often beneficial from scanners that capture bitmap data to help transfer and manipulate textures in a virtual environment.

The simplest forms of export format are polygon meshes, such as the commonly used STL file format (the STL file format is described in Chapter 3). Many modern scanners will output data directly into the STL file format rather than point clouds, and typically a number of export options will be available. However, the export data format used will depend very much on the anticipated use of the data; this area is explored in greater depth in the next chapter. Some techniques can be applied directly to the point cloud data. These are typically done to remove erroneous points, decrease noise, or simply reduce the number of points (so to reduce computational load). This has the effect of cleaning up the dataset and reducing the file size.

There are many open-source point cloud manipulation software tools available (e.g., CloudCompare, https://www.cloudcompare.org/main.html), which enable point cloud processing functions such as point and noise reduction, alignment (typically referred to as *registration*) of multiple point clouds, and deviation of multiple scans and mesh data.

2.8 Media

In combination with the data format, the output media type is also a part of the whole system. Advances in computer networks and telecommunications have enabled the phasing out of storage media such as magnetic optical disks previously used for archiving large data sets. These were expensive and required specialist hardware and software to translate data. In addition, there has been widespread adoption of Windows operating systems, PC hardware, compact discs (CDs), digital versatile discs (DVDs), and portable USB drives as data storage media. With teleradiology now commonplace, most hospitals put medical image data onto the hospital local area network. PACS is commonly used to store and manage medical image data within hospitals. This enables clinicians to remotely access medical images data in their own departments. Once the data are available over the hospital network, they can be transferred over secure, internet-based systems to external parties. Given that DICOM data contain sensitive patient information, there are tight regulations on the security and transfer of scan data. For example, information on patients within the European Union should not be stored outside the European Union, and there should be compliance with standards such as ISO/IEC 27001:2005, which specifies a management system that is intended to bring information security under explicit management control. Data that are archived to a CD/DVD can be password protected if being sent by postal systems, but this is becoming less desirable in favor of internet-based transfer.

When dealing with point cloud data, media formats are normally those typically used in the design and engineering community, and translation does not pose the problems associated with radiological data; it is far simpler to strip data of sensitive information that could relate it to an individual patient.

2.9 Summary

This chapter has covered the essential steps needed to select appropriate medical imaging and three-dimensional data capture. The fundamentals of CT, CBCT, and MRI imaging have been explained showing that CT and CBCT are well suited to the capture of bony anatomy, while MRI enables the capture of specific soft tissues and internal organs as long as the complex parameters and limitations are well understood. The manner in which the grayscale pixel images resulting from these imaging methods can be manipulated through concepts such as segmentation and region growing that enable the isolation of individual organs or structures have been introduced with worked examples shown using popular available software tools. The fundamentally different approach and characteristics of optical surface scanning have been introduced along with the basic approaches to the creation of useful digital three-dimensional models from point cloud data such as polygon meshing. The considerations and implications of data files type, size, and media have been introduced, but it remains crucial to recommend the early involvement of those experts responsible for the capture of the data to ensure the resulting data are appropriate and of the highest quality. The next chapter discusses how these resulting data can be manipulated further in the design and or preparation of virtual anatomic models and the data needed to subsequently produce physical models via AM.

References

[1] Lawrence-Zeng G. Image reconstruction: applications in medical sciences. Walter de Gruyter GmbH; 2017. ISBN: 9783110500486.
[2] Salditt T, Aspelmeier T, Aeffner S. Biomedical imaging: principles of radiography, tomography and medical physics. Walter de Gruyter GmbH & Co KG; 2017.
[3] Jayaratne YSN, McGrath CPJ, Zwahlen RA. How accurate are the fusion of cone-beam CT and 3-D stereophotographic images? PLoS One 2012;7(11). www.plosone.org.
[4] Erovic BM, Chan HH, Daly MJ, Pothier DD, Yu E, Coulson C, et al. Intraoperative cone-beam computed tomography and multi-slice computed tomography in temporal bone imaging for surgical treatment. Otolaryngology—Head and Neck Surgery January 2014;150(1):107—14. https://doi.org/10.1177/0194599813510862.

Further reading

[1] Wake N, editor. 3D printing for the radiologist. St. Louis, Missouri, USA: Elsevier; 2022. ISBN: 978-0-323-77573-1.
[2] Kalander W. Computed tomography. Wiley-VCH; 2000. ISBN: 3895780812.
[3] Hofer M. CT teaching manual. Thieme-Stratton Corp; 2000. ISBN: 0865778973.
[4] Henwood S. Clinical CT: techniques and practice. Greenwich Medical Media Ltd.; 1999. ISBN: 1900151561.
[5] Gibbons AJ, Duncan C, Nishikawa H, Hockley AD, Dover MS. Stereolithographic modelling and radiation dosage. British Journal of Oral and Maxillofacial Surgery 2003;41:416.
[6] Swann S. Integration of MRI and stereolithography to build medical models: a case study. Rapid Prototyping Journal 1996;2:41—6.
[7] Loubele M, et al. Comparison between effective radiation dose of CBCT and MSCT scanners for dentomaxillofacial applications. European Journal of Radiology September 2009;71(3):461—8.

[8] Li G. Patient radiation dose and protection from cone-beam computed tomography. Imaging Science in Dentistry 2013;43:63—9. https://doi.org/10.5624/isd.2013.43.2.63.

[9] Drage NA, Sivarajasingam V. The use of cone beam computed tomography in the management of isolated orbital floor fractures. British Journal of Oral and Maxillofacial Surgery 2009;47:65—6. https://doi.org/10.1016/j.bjoms.2008.05.005.

[10] Bianchi S, Anglesio S, Castellano S, Rizzi L, Ragona R. Absorbed doses and risk in implant planning: comparison between spiral CT and cone beam CT. Dentomaxillofacial Radiology 2001;30:S28.

[11] Scarfe WC, Farman AG. What is cone-beam CT and how does it work? Dental Clinics of North America 2008;52(2008):707—30. http://endoexperience.com/userfiles/file/unnamed/New_PDFs/cbCT/CBCT_how_does_it_work_Scarfe_et_al_2008.pdf.

CHAPTER 3

Working with medical scan data

3.1 Image segmentation

As described in earlier chapters, both computed tomography (CT) and magnetic resonance (MR) images are made up of greyscale pixels. In CT, the greyscale is proportional to the X-ray density. In MR, the greyscale will be proportional to the magnetic resonance of the soft tissues. In many cases, it is advisable to work with the original data rather than any three-dimensional reconstruction derived from it. Therefore, CT and MR image data are often manipulated in the pixel format. Much of these data manipulation are similar in concept to popular photo-editing software such as Adobe Photoshop. Many software packages are available that utilize such pixel manipulation to allow specific individual anatomical structures to be isolated form a CT or MR data set and exported in an appropriate format. Many of these packages operate in a similar manner to the software that radiographers routinely use to generate images in radiology departments.

3.1.1 Thresholding

Thresholding is the term used for selecting anatomical structures depending on their density or greyscale value. By specifying upper and lower density thresholds, tissues of a certain density range can be isolated from surrounding tissues. Due to the partial pixel effect described in Chapter 2, small variations in the thresholds may affect the quality of the anatomical structures isolated. The effect may be to make them slightly larger or smaller as illustrated in Figs. 3.1—3.3. However, thresholding will select all pixels within the specified density range regardless of their relationship to individual anatomical structures. This may be overcome using region growing.

The effect can be clearly seen in the real example shown in Fig. 3.4. In this example, bone is selected by setting a high upper threshold and an appropriate lower threshold, with the resulting region shown in Fig. 3.5.

Medical Modeling
ISBN 978-0-323-95733-5
https://doi.org/10.1016/B978-0-323-95733-5.00003-X

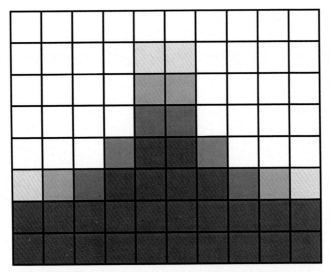

Figure 3.1 Original computed tomography image.

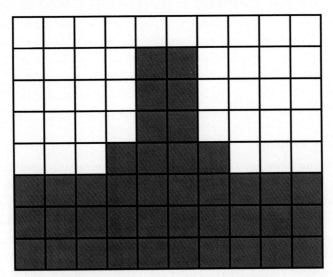

Figure 3.2 The effect of a low threshold.

However, the effects of varying the lower threshold can be seen in Figs. 3.6 and 3.7. Notice that in Fig. 3.6, it is clear that bone is present beyond the boundary of the selected region. However, in Fig. 3.7, we can see those other areas, unconnected to our region of interest, have been selected. It is

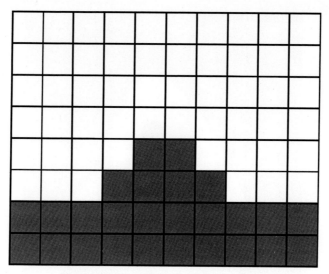

Figure 3.3 The effect of a high threshold.

Figure 3.4 Original computed tomography image.

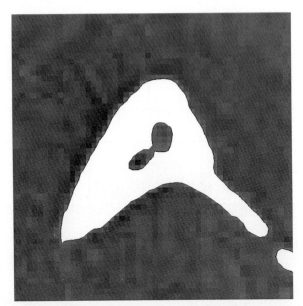

Figure 3.5 Region selected using appropriate threshold values.

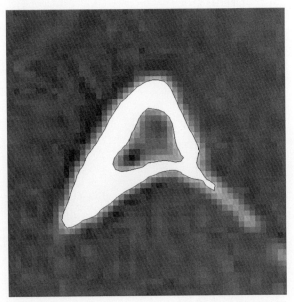

Figure 3.6 Region selected using an increased lower threshold value.

Figure 3.7 Region selected using a decreased lower threshold value.

therefore essential that thresholds be accurately selected where accuracy is of high importance. This becomes particularly critical when very thin or narrow objects are of interest as small changes in threshold can result in these areas not appearing in the selected region.

3.1.2 Region growing

To select single anatomical structures from all of those present within the specified thresholds, a technique called region growing is typically used. This works by allowing the user to select a single pixel within a region already specified by thresholding. The software then automatically selects every pixel within the specified thresholds that is connected to the one selected by the user. This results in single anatomical structures being isolated from neighboring, but unconnected, structures.

However, this is not a simple as it might first appear. It only requires one single pixel (representing perhaps 0.25 mm) to connect two regions for the software to assume that they are the same structure. Therefore, structures that are separate but in close proximity or contact may need to be separated manually before region growing will be successful. This may occur, for example, in joints.

3.1.3 Other techniques

Many other techniques may be available depending on the software being used. Manual techniques can be used and are similar to those found in photo editing software, such as draw, delete, cavity fill, etc. These allow the user to edit data to remove artifacts or connect neighboring structures.

Other techniques are sophisticated variations on the thresholding and region growing functions. These may incorporate local variations to thresholds or add or remove pixels from selected regions to alter the boundaries. The exact nature of these functions will depend on the software being used; therefore, it is not appropriate to attempt to describe them all here.

3.2 Using CT data: Worked examples

Workflows will vary depending on the software being used to segment medical scan data. These examples illustrate some of the most common software available at the time of writing: Mimics (Materialise N.V., Technologielaan 15, 3001 Leuven, Belgium), D2P (3D Systems Inc, Rock Hill, South Carolina, USA) and itk-SNAP [1].

3.2.1 Mimics worked example

Mimics is one of the earliest examples of software developed to segment DICOM format medical data into surface tessellated models ready for fabrication. It is also one of the most widely used in industry and within hospitals around the world. It is certified for medical use being CE marked in Europe and FDA510(k) in the USA.

The first step is to import the data. In most cases, the data will be recording in a format that is compliant with the internationally recognized DICOM format. If this is the case, the software has an automatic import function. During importing, the software converts the images into a format it recognizes as its own and displays the resulting axial images on the screen. The axial images are the original scan images from the CT data. Mimics software also uses the axial pixel data and the slice distance to calculate images in the sagittal and coronal planes.

As described earlier, in CT images, the greyscale is proportional to the density of the tissue. Therefore, the denser the tissue is, the lighter the shade of gray that corresponds to it will be, see Fig. 3.8. Mimics, and software like it, use these greyscale values to differentiate between different tissue types.

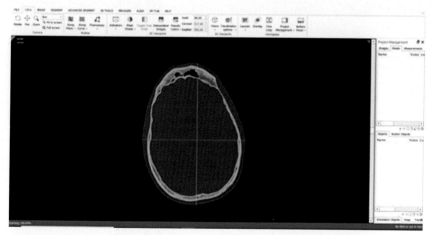

Figure 3.8 Greyscale axial computed tomogrpahy slice image after import into Mimics V24 software.

By selecting upper and lower greyscale values, specific tissue types can be selected. These levels are typically referred to as thresholds. On importing a new data set, Mimics displays the images using a default threshold for bone, which is shown in green in Fig. 3.9. Selecting the desired tissue type is

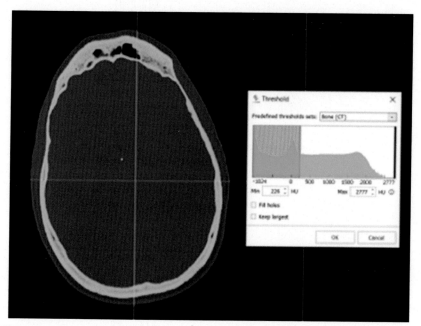

Figure 3.9 Thresholding to select bone density anatomy as a "mask" in Mimics V24.

accomplished by varying the upper and lower thresholds until the required tissue type is isolated. This process is usually referred to as segmentation. The fill holes option can be used to automatically fill in small, trapped volumes early in the segmentation process.

Once the desired tissue type, in this case bone, has been segmented, it may be necessary to limit the selected data to one particular structure. Region growing allows the user to select a certain pixel within the desired structure and the software then automatically selects all other pixels that are connected to it. As this function operates in all three dimensions, single structures can quickly be segmented from the whole data set. This can be seen in the yellow structure highlighted in Fig. 3.10.

The software can also be used to view three-dimensional shaded images of the selected mask data using the "Calculate Part" tool. Many other segmentation and manipulation tools are available to refine the 3D model ahead of manufacturing a physical replica, designing a custom device or using for visualization. The segmented data are then exported in the format used to create the computer files necessary to build the medical model, see Fig. 3.11.

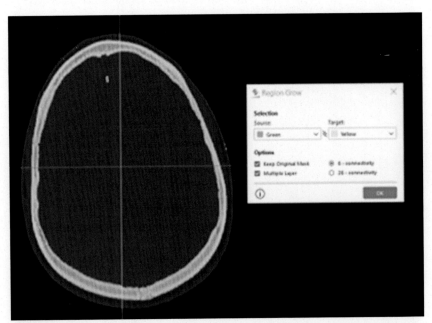

Figure 3.10 Region growing to select only structures attached to the anatomy of interest.

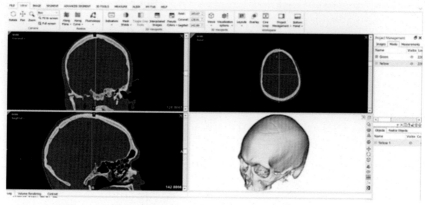

Figure 3.11 Creating a 3D rendering of the anatomical region.

Scans are usually performed in slices or a spiral form in a plane perpendicular to the long axis of the patient. The interval between the slices may be in the order of a millimeter or two, whereas the models are built in layers, which are typically 0.1—0.2 mm in thickness. Therefore, additional software is used to interpolate intermediate slices between the scan data slices. Another interpolation is carried out within the plane of the scan to improve resolution. Together, these operations result in the natural and accurate appearance of the model. This interpolated data can then be exported in several formats that may be transferred to computer-aided design packages or rapid prototyping preparation software.

Some of the different output data formats are described later in this chapter. The same example as shown here is used to enable comparison.

3.2.2 D2P worked example

D2P is another commercially developed software dedicated to segmenting, visualizing, and creating replica anatomical structures. Like Mimics, it is available in different modules that are expandable depending on user needs. It is also certified, in this case FDA 510(k), meaning it is recognized in the USA for application in the custom medical device workflow.

The D2P workflow begins adding DICOM files from a CT or MR scan to either a new patient, or an existing patient folder. When the data are selected, there is an option to automatically open a 3D visualization of the anatomy (Fig. 3.12), which can be adjusted to show different ranges of tissue density. Using automatic options for bone segmentation picks up everything within the predetermined Hounsfield Unit threshold range (Fig. 3.13).

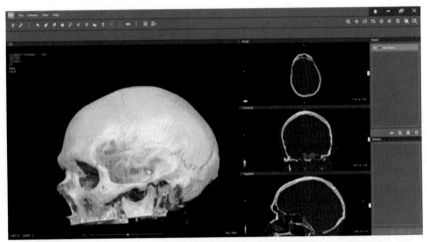

Figure 3.12 Automatic 3D visualization in 3D-Systems, D2P software.

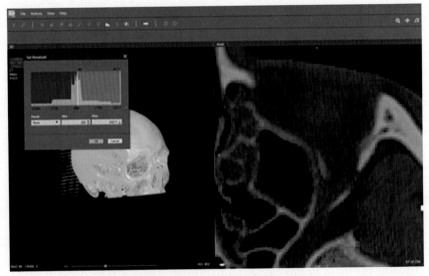

Figure 3.13 Selecting a Hounsfield unit range.

"Path tracking" is used to select portions of anatomy more selectively (Fig. 3.14). Seed points called "Markers" are placed in the axial, sagittal, or coronal views. Areas of anatomy within the threshold range are selected from these marker points, but the selection stops where a boundary of tissue outside of the range is met. This method is best when building up a selection gradually. A "Thin Bones" option can be used to ensure areas of

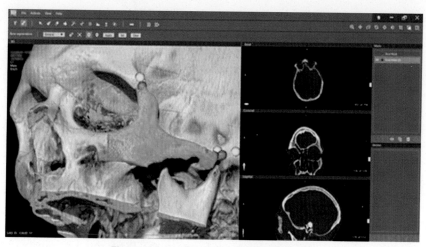

Figure 3.14 Using the path tracking tool.

anatomy such as orbital floors are included in the selection. The "Set" tool is used to create a mask of the selected area ready for further editing. A "Fill" tool is used to close cavities.

As with Mimics, manual tools are used to add in areas of anatomy that were missed during automatic or semi-automatic segmentation. For example, a "Paint" tool can be used to select areas of anatomy using a wider threshold range (Fig. 3.15). This is useful for areas affected by the partial pixel effect, such as orbital floors.

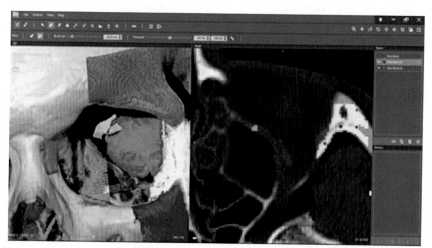

Figure 3.15 Adding thin areas of orbital floor into masks within D2P.

D2P offers visualization in a virtual reality (VR) environment. VR is only used for visualization, education, and training (there are warnings stating that it is not suitable for diagnostic or measurement purposes). The environment enables the user to navigate through anatomy, introduce cut planes, adjust the visualization to show different tissue densities, and make basic measurements. The immersive nature of visualization makes it a potentially useful tool when communicating complex anatomy within a multidisciplinary team.

3.2.3 Itk-SNAP worked example

Itk-SNAP was developed through a collaboration between Paul Yushke-vich, Ph.D., of the Penn Image Computing and Science Laboratory (PICSL) at the University of Pennsylvania, and Guido Gerig, Ph.D., of the Scientific Computing and Imaging Institute (SCI) at the University of Utah, USA. It is a freely available software, which makes it an appealing option for individuals or organizations with limited budget. It is not however, recognized as a medical device according to the FDA, Europe or elsewhere. There is no formal support from a company, but tutorials are available online and linked to through the official website. Features are pared down compared to Mimics or D2P, but still enable DICOM data to be processed into 3D replicas of anatomy. This example illustrates using version 3.8.0.

Importing the data is carried out by selecting the "main" image from the set, using the "DICOM Image Series" file format option. Selecting the first image in the sequence results in the entire set being imported. Axial, sagittal and coronal views are presented automatically (Fig. 3.16). These can be scrolled through and zoomed into.

Segmentation is different to Mimics and D2P. It is semi–autonomous. The first step uses the "Active Contour" segmentation mode to select the region of interest you wish to segment. A segmentation box can be dragged around to isolate the area of interest, in this case, the anterior portions of a cranium (see Fig. 3.17). Selecting "Segment 3D" focusses the views on the area of interest, also opening the next stage, thresholding. The principle of segmentation is similar to Mimics and D2P, but with key differences. A tissue density range is selected, with a preview provided in a "speed" image, which illustrates the results. Areas that will not be selected are colored blue by default and areas that will be selected are left unco-lored (Fig. 3.18).

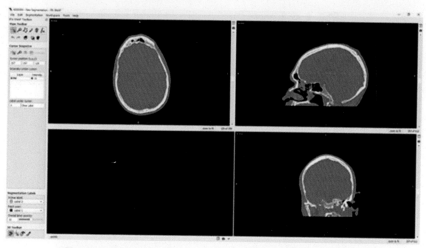

Figure 3.16 The itk-SNAP interface following DICOM import.

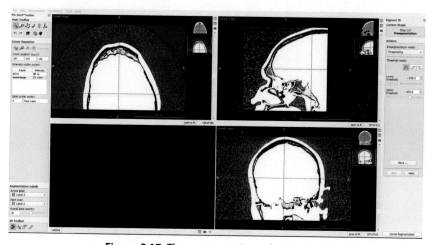

Figure 3.17 The segmentation selection box.

Step 2 of segmentation requires the user to add "Bubbles" in areas of desired segmentation, from which the selection can grow. This is shown in Fig. 3.19. The size of the bubbles can be adjusted so that they do not extend into surrounding areas in three dimensions. Creating multiple "Bubbles" gives more areas to start growing the selection.

Clicking next moves to the "Evolution" stage of segmentation. Parameters that control the segmentation process can be manually adjusted or

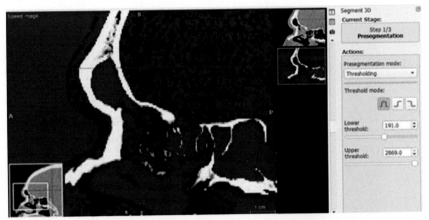

Figure 3.18 Presegmentation selected areas.

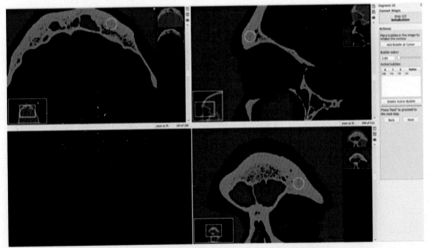

Figure 3.19 Step 2 of segmentation using bubbles.

left as default. Selecting "play" starts the contour, "snake" evolution process, which grows pixel selections throughout the slice series from the bubbles placed previously. This process takes a few minutes depending on the power of your computer. A preview is provided as the selection is grown, so the user can determine when all required connected portions of anatomy are selected. The growing operation does not extend beyond gaps in the tissue density range selected and where a "Bubble" has not been placed. For example, the temporomandibular joint has an articular disk

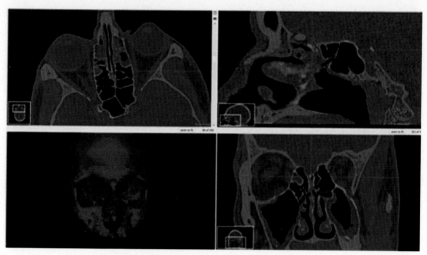

Figure 3.20 Results of a sample segmentation.

separating the mandibular head and fossa. If the mouth were scanned open so teeth were not touching, the growing operation would be unlikely to cross the temporomandibular joint. This can be a useful feature that prevents the need for subsequent operations required to separate structures. The "update" option in the 3D viewer window can be used to show the results of the operation. The results of a sample segmentation are illustrated in Fig. 3.20.

In this case, thin bones around the orbital floors, ethmoid, and maxillary sinus were not captured during the contour evolution process, so adjustments in the threshold range and options within the parameters of the contour evaluation would be required if those areas were of interest. Alternatively, the "paintbrush mode" could be used to selectively draw areas of missing tissue in each image.

There are far fewer complex editing tools and options to link the outputs from itk-SNAP to analysis software or manufacturing processes when compared with Mimics or D2P. There are also few options to refine the 3D results to ensure they are a fair representation of the anatomy. A basic plane cut tool called 3D scalpel can be used to cut and separate areas of anatomy, but in practice, this is rudimentary and more complex segmentation tools are available in other software packages. Additional software tools would also be needed to fill trapped volumes ahead of stereolithography fabrication, for example. The results can, however, be exported as an STL file prior to further software being used to improve quality ahead of manufacture. Another option missing is the ability to import STL or other 3D representations. This can be an important way to check the fit of a custom implant or guide against the original data to check fit and other features required for safe use.

3.3 Point cloud data operations

Noncontact scanning typically produces a large number of points that correspond to three-dimensional co-ordinate points on the surface of the target object. The collection of points taken of an object is usually referred to as a "point cloud." The nature of point cloud data is completely different to the pixel image data we obtain from CT and MR scanning and therefore requires completely different software.

Point clouds in themselves are of little use. They are therefore converted into three-dimensional surfaces, usually called meshes. The simplest method is polygonization. This involves taking points and using them as vertices to construct polygon facets. The collection of facets is usually known as a polygon mesh. The simplest form of polygon is the triangle, so it is frequently used. The steps from point cloud data to polygon surface are shown in Fig. 3.21. Triangular faceted meshes can be easily stored in the STL file format. Applying other shapes of polygon mesh, such as square or hexagonal may be more appropriate, particularly for use in finite element analysis techniques.

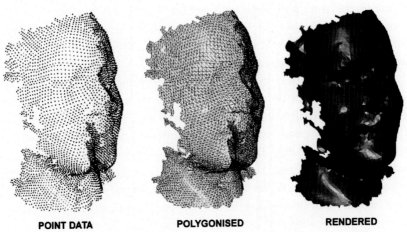

POINT DATA   POLYGONISED   RENDERED

Figure 3.21 A triangular polygon mesh created from a point cloud.

Often scan data of an artifact's topography will extend beyond the principal area of interest, and unnecessary points should be removed immediately. This reduces the memory requirement for any subsequent operations. Once a mesh has been generated from the available points, any gaps will become apparent. At this point, holes and gaps can be filled or corrected to form a single coherent or "manifold" mesh.

As an alternative or extension to mesh modeling, the user may wish to use the scan data to act as a template to construct a series of mathematically defined curves and surfaces. This is typically done in reverse engineering, enabling the artifact's topography to be taken into traditional engineering CAD systems. This requires a great deal of skill and expensive software and is only necessary when it is desirable to perform precise changes to the surfaces. If the ultimate aim is to produce a physical model by AM techniques, then this step often is redundant, as the data will only have to be converted into an appropriate format (e.g., STL) again.

Whichever route is taken, if a physical model is to be made, the surface has to be turned into a single bound volume. This can be achieved by offsetting the surface and filling the gap, resulting in a model with a definite thickness. This can be seen in Fig. 3.22, which shows two views of a three-dimensional model created from scan data of a face. The oblique view clearly shows the thickness of the model. To reduce file size, the rear surface may be simplified as it is of no consequence except to give the model bulk.

There are many open-source point cloud software programs available now, such as CloudCompare (https://www.cloudcompare.org/). Libraries

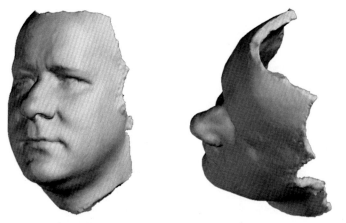

Figure 3.22 A bound volume STL file created from an offset surface mesh.

such as PointClouds (https://pointclouds.org/) are also available to enable developers to integrate existing functionality into custom programs/APIs. Many of these open-source programs and libraries now offer functionalities such as manual and global registration (i.e., alignment of multiple scans), segmentation, and key point reference target positioning for critical points of interest.

3.3.1 Data clean up

Clean-up tools for scan data have improved significantly over recent years, with new algorithms for more complex functions. Often the simplest way to filter and clean scan data is to first ensure the scan has been through the polygonization process, that is, advanced from a point cloud into a mesh, as the visualization of the scan is easier to comprehend to the user in virtual space. There are several tools that the user may decide to use in addition to filtering to enhance the mesh for the intended application. Furthermore, smoothing of meshes may also be necessary through filtering or even relaxation of the overall topography.

Decimation may be required to reduce the number of facets of an overall mesh or a previously selected portion of a mesh. Subsequently, the size of the file reduces due to fewer entities (points, edges, and facets), and future data processing may be accelerated as a result. However, one must carefully consider an acceptable level of accuracy required to balance the file size. Fig. 3.23 shows the effects of decimation, by a 50% and 75% reduction in triangular facets, respectively.

Conversely, the user may wish to increase the number of facets around topographies with high curvature and surface detail to increase surface

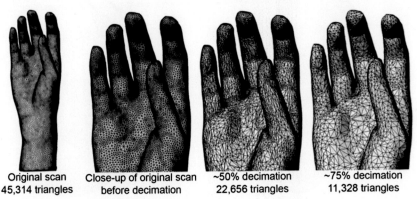

| Original scan | Close-up of original scan | ~50% decimation | ~75% decimation |
| 45,314 triangles | before decimation | 22,656 triangles | 11,328 triangles |

Figure 3.23 Showing the effect of triangle reduction.

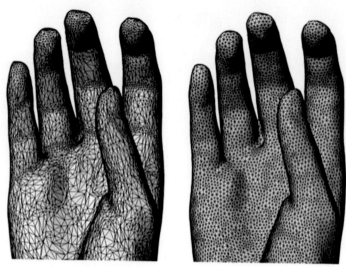

Figure 3.24 Effect of remeshing, decimated scan (*left*) and remeshed mesh with maximum triangle edge length of 2 mm (*right*).

quality. This is achieved by subdividing the existing mesh by a given factor (e.g., 4). The tessellation of meshes may also be adjusted to suit preference; with relatively consistent curvature, a mesh may be re-triangulated to make facet size more uniform (relative to edge length of triangles as a guideline), or by varying the spacing between points. Fig. 3.24 shows effects of re-triangulation or "remeshing," with the goal of creating equilateral triangular facets.

Depending on the data capture method used or the data manipulation performed, a mesh may present several holes; this is particularly prevalent for laser scanning artifacts, which have overlaps and areas, which may not be captured due to line-of-sight limitations. Therefore, several tools are available to fill holes within suitable clean up software. The user may specify varying levels of continuity when filling holes, from flat fill to tangency and curvature. Parameters may also be set to automatically fill holes below a specified diameter, to increase efficiency, and to allow the user to filter out larger more problematic holes, which may require more attention to fill.

3.3.2 CAD data generation

There are several options to generating alternate CAD data from point clouds but, in most cases, require the initial conversion to mesh data to ensure the desired topography is achieved. Visualization of a point cloud

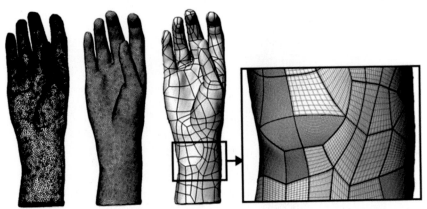

Figure 3.25 Automated creation of NURBS surface patches from a polygon mesh.

can be misinterpreted unless a highly dense point cloud can be achieved. Misinterpretation is often a result of misunderstood point placement and subsequently may create an undesirable output once converted to mathematical surfaces known as NonUniform Rational B-Spline (NURBS) surfaces. Characteristics of NURBS surfaces can be found later in this chapter. Most data editing software packages now feature auto-surfacing tools, which create a series of four-sided NURBS patches over the topography of the artifact. Fig. 3.25 shows the automated creation of NURBS surface patches from a polygon mesh; the close up shows each patch is four sided. Various parameters are available to the user, such as patch count and number of control points. The benefit of this approach is that NURBS patches present a smoother appearance due to their mathematical construction and therefore may offer a more realistic appearance compared to mesh modeling, which may offer an approximate resolution to the original artifact.

3.4 Two-dimensional formats

Typically, in radiography, the output of medical scanning modalities are two-dimensional (2D) images. These are usually prepared from the scan data by the radiographer according to instructions from doctors and surgeons. These images may be from the slices taken through the body or three-dimensional reconstructions. Often, these images are printed on film and treated in much the same way as X-ray films. As medical scan data images are made up of pixels, the images can be exported in familiar computer graphics formats such as bit maps or JPEGs.

Sonography can also be presented with 2D images; using captured imagery in the context of fetal growth, key anatomical landmarks are used to check a range of anthropometric measurements including head circumference, femur length, as well as heart chamber arrangement.

3.5 Pseudo 3D formats

Data can be exported in formats that allow three-dimensional operations to be undertaken without being true three-dimensional forms. The objects are defined by a series of two-dimensional contours arranged in increments in the third dimension. These types of files are often referred to as $2^1/_2$ D data or "slice" formats. More typically, however, these formats are used as an intermediate step in creating true three-dimensional CAD representations.

The formats typically are in the form of lines delineating the inner and out boundaries of structures isolated by thresholding and region growing techniques. The lines are usually smooth curves or polylines that are derived from the pixel data. This technique results in smooth contours that more closely approximate the original anatomical shape that the pixelated data. For example, if we consider the original CT data that are shown in Fig. 3.26, we can see there is a high-density bone structure surrounded by

Figure 3.26 Original pixelated CT image.

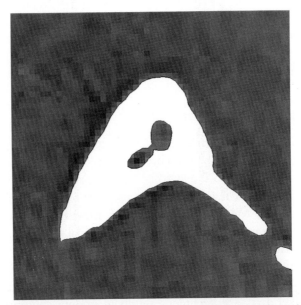

Figure 3.27 Segmented region showing smooth polyline boundaries.

lower density soft tissue. The effect of specifying upper and lower threshold and region growing is shown in Fig. 3.27. The inner and outer boundaries of the selected region are smooth polylines.

The pseudo-3D effect arises when the 2D polylines are stacked in correct orientation and spacing to provide a layered model, similar in effect to a contour map. Fig. 3.28 shows a 3D rendering derived from CT data of the proximal tibia alongside the same data exported in a $2^{1}/_{2}$ D polyline or "slice" format. When such formats are used in CAD, it is common to create surfaces between the slices to generate true 3D surfaces. However, when these formats are used to interface directly with rapid prototyping machines it is common to interpolate intermediate layers between the original slices so that data exists at layer intervals that correspond with the build layer thickness of the machine being used.

The formats that follow are essentially the same and appear similar to that shown in Fig. 3.28. The differences between them are concerned with the order and amount of information stored in them.

3.5.1 IGES contours

Initial Graphics Exchange Specification (IGES) is an international standard CAD data exchange format that has been used for many years. More

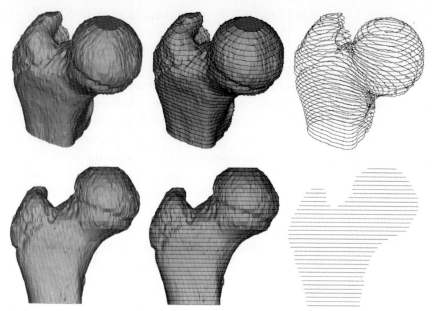

Figure 3.28 Two-and-a-half-dimensional polyline data compared to 3D rendered model.

information about the standard can be found at http://www.nist.gov/iges/. The standard consists of lists of geometric entities, which can be assigned with size and position properties in three dimensions. When dealing with layer data, two-dimensional contours can be described as simple polylines at increments in the third dimension. These contours can be imported into CAD packages that can then form true three-dimensional surfaces on them. IGES files have the three-letter suffix IGS.

3.5.2 Slice file formats

Derived from the word "slice," the SLC is a relatively simple file format that describes the perimeter of individual slices through three-dimensional forms. The inner and outer boundaries are expressed as a set of vectors. The format therefore is an approximation but in practice, the vectors can be generated with sufficient resolution, such that curved surfaces on RP models derived from them appear smooth.

The format was developed by 3D Systems as an alternative input file for their stereolithography technology, but some other RP machines are able to accept the format as input. If an SLC file can be generated from the original source data, it effectively bypasses the need to create STL files. For a given data set, an SLC file will be much smaller than an STL file of comparable resolution. The SLC therefore provides a highly efficient data transfer format for medical modeling using stereolithography. The format is little used compared to the typical STL file.

Other varieties of slice format are available depending on the software being used but are increasingly obsolete. The common layer interface (CLI) is a contour file format that describes cross sectional slices through an object in much the same way as the SLC file. However, the format is commonly available to many software developers. Also derived from the word "slice," the SLI file format is similar to the SLC file format, but rather than being an input file, it is an intermediate file format created during the preparation of stereolithography builds. The file, developed by 3D Systems, not only describes the perimeter of the slices but also includes raster lines that make up the cross-sectional area within the boundary. These raster scan lines are usually referred to as hatches. The file format also includes different types of hatches for up facing and down facing layers. This is important in stereolithography because up-facing and down-facing layers are built differently to optimize accuracy. As with the SLC file format, the ability to generate this file format directly from medical scan data provides an efficient data transfer when generating stereolithography files. The extra hatch information contained in the file increases its size in comparison to the SLC.

3.6 True 3D formats

3.6.1 Polygon-faceted surfaces

Unlike the contour or slice-based formats described previously, true three-dimensional data formats generate computer models that have a surface. If a computer model has a surface, then it can be rendered and visualized on a screen and be manipulated with a higher degree of sophistication than $2^1/_2$ D data. However, file sizes are typically higher.

One of the simplest methods of generating a three-dimensional computer model is to create a polygon faceted mesh. This is achieved by approximating the original data as a large number of tessellating polygon facets. As the triangle is the simplest polygon, it is frequently exploited in polygon faceted representations. However, other polygons are also used, particularly in Finite Element Analysis (FEA).

3.6.2 Finite element meshes

Finite Element Analysis packages rely on breaking down three-dimensional objects into a large number of small discrete elements. These elements may make up only the surface of the object; for example, the STL file described above may be considered a triangular surface mesh. However, some meshes may break the whole object into discrete three-dimensional elements. These may be tetrahedral, cuboids, or other three-dimensionally tessellating polygons.

Depending on the software package used, the voxel data of a CT or MR scan may be exported or translated into a solid or surface mesh suitable for use in FEA. There are usually variables to set when conducting the translation that affect the quality of the mesh and the resulting file size. This is similar to specifying the quality versus file size compromise for STL files described previously.

Many FEA packages have their own formats for meshing and some medical software packages can produce and export the correct format for a given analysis package. However, this may require the purchase of specific translators or additional modules for the software package.

3.6.3 Mesh optimization

A collection of manifold facets is referred to as a mesh. There are several methods to optimize a mesh in order to balance surface representation (accuracy and resolution) with file size. These optimization methods have been described in more detail later in this chapter.

3.6.4 Mathematical curve based surfaces

Unlike faceted polygon surfaces, curve-based surfaces use complex mathematical routines to produce smooth curves in three-dimensional space, such as NURBS curves (Fig. 3.29). NURBS curve construction involves control points, knots, and knot vectors.

NURBS curves with more than two control points have different geometric continuity (G) states defined by a degree; G0, G1, and G2 states can be seen in Fig. 3.30. Assuming the user creates a curve with seven control points, placed in the same position (Fig. 3.30a); a G0 curve has no degree of continuity and thus appears as six distinct sections (Fig. 3.30b). G1 displays tangent continuity, where sections are perpendicular but may still display sharp changes in trajectory; this can be seen in the curvature map on Fig. 3.30c, where map direction can have a sudden stepped change. G2, however, has perpendicular continuity but also ensures a smoother trajectory across sections, and this can be seen with the curvature map in Fig. 3.30d, which ensures a smooth transition across sections without stepping.

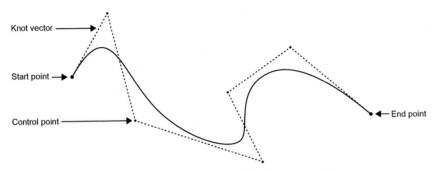

Figure 3.29 Nonuniform rational B-spline curve features.

Some of these routines also operate in three dimensions in order to create complex curved surfaces called "patches." NURBS patches are typically formed from four joined curves with G0 continuity, and principles of curve continuity may be extended to joining surfaces for boundary representation (B–Rep) modeling, for example. An object may require a number of patches to cover the whole object surface. The patches differ according to the complexity of the mathematical curve routine that is used.

This kind of surface modeling produces highly sophisticated surface models that are typically used in the automotive, motor sports, and aerospace industries. Usually, objects are designed using surface modeling packages. However, surface patches are also often used in reverse engineering to create useful CAD geometry from digitized physical objects.

The modeling behavior of NURBS surface modeling is very different to that of polygon modeling. As described previously, NURBS surfaces are typically constructed from four boundary curves or edges from adjoining surfaces. B–Rep models are constructed of a series of trimmed NURBS surfaces, but each surface still retains its underlying data, as shown in Fig. 3.31.

Curves and surfaces are mathematically described and therefore typically relatively smooth and simple surfaces. As such, they are often used to define the smooth but accurate surfaces of objects in product, aerospace, and automotive design. This makes them less well suited to the highly complex surfaces of human anatomy. However, there are many cases where it may prove to be a useful approach, particularly when attempting to integrate human anatomy with the design of products that must accommodate or fit around people.

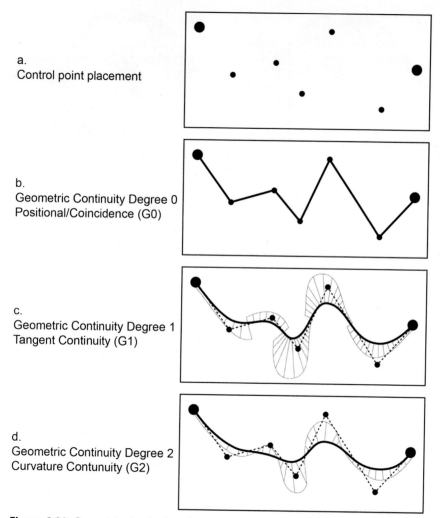

a.
Control point placement

b.
Geometric Continuity Degree 0
Positional/Coincidence (G0)

c.
Geometric Continuity Degree 1
Tangent Continuity (G1)

d.
Geometric Continuity Degree 2
Curvature Contunuity (G2)

Figure 3.30 Geometric continuity along a nonuniform rational B-spline curve with seven control points, (a) control point placement, (b) geometric continuity degree 0 positional/coincidence G0, (c) geometric continuity degree 1—tangent continuity G1, (d) geometric continuity degree 2—curvature continuity G2.

The curves and surfaces are created by positioning the control points and boundary lines that define the surface patch onto the surface of the source data. The surface patch itself is then mathematically created according to the type of surface the software uses. The most complex type of surface patch is defined by NURBS surfaces.

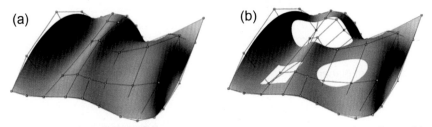

Figure 3.31 Nonuniform rational B-spline geometry, (a) untrimmed surface with construction features (control points and knot vectors) and (b) trimmed surface showing maintained underlying control features.

The structure of mathematically defined surfaces can be communicated with isometric parameters or "isoparms"; integrated curves which assist in defining the topography of the surface. The isoparms typically travel in two opposite directions, termed "u" and "v" (Fig. 3.32). While the surface patch may have a given number of isoparms in u and v directions, the user can also adjust the visual count to help with communicating the form in wireframe; in this instance, the user views "isocurves".

If used to replicate the topography of a scan, for example, then the degree to which this kind of surface patch matches the source data can be controlled by altering the number of control points in the surface. More control points enable the surface patch to be more complex and therefore follow the original data more closely. The number of control points is a variable set by the user when creating the patch.

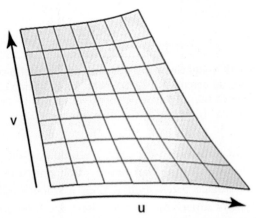

Figure 3.32 Nonuniform rational B-spline surface with isoparms defining the surface topography.

The effect of altering the number of control points can be seen in Fig. 3.33, which shows an IGES surface patch created from noncontact scan data of a human hip with control points varying from 6 to 100. The areas of darker gray show where the IGES surface patch closely approximates the scan surface, while the lighter areas show that the IGES surface patch is below the scanned surface. The surface patch is also shown as a mesh

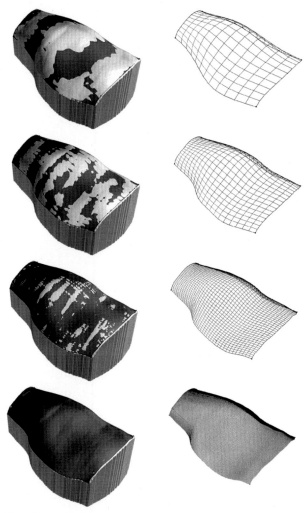

Figure 3.33 The effect of the number of control points on patch quality (from top), six control points, 10 control points, 20 control points, and 100 control points (bottom).

showing the complexity of the surface as the number of control points increases.

Depending on the software being used, the quality of the fit between the surface and the original data can be visually inspected on screen as shown in Fig. 3.33 or numerically quantified. For the surface with only six control points, the average gap between original surface and the patch is 0.818 mm. In comparison, the surface shown with 100 control points is a much closer fit with the average gap almost negligible at 0.00678 mm. As NURBS surfaces are controlled by mathematical equations, the patches themselves have to obey certain criteria in order for the equations to solve. Failure to obey these constraints will result in no surface patch being generated or a patch that flips creases or twists. Typically, surfaces that exhibit these faults cannot be physically manufactured or they give rise to other problems.

To create solid models from surface patches, all of the surface area of the object must be covered. In addition, the patches must meet at common edges and not overlap. These surfaces can then be "stitched" together to form a true three-dimensional solid computer model of the object.

Advances in surface generation tools now allow for automation of surface patches, the user may specify a patch count and topography may then be created. Auto-surfacing is described in more detail later in this chapter.

3.6.5 Subdivisional modeling

Subdivisional modeling is a novel modeling approach, which bridges polygon and NURBS modeling strategies. With typical functions such as face, edge and vertex pulling and face bridging, subdivisional modeling seeks to make the CAD workflow of conceptual and visual design faster and easier and is ideal for creating and manipulating digital representation of organic forms. Often used in animation and visual effects, significant benefits include simple transfer between meshes to NURBS surfaces, as well as a combination of unique modeling capabilities not previously possible with NURBS or polygon modeling; star points can be included which involve more than four facets or patches aligning at a single point which can be inherently problematic to capture smooth surface transitions using NURBS modeling. T-junctions are also possible, as well as localizing subdivision for focusing complex features. Finally, surface quality is not compromised unlike polygon modeling, which can result in angular

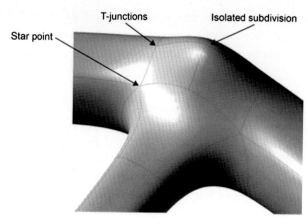

Figure 3.34 Close up view of a subdivisional model showing a star point and T-junction.

aesthetics if the geometry is represented with a low facet count mesh. Fig. 3.34 shows a close-up view of a subdivisional model showing a star point and T-junction.

However, subdivisional modeling is still an approximate modeling topology. For high accuracy, high tolerance components, or to ensure accurate fit to anatomical modeling, both polygon and NURBS topologies are still typically used.

The use of subdivisional the medical sector is still limited but increasingly used in product design through widened adoption in many CAD programs such as Fusion 360 (Autodesk), Creo (PTC), and Rhinoceros (McNeel).

3.6.6 Voxel modeling

A major impediment to the application of CAD in medical applications, such as prosthetics and implant design, is the fact that it requires the integration of existing anatomical forms with the creation of complex, naturally occurring free form shapes. Until recently, CAD has been driven and developed specifically to define geometry for engineering processes using techniques described in the previous sections of this chapter. Consequently, the way they operate makes it extremely difficult to integrate human anatomy and create similar forms. In addition, the methods used to define shapes in CAD are based almost entirely on mathematical geometry (straight lines, angles, arcs, splines, etc.). In contrast to engineers, prosthetists have highly developed visual and tactile skills that allow them to handle materials to create accurate free form shapes. This creates a significant barrier to the application of current engineering-based CAD software by experienced prosthetists or others designing bespoke-fitting medical products.

Voxel-based modeling software has the potential to overcome many of the geometric constraints of other, more engineering-focussed CAD software. Unlike traditional engineering-based CAD systems, voxel-modeling software has been developed to visualize and manipulate solid, complex, unconstrained three-dimensional shapes and forms by manipulating many thousands of tiny cubes (akin to volumetric pixels). The effect is more like manipulating clay or wax but in a digital environment. This provides the opportunity to add higher levels of complex details, such as physical textures, and enables more precise blending of designed structures into surrounding anatomy.

There is a limited range of voxel modeling software. Three of the foremost are Z-Brush (Pixologic Inc, USA), 3D-Coat (http://3d-coat.com/) and FreeForm (Geomagic, 3D-Systems, USA). Of these, FreeForm has the most widely reported use in medical applications. FreeForm has a unique interface (Touch Touch X haptic interface); the hardware consists of a three-dimensional stylus developed from research carried out at Massachusetts Institute of Technology. The device gives all six axes of movement via a handheld stylus, similar to a pen or sculpting tool. Crucially, this device not only has six axes of freedom but also incorporates tactile-feedback. Thus, when moving the cursor on screen in three-dimensions when the cursor comes to the surface of an object on screen, the user feels the contact through the stylus. The resistance can be varied to simulate materials of different hardness and consistency.

Unlike traditional engineering-based CAD systems, FreeForm and the other voxel modeling software noted has been developed to visualize and manipulate solid, complex, unconstrained three-dimensional shapes and forms. Material can be ground away, drilled, stretched, pushed, or added in a manner analogous to sculpting with clay or wax.

3.6.7 STL modeling

It is often rare for CAD users to start their initial CAD modeling strategies with triangular mesh modeling in the form of STL data; instead, the user may start with other forms of 3D data, such as a point clouds, quadrilateral mesh modeling, or B-Rep modeling, before progressing onto STL. In most cases, STL data are normally exported with the intent to produce valid data for Additive Manufacture. In this case, it is vital that STL data is "water-tight" or closed prior to transferring to an AM system. Therefore, STL modeling is often necessary to produce adequate files in support of AM. Software such as Geomagic Studio (Geomagic Solutions, 430 Davis Drive, Suite 300, Morrisville, USA) and Materialise Magics (Technologielaan 15, 3001 Leuven, Belgium) are specifically tailored for reverse engineering tasks, such as point cloud manipulation and STL modeling. Clean-up tasks including polygonization and decimation are described in more detail within Section 3.3.1.

Some CAD packages enable modeling directly in mesh formats and can be used for modeling arbitrary forms for artworks, sculpture, jewelery, or medical modeling. Examples include 3D Studio Max (Autodesk) and 3-Matic (Materialize).

3.7 File management and exchange

3.7.1 STL

Derived from the word "stereolithography," the STL file is a simple file format that describes objects as a series of triangular facets that form its surface. For example, if we view a simple object as an STL file, we can see the triangles. It can be seen that large flat areas require few facets, whereas curved surfaces require more facets to approximate the original surface closely, see Fig. 3.35. The format was originally developed by 3D Systems to provide a transfer data format from CAD systems to their stereo-lithography technology, but it has subsequently been adopted as the de facto standard in the RP industry.

The STL file simply lists a description of each of the triangular facets, which make up the surface of a three-dimensional model. Fig. 3.36 shows the beginning and end of an STL file in text format. The first line describes the direction of the facet normal. This indicates which surface is the outside of the facet. The next three lines give the co-ordinates of the three corners, or vertices, of the facet.

The simplicity of the triangle makes mathematical operations such as scaling, rotation, translation, and surface area and volume calculations straightforward. The format also allows the angle of facets to be identified, which is necessary for stereolithography.

STL files can be in binary or text (ASCII) format. Binary format files are much smaller and should be used unless there is a specific reason the text

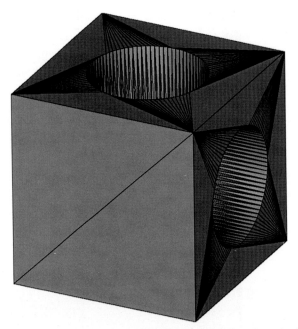

Figure 3.35 STL file of a cube with holes showing the triangular facets.

```
solid FILENAME
   facet normal 1.000000e+00  0.000000e+00  0.000000e+00
      outer loop
         vertex 0.000000e+00 -1.204845e+00 -1.658504e+00
         vertex 0.000000e+00 -1.235913e+00 -3.804270e+00
         vertex 0.000000e+00 -4.000000e+00
      endloop
   endfacet

and so on...

   facet normal 1.000000e+00  0.000000e+00  0.000000e+00
      outer loop
         vertex 1.000000e+00  4.000881e+00  1.221143e-04
         vertex 1.000000e+00  3.535500e+00  3.535500e+00
         vertex 1.000000e+00  3.999953e+00  0.000000e+00
      endloop
   endfacet
endsolid FILENAME
```

Figure 3.36 The beginning and end of an STL file in text format (ASCII).

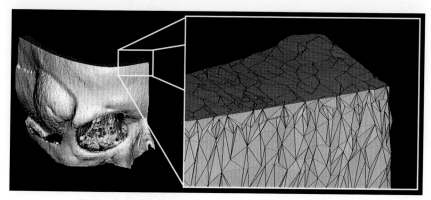

Figure 3.37 Close up view of an STL file showing facets.

format is required. STL files can vary in size from around 50K to hundreds of megabytes. Highly complex parts may result in excessively large STL files, which may make data operations and file transfer difficult. Fig. 3.37 shows a close-up view of an STL file illustrating the vast number of triangles required to provide accurate representation of an anatomical form. However, highly effective compression software is freely available that can reduce an STL file to a small fraction of its normal size to enable easy transfer of files over the Internet or by email.

To be used successfully in AM, the STL file must form a single enclosed volume; meaning that it should have no gaps between facets and all the facets should have their normals facing away from the part (i.e., identifying which is the inside and outside surface). Usually, small problems with an STL file can be corrected with specialist software or the AM machine preparation software. STL files can be generated from practically all three-dimensional CAD systems. Solid modeling CAD systems rarely have problems creating STL files, but surface modellers can pose problems if the surfaces are not all properly stitched and trimmed.

When exporting an STL file from a CAD system, the user will normally specify a resolution or quality parameter to the STL file. This is normally achieved by specifying a maximum deviation. The deviation will be the perpendicular distance between a facet and the original CAD data where the facet forms a chord at a curved surface, as shown in Fig. 3.38. In essence, a smaller deviation will give a more accurate representation of the CAD model, but this will result in an STL file with a greater number of smaller facets. The file size depends only upon the number of facets and so a smaller deviation will give rise to a larger file. The effect is illustrated in Fig. 3.39.

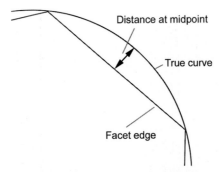

Figure 3.38 Facet deviation.

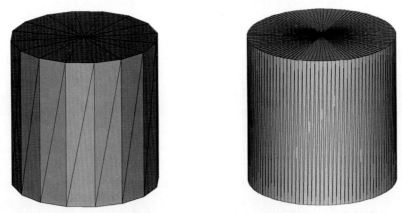

Figure 3.39 The effect of deviation settings (*left*) large deviation = small file but poor quality and (*right*) small deviation = large file but better quality.

When creating STL files from medical scan data a similar effect can be achieved. Typically, rather than setting a maximum deviation, the STL file resolution is determined by relating the triangulation to the voxel size of the CT data. If every voxel is triangulated the maximum resolution will be produced, however, the file size will be correspondingly large. By triangulating multiple voxels, a simpler but smaller STL file will be produced. Typically, a compromise is achieved between file size and surface quality. Some software will simplify this by allowing the user to specify low-, medium-, or high-quality settings.

For example, compare the results of software settings that range from "low" at 4.3 MB, "medium" at 11.7 MB and "high" at 34 MB, with the highest possible resolution settings resulting in an STL file of over 94 MB.

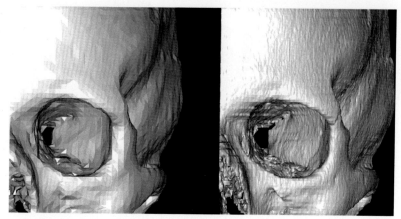

Figure 3.40 The effect of STL file resolution settings on medical scan data showing (*left*) low resolution and (*right*) the highest possible resolution.

The two extremes of the effect of these settings can be seen in the close-up views of the resulting STL files shown in Fig. 3.40.

STL files can be postprocessed in several ways usually to enable them to be produced by AM methods more efficiently. Many software applications are available that allow STL files to split into separate parts or re-oriented in space. In addition, the quality of the STL file can be manipulated somewhat. File size can be reduced by triangle reduction techniques and smoothing algorithms can be applied.

3.7.2 OBJ

Capable of transferring NURBS geometry (curves and surfaces) and polygon meshes, the Object (.obj) file format is one of the most powerful and versatile transfer methods. Furthermore, properties relating to a file may also be included such as object names, materials and color.

3.7.3 VRML/X3D

Early versions of Virtual Reality Modeling Language (VRML) are also triangular faceted surface representations. They are in essence very similar to STL files but are typically created at very small file sizes and are coded in a more efficient manner to make them suitable for transfer over the Internet. VRML also allow the inclusion of color and texture mapping to models (which can be applied to an object, mesh, shell, individual polygon facets, vertices, or using a texture/UV map) and may even contain animation/

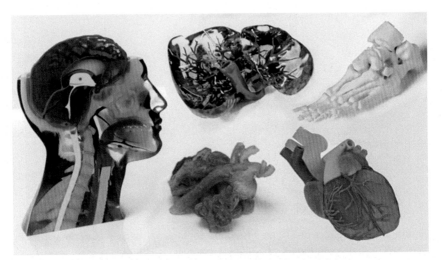

Figure 3.41 Full-color 3D-printed medical models.

sound data. An extended version of VRML, termed X3D, may include additional data such as geo-locations as well as improved rendering. Applications in the medical sector include interactive teaching aids and procedural training/planning visualisations, which may be beneficial to healthcare professionals with varying levels of expertise, from medical students to consultants (See https://www.stratasysdirect.com/resources/design-guidelines/polyjet-multi-color) and https://grabcad.com/tutorials/how-to-3d-print-in-full-color-part-1) (Fig. 3.41).

3.7.4 STEP

The Standard for the Exchange of Product Model Data format (ISO10303) or STEP format is used to transfer solid models between high-end CAD packages. The advantages of the step format are that includes not only the geometrical features that make up the model, but it also includes the history tree—that is the list of operations or construction stages that led to the model. This enables a model developed in one CAD package to be edited, revised, or redesigned in another CAD package. http://en.wikipedia.org/wiki/ISO_10303-21.

3.7.5 IGES

As stated previously, the IGES format is an international standard that describes computer aided design data as mathematically defined geometries positioned in the three-dimensional space (see http://www.nist.gov/iges/). By converting data from the original source through an intermediary three-dimensional format, such as the STL file some CAD packages may be able to generate geometry such as vertices, curves, and surfaces based on the original data. The nature of the surface and the degree to which it accurately reproduces the original anatomy will depend greatly on the data formats and CAD packages used. IGES file exchange also allows for the transfer of model properties such as labeling, notes, and color. Some CAD packages also enable the user to create curves and surfaces from point cloud data obtained using touch probe or noncontact surface scanners.

3.7.6 AMF/STL2.0

Fundamental limitations of the STL format for file management are that STL only communicates an approximate form of a mesh relative to its resolution and cannot include additional build properties or metadata. Given the expansion of the AM market leading to development of new technologies and supporting software, a data management system was needed to communicate characteristics of CAD models such as multi-material and multicolor constructs as well as complex lattices and scaffolds. Intellectual property has also been a concern since STL formats do not permit limited access, leading to unauthorized sharing and access of the file content.

The Additive Manufacturing Format (AMF) or STL 2.0 (ISO/ASTM52915) overcomes these limitations. In tackling lack of detail through approximation, the AMF triangles no longer remain planar like STL facets; AMF triangles effectively become curved patches. This adaptation encompasses tangency in facets by noting the normal direction at mesh vertices (Fig. 3.42) and subsequently subdividing facets (Fig. 3.43). This reduces the file size, since larger triangular facets may be used to communicate a larger surface area while also increasing the accuracy of the part relative to the origin of the data. Uptake of the AMF format by software developers for 3D CAD software has been limited to date but shows much promise in future applications due to its versatility.

3.7.7 3MF

Led by an international consortium of companies such as Autodesk EOS and Microsoft [2], the open-source format is generated with XMF, meaning the contents are easily readable, interpreted, and modified in a

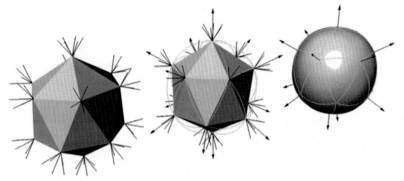

Figure 3.42 STL to AMF conversion, showing (*left*) triangular polygon mesh of a sphere (24 faces) with individual vertex normals, (*middle*) combined normal at vertices (*arrowed*) and approximated curves and (*right*) curved facets through combining the vertex normals of surrounding facets.

Figure 3.43 Mesh face subdivision (*left to right*) subdivided by a factor of four, a maximum of four times resulting in 256 triangles.

text editor if necessary. Furthermore, the files include additional metadata, such as model color, multimaterial properties, creator, and copyright information as well as encryption (using the Secure Content Extension). One of the most complex demands of design for AM is the development of multimaterial parts, particularly with functional grading (the transition from one or more material properties to another). There are various CAD methods to enable the transition of materials, including dithering patterns of voxels to uniformly or stochastically, but often the transfer of these data to AM machines has been troublesome and time-consuming using conventional file formats such as STL, as each volume must be defined as a separate shell, resulting in extremely high file sizes. 3 MF aims to solve complex design-to-AM limitations such as this, with the volumetric extension (see https://3mf.io/announcement/2021/11/3mf-launches-volumetric-extention-work/); the ability to communicate forms with volumetric mathematical formulae (see, https://github.com/3MFConsortium/spec_volumetric/blob/master/3MF%20Volume

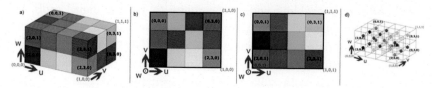

Figure 3.44 Voxel distribution for multiple material AM.

tric%20Extension.md; Fig. 3.44) as opposed to defining separate shells, thereby enabling enhanced functions such as gradual color/material blends. Data are transmitted to AM machines using image stacks representing voxel grid distribution.

It is likely that 3 MF will continue to evolve as AM technologies adapt to offer advanced functionality (e.g., more material choice, reduced layer thickness, higher resolution material deposition, etc.), and demands of end users evolve in response to technological developments. Adoption of the volumetric extensions in CAD programs will also be necessary to enable users to gain the benefits of these developments.

3.8 Summary

This chapter covers the methods, tools and key considerations for working with medical scan data in various forms prior to file export in preparation for downstream activities, including (but not limited to) visualization, simulation, and digital fabrication (i.e., AM). The workflows required to transition data from imaging to file formats suitable for downstream activities vary depending on the imaging methods used, software available, and the selection of digital fabrication process. It is important to consider the key requirements of the intended outcomes, whether they be virtual visual aids or physical artifacts to be fabricated (e.g., material physical properties, cost, color, etc.) early in the design and fabrication workflow, as this has an impact on the imaging, software, and hardware used throughout the workflow.

It is vital to choose appropriate image processing and CAD software to fully define and capture the user's design intent of the intended artifact(s). Effective software tools enable the quick, effective exploration, and design stages required, based on the imported data, to deliver on key design requirements of the artifacts (often determined through design specifications) to help reduce lead times; while enhanced digital tools often come with additional cost, advances in low-cost/open source image processing, and

CAD software is enabling wider accessibility to suitable digital tools, but fundamentally, many software programs regardless of price rely on similar principles of topology, that is, mesh/polygon, NURBS, subdivisional, and voxel-based geometry. The effective communication of such geometry is then only realized with suitable file management and exchange, with increased availability of more dynamic file formats enabling a wider selection of fabrication opportunities. The next chapter covers physical reproduction of artifacts using AM and related manufacturing technologies.

Acknowledgments

With thanks to Jason Ingham, Prince Charles Hospital, Cwm Taf Morgannwg University Health Board for access to the 3D-Systems D2P software.

References

[1] Yushkevich PA, Joseph P, Heather CH, Rachel GS, Ho S, Gee JC, et al. User-guided 3D active contour segmentation of anatomical structures: significantly improved efficiency and reliability. NeuroImage July 1, 2006;31(3):1116—28.
[2] https://docs.microsoft.com/en-us/windows/uwp/devices-sensors/generate-3mf.

CHAPTER 4

Physical reproduction

4.1 Introduction to additive manufacturing

4.1.1 Introduction

Rapid prototyping is a phrase that was coined in the 1980s to describe innovative technologies that produced physical models directly from a three-dimensional computer-aided design (CAD) of an object. Many other phrases have been used over the years including solid freeform fabrication; layer additive manufacturing, 3D printing, and advanced digital manufacturing. In the late 1990s, the application of these technologies to tooling was investigated, and the phrase rapid tooling was used to cover these direct and indirect processes. Concurrently, these technologies were applied to the manufacture of components and products and not just prototypes, and the phrase rapid manufacturing was used. As the importance of manufacturing overtook prototyping, the term additive manufacturing (AM) has become accepted as the term that encompasses all layer additive technologies, and therefore, it is the one used here. Although, some of the case studies may use a variety of terms depending on when they were first written.

The objects created by AM processes may be used as models, prototypes, patterns, templates, components, and end-use components or complete products. However, for simplicity, the objects created by the AM processes described in this chapter will be referred to as models or parts, regardless of their eventual use.

Due to the technology-driven nature of AM companies and their products, the industry is awash with trade names, abbreviations, and acronyms for the various processes, software, hardware, and materials. Many of these are registered or recognized trademarks, and these have been indicated where possible. The consensus terminology is covered in the next section, and a glossary of terms can be found in Chapter 7.

Medical Modeling
ISBN 978-0-323-95733-5
https://doi.org/10.1016/B978-0-323-95733-5.15006-9

The most common AM processes are described later in this chapter. Each major AM process type is covered in principle. While it is not practical to describe every single aspect of every machine available, the sections provide an overview of the technologies, their pros and cons, and their appropriateness for medical applications.

4.1.2 AM terminology

The categorization and terminology of AM took many years to reach a consensus view. As these technologies moved from modeling and proto-typing into the manufacture of end-use components and products, the need for internationally recognized standards led to the ISO defined terminology found in *ISO/ASTM 52900 Additive Manufacturing—General principles—Terminology* initially in 2015 and revised in 2021. The standard defines seven categories of AM given below (Table 4.1). This chapter focuses on the processes most commonly used in medical applications. Directed energy deposition is omitted as it typically refers to metal materials only and is typically used for larger structural and mechanical parts. Sheet lamination and binder jetting are omitted as they are diminishing in use, few systems are currently available, and they are little used in medical applications compared with the other processes.

Table 4.1 ISO-defined AM processes.

Category	Description
Binder jetting	A liquid bonding agent is selectively deposited to join powder materials
Directed energy deposition	Focused thermal energy (e.g., laser, electron beam, or plasma arc) is used to fuse (melt) materials by melting as they are being deposited
Material extrusion	Material is selectively dispensed through a nozzle or orifice
Material jetting	Droplets of build material are selectively deposited
Powder bed fusion	Thermal energy selectively fuses regions of a powder bed
Sheet lamination	Sheets of material are bonded to form a part
Vat photopolymerization	Liquid photopolymer in a vat is selectively cured by light-activated polymerization

4.1.3 Layer additive manufacturing

AM systems work by creating models as a series of contours or slices built in sequence, often referred to as layer additive manufacturing. The different AM systems vary in how they create the layers and in what material.

By convention, and subsequently adopted in *ISO/ASTM 52921:2013 Standard Terminology for Additive Manufacturing—Coordinate Systems and Test Methodologies*, the axes X and Y represent the plane in which the layers are formed, and the Z axis is the build direction, usually referred to as the height. This is shown in Fig. 4.1. While the ISO standard defines the origin (0, 0, 0) to be at the center of the bed, most machines define their build volumes in positive space and therefore have their origin in one corner of the build volume as shown. Consequently, the number of layers needed for a given object is a function of the layer thickness and height of the model. The accuracy and resolution in the XY plane are therefore dependent on this mechanism. Most are accurate to fractions of a millimeter so that any geometry in the XY plane should be faithfully reproduced.

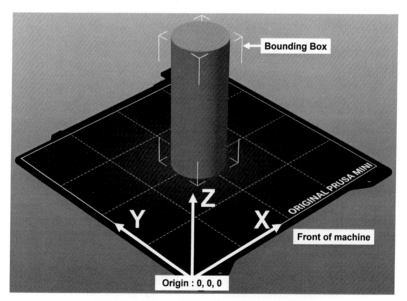

Figure 4.1 Conventional coordinate system.

The layer thickness is dictated by the mechanical process that adds the build material and may vary according to the material being used. Layer thickness is usually in the order of 0.05−0.30 mm, although some material jetting technologies are capable of depositing thinner layers. The addition of layers leads to a stepped effect in geometry perpendicular to the XY plane.

As an example, consider a cube with two perpendicular holes through it, as shown in Fig. 4.2. The hole in the top of the cube, as viewed from above, formed by the scanning mechanism in the XY plane will be formed perfectly (within the capability of the drawing mechanism). However, the hole in the side of the cube, as viewed from the side, formed by the addition of layers will display a stepped effect (shown in Fig. 4.3). Consequently, when building objects, careful consideration should be given to the orientation so that the optimum features are formed in the XY plane and stepping is avoided. For example, a cylindrical shape should be built upright if the circular section is to be faithfully reproduced.

The extent of the stepping can be diminished by creating thinner layers, but as the overall build time tends to be more dependent on the layer addition process, a larger number of thinner layers leads to longer build times. Therefore, a compromise between surface finish and speed is established for each AM process.

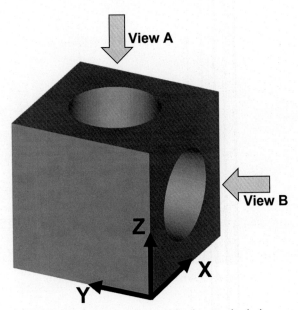

Figure 4.2 Cube with perpendicular circular holes.

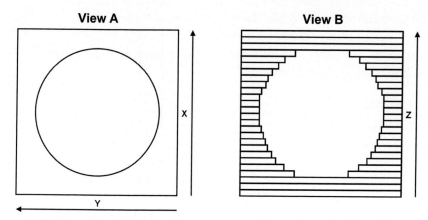

Figure 4.3 Top view (a) and side view (b) showing the stair-step effect.

4.1.4 Boundary compensation

As AM is an additive process, the method used to create the layers has a finite minimum size. In the case of the material jetting, this is usually expressed as the pixel size that the print head is capable of reproducing and expressed in the same terms as paper printers in dots per inch or dpi. However, with laser-based technologies such as stereolithography (SL) and laser sintering (LS), the diameter of the laser beam provides the minimum capability. For maximum accuracy, the software that controls this process offsets the path of the laser by half of its diameter so that it draws a path within the boundary of the object. This is illustrated schematically in Fig. 4.4 (left). However, this means that small, thin features below that minimum size cannot be reproduced. A similar effect is shown by material extrusion where the minimum thickness of the deposited bead of extruded molten plastic is compensated for in the same manner.

In practice the size of offset is very small, typically small fractions of a millimeter; however the effect should be considered separately to

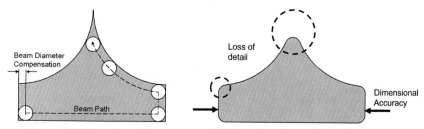

Figure 4.4 Boundary compensation (left) and resulting loss of detail (right).

dimensional accuracy. While dimensional accuracy may remain very good, large offset values will have a detrimental effect on the ability of the AM system to reproduce intricate detail, crisp edges, and sharp corners, as illustrated in Fig. 4.4 (right).

4.1.5 Data input

All AM systems require a computer model of the object to be built. In all cases, software is used to take a digital three-dimensional model of the object to be built and slice it into a vast number of very thin cross-sections. These cross-sections then are translated into an appropriate format to direct the layer manufacturing process of the AM system.

As there are a wide variety of CAD software programs and AM systems available, each with different requirements, industry required a standard translation format. SL was the first widely commercialized AM process, and it employed a mathematically simple approximation of three-dimensional CAD models, called the STL file, which is fully described in Chapter 3. The format describes models by closely approximating their shape with a surface made up of many triangular facets. As it was the first convenient transfer format to be offered by CAD software developers, the STL was widely adopted by other AM manufacturers and has since become the de facto standard AM file format. Consequently, despite some shortcomings, all AM machines can use the STL file. However, other formats are also available, and a description of some of the most relevant is provided in Chapter 3.

In practice, AM software slices the STL file into cross-sections. The software then creates control files that will instruct the AM machine how to deposit each layer and how to deposit subsequent layer material. Depending on the process, there may be a degree of user interaction at this point that can help to optimize the build. The result is one or more files that are transferred to the AM machine itself. Typically, the build file is then checked, and the machine is prepared for a new build.

Although the models may take many hours to build, AM machines typically operate unattended, nonstop. Therefore, a build will frequently be started last thing in the working day, and the machine will run overnight; often the model will be completed by the next morning. Large industrial machines may operate unattended for several days.

All AM technologies require some method of supporting the build in progress, which will subsequently require some degree of postprocess cleaning and finishing after the model is built. Cleaning and finishing are usually done by hand to remove residual material, remnants of support structures, and the stepped effect. Some processes require secondary processes. The level of cleaning and finishing employed depends on the end use of the model.

AM technologies are constantly being developed and improved, so the descriptions here are intended to provide an overview of how the different technologies work and compare their relative strengths and weaknesses. However, when assessing AM technologies from manufacturers or service providers, it is advisable to obtain the latest specifications. AM manufacturer websites are the best source of such information.

4.1.6 Basic principles of medical modeling: orientation

Every medical model is unique, and the characteristics of each model need to be considered carefully when selecting and utilizing a particular AM system. Compared with engineering products that have been designed for manufacture, models of human anatomy may be much more challenging to prepare for a successful build, even for those experienced in the operation of their AM machines.

The following sections describe some of the most common AM processes and highlight their key technical considerations. However, there are some basic principles that apply to nearly all AM technologies when building medical models. An important consideration is the orientation of the build. The orientation of the build will have an influence on the surface finish of the model, the time it takes to build, the cost of the model, the amount of support required, the risk of build failure, and even its mechanical integrity.

These factors are interdependent; the key to producing a high-quality medical model is a thorough understanding of them, correctly identifying the priorities, and reaching a compromise solution that best meets the requirements of the model. The effect of orientation on these factors is explored below.

4.1.6.1 Build time and cost

As all AM processes work on a layer-by-layer basis, the builds consist of two repeated stages: drawing or creating the layer and recoating or depositing material for the next layer. Generally, the material deposition stage takes longer and often poses the greatest risk to build failure. Therefore, orienting a model so it minimizes the number of layers will reduce the build time and risk.

Generally, costs are directly related to the build time. Therefore, the longer the build time is, the more the model will cost. In many cases, the automatic option is to orient a model for minimum height and therefore minimum layers and minimum cost. However, as described below, this may have an undesirable effect on the quality of the model.

4.1.6.2 Surface finish and model quality

As described previously, the layer-by-layer building process results in a stair-step effect on sloping or curved surfaces. Depending on the shape of the object, the orientation may have a great effect of the degree of stair stepping on the model surface. However, the mechanisms that create the layer geometry usually offer better resolution. When considering engineering parts, the most important feature is identified, and the build is oriented to provide the optimum surface quality for that feature. However, human anatomy usually possesses curved surfaces in all directions, and the optimum orientation may depend on other, more important, factors. However, there are some obvious examples where orientation makes a considerable difference to surface quality. The long bones of the arms and legs, for example, are essentially cylindrical in form. Orienting a model of a long bone so it lies flat will minimize build time and cost, but the layers will be readily apparent in the model. Orienting the build in an upright sense will increase build time and cost considerably, but the layered effect would be drastically reduced in comparison, leading to much better model quality. Fig. 4.5 illustrates the effect of stair stepping on a model of a proximal tibia (the relative thickness of the layers has been increased to exaggerate the effect). The upper model was built lying horizontally; while this minimized build time, the thickness of the layers has had a negative effect on the reproduction of the contours of the model. The lower figure shows a model that was built upright, and consequently, the layer thickness has not had such a detrimental effect, at the cost of increased build time.

4.1.6.3 Support

All AM processes provide support to the model as it is built. Some processes build supports concurrently with the model, while others use the unused build material to support the part as it is built. Usually, parts that are supported by the unused material may be oriented to provide the minimum cost or to fill the available build volume with the greatest number of parts.

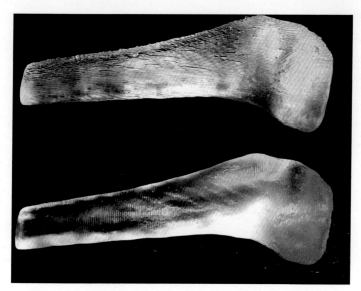

Figure 4.5 Model showing the stair-step effect of orientation.

However, for those processes that construct supports or deposit a second supporting material, their effect on model quality and their subsequent removal should be considered.

Typically, the more supports or supporting material that is present, the longer the build will take, the more materials costs will be incurred, and the more time will be spent manually removing them, which can add to the overall time to delivery and labor costs. Where supports are constructed from build material, the surface finish and quality will also be affected, as most supports leave a witness mark on the surface after their removal. For example, the effect of support removal from SL models is shown in Fig. 4.6.

Another consideration when building medical models is the presence of internal cavities. The human skull, for example, contains cavities, the sinuses, and supports within them may prove difficult or impossible to remove by hand. Some AM processes use soluble supports, which can be removed by immersing the model in a solvent or can melt away supporting materials with a lower melting point (for example wax). While this still adds to the overall time to delivery, labor costs are reduced, and supports inside cavities are more easily removed. Orientation may be optimized to minimize support materials as they represent a considerable materials cost (and of course waste). In some circumstances, the support material might even

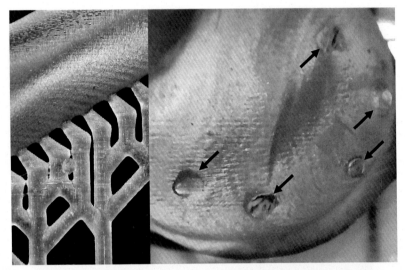

Figure 4.6 Close-up showing witness marks after support removal.

prove to be a significantly higher volume than the actual model. It is therefore advantageous to orient a model to minimize the amount of support but not to the degree where the build process is threatened.

4.1.6.4 Risk of build failure

While AM machines are generally reliable and able to operate for long periods unattended, the build process can be threatened by pushing parameters to their operational limits. As has been stated previously, building models of human anatomy poses challenges in AM due to the complex nature of the forms being built. This makes the risk of build failure higher when attempting medical modeling compared with engineering parts. Often the risks to build failure depend very much on the specific AM process being used, but some general principles apply. The overall stability of the model is important in most AM processes. It therefore makes sense to orient the model so it is most stable, i.e., wider at the bottom that the top. This will certainly also directly affect the supports required and the stair–step effect. The principle of stability and the subsequent effect on the supports are shown clearly in the example in Fig. 4.7, where the orientation on the right offers a more stable build and a much reduced amount of support.

Economic factors are also important. Most AM parameters are set to provide the fastest, and therefore cheapest, build possible. However, setting parameters for speed may increase risk of build failure. The complex nature

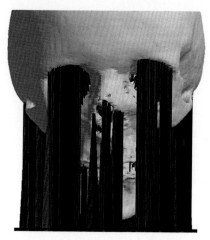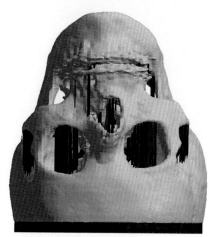

Figure 4.7 The effect of orientation on supports and stability.

of medical modeling means that parameters may have to be altered to lower risk, and consequently, build time and cost may increase compared with an engineering model of comparable size.

4.1.6.5 Data quality

While most orientation decisions are arrived at by considering the final model's shape and size, the computer data used to build it may also have an influence on the choice of build orientation. The choice of data format could limit the options available for orienting the model. For example, if a $2^{1}/_{2}$ D format is used, such as an SLC file, then the choice of orientation is fixed when the data are created. Usually, these formats are created in the same orientation as the original scan data, and this may provide a higher quality data file. The advantages of the data format in terms of quality and efficiency may override the other orientation considerations. The data formats and their various advantages and disadvantages are described in Chapter 3.

4.1.6.6 Illustrative example

Fig. 4.8 shows three instances of the same model at three orientations, A upright, B at an angle, and C flat to the build platform. Orientation A would provide the best surface finish and suffer the least from the stair-step effect. It would increase build time and be at greatest risk of build failure due to instability. Orientation C is the opposite, providing the greatest stability, faster build time, and least risk. However, the stair-step effect

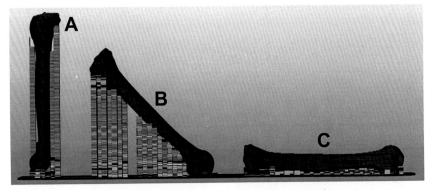

Figure 4.8 Orientation considerations example.

would have the greatest effect on surface quality. Orientation B provides a good compromise but creates the most supports so wastes material.

4.1.7 Basic principles of medical modeling: sectioning, separating, and joining

Before preparing an AM build, it is worth considering the nature of the model and its intended use before finalizing the data. There are several considerations to address relating to the overall size of the model, whether it is one structure or several related structures and whether their positional relationship is important. For these reasons, it may become necessary to section, separate, or join parts to make the model or models required.

4.1.7.1 Sectioning

Sectioning may be done when selecting the data from the original scan data or could be done by splitting the export data files, such as STL files. The most obvious reason for sectioning a model is to reduce the extents of the model to only those areas that are needed. This reduces build time and cost. However, it may also be used to split large models into parts that can be accommodated within the build volume of a given AM machine. Sectioning may also be used to gain access to trapped volumes (internal cavities). This may be necessary to remove waste material or supports but may be because the internal anatomy is also of interest to the clinicians.

When sectioning models, it is advisable to incorporate a stepped or keyed section so the two separate parts are easily located and aligned when they are put back together. AM preparation software packages provide functions to achieve this. An example is shown in Fig. 4.9 where a long

Figure 4.9 A long bone model with keyed section.

bone has been cut into two pieces. The keyed section helps to align the two pieces when they are joined together. It also provides a larger area for adhesives to be applied, forming a stronger bond than a simple cut through the shaft.

4.1.7.2 Separating

As opposed to sectioning models at convenient locations, it may be desirable to separate different adjacent anatomic structures. Often when preparing data from the original scan data, different anatomic structures are touching or sufficiently close to one another that the data become a single object. For example, close joints can become closed, effectively creating a single object from two distinct bones. It may therefore be desirable to edit the data to separate the different anatomic structures so that they can be built individually. This may be so that the parts can be built separately to save time or cost but may also be because the clinicians wish to be able to articulate the two bones.

4.1.7.3 Joining

Just as it may be desirable to separate adjacent anatomic structures, the opposite may be true. When building separate anatomic structures, their final use should be considered. If the spatial relationship between the structures is important, then that will have to be created physically in the resultant model, for example where bones have been fractured and displaced. In some cases, this may be achieved by leaving supports in place between the separate parts. However, supports are typically not strong enough to achieve this reliably. Instead, it may be necessary to create bridges that join the separate parts together to maintain their relative spatial relationship. When creating such bridges, it is important to make them clearly artificial in appearance and locate them away from critical areas, so they do not obscure or become confused with the anatomy. An example of a bridge between two separate bones is shown in Fig. 4.10.

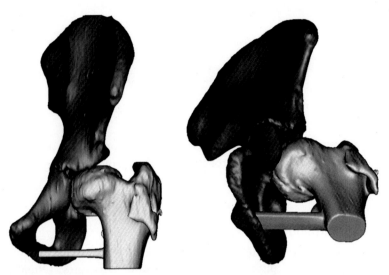

Figure 4.10 An example of a bridged model connecting the proximal femur to the pelvis.

4.1.8 Basic principles of medical modeling: trapped volumes

Compared with engineering parts, the presence of trapped volumes (internal cavities) can pose problems in medical modeling. As mentioned above, support materials need to be removed, and this may prove difficult where physical access to the cavity is limited. Those processes with soluble or low-melting-point support materials have an advantage here, but they still need adequate access to allow the solvent into the cavity and drain the liquid materials out. The "trapped volume" problem is particularly associated with SL, where the effect can lead to build failure (see Section 4.2 for more detail on the trapped volume effect). However, the presence of trapped volumes can cause problems for many AM processes. To produce the unwanted effects of a trapped volume, the cavity does not need to be entirely closed. If the openings to such a cavity are small enough, the effect can be as bad as a fully closed cavity. Examples of cavities are the cranium, the sinuses in the skull and face, and the marrow space inside large bones.

The main problem with trapped volumes is removing the waste material and/or supports from within them. Various techniques can be used to address trapped volumes. If the cavities are small, totally closed, and of no interest to the clinician, the data can be edited to fill in the cavity. For example, the marrow space inside long bones might be of no interest, in which case, the cavity can be filled at the data segmentation stage, resulting in a solid model. It may be tempting to leave such cavities in place when they are fully closed; however, this poses risks. If the model is broken, sawn, or drilled into later, the unused material may leak out. This could be in the form of loose powder, solvent, or liquid resin, which may damage the model or prove a nuisance (or depending on the material being used even a minor health risk) to the user or patient.

Where cavities are larger or deemed important, artificial openings can be created to enable the waste material to be removed manually or drained out. When creating such openings, the location and shape should be created to make them obvious so that they are not confused with the anatomy. For example, drain holes should be made square, so they are less likely to be confused with naturally occurring holes (called foramen). Alternatively, as described above, the model can be sectioned into parts that are built separately to either enable access to the cavity or eliminate it altogether.

4.2 Vat polymerization

According to the ISO definition, vat photopolymerization describes AM processes in which liquid photopolymer in a vat is selectively cured (solidified) by light-activated polymerization. There are now a variety of AM processes that work on the principle of vat polymerization. The main differences are based on the light source, which ranges from liquid crystal display (LCD) screens (like mobile phone screens) to digitally projected light and lasers. The light is typically ultraviolet but can be visible (white) light.

4.2.1 Stereolithography
4.2.1.1 Principle
Liquid resin is selectively cured to solid by ultraviolet (UV) light accurately positioned by a laser. Layers are created by placement of the build platform immediately above or below the surface of a photocurable resin placed within a vat. A laser then scans the layer pattern at the interface, generating the initial layer on the platform. Successive layers are cured by lowering this platform and applying an exact thickness of liquid resin. When the model is complete, the platform rises out of vat and usually is left to drain for a time before being removed and washed.

4.2.1.2 Detail

The first SL apparatus was developed and commercialized by 3D Systems Inc. in the late 1980s. Models are made by curing a photopolymer liquid resin to solid using a UV laser. Models are built onto a perforated platform that lowers into the vat by a layer thickness after each layer is produced. A waiting phase after the platform lowers allows the liquid to flood over the previous layer and level out. Then a recoater blade will level the resin and remove bubbles or debris from the resin surface. The length of the wait states and speed of the recoater blade will depend on the viscosity of the resin. This method also means that there are problems with building objects with trapped volumes as the liquid in these areas is not in communication with the resin in the vat and does not level out, leading to build failure. To remedy this, typically the recoater blade has a U-section that picks up a small amount of resin with a vacuum and deposits it over the previous layer. As the platform lowers into the resin, a deep vat is needed to accommodate the height of the build.

More recently machines, such those by FormLabs, have opted for the rising platform and a shallow vat arrangement. Unlike deep vat machines that project the laser down onto the free resin surface, these machines direct the laser through the transparent bottom of a shallow vat with the build platform rising upward out of the vat. The model remains attached to the rising platform (like most DLP and LCD machines described in the next section). The shallow vat makes changing materials quicker and easier as the shallow vat is easily removed, and no moving parts need to be cleaned.

Overhanging or unconnected areas must be supported. Supports are generated by the slicing software and built concurrently with the model. When a model is complete, excess resin is washed off using a solvent or cleaning agent (e.g., isopropanol), and the supports are removed. It should be noted that cured resin materials are thermoset polymers (crosslinking is a one-way process). Consequently, postprocessing and support material removal are only possible through mechanical methods. This limits the sustainability of these materials, as unlike thermoplastics, they cannot be recycled or reprocessed. Filtration of set polymer fragments from the liquid resin is required when builds fail.

The model is then postcured in a special apparatus by a UV light source ensuring all resin is fully solidified. Postcuring may also involve heat, which will help strengthen some resins. Postcuring is important as any remaining liquid resin would leave a sticky residue on the surface of the model and may be a skin irritant.

All lasers used in SL emit in the UV spectrum and are therefore not visible to the naked eye. However, the laser spot can usually be seen on the resin surface as a bright glow that is briefly emitted by the reaction in the resin. The laser and the precision optics required lead to comparatively high machine costs compared with other forms of AM.

The speed of the machine depends on how much energy the resin requires to initiate polymerization, as the power of the laser is constant; if more energy is required, the laser must travel slower. Material properties and accuracy also depend on the resin characteristics. As the material polymerizes, there will be some degree of shrinkage; this can be compensated for in the build parameters but may also lead to other problems, most notably curl. This was especially true early in the development of SL when most systems used acrylate-based resins. The development of epoxy-based resins largely eliminated these problems, resulting in very accurate models. New materials are becoming available with a range of physical properties that enable a range of applications, from communication models to surgical guides.

More broadly, four distinct material categories have emerged and comprise the following:

- standard photopolymer: resins that produce parts with high detail reproduction and transparency, with applications being primarily for prototyping or communication purposes;
- engineering resins: resins that produce parts with material properties that approximate those of common thermoplastics, such as acrylonitrile butadiene styrene (ABS) or polypropylene, or elastomeric materials, such as rubber and silicone;
- dental and medical: resins that produce high-resolution, durable parts with varying degrees of biocompatibility from class I to IIa. Parts can be sterilized and used in surgery (short term) or longer term contact with healthy skin;
- castable: resins that produce sacrificial casting patterns with low levels of ash residues.

Resins are typically transparent or translucent, but colored and filled resins are now available. More recently, flexible resins have become available that provide soft, flexible parts, which might be desirable for some medical applications. However, elasticity remains limited, and the material properties degrade with age, especially exposure to UV light.

Typical SL medical models are shown in Figs. 4.11 and 4.12. In medical modeling terms, SL is in many ways ideal. SL models show excellent

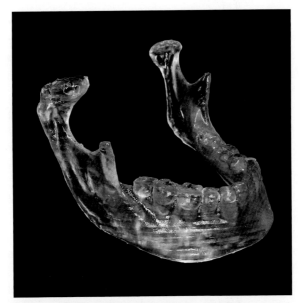

Figure 4.11 SL model of the mandible.

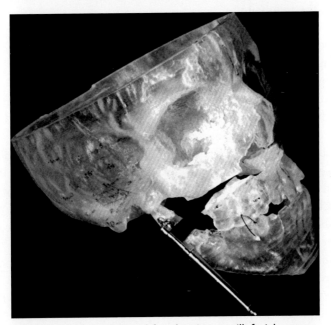

Figure 4.12 SL model used for planning maxillofacial surgery.

accuracy and very good surface finish. The transparency of most SL materials enables internal details such as sinuses and nerve canals to be seen clearly. The fact that unused material remains liquid also means that it can be easily removed from internal cavities (as long as the resin can drain freely). This is especially beneficial when modeling the human skull, which possesses many such internal features as well as the cranium itself. These advantages can be clearly seen in the examples illustrated here.

The solid, fully dense finished models lend themselves well to cleaning and sterilization, and the development of medical standard materials is an advantage. A variety of resins are available that have been tested to internationally recognized standards (for example ISO 10993 and USP 23 class VI), making parts suitable for handling in the operating theater as a surgical aid or to be used as surgical templates and guides.

Some resins can also be selectively colored. A single pass of the laser will solidify the resin, and a subsequent high-power pass will cause the solidified resin to change color. This allows internal features to be made visible through the thickness of the transparent model. This can be used to show, for example, the roots of teeth in the jaws or tumors, as illustrated in Figs. 4.13 and 4.14. These factors combined with the inherent accuracy of SL make it highly appropriate for medical modeling.

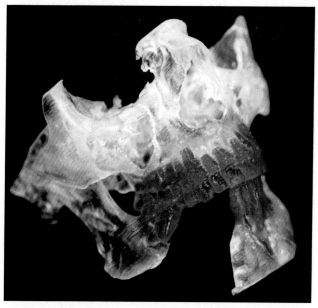

Figure 4.13 Selectively colored model showing the roots of teeth.

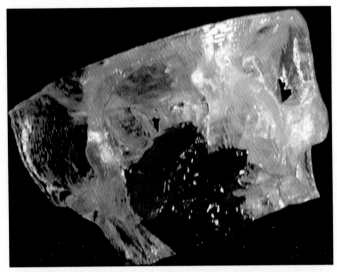

Figure 4.14 Selectively colored model showing a tumor.

SL remains a specialized market, but machines are available from a variety of manufacturers most notably 3D Systems, who continues to produce large-capacity industrial specification machines (with platform sizes up to 650 × 750 × 550 mm), and FormLabs, who have developed a range of highly successful and comparatively affordable benchtop machines. The dental technology market has widely adopted SL, and dedicated machines and materials have been developed for these applications.

Advantages	Disadvantages
Well-developed, reliable machines and software	Relatively high cost of machine and materials
Established sales, support, and training	High maintenance costs
High accuracy, good surface finish	Resin handling requirements
Little material waste	"Trapped volumes" problematic
Medical standard materials	Long laser scanning time (compared
Can be sterilized and selectively colored	with DLP [digital light processing]/ LCD)
Transparent models	

4.2.2 Digital light processing

4.2.2.1 Principle

Liquid resin is cured to solid by light (UV or visible) accurately projected as a two-dimensional mask using a digital micro-mirror device (DMD). The cross-sections are projected as images onto the bottom of a shallow transparent vat, curing a layer of resin between the base of the vat and the build platform. Successive layers are cured by raising the build platform by an exact layer thickness and allowing resin to flow under the previous layer before reexposure of the resin. The model remains suspended from the build platform, rising away from the shallow vat.

4.2.2.2 Detail

This process was initially developed by EnvisionTEC GmbH, resulting in their Perfactory machine. These early machines relied on DMDs that could project visible light images through the transparent base of a vat of resin. These machines relied on high-power light bulbs as found in projectors, and these expensive bulbs had a limited service life.

The cured layer is trapped between the flat, transparent base of the vat and the build platform; subsequently, the previous layer of the build results in perfectly flat layers created at the exact layer thickness. Creating this fixed gap eliminates any of the settling or leveling procedures required in SL, which in turn accelerates the process. This process also means that the layer thickness can be reduced, leading to models built from very thin layers, which improves surface finish. A range of machines are available from small desktop machines to larger free-standing machines with envelopes as large as 510 × 280 × 350 mm. The smaller machines are popular in dental applications, while larger machines compete with SL.

As the cross-section is projected in one instant, the exposure time for each layer is fixed regardless of the geometry of any given cross-section. This provides rapid and predictable build rates and means that a full platform can be cured as quickly as a small part.

As in SL, overhanging or unconnected areas must be supported. Supports are generated by the build software and built along with the model. When a model is complete, excess resin is washed off using a solvent, and the supports are manually removed. Depending on the material, postcuring with light and or heat might be needed. Resins cured by visible light can be postcured in daylight. Unlike SL, there is no laser and very few moving parts, which leads to considerable savings in manufacture, maintenance, and running costs.

The accuracy of the process is a function of the projected image of the cross-section. The DMD will have a fixed pixel array (e.g., 1920 × 1200), but on some machines, the image can be scaled up or down using optics. Therefore, the resolution of the model will be defined by the overall size of the build area divided by the number of pixels (i.e., pixel size) and the layer thickness. This resolution (pixel size) in the X—Y plane can range from as little as 0.016 to 0.069 mm. The layer thickness may range from 0.015 to 0.150 mm depending on the machine specification and optics used.

The materials used are very similar to SL resins and produce models that are solid, reasonably tough, and can be transparent. UV curing resins are typically transparent or translucent, whereas visible light curing resins are usually amber or red in color. An example is shown in Fig. 4.15. Materials have been developed and approved for specific medical applications in dental technology and hearing aid manufacture (see Fig. 4.16), and castable materials are also available.

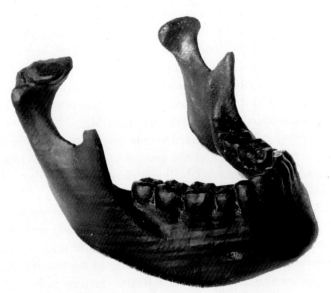

Figure 4.15 Perfactory model of a mandible.

Figure 4.16 A hearing aid shell manufactured using DLP.

Advantages	Disadvantages
High accuracy, excellent surface finish	Resin handling requirements
Little material waste	Limited material choice
Lower purchase, maintenance, and running costs	Limited build size
Transparent models	
Medical standard materials	

4.2.3 Liquid crystal display

4.2.3.1 Principle

Liquid resin is cured to solid by visible/UV light generated from individual pixels on an LCD. A layer is generated by exposure of a photopolymer resin contained within a vat to light from an LCD screen position at the underside of the vat. The individual pixels of the screen form the details of a given layer. The first layer is generated by placing a build platform a print layer distance away from the LCD screen, before exposure solidifies the resin on the build platform. Successive layers are cured by raising the build platform by an exact layer thickness and allowing resin to flow under the previous layer before subsequent layer exposure.

4.2.3.2 Detail

LCD SLA (stereolithography apparatus) is a technologic derivative of DLP where models are formed by curing a photopolymer liquid resin to solid using visible light. Typical machines work almost entirely like DLP machines as described previously, with the exception that exposure of the photopolymer is achieved using an LCD screen over the projection of light controlled by a DMD. A range of machines are currently available ranging from desktop machines with a build envelope of approximately 115 × 65 × 155 mm to larger variants with build envelopes of approximately 510 × 280 × 350 mm. The resolution and reproduction quality are largely dependent on the pixel density of the LCD screen, with early variants comprising a 2K screen (2560 × 1440 pixels) and lower layer resolution of 25 μm. Modern LCD SLA systems now comprise 8K screens (7500 × 3240 pixels) and lower layer resolutions of 10 μm. Current systems are now providing a cost-effective solution that matches the performance of comparative SLA and DLP systems but at a lower price point, making SLA 3D printing more accessible to researchers and medical professionals.

As with DLP systems, LCD machines expose the resin to an entire layer at any instance, meaning layer time is generally fixed regardless of geometry; however, exposure time may vary depending on a specific resin's photo initiation characteristics. Consequently, reproduction of a part is considerably quicker than conventional SLA machines.

As with all SLA systems, overhanging and unconnected part regions require the use of supports, which are generated in the respective slicing software. Postprocessing of parts requires the washing of excess resin using solvents or specialist cleaning agents, while supports are mechanically removed. Typically, a final phase of UV exposure is required to ensure all remaining monomers are fully cross-linked before a part is ready for use. Unlike SLA machines, no lasers are required or additional optics as with DLP systems, while the power requirements of the LCD are low. Therefore, LCD systems realize low running and maintenance costs, and the LCD is considered a consumable part, with the potential to be easily replaced or upgraded over time.

Again, as with DLP, the materials used are similar to SLA resins and produce parts that are robust, with a range of material characteristics and being potentially biocompatible. Following advances in photopolymer formulations, resins now exist with excellent transparency and in a range of colors. Fig. 4.17 shows an example of a wrist splint produced using a large

Figure 4.17 A wrist splint with supports attached the build platform.

build envelope LCD system with the supports clearly seen and the parts still attached to the build platform. Fig. 4.18 shows the completed parts with supports removed.

Advantages	Disadvantages
High accuracy, excellent surface finish	Resin handling requirements
Little material waste	Postcuring often required
Low maintenance and running costs	
Transparent models	
Large build size	

Figure 4.18 The wrist splint from Fig. 4.17 cleaned with supports removed.

4.3 Material extrusion

The ISO definition describes material extrusion as a process in which material is selectively dispensed through a nozzle or orifice. Typically, this is a thermoplastic material that is heated and extruded in a semimolten state, cooling rapidly when deposited. The process is commonly known as fused deposition modeling (FDM) or fused filament fabrication.

4.3.1 Principle

Thermoplastic material is fed in filament form to a heated extrusion head. Layers are made by semimolten material deposited as a fine bead through a fine nozzle (typically 0.4 mm). The build table lowers an exact amount, and the next layer is deposited onto the previous layer. The heat emitted by the deposition is sufficient to cause local melting to the previous layer and adjacent deposited beads, ensuring a strong bond is formed.

4.3.2 Detail

This process was originally developed in the United States by Stratasys Inc., and the technology is used in their Fortus branded machines (FDM is a Stratasys trademark). In recent years, the expiration of key patents and the RepRap open-source movement (www.reprap.org) led to an explosion in the market for cheap, small "3D printers." These machines range in quality from home hobbyists' kits costing a few hundred dollars to substantial industrial machines. The use of low-cost machines in medical applications is possible, but care must be taken to ensure the materials' properties, quality, and accuracy are adequate for the medical purpose. Models made on low-cost machines are likely to be inferior to models built on industrial standard equipment, and quality control and certification will be difficult to achieve for mainstream clinical use. Consequently, this section will focus on industry standard machines.

Models are made by extruding thermoplastic materials fed in the form of a filament from a spool through a heated nozzle. The nozzle moves in the X and Y axes to produce layers, and then the build platform lowers by a layer thickness, and the next layer is deposited on top of the previous layer. The heat of the newly deposited layer causes local melting to adhere the layers together. Supports for overhangs are built concurrently from either the build material in low-cost machines or a second different material in industrial machines and are removed when the model is complete. Separate support materials are typically soluble, making removal easy and labor free, even from cavities (assuming the liquid can flow freely). The commonly used materials are thermoplastics such as polylactic acid (PLA) and ABS, but other materials include Nylon and polyethylene terephthalate glycol. More sophisticated machines that can operate at higher temperature can process tougher materials such as carbon fiber—reinforced materials, polycarbonate, and polyphenylsulfone. The models produced therefore can be handled directly and require no special cleaning or curing. The physical properties can be close to injection-molded plastic components and are typically very tough and durable compared with other AM processes.

The process also means that machines are typically quiet and clean. The technology is very well established and therefore reliable and comparatively easy to maintain. Surface finish and accuracy are not as good as SL, for example, but the machines cost less than similarly sized SL machines.

Material extrusion models are hard but typically have a lattice internal structure surrounded by solid walls. Models are usually opaque, which hides internal details that could be visible in a transparent SL model. A typical material extrusion medical model is shown in Fig. 4.19.

Material extrusion is well suited to medical modeling. The models show reasonable dimensional accuracy and surface finish. The models are tough and can withstand repeated handling, making them ideal for teaching models but also potentially for functional items such as orthoses. The finished models lend themselves well to cleaning, and materials are available that have been tested to recognized standards to show low toxicity. Some manufacturers have produced appropriate guidelines for sterilization of materials extrusion parts (e.g., Sterilization of FDM-Manufactured Parts, Stratasys, www.stratasys.com/en/resources/whitepapers/sterilization/), and

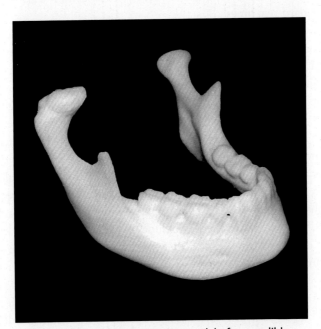

Figure 4.19 Material extrusion model of a mandible.

there has been research showing that parts can be successfully sterilized [1]. Some materials can withstand elevated temperatures so in theory could be sterilized, although autoclave poses a challenge to the commonly used thermoplastic materials such as PLA, ABS, which easily distort due to the process exceeding the low glass transition temperatures (T_g). On low-cost machines, the support material is solid and can be difficult and sometimes impossible to remove from the internal cavities often found in human anatomy, making soluble supports essential for modeling the skull for example.

The white material typically used represents bones in a familiar manner, and models of long bones and joints can be well modeled. These advantages can be clearly seen in the examples illustrated in Figs. 4.20 and 4.21.

Advantages	Disadvantages
Relatively cheap to buy and run	Poorer accuracy and surface finish
Reliable and well proven	compared with SL
Clean and safe process	Small models and small features
Strong, tough models	difficult
Medical standard materials available	Opaque material
Some materials can be sterilized	

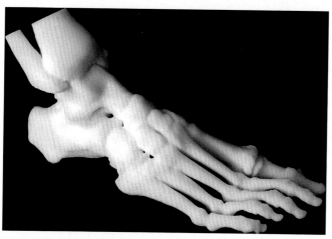

Figure 4.20 Material extrusion model of the foot.

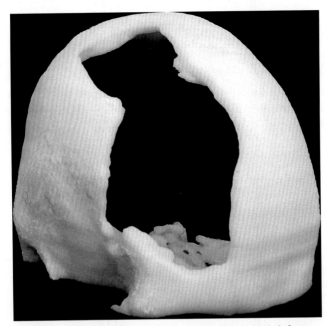

Figure 4.21 Material extrusion model of a cranial defect.

4.4 Powder bed fusion

The ISO definition describes powder bed fusion as a process in which thermal energy selectively fuses regions of a powder bed. Heat causes particles to partially or fully melt, so they adhere to each other, forming a solid layer. The heat penetrates through more than one layer, ensuring interlayer bonding. The heat can be supplied selectively by a laser or to the whole layer by radiant heat, in which case, special liquids printed into the surface of the powder bed absorb more heat, enabling only the printed area to bond.

4.4.1 Laser sintering
4.4.1.1 Principle
LS is like SL except using powders instead of liquid resins. A powerful laser locally fuses or sinters thin layers of fine-particulate thermoplastic material. The build platform is lowered, fresh powder is applied, and the next layer is scanned on top of the previous layer. Heat penetration causes local melting to form the interlayer bond.

4.4.1.2 Detail

Models are made by selectively sintering thermoplastic powder material using a laser. The build materials are heated to a level below their melting point. The material is heated further to near melting point by laser scanning the cross-sections, locally heating the powder enough to fuse the particles together. Note that in LS of thermoplastics, the particles are fused together but do not fully melt into the liquid state. The build platform lowers each layer, and fresh powder is spread across the build area by a roller. The inherent dangers of handling fine powders are controlled by purging the build volume with nitrogen gas. Models are supported by the unused powder. Overall, build times are comparatively slow to allow for heating up the build chamber, around 1.5 hours, and cool down of the chamber and the powder bed, around 2 hours. When completed, the model is dug out of the powder and bead blasted to remove any powder adhering to the model's surface. The machines are large and heavy and require water cooling, extraction, and nitrogen supply. Consequently, the operating costs are considerable, and a high throughput is required to justify the investment in the technology and its associated infrastructure.

The market is dominated by two manufacturers, 3D Systems and EOS GmbH. The most commonly used glass-filled Nylon material gives relatively low porosity, resulting in strong, robust models compared with SL for example. Surface finish is relatively rough compared with SL as may be expected from a powder, but this tends to hide the stair-step effect. Accuracy is almost as good as SL and comparable with material extrusion. Specific materials allow LS models to be used as sacrificial patterns in investment casting. Elastomeric materials are also available for prototyping flexible components, and aluminum-filled materials are available for enhanced physical properties. When installed, it takes quite a while to set up the machine parameters, involving a certain amount of trial running. However, once correctly set up, the machines are well proven and reliable.

The models are reasonably accurate and tough but remain porous. Although the materials themselves pose no inherent medical problems, the porosity makes them challenging to clean and sterilize effectively, although it can be achieved. LS models are opaque, which may hide internal details. A typical LS medical model is shown in Fig. 4.22.

In medical modeling terms, LS models have proven useful. LS models offer reasonable accuracy and surface finish comparable to material extrusion models. The strength and toughness of the models is a definite advantage, but the powder nature of the material leads to a rough surface,

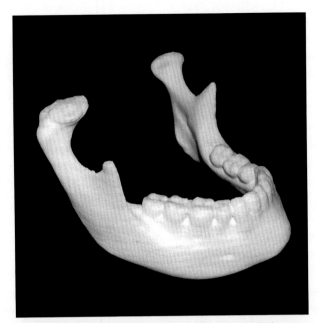

Figure 4.22 Laser sintered model of a mandible.

which can trap dirt and grease in repeated or continual handling. Despite these issues, LS parts are now routinely used to make surgical guides (single use) and end-use devices such as foot orthoses. The porosity of the models poses concerns where long-term skin contact is concerned. Unused material remains as loose powder, so it can be removed from internal spaces and voids, though this may sometimes be difficult unless there are apertures large enough for the powder to flow freely.

Advantages	Disadvantages
Strong, tough models	High cost of machine and materials
Reasonably accurate	Large machine and infrastructure
Good capacity and	requirements
throughput	Heat-up and cool-down time
	Poor surface finish

4.4.2 Laser melting

A number of manufacturers offer machines that can process metals. Although similar in principle to LS, these processes fully melt the powder to form fully dense metal parts, so this is referred to as laser melting. All of the

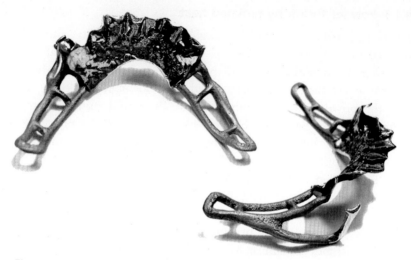

Figure 4.23 Partial removable denture framework made using laser melting.

machines are similar in principle, utilizing high-power fiber lasers to melt very fine metal powders in inert gas atmospheres (argon or nitrogen). The full melting enables the production of solid, dense metal parts in a single process (i.e., not using binders or postprocess furnace operations to make metal parts indirectly). A variety of metals can be used including stainless steels, cobalt—chrome, and titanium. While these processes are not suited to the production of anatomic models, they are being used to produce dental appliances, surgical guides, custom-fitting implants, and prostheses as described in the case studies in Chapter 5. For example, Fig. 4.23 shows a partial removable denture framework made using laser melting.

The process offered by Arcam also uses a powder bed, but an electron beam directed by magnetic fields supplies the melting energy. Parts are commonly made from titanium, but other metals can be used. In comparison with laser melting processes the Arcam process typically has faster deposition rates and thicker layers. The build chamber is a vacuum rather than using an inert gas.

Advantages	Disadvantages
Very strong, very tough metal parts	High cost of machine,
Accurate	infrastructure, and materials
Materials can be sterilized easily (using autoclave)	Inert gas supply required
Wide range of metal materials	

4.4.3 Powder fusion by radiated heat

Developed from research at Loughborough University and initially referred to as high-speed sintering, powder bed fusion by heat involves exposing the deposited layer of powder to radiated infrared heat. To control where the powder fuses together or not, special inks are printed into the powder to both enhance heat absorption (fusing agent) and inhibit it (detailing agent). The cross-section to be fused is printed with a heat-absorbing ink, and the powder immediately surrounding the perimeter is printed with an inhibiting ink. The process commercialized by HP is called   Multi Jet Fusion (MJF). Although the ink–printing element has similarities with ISO definition of binder jetting, in this case, it is the thermal energy that fuses the powder; we will therefore consider it to be a powder bed fusion process.

The machines are similar to LS machines and are supplied with associated machines that prepare powder materials and build chambers as well as facilitate cooling-finished build chambers and removing the built parts from the powder and recycling unused powder. The machines and associated facilities and infrastructure are on a par with smaller LS machines but could offer greater throughput.

Similar to LS, materials are typically formed of Nylon (such as polyamide 11 and 12), but a thermoplastic polyurethane is also available for flexible parts. The completed models are similar to LS models in terms of physical properties and accuracy. An MJF jaw model is shown in Figs. 4.24 and 4.25 shows a typical MJF part featuring a built-in hinge, which is possible with powder processes.

Advantages	Disadvantages
Very strong, very tough metal parts	High cost of machine, infrastructure, and materials
Accurate	
Materials can be sterilized easily (using autoclave)	Inert gas supply required
Wide range of metal materials	

4.5 Material jetting

The ISO definition of material jetting describes it as a process in which droplets of build material are selectively deposited. This is typically achieved using technology similar to that found in office inkjet printers.

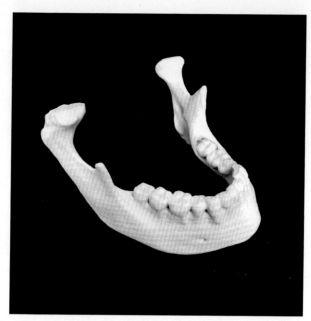

Figure 4.24 MJF model of the jaw. *(Model kindly provided by Prof. Carl Diver and Mark Chester, PrintCity, Manchester Metropolitan University, Manchester, UK.)*

Figure 4.25 Typical MJF model featuring built-in hinge. *(Provided by Yigit Basar from 3DGBIRE.)*

4.5.1 Principle

Build material is deposited in the form of extremely small droplets, discretely, by an array of jetting heads. The material is ejected as a liquid, solidifying on contact with solid material or the build platform by cooling, drying, or curing to solid by UV light. The head moves in the X and Y axes to build the layers. The build platform lowers by a layer thickness, and material is deposited onto the previous layer. Separate supporting materials are concurrently deposited by similar print heads.

4.5.2 Detail

There are several companies producing machines that utilize deposition processes similar to those found in ink jet printers, such as Objet and 3D Systems. The 3D Systems ProJet machines and the Objet PolyJet machines use photo-polymerizing resins that are simultaneously solidified by UV light as they are printed. A ProJet model is shown in Fig. 4.26, and an Objet model is shown in Fig. 4.27.

Generally, the machines and materials are clean and safe and do not require extraction or special handling. ProJet machines surround the model with a secondary wax support material that is melted away from the

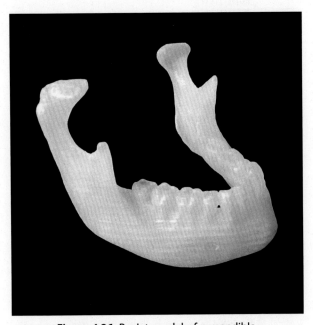

Figure 4.26 ProJet model of a mandible.

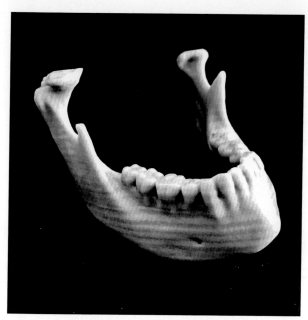

Figure 4.27 Objet mandible model.

completed model, before being soaked in cooking oil and de-greased. Objet systems deposit a gel-like support material, which can be removed easily by hand or with brushes, although pressurized water jets are typically used. A caustic soda/hydrochloric acid (2% concentration) solution can also be used for model cleaning.

As might be expected from technologies that rely on printing technologies, the resolution of these machines is often quoted in terms of dpi. Machine specifications vary, but generally, the models are comparable to SL models.

One of the advantages of printing technologies is the ability to produce very thin layers. This reduces the stepping effect, and models are very smooth compared with other layer manufacturing techniques. Layer thicknesses of 0.030 mm are typical but can be as thin as 0.016 mm. Another advantage is that each layer is created during a pass of the print head regardless of the geometry of any given cross-section, which makes prediction of build times relatively simple and accurate.

Most of the machines use at least two print heads, with the Objet Connex featuring up to eight print heads in total: four for distributing support material, while the remainder are used to distribute model material.

The print heads traverse across the build platform on a single block to strategically deposit photopolymer resin. The benefit of having large numbers of print heads is accelerated building speed.

The relative accuracy of the Objet process and the transparency of the materials make them ideal for medical modeling and comparable in most respects to SL models. In addition, an Objet build material has been tested to an internationally recognized standard for plastics (USP 23 Class 6) and therefore may be sterilized and used in theater.

Uniquely, softer, rubber-like materials have also been developed for these printers. These technologies enable multiple material builds. The Objet Connex machines for example can print two different build materials—typically one will be a rigid plastic material and the other a soft, rubber-like material. In addition, intermediate materials are possible by printing differing combinations of the two materials. These intermediate materials are referred to as "digital materials." This enables models to be made that exhibit a range of physical properties in a single build. To create multimaterial models, a separate volume (STL file or shell) must be created to define each material region. This is shown in Fig. 4.28, which shows a CAD model of a wrist orthosis featuring multiple shells (see color section), and Fig. 4.29, which shows an Objet Connex model featuring a range of soft black features, hard white features, and intermediate gray features.

Figure 4.28 Assembly of six STL files for different build materials.

Figure 4.29 Objet Connex build showing multiple digital materials.

More recently the technology has developed to enable color to be combined with multiple materials enabling life-like representations of human anatomy. Manufacturers such as Stratasys and Mimaki are specifically addressing the anatomic modeling market (see later case studies and Fig. 4.30).

Advantages	Disadvantages
Easy to use	Expensive materials
Suitable for office environment	Expensive machines
Thin layers	Replacing jetting heads can be
High resolution, good accuracy	expensive
Transparent models possible	Materials degrade with age and
Medically appropriate materials available	UV exposure
Multiple material builds	
Color builds	

4.6 Computer numerical controlled machining

Unlike AM techniques, milling has always been of limited use when producing shapes with undercuts, re-entrant features, and internal voids. Yet, it proves economical and rapid when forming simple, solid shapes and forms. The principal advantages of machining are its availability and versatility. A large number of machines exist from very cheap desktop routers to very large factory-based five-axis machine tools. Unlike most AM technologies, machining is not restricted to a limited range of materials.

Figure 4.30 Add a full color anatomic model.

Almost any material can be machined ranging from the hardest metals to soft foams. Computer numeric controlled machining (CNC) allows machining to be carried out under computer control based on CAD data. Depending on the machine and material configuration, CNC allows complex forms to be machined rapidly and accurately.

The characteristics of machining make their use in medical modeling particularly suitable to producing larger models of the external anatomic topography. Typical applications may be for molds and formers or custom-fitting supports and wearable devices.

The artificial hip model shown in Fig. 4.31 was made to replicate the human body in impact testing of hip protection devices for the elderly. The model is made of dense, closed-cell foam and shaped using CNC machining to replicate not only the external anatomic topography but also an internal pocket in which an artificial femur can be located. The materials

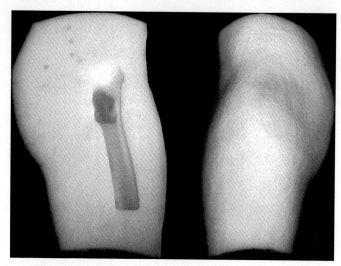

Figure 4.31 CNC-machined closed-cell foam hip model.

properties, size, and volume would be unsuitable for 3D printing but are ideal for CNC machining.

Advantages	Disadvantages
Very wide range of machine size and cost	Time-consuming preprocessing
	Set-up time
Very wide range of materials	Clamping the part during cutting
Clean, safe materials	Poor for thin walls and small features
Good for large, thick, simple models	Poor for internal detail, hollow parts,
Good accuracy possible	or undercuts

4.7 Cleaning and sterilizing medical models

4.7.1 Introduction

The application of AM technologies in the medical sector has led technology and material vendors to develop processes and materials that meet prerequisite standards that can enable products to be used in direct contact with skin or as temporary or permanent implants. Common standards that developers cite are USP 23 class VI and/or specific parts of ISO 10993. However, while some materials have undergone testing to demonstrate levels of biocompatibility, it is the responsibility of the device manufacturer to ensure that it is fit for purpose and complies with the guidance set out in the Medical Device Regulations (MDRs), which have replaced the Medical Device Directive in Europe and are more widely adopted. Cleaning and sterilizing become an increasingly important part of the production process as the risk to the individual using the device or the patient receiving treatment increases. MDR classification is related to many factors, including the duration of exposure, the invasiveness, and whether it is passive or active. The higher the MDR classification is, the more risk it poses to the health of the patient and the more rigorous are the controls required in the specification and testing.

When compared with alternative manufacturing processes, such as CNC machining, AM processes can involve the use of far fewer non-biologic contaminants, such cutting fluids, oils, and swarf. However, AM materials themselves and the processing and postprocessing of them still provide significant opportunity to introduce contamination that has the potential to harm when used for medical purposes. Many AM manufacturing processes have prescribed methods of postprocessing, but there is still significant opportunity for deviation in part cleanliness, especially when dealing with complex, anatomic-based structures that include areas that are difficult to access with fluids and cleaning implements. The suitability of various AM process for any given medical application will vary based on factors such as the following:

- the geometry being produced (will it have any deep crevices or features that may prevent support structures being removed or adequate cleaning?);
- the intended application (classification according to the MDRs; this will dictate the necessary level of material testing, cleaning, and sterilizing parts should be subjected to).

When choosing an AM process for fabricating end-use medical devices, the ability to rigorously clean and/or sterilize parts should be considered. Some processes, such as powder bed 3D printing do not lend themselves well to cleaning and would not withstand most forms of sterilizing. Other processes, such as SL, SLS, DLP, some material jetting technologies, and most metal-based technologies, have demonstrated their suitability for producing end-use medical devices. The common characteristic of these processes is the availability of materials that have undergone prerequisite testing to recognized standards such as USP/23 class VI or ISO 10993 and the ability to produce parts that can withstand rigorous cleaning (i.e., they will not degrade with handing and scrubbing for example). With the exception of SLS, processes that have a demonstrated track record in medical applications also tend to be those capable of producing 100% dense components, which makes them easier to clean and sterilize. Further advantages are also offered by processes that either do not require a support structure (such as SLS) or that utilize the same material for support structure and part (such as SL or DLP), which means that only one material requires validation. However, one drawback of using a support structure in the same material as the part is that they can be difficult to fully remove from hard-to-reach areas (such as sinuses in anatomic models) and leave "witness" marks (as shown in Section 4.2). Material Jetting technologies, such as Objet and ProJet, use support materials that are dissimilar and encase the part, further compounding the difficulty in fully removing residues or unwanted contamination.

Despite the wide range of medical applications where AM technologies are used, there is a limited range of academic literature describing how parts should be cleaned and sterilized. The following sections will describe general procedures used to clean and sterilize parts that are used as medical devices and put this in context of AM production.

4.7.2 Cleaning

The purpose of cleaning can generally be classified as twofold: the removal of nonbiologic (manufacturing) contamination and minimization of biologic contamination (bioburden) to controlled levels.

Postprocessing and cleaning regimes are particular to each rapid prototyping process. A limited number of material suppliers and equipment manufacturers have published guidelines that describe both build parameters and postprocessing procedures that should be adhered to if cleaning for medical applications. However, it may also be necessary for the device manufacturer to adapt processes to suit the final application and demonstrate through testing that their processes meet the requirements defined in the MDRs.

In general, any cleaning method should be undertaken in a controlled environment, where the risk of contamination or bioburden is reduced. This means carefully considering sources of contamination and implementing controls to avoid it. Further consideration must also be given to how the cleaning process may affect the mechanical and chemical properties of the AM material. For resin-based SL and DLP processes, solvents such as isopropyl alcohol are used to remove residual monomer.

Ultrasonic cleaning methods are commonly used to remove contaminants from metal components but are less effective on polymers. Typical cycles will include multiple-stage ultrasonic cleaning (with each stage gradually reducing the contamination) or passivation (for titanium implants), followed by rinsing and drying.

4.7.3 Sterilization

The purpose of sterilization is to remove all forms of microbial life present on a surface, in a fluid, medicine, or other compound. The aim is to achieve a high sterility assurance level, which is a term used to describe the probability of a single unit remaining nonsterile after the sterilization process. There are numerous methods of sterilization, some more compatible with AM processes than others. Three common methods are as follows:

- autoclave
- gamma radiation
- ethylene oxide

Autoclaving is a widely used and highly accessible method of sterilization available in the vast majority of hospitals and clinics. It involves the use of disinfecting/cleaning stages, combined with high-temperature steam and pressure to remove debris and sterilize the part. A typical cycle will involve disassembly and labeling of equipment, which is then placed on a metal tray; pressure washing in a detergent to remove debris; inspection; sorting according to the operating theater lists; wrapping in a linen bag or packing in a special bag; labeling with a sterilization indication marker; autoclaving on a specified cycle at 134°C; and cooling. The entire cycle can take around 4 hours. If at any point during the cycle there is an indication that the parts are not sterile, the cycle is repeated. The high temperatures and pressures may make autoclaving an unsuitable method for sterilizing many polymer parts, especially if it necessary to sterilize those multiple times.

Gamma radiation sterilization is a method commonly used for sterilizing disposable medical devices and those that are sensitive to heat and steam. This method of sterilization is typically found in medical device factories and is less common in hospitals. Irradiation can also have a detrimental effect on some polymer materials.

Ethylene oxide (EO or EtO) is used to sterilize objects sensitive to temperatures greater than 60°C and/or radiation, such as polymers. It is commonly used for large-scale sterilization of disposable devices but is less common in hospitals; it is often outsourced to private companies. One reason for this is that EtO gas is highly flammable, toxic, and carcinogenic with a potential to cause adverse reproductive effects. Following initial cleaning and preconditioning phases, the sterilization process can take around 3 hours. This is followed by a stage to remove toxic residues.

In general, the physical and chemical effect of sterilization processes on many AM materials, particularly polymers, remains the subject of research, although some suppliers and manufacturers have developed acceptable standard operating procedures that have passed relevant approvals.

4.8 Summary

This chapter has introduced the four main types of AM processes that are most commonly used in medical modeling. The agreed standard definitions have been explained, and the four process groups of vat polymerization, powder bed fusion, material extrusion, and material jetting have been described in principle and in detail. The main types of machines in each category have been explained, and the fundamental pros and cons of the processes have been described. The chapter concludes with recommendations about the uses of AM processes for medical devices and introduces some key considerations relating to cleaning and sterilization. The materials and technologies remain in constant development, but the fundamental characteristics are established in this chapter, and there are also many other processes that cannot be full described here and increasingly hybrid approaches that seek to combine different processes to gain the relative benefits of multiple processes. For example, the combination of AM and CNC milling into one machine or the use of five axes of movement. Such developments can be reviewed by considering the fundamental characteristics of each process step. The following chapter provides a wide variety of case studies that describe applications in greater detail.

Reference

[1] Frizziero L, Santi GM, Leon-Cardenas C, Ferretti P, Sali M, Gianese F, Crescentini N, Donnici G, Liverani A, Trisolino G, Zarantonello P. Heat sterilization effects on polymeric, FDM-optimized orthopedic cutting guide for surgical procedures. Journal of Functional Biomaterials 2021;12(4):63.

CHAPTER 5

Case studies

5.1 Introduction

This chapter contains a large number of case studies covering a wide range of applications of medical modeling in a variety of different disciplines and research projects. The chapter is made up of individual case studies based on the authors' research over the last 25 years with some invited contributions from colleagues.

The chapter is divided into six sections covering

1. Implementation
2. Surgical applications
3. Prosthetic rehabilitation applications
4. Orthotic applications
5. Dental applications
6. Research applications

The case studies go into detail about the technologies and techniques employed based on the concepts introduced in Chapters 1 through 4. The majority of the work has been previously published in international peer-reviewed journals and conference proceedings, and in this case, full acknowledgments and citations to the original publications are provided including digital object identifiers where available.

The first section, implementation, covers the issues to consider when collaborating between designers, engineers, service providers, radiographers, and clinicians. The cases highlight possible problems and suggest ways of avoiding or overcoming them.

The second section, surgical applications, covers the use of medical modeling techniques in the support of surgical planning and the design and manufacture of surgical guides and implants.

The third section, prosthetic rehabilitation applications, focuses on the use of medical modeling techniques to support the provision of prosthetics (typically facial prostheses but also including burns conformers and breast prostheses).

Medical Modeling
ISBN 978-0-323-95733-5
https://doi.org/10.1016/B978-0-323-95733-5.00015-6

The fourth section, orthotic applications, describes the use of scanning, CAD, and AM (additive manufacturing) in the development of custom-fitting, wearable orthotic devices such as wrist splints.

The fifth section, dental applications, describes the development of CAD and AM technologies in the design and manufacture of dental appliances such as removable partial denture (RPD) frameworks and custom-fitting sleep apnoea devices.

Finally, the sixth section, research applications, covers a variety of applications where medical modeling techniques have been used to support research projects including the validation of finite element analysis, the appearance of AM materials in CT images, texture reproduction, archaeology, bone surrogate materials, and training devices.

PART 5.1

Implementation

CHAPTER 5.1

Implementation case study 1: Computed tomography guidelines for medical modeling using additive manufacturing techniques*

5.1.1 Introduction

Medical modeling is the term for the production of highly accurate physical models of anatomy directly from 3D medical image data utilizing computer-controlled manufacturing machines commonly referred to as rapid prototyping (RP). Medical modeling involves acquiring three-dimensional image data of human anatomy, processing the data to isolate individual tissues or organs of interest, optimizing the data for the RP technology, and finally building the physical model in an RP machine. These machines have been primarily developed to enable designers and engineers to build exact models of products that they have designed using computer-aided design (CAD) software. Consequently, RP technologies have developed to produce models of very high accuracy as rapidly as possible.

---

* The work described in this chapter was first reported in the reference below and is reproduced here, in part or in full, with the permission of Elsevier Publishing. **Please note the term** additive manufacturing **appears in the chapter title for consistency but the original title and text uses the previously common term of** rapid prototyping.

Bibb R, Winder J. A review of the issues surrounding three-dimensional computed tomography for medical modeling using rapid prototyping techniques. Radiography 2010;16:78—83. https://doi.org/10.1016/j.radi.2009.10.005

Medical Modeling
ISBN 978-0-323-95733-5
https://doi.org/10.1016/B978-0-323-95733-5.00017-X

Medical modeling has many applications, and the most common has been in head and neck reconstruction including neurosurgery, craniofacial/maxillofacial surgery, and manufacturing prosthetics and implants. Medical models are not only being used for diagnosis, communication, and pre-surgical planning, but also they are increasingly used in the design and manufacture of implants and prosthesis [1−9], creating surgical guides [10−15], making imaging phantoms and teaching. Clear demonstration of benefits published in case studies and review articles has led to increasing interest in medical modeling [16−18]. Hundreds of medical models are produced in the UK each year, and the numbers are growing rapidly making it increasingly likely that radiographers will be asked to provide computed tomography (CT) images for medical modeling.

Usually, a surgeon or clinician who requires a medical model will request a 3D scan of the area of interest. The medical modeling process requires the relevant anatomy to be captured in a three-dimensional format, and although a number of medical imaging technologies have been successfully employed to make models, including MRI [19] and ultrasound [20], volumetric CT (3D CT) is by far the commonest imaging modality. The 3D medical image data is processed, mathematically modeled and subsequently transferred to a rapid prototype model provider for manufacture. Useful reviews of the RP methods and clinical applications are available [21,22]. After acquisition and transfer, the images are imported into specialist RP software, and techniques such as image thresholding and region growing are used to isolate the desired anatomical structure [23]. These data are then exported in a format called stereolithography (STL) that can be utilized by RP machines. The process is illustrated in Fig. 5.1.1.

The software for the RP machine then slices the three-dimensional data to produce a cross-section for each layer the machine will build. It is not

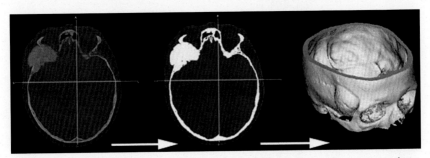

Figure 5.1.1 Steps from computed tomography (CT) image to 3D reconstruction.

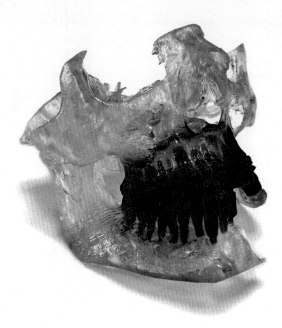

Figure 5.1.2 Example of a medical model.

possible to describe each RP technology in detail here, but more information can be found in reference texts [21]. A typical medical model of the mid-face and mandible is shown in Fig. 5.1.2. The model is multicolored to help the clinician distinguish between different tissue types identified within the CT images.

This technical note aims to provide radiographers with recommendations that will enable them to provide good-quality 3D CT images that can fulfill the requirements of medical modeling quickly and efficiently. There have been many case studies describing the use of medical models, but comparatively few have addressed the issues encountered when attempting to utilize medical modeling [24,25]. Potential problem areas in data acquisition and transfer are addressed in this note, and the guidelines described here aim to help avoid these errors.

5.1.2 CT guidelines for medical modeling

5.1.2.1 Anatomical coverage

An ideal CT acquisition should be free from image artifacts, have isotropic voxel resolution, and high image contrast between the anatomy of interest and neighboring tissues and low noise. It is clear that the series of axial images must begin and end either side of the anatomy of interest, but it is important to begin and end the scan some distance either side of the region/anatomy of interest. It is better to include anatomy (perhaps a few centimeters either side up to a maximum of 5 cm) above and below the area of interest, the amount being dependent on the region being scanned. In some cases, anatomy beyond the area of direct interest is used to help form or shape a repair to a bone of soft tissue defect [4,5,26—28]. The data volume should be continuous as noncontinuous data may contain areas where the patient has shifted position slightly, and the separate series will not align perfectly.

5.1.2.2 Patient arrangement, positioning, and support

Movement during the CT acquisition will result in movement artifact and distort the image data, which will translate directly through to the finished model. Anything greater than a 1-mm movement may render a model unusable. Involuntary movement of the chest, neck, head, or mouth can occur through breathing or swallowing. Gated acquisition for respiratory motion compensation as well as cardiac compensation is becoming increasingly common in multislice CT scanning [29], and their implementation would lead to significant movement artifact reduction. Other sources of movement like swallowing or talking should be controlled as far as possible.

It is increasingly common for 3D CT scans to be utilized in the multidisciplinary management of a patient, where the image data are not only used for diagnosis but also for surgical planning, computer-guided surgery, medical modeling, and prostheses design. The 3D CT data may be used to represent both hard and soft tissue internally but may also find subsequent application in tissue reconstruction or prosthetic rehabilitation. It is therefore important to consider the positioning and support of soft tissues to eliminate or reduce unwanted deformation of soft tissue. The use of positioning pads should be considered so that the anatomy of interest does not become distorted prior to scanning. Patient immobilization techniques such as vacuum pillows or simple foam pads may be used to support the patient, even to the extent that the surgical position of the patient may be replicated within the scanner. Examples of unwanted soft tissue deformations caused during CT acquisition can be seen in Fig. 5.1.3 where the images were acquired for the soft tissue information. Note the use of a hand in supporting a child's chin in Fig. 5.1.3. The use of the data should be clarified with the referring clinician to ensure appropriate positioning and support is used.

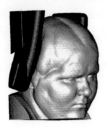 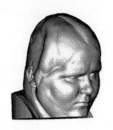 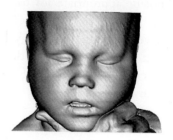

Figure 5.1.3 Soft tissue deformations caused by support.

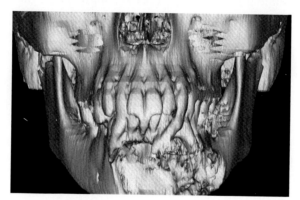

Figure 5.1.4 Overlap of the occlusion.

When using CT data for surgical planning, it is often necessary for different bones to be manufactured individually so that they can be moved independently to simulate surgical techniques. It is common for example when planning maxillofacial surgery to perform osteotomies and move parts of the mandible or maxilla. Often the patient is scanned with a closed bite, which causes the data for upper and lower teeth to merge. This makes subsequent separation of the mandible and maxilla using image-processing techniques very difficult. The effect of this overlap can be seen in Fig. 5.1.4. For maxillofacial cases, it is recommended that a slightly open bite or spacer be used that will enable the different jaws to be separately segmented.

5.1.2.3 CT parameters

There is a very wide range of CT technology available in the Health Service (from single slice helical to 64 slices or greater), and the variation in clinical practice throughout the UK and Ireland is very wide. It would not be appropriate for us to provide a definitive CT scanning protocol for any or all regions of the body, given that different centers will have different technology and levels of experience. It is recommended that radiographers use their routine 3D volumetric protocol for the given anatomical area also taking into consideration some of the issues below.

5.1.2.4 Slice thickness

To maximize the data acquired, the reconstructed slice thickness should be minimized. Some scanners can produce 0.5 mm slices, which gives excellent results, but this must be balanced against increased X-ray dose. Typically, slice thicknesses of 0.5–1 mm produce acceptable results. A slice thickness of 2 mm may be adequate for larger structures such as the long bones or pelvis. A slice thickness greater than 2 mm will give poorer results, and it is not recommended for rapid prototype models.

Consideration may also be given to collimation and overlap. In most circumstances, the scan distance and collimation should be the same. However, using a slice distance that is smaller than the collimation gives an overlap. When scanning for very thin sections of bone that lie in the axial plane, such as the orbital floor or palate, an overlap may give improved results [7].

5.1.2.5 Gantry tilt

Gantry tilt is commonly used it CT to provide the appropriate angle of slice relative to the anatomy of interest and also to reduce the radiation dose to the orbits in head scanning. However, for the purposes of medical modeling, gantry tilt should be set to zero (0 degree) as it does not significantly improve the quality of the acquired data and provides an opportunity for error when the service provider imports the images. Large gantry tilt angles are clearly apparent on visual inspection of the data and may be corrected. However, small angles may not be easy to check visually and may remain undetected or be compensated for incorrectly. Even the use of automatic import of the medical image standard DICOM is no guarantee as although the size of angle is included in the format, the direction is not. Failure to compensate for the direction correctly will lead to a distorted model, wasting time and money, and potentially leading to errors in surgery or prosthesis manufacture.

5.1.2.6 X-ray scatter

Dense objects such as amalgam or gold fillings, braces, bridges, screws, plates, and implants scatter X-rays resulting in a streaked appearance in the image. It is common practice in radiography to remove all metal to reduce artifacts where possible. However, where metal implants cannot be removed the effects of these can be manually edited using medical imaging software to produce a better medical model. However, this does depend on the expertise of the service provider, and consequently, the accuracy of the model in the affected areas cannot be guaranteed.

5.1.2.7 Noise

Noise is a fundamental component of a CT image and is especially prevalent in dense tissues. Medical modeling depends on identifying smooth boundaries between tissues. Noise interrupts the boundary, resulting in poor 3D reconstructions and consequently poor medical models, typically appearing rough or porous. This is shown in Fig. 5.1.5. Efforts should be made to reduce noise where medical models are required.

5.1.2.8 Image reconstruction kernels

During the reconstruction, digital filters (kernels) are applied which enhance or smooth the image depending on the clinical application. Typically, the options will range from "sharp" to "smooth." Sharpening filters increase edge sharpness but at a cost of increasing image noise. Smoothing filters reduce noise content in images but also decrease edge sharpness. In general, when building medical models, smooth filters tend to give better results and

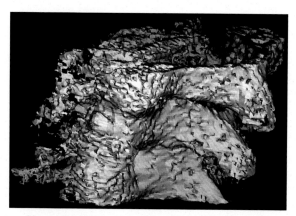

Figure 5.1.5 Porous appearance caused by noise.

are easier to work with. Although the smooth image contrast appears poor on screen (Windows computers can only display 256 shades of gray and the human eye can only perceive about 70 gray levels), density profiles show that the actual contrast is good and allows a lower threshold to be used.

5.1.2.9 Data transfer

3D medical image processing and medical RP software (e.g., Mimics, Materialise N.V, Technologielaan 15, 3001 Leuven, Belgium, AnalyzeDirect, Inc. 7380 W 161th Street, Overland Park, KS, 66085 USA) require DICOM V3.0 data format and are usually sent to a medical modeling service provider on CD-ROM. In nearly all cases, only the reconstructed axial/transverse images are required as further image processing and modeling will be carried out by the service provider. CT images should be written without image compression or the automated viewing software. From a patient confidentiality point of view, the exclusion of the manufacturer's viewing software means that images on a lost or misplaced CD cannot be easily viewed without specialist knowledge and software. The images do not need to be "windowed" prior to storage or transfer as access to the original DICOM images allows the service provider to view them with their own settings. Careful consideration should be given to patient confidentiality and data security, and procedures should be agreed with the service provider to ensure all data are securely and ethically treated. For example, data should be sent by registered post, the service provider should be requested to store data in a locked cabinet, and access to any data should be password protected. Arrangements should also be made to return or destroy the CD on completion of the model.

5.1.3 Conclusion

Following the general guidelines above will help to improve the source volumetric CT image data used for medical RP and subsequently improve on model quality.

Acknowledgments

Fig. 5.1.2 is reproduced with the permission of Queen Elizabeth Hospital Birmingham, Fig. 5.1.3a was kindly provided by Dr. Jules Poukens, University Hospital Maastricht, the Netherlands. Figs. 5.1.3b and 5.1.4 were kindly provided by Carol Voigt, Brånemark Institute, South Africa.

References

[1] Eufinger H, Wehmoller M. Individual prefabricated titanium implants in reconstructive craniofacial surgery: clinical and technical aspects of the first 22 cases. Plastic and Reconstructive Surgery 1998;102(2):300—8.

[2] Heissler E, Fischer FS, Bolouri S, Lehmann T, Mathar W, Gebhardt A, et al. Custom-made cast titanium implants produced with CAD/CAM for the reconstruction of cranium defects. International Journal of Oral and Maxillofacial Surgery 1998;27(5):334—8.

[3] Joffe J, Harris M, Kahugu F, Nicoll S, Linney A, Richards R. A prospective study of computer-aided design and manufacture of titanium plate for cranioplasty and its clinical outcome. British Journal of Neurosurgery 1999;13(6):576—80.

[4] Winder J, Cooke RS, Gray J, Fannin T, Fegan T. Medical rapid prototyping and 3D CT in the manufacture of custom made cranial titanium plates. Journal of Medical Engineering and Technology 1999;23(1):26—8.

[5] D'Urso PS, Earwaker WJ, Barker TM, Redmond MJ, Thompson RG, Effeney DJ, et al. Custom cranioplasty using stereolithography and acrylic. British Journal of Plastic Surgery 2000;53(3):200—4.

[6] Bibb R, Bocca A, Evans P. An appropriate approach to computer aided design and manufacture of cranioplasty plates. Journal of Maxillofacial Prosthetics and Technology 2002;5(1):28—31.

[7] Hughes CW, Page K, Bibb R, Taylor J, Revington P. The custom-made titanium orbital floor prosthesis in reconstruction for orbital floor fractures. British Journal of Oral and Maxillofacial Surgery 2003;41(1):50—3.

[8] Evans P, Eggbeer D, Bibb R. Orbital prosthesis wax pattern production using computer aided design and rapid prototyping techniques. Journal of Maxillofacial Prosthetics and Technology 2004;7:11—5.

[9] Singare S, Dichen L, Bingheng L, Zhenyu G, Yaxiong L. Customized design and manufacturing of chin implant based on rapid prototyping. Rapid Prototyping Journal 2005;11(2):113—8.

[10] Bibb R, Bocca A, Sugar A, Evans P. Planning osseointegrated implant sites using computer aided design and rapid prototyping. Journal of Maxillofacial Prosthetics and Technology 2003;6(1):1—4.

[11] Di Giacomo GAP, Cury PR, De Araujo NS, Sendyk WR, Sendyk CL. Clinical application of stereolithographic surgical guides for implant placement: preliminary results. Journal of Periodontology 2005;76(4):503—7.

[12] Goffin J, Van Brussel K, Vander Sloten J, Van Audekercke R, Smet MH, Marchal G, et al. 3D-CT based, personalized drill guide for posterior transarticular screw fixation at C1-C2: technical note. Neuro-Orthopedics 1999;25(1—2):47—56.

[13] Goffin J, Van Brussel K, Vander Sloten J, Van Audekercke R, Smet MH. Three-dimensional computed tomography-based, personalized drill guide for posterior cervical stabilization at C1-C2. Spine 2001;26(12):1343—7.

[14] Sarment DP, Sukovic P, Clinthorne N. Accuracy of implant placement with a stereolithographic surgical guide. The International Journal of Oral and Maxillofacial Implants 2003;18(4):571—7.

[15] Sarment DP, Al-Shammari K, Kazor CE. Stereolithographic surgical templates for placement of dental implants in complex cases. The International Journal of Periodontics and Restorative Dentistry 2003;23(3):287—95.

[16] Petzold R, Zeilhofer H-F, Kalender WA. Rapid prototyping technology in medicine - basics and applications. Computerized Medical Imaging and Graphics 1999;23(5):277—84.

[17] Bibb R, Brown R. The application of computer aided product development techniques in medical modelling. Biomedical Sciences Instrumentation 2000;36:319–24.

[18] Sanghera B, Naique S, Papaharilaou Y, Amis A. Preliminary study of rapid prototype medical models. Rapid Prototyping Journal 2001;7(5):275–84.

[19] Swann S. Integration of MRI and stereolithography to build medical models. Rapid Prototyping Journal 1996;2(4):41–6.

[20] D'Urso PS, Thompson RG. Fetal biomodelling. The Australian and New Zealand Journal of Obstetrics and Gynaecology 1998;38(2):205–7.

[21] Bibb R. Medical modelling: the application of advanced design and development technologies in medicine. Cambridge: Woodhead; 2006.

[22] Webb PA. A review of rapid prototyping (RP) techniques in the medical and biomedical sector. Journal of Medical Engineering and Technology 2000;24(4):149–53.

[23] Gonzalez RC, Woods RE. Digital image processing. 3rd ed. Addison Wesley; 2007. p. 443–58.

[24] Winder RJ, Bibb R. Medical rapid prototyping technologies: state of the art and current limitations for application in oral and maxillofacial surgery. Journal of Oral and Maxillofacial Surgery 2005;63(7):1006–15.

[25] Sugar A, Bibb R, Morris C, Parkhouse J. The development of a collaborative medical modelling service: organisational and technical considerations. British Journal of Oral and Maxillofacial Surgery 2004;42:323–30.

[26] Winder J, McRitchie I, McKnight W, Cooke S. Virtual surgical planning and CAD/CAM in the treatment of cranial defects. Studies in Health Technology and Informatics 2005;111:599–601.

[27] Vander Sloten J, Van Audekercke R, Van der Perre G. Computer aided design of prostheses. Industrial Ceramics 2000;20(2):109–12.

[28] Verdonck HWD, Poukens J, Overveld HV, Riediger D. Computer-assisted maxillofacial prosthodontics: a new treatment protocol. The International Journal of Prosthodontics 2003;16(3):326–8.

[29] Li G, Citrin D, Camphausen K, Mueller B, Burman C, Mychalczak B, et al. Advances in 4D medical imaging and 4D radiation therapy. Technology in Cancer Research and Treatment 2008;7(1):67–81.

CHAPTER 5.2

Implementation case study 2: The evolving development of a collaborative service: Organizational, technical, and regulatory considerations*

5.2.1 Introduction

Technologies, organizational challenges, and regulatory conditions have evolved significantly since earlier editions of this book. The scope of in-hospital medical modeling services has changed in parallel. The production of medical models using additive manufacture (AM) has been a valuable aid in the diagnosis of medical conditions, the planning or surgery, and production of prostheses for many years [1–4]. AM techniques have shown significant advantages over previous milled models used in very early

* Early sections of this case study were, in part, first reported in the reference below and are reproduced here, in part or in full, with the permission of the British Association of Oral and Maxillofacial Surgeons. Significant updates have been added to reflect developments in technology and regulations.

Sugar A, Bibb R, Morris C, Parkhouse J. The development of a collaborative medical modeling service: organisational and technical considerations. British Journal of Oral and Maxillofacial Surgery 2004;42(4):323–30.

Substantial additions have been made, which reflect the evolution of collaboration.

This work has only been possible because of the enthusiasm and hard work of the entire collaborating team from PDR and Morriston Hospital. In particular, the authors gratefully acknowledge the work of Peter Evans, Alan Bocca, Steven Hollisey-McLean, and Lawrence Dovgalski of the Maxillofacial Unit and Dr. E Wyn Jones, Rose Davies, and Sian Bowen of the Radiology Department of Morriston Hospital. We are also grateful to our neurosurgical colleague, Tim Buxton, who has supported this project physically and financially since its inception.

Medical Modeling
ISBN 978-0-323-95733-5
https://doi.org/10.1016/B978-0-323-95733-5.15007-0

years [2]. As software and AM technologies have evolved and become more affordable, in-hospital use has grown in scope. For example, it is now relatively commonplace for hospitals to incorporate services to produce transient use custom surgical cutting and drilling guides [5]. A limited number of hospitals now also include services for the AM of long-term custom implants [6]. This is, in part, due to the clinical benefits of using AM in production of custom transient surgical devices and long-term implants being more widely accepted. Despite the growing use of design and AM technologies within hospital environments, barriers to safe and efficient use remain, but with a different emphasis. Close cooperation between healthcare providers, researchers, industry partners, and regulatory bodies is still required to ensure sustainable development of in-hospital custom medical device design and manufacturing service.

This case study describes the evolving relationship of a Wales-based National Health Service (NHS) hospital and design/research organization that developed the use of design in healthcare and medicine. This collaboration was established in 1999, but the nature and balance of roles and responsibilities have changed significantly since then. Charting these changes and the organizational, technical, and regulatory considerations that informed them will provide insight into how in-hospital services can meet clinical needs more effectively.

5.2.2 The early years of collaboration—Establishing a joint medical modeling service

The production of medical models was, only 2 decades ago, dominated by commercial services with few being close to hospitals needing them. In most cases, scan data on archive media (e.g., magnetic optical disk, DAT tape, or CD ROM) were posted from the radiology department to the specialist service supplier. The supplier translated and segmented the data from which the medical model build files were created. The transfer of medical scan data is described in Chapter 3. This required careful communication so that the correct parameters, tissues, and extents were used. Once translation was complete, the build files were returned to the service provider who built and delivered the model to the hospital. This procedure, schematically shown in Fig. 5.6 and called the "disconnected" procedure, was time consuming and removed potentially important clinical decision-making opportunities from medical staff. A limited number of specialist suppliers dominated this service. One of the principal reasons for

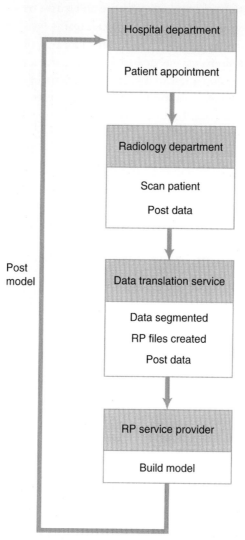

Figure 5.6 The "disconnected" procedure.

this was the dependence on relatively expensive and technically difficult to use hardware and software. This meant that high volumes of cases were needed to justify the expense. Furthermore, the need for secure data handling and training staff from engineering, manufacturing, or design backgrounds in medical terminology also limited more widespread commercial service expansion. During this time, in-hospital services were also constrained for the same reasons; software and hardware was expensive, and the skillsets did not exist as part of hospital teams.

A collaborative approach between organizations close to each other in the United Kingdom was needed to improve medical model production efficiency and develop a research relationship. One of the first hospital and AM service provider relationships developed was between Morriston Hospital in Swansea and International Centre for Design Research (PDR) at Cardiff Metropolitan University [7—9]. This process involved sharing the necessary procedures and resources between the medical staff and the AM service provider. Tasks required to produce medical models were accomplished by the most appropriate staff, improving decision-making opportunities while reducing cost and turnaround time. This was made possible due to increased use of over-the-internet data exchange and expansion of data storage capability in both partners.

Due to the modular nature of the software used, the initial integrated procedure shown in Fig. 5.7 was first attempted [8]. Experienced radiographers in a location accessible to the surgeons conducted the segmentation of the image data within the radiology department. The segmented data were then sent to the AM service provider who used the remaining software modules to prepare and manufacture the models. Once such a route was established, the turnaround time from initial scan to finished model was reduced from weeks to days.

The hospital partners used specific modules of specialist software (Mimics, Materialise NV, Leuven, Belgium) for the import and segmentation of scan data. A separate module (CT Modeler, CTM) generated the rapid prototyping build files from data exported from Mimics. This allowed the segmentation software (Mimics) to be installed in both the radiology and the clinical departments, while the other module (CTM) remained at the AM service provider. The segmentation module was installed on well-specified PCs connected to the hospital network allowing direct access to the scan data. The basic principles of data segmentation are more fully described in Chapter 3.

When using this software, there was a convenient intermediate step between the different modules. The export file format from Mimics to CT Modeler (called a .3dd file) was highly compressed, typically in the region of 1 MB. Files this small were easily transferred as an attachment to an ordinary email. This was an important consideration in the days before low-cost, easy to access, over-the-internet data transfer. The use of existing

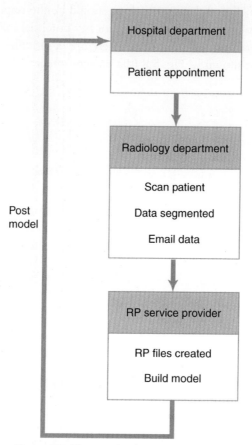

Figure 5.7 The initial integrated procedure.

email protocols including the firewalls at the hospital and the service provider eliminated security and network access issues.

After trials, it became clear that the file import, preparation, and segmentation were too time consuming to be undertaken by the radiographers. Although highly trained in the operation of the scanner, the radiographers were unfamiliar with other computer formats and network procedures. It was felt that this was not the best use of their time. It was originally envisaged that radiographers would be better able to segment the scan data. However, in practice, the simplicity of the software allowed accurate segmentation to be accomplished by any adequately trained user, such as clinicians. Furthermore, clinicians had a much better idea of what they wanted to achieve by the segmentation. It was also felt that the majority of the clinical decision-making should be the responsibility of the clinical department. This would eliminate any potential misunderstandings between medical and technical staff. Therefore, a modified procedure was implemented as shown in Fig. 5.8 (the current integrated procedure). This

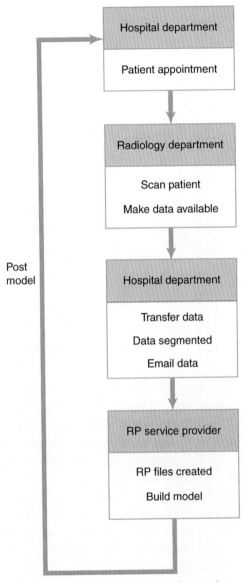

Figure 5.8 The current integrated procedure.

procedure involved the clinical department having the software to enable them to segment the data and email the compressed file format (a 3D image) to the AM service provider.

This method also kept as much of the workload (and therefore cost) within the clinical department and minimized the workload transferred to radiology. The added benefit of this method was the capability of the clinical department to produce 3D reconstructions on screen. The surgeon could then view and move these images at will rather than relying on a selection of fixed views produced on film by the radiology department. Axial slices could be viewed simultaneously with coronal and sagittal reformats, and the slices can be run through in seconds. Areas of specific interest could be generated in three dimensions and viewed from any angle. The increased access to 2D and 3D data in various forms may also eliminate the need to produce a physical model in many cases, with a potential saving of time and money. The images produced could also be saved on hard disk, printed on paper for storage in patients' notes, and exported into standard image software and, from there, into slide presentation formats for teaching/lecturing/demonstrations. Hard copy could also be given to patients to help them understand their condition.

When following this procedure, the radiology department obtained the scans in the usual manner and saved the data into a secure directory on a PC hard-wired to the scanner. This PC enabled storage of data in the radiology department in their preferred format. The PC was connected to the hospital's local intranet, which allowed sharing with a registered user in the maxillofacial unit using File Transfer Protocol (FTP) to another PC in the maxillofacial unit also connected to the intranet. Other registered users in other hospital departments could also be included if necessary. The data were then burned in the maxillofacial unit onto CD, saved on the PC's hard disk, and imported into the specialized imaging software (Mimics). Image manipulation was then carried out by clinicians. Sections could be mirror-imaged and manipulated, and areas were measured without interfering with the schedule of a busy radiology department. The file for AM model construction, if required, could then be emailed to the service provider as an attachment.

Of course, when CT scanners were changed, it was necessary to check that existing procedures still worked and that the imaging software (e.g., Mimics) was configured appropriately to receive the new scan format. Current scanners almost invariably now comply with a standard digital format (DICOM), and the importation of such scans into Mimics is in fact made much easier in this format. When Morriston Hospital acquired a new Multislice CT scanner in 2002, all its raw CT data was saved on a server within the hospital intranet. The maxillofacial unit was then given direct access to this server through the intranet, and CT scans can be picked up directly and saved. This method replaced the FTP arrangement.

In these early years, technical constraints revolved around challenges of sharing large data sets efficiently and securely. The hospital was constrained to using now superseded ways of transferring DICOM format CT data within the hospital intranet to the maxillofacial team, who then had to duplicate it to physical media. At the time, many radiology departments were not accustomed to exporting data and may only have such archiving formats available as DAT or Optical Disk. In addition, some radiology departments were still not networked. Data exchange between the hospital and AM service provider were also highly constrained, only allowing file sizes in the order of <5 mb to be transferred without the use of specialist FTP. Although transferring data to a CD ROM was possible, this reduced joint service efficiency significantly and prevented the technology from being used in urgent cases.

From an organizational perspective, creating a local collaboration reduced the timescale of medical model production, but it required cooperation of the surgical departments, the radiology department, the hospital IT department, and the service provider. Consequently, the organizational considerations related to economics and budgeting, staff workload, and responsibility management. It is widely acknowledged that NHS hospital departments in the United Kingdom operate within strictly controlled budgets. Medical models were perceived to improve quality of patient care, and cost-effectiveness incurred an additional charge on tight budgets. It was therefore crucial to obtain the support and commitment of senior hospital management at the outset of the project in this case study. At the time, the university-based AM service provider could make cost-effective use of AM equipment by providing medical models across the United Kingdom. This helped to offset the high purchase, material, and running costs of the AM equipment and software. The hospital themselves needed only to spend time and resources segmenting data and planning for the medical model. Issues around in-hospital financial management were evident. The prescription of a medical model also incurred costs and required precious resources from radiology through to the maxillofacial department. In many cases, clinical departments will want to maintain the maximum amount of responsibility and control while minimizing workload and expenditure. Interdepartment budgeting and planning was required to ensure appropriate resource and effort allocation.

During the early years of collaboration, regulatory considerations were minimal since the organizations were operating under the Medical Device Directive. This gave relative leniency to custom-made devices. Medical models were also not considered medical devices, so they were not affected as significantly by the directive. The AM service provider did, however, have an internationally recognized ISO 9001 [10] quality management system in place, which enabled control of AM production. Issues around regulations have, however, become much more prominent in parallel with AM technology development and hospital service evolution.

5.2.3 Service evolution toward greater in-hospital capability

Establishing a geographically close joint medical modeling service led to greater research collaboration. By c.2016, AM technologies and associated software had evolved significantly. Advances in the way radiological images were stored and shared within hospitals had overcome many of the technical issues identified in the early stages of collaboration. Sharing of large datasets outside the UK NHS was also possible, but it still required use of specialist software and tightly controlled security protocols needed to protect sensitive information. Staff in the Maxillofacial Unit at Morriston Hospital could now pull CT data directly from hospital servers into the segmentation software, using techniques described in Chapter 3 to create 3D computer models of patient anatomy. Data exchange between the partners was facilitated by a secure, dedicated server space, hosted by NHS Wales. This provided the freedom to exchange information such as 3D computer-aided design models, images, and documents that were helpful to support communication on complex cases. Data exchange between the university-based organization and other hospitals in the United Kingdom was enabled by a specialist service (Image Exchange Portal, IEP, Sectra Systems). IEP was, however, constrained to the exchange of DICOM format data from CT scans.

Medical models were still required to support clinical decision-making and procedure planning, but greater focus was placed on the production of end-use custom, transient surgical guides (see case studies). There was increased evidence that using custom surgical drilling/cutting guides and implants improved surgical procedure accuracy [11]. It was also viewed as more efficient since it was only necessary to produce a small device using AM, rather than employ multiple stages of indirect fabrication on a medical model. This was considered important; although there was a greater range of AM materials suitable for producing transient surgical devices, the technology was still expensive to own and operate. Both metal and polymer AM methods were used to produce custom surgical guides. Despite the high costs, metal AM was appealing due the ability to produce stronger, smaller devices that were more resistant to wear [12]. Polymer and, in a small number of cases, metal guide production was undertaken by the AM service provider who had invested in technologies and materials that were well suited to the application and were deployed across the United Kingdom.

However, guide design and fabrication amplified issues around the disconnect between surgeons, in-hospital services to plan procedures, and AM service providers. Designing end-use custom surgical devices introduced more stages, which required greater levels of collaboration and verification to ensure safety and suitability for AM fabrication. The busy nature of hospital environments made external collaboration on complex case planning, design, and manufacture a challenge. Cooperation between the partners had led to a high degree of knowledge exchange; the maxillofacial laboratory team had learned how to design custom devices in parallel with the service provider through their research and application. Even with limited dedicated resource, having in-hospital custom device design knowledge, supported by the external collaborator, helped to overcome the disconnect between prescribing surgeons and designers.

5.2.4 Bringing greater design expertise and guide production capability in-hospital

Investment in design knowledge to support expansion of 3D surgical planning and custom guide design was viewed as the next important step to improve process efficiency. Collaboration with the university-based partner had emphasized the importance of disciplines outside traditional maxillofacial laboratory roles; PDR employed experts in design, engineering, and manufacturing, who then learned about medical procedures to inform commercial and research projects. Conversely, the Maxillofacial Lab at Morriston Hospital employed expertise in dental technology, facial prosthetics, and medical procedures. Bringing expertise in 3D surgical planning and design software within the hospital setting would help to overcome the disconnect between device prescription, design, and fabrication. Approval was therefore sought for the maxillofacial unit to employ a dedicated 3D biomedical technician, who would be responsible for realizing surgical plans and guide designs within the hospital and who would maintain links with the university-based collaborator where needed.

During the period from c.2017–18, numerous surgical planning and design software tools were used by Morriston Hospital, including Mimics, Simplant (Materialise NV, Leuven, Belgium), and Geomagic Freeform Plus (3D-Systems, USA). Less emphasis was placed on in-hospital AM production hardware, but falling prices and greater levels of industry support meant that it became economically feasible for the hospital to invest in modest technology. Market forces and expiration of patents meant that fused deposition modeling (FDM)-based machines (see Chapter 4) became increasingly affordable. Morriston purchased a Wanhao Duplicator i3 and i5 (Wanhao Zhejiang, China) for the purposes of medical model production. Material prices of these machines were about 10 times less than previously used stereolithography resin. Maintenance costs were also significantly cheaper, and they were not locked into industry support packages. Although they were not suitable to produce surgically invasive custom devices, the low-cost machines were used extensively for medical models and mold tool production for face and body prostheses (see case studies), where high-fidelity and high-performance materials were not needed. The hybrid approach where surgical plans and device designs were undertaken in-hospital, with surgical guides produced by the external provider, continued until around late 2018.

The next major development opportunity presented with the advent of relatively low-cost stereolithography technologies. The primary limitations of FDM extrusion AM equipment are materials (which are not suitable for sterilizing) and geometry resolution (details required for intricate custom surgical devices cannot be replicated). The stereolithography apparatus, used by the university partner to produce homogeneous, solid guides in a material that met prerequisite requirements for biocompatibility, was more suitable for sterilizing and had a proven history for the production of custom surgical guides. It was, however, prohibitively expensive and was better suited to higher volume production where costs could be recouped more effectively. Formlabs (Massachusetts, USA) was one of the first manufacturers to offer an affordable SLA technology that was suitable for in-hospital production of detailed custom surgical devices. Morriston invested in two Formlabs, Form 2 machines, which enabled them to fabricate their own surgical guides. This was a turning point that demonstrated commitment to a total in-hospital custom device solutions, but also one that presented new challenges. For a hospital lab already familiar with handling chemicals and other materials, the practicalities and technical issues associated with production and postproduction of stereolithography parts were not too challenging. Adequate hazardous material storage and disposal were supported, and the environment was designed with appropriate ventilation and temperature control. The small footprint of the machines also helped to make this relatively simple. From an organizational perspective, financial arguments for introducing further AM equipment centered around the high costs associated with industry providers; it was perceived that having the production in-hospital would almost eliminate the need for contract production and maintain greater control, and agility, of the process. Techniques for surgical guide production evolved and are described in more detail in the case studies.

The way custom implants were designed and fabricated also evolved with technology evolution and changes in regulations. Morriston's Maxillofacial Lab had historically manufactured custom implants using indirect methods, such as plate pressing and bending around anatomic models. The transition of using 3D CAD and direct metal AM for the production of custom implants began c.2016. Between then and 2019, the team transitioned from design being a shared activity between Morriston's Maxillofacial Lab and PDR to being led entirely in-hospital. By this point, knowledge exchange and collaboration with industry partners who fabricated the final custom implants (typically using metal AM) had enabled the embedded 3D biomedical technician, supported by the wider laboratory team. At this time, organizational challenges centered on demonstrating value for money that a dedicated 3D biomedical technician, numerous software packages, and production technology provided. The argument was that bringing technology in-hospital almost eliminated subcontracting costs. Perhaps a stronger argument was that it provided a faster response time to urgent cases and gave greater control over the process, which is valuable given the fluid nature of interactions between surgical and maxillofacial laboratory staff. It completely overcame the disconnect between prescribing surgeons and designers but did create dynamics whereby the demand for ad hoc case planning meetings were more common. There was a need for more rigorous structure to ensure essential case details that informed design and manufacture decisions were captured during surgeon/planning team meetings.

5.2.5 Evolving regulatory considerations

The Medical Device Regulations (MDRs) came into force in May 2021 following a 3-year transition period from the Medical Device Directive (MDD) that covered the European Union. Under the MDD, custom medical devices were given a degree of relative leniency for in-hospital producers, with rules around design and production relaxed compared with industry-produced medical devices. Along with other changes, the MDR stipulated that hospitals, like industry, must have an appropriate quality management system in place, together with statements of conformity and other information that help to demonstrate safety and regulatory compliance. The team at PDR had developed an ISO 13485:2016 [13] quality management system covering the design of custom implants and guides. The team at Morriston understood the importance of a more robust quality management system and started working on their own in 2019. A dedicated person was employed to develop and maintain an ISO 13485:2016 quality management system for the Maxillofacial Lab and Rehabilitation Engineering Unit within Morriston Hospital. The system developed covered production of custom-made medical devices, which was a major step in addressing regulatory requirements.

Introduction of the MDR also led to wider changes in previous practices used in the production of custom-made implants. Industry had started to provide stock plate systems in a sterile state, which meant that in absence of resterilization guidelines, they should only be handled in a sterile environment. Whereas, previously, plates would be prebent using anatomic models as a guide in a lab environment, Morriston had to develop stringent controls for handling and bending stock plates prior to surgery. These processes were included within their quality management system together with those designed to control production using AM equipment. Despite many years of success, previously used methods of pressing custom implants from sheet titanium were also stopped due to concerns over the difficulty in demonstrating the appropriate levels of quality control. Plate pressing was replaced by 3D CAD design and direct AM production by an industry supplier. This was perceived as a regressive step in cases that required an extremely fast turnaround. For example, custom orbital floor implants benefit from being used within 7 days of injury, which places pressure on designing, signing off the design, and manufacturing a plate in a very short timescale, which is not always possible using metal AM.

Organizational challenges also emerged around the compatibility of different hospital quality management systems. Morriston Hospital's Sterilisation and Disinfection Unit also had a stringent ISO 13485 quality management system that required regulatory compliant instructions for disinfecting and sterilizing products that entered their process. This meant that units had to work together to establish safe practices or eliminate those that could not be sufficiently controlled.

At the time of writing, there is still uncertainty around how the MDR and UK MDR affect some elements of in-hospital design and production of custom-made medical devices in the United Kingdom. As others have identified, the trend is certainly toward greater levels of quality management that align to what is expected in industry and greater harmonization of international standards [14,15]. Morriston Hospital's quality management system and other processes related to regulatory compliance are designed to continuously improve based on evolving evidence and continued collaboration with the academic and industry partners. This same model of collaboration is likely to emerge in many other hospitals.

References

[1] Klein HM, Schneider W, Alzen G, Voy ED, Gunther RW. Pediatric craniofacial surgery: comparison of milling and stereolithography for 3D model manufacturing. Pediatric Radiology 1992;22:458−60.

[2] Swaelens B, Kruth JP. Medical applications of rapid prototyping techniques. In: Proceedings of the fourth international conference on rapid prototyping. Dayton, Ohio, USA; 1993. p. 107−20.

[3] Greenfield GB, Hubbard LB. Computers in radiology. New York: Churchill-Livingstone; 1984. p. 91−130.

[4] Jacobs PF, editor. Stereolithography and other RP&M technologies. Dearborn, MI, USA: Society of Manufacturing Engineering, One SME Drive; 1996.

[5] Calvo-Haro JA, Pascau J, Mediavilla-Santos L, Sanz-Ruiz P, Sánchez-Pérez C, Vaquero-Martín J, et al. Conceptual evolution of 3D printing in orthopedic surgery and traumatology: from "do it yourself" to "point of care manufacturing". BMC Musculoskeletal Disorders 2021;22(1):1−10.

[6] Sharma N, Aghlmandi S, Dalcanale F, Seiler D, Zeilhofer H-F, Honigmann P, et al. Quantitative assessment of point-of-care 3D-printed patient-specific polyetheretherketone (PEEK) cranial implants. International Journal of Molecular Sciences 2021;22(16):8521.

[7] Bibb R, Freeman P, Brown R, Sugar A, Evans P, Bocca A. An investigation of three-dimensional scanning of human body surfaces and its use in the design and manufacture of prostheses. Proceedings of the Institute of Mechanical Engineers Part H, Journal of Engineering in Medicine 2000;214(6):589−94.

[8] Bibb R, Brown R. The application of computer aided product development techniques in medical modelling. Biomedical Sciences Instrumentation 2000;36:319−24.

[9] Bibb R, Brown R, Williamson T, Sugar A, Evans P, Bocca A. The application of product development technologies in craniofacial reconstruction. In: Proceedings of the ninth European conference on rapid prototyping and manufacturing. Athens, Greece; 2000. p. 113—22.

[10] International Organization for Standardization. ISO 9001:2015—quality management systems—requirements. Available at: https://www.iso.org/standard/62085.html; 2015.

[11] Di Giacomo GAP, Cury PR, de Araujo NS, Sendyk WR, Sendyk CL. Clinical application of stereolithographic surgical guides for implant placement: preliminary results. Journal of Periodontology 2005;76(4):503—7.

[12] Bibb R, Eggbeer D, Evans P, Bocca A, Sugar A. Rapid manufacture of custom-fitting surgical guides. Rapid Prototyping Journal 2009;15(5):346—54.

[13] International Organization for Standardization. ISO 13485:2016—medical devices— quality management systems—requirements for regulatory purposes. Available at: https://www.iso.org/standard/59752.html; 2016.

[14] Tel A, Bordon A, Sortino M, Totis G, Fedrizzi L, Ocello E, et al. Current trends in the development and use of personalized implants: engineering concepts and regulation perspectives for the contemporary oral and maxillofacial surgeon. Applied Sciences 2021;11(24):11694.

[15] Carl A, Hochmann D. Comparison of the regulatory requirements for custom-made medical devices using 3D printing in Europe, the United States, and Australia. Biomedical Engineering/Biomedizinische Technik 2022;67(2):61—9.

CHAPTER 5.3

Implementation case study 3: Medical additive manufacturing technologies: State of the art and current limitations for application in oral and maxillofacial surgery[*]

5.3.1 Introduction

Medical rapid prototyping (MRP) is defined as the additive manufacture (AM) of dimensionally accurate physical models of human anatomy derived from medical image data using a variety of technologies. It has been applied to a range of medical specialities including oral and maxillofacial surgery [1–7], dental implantology [8], neurosurgery [9,10], and orthopedics [11,12]. The source of image data for 3D modeling is principally computed tomography (CT), although magnetic resonance imaging and ultrasound have also been utilised. Medical models have been successfully built of hard tissue such as bone and soft tissues including blood vessels and nasal passages [13,14]. MRP was described originally by Mankowich et al. in 1990 [15].

---

[*] The work described in this chapter was first reported in the reference below and is reproduced here, in part or in full, with the permission of the American Association of Oral and Maxillofacial Surgeons. Please note the term "additive manufacturing" appears in the chapter title for consistency, but the original title and text uses the previously common term of "rapid prototyping."
Winder RJ, Bibb R. Medical rapid prototyping technologies: state of the art and current limitations for application in oral and maxillofacial surgery. Journal of Oral and Maxillofacial Surgery 2005;63(7):1006–15. https://doi.org/10.1016/j.joms.2005.03.016.

Medical Modeling
ISBN 978-0-323-95733-5
https://doi.org/10.1016/B978-0-323-95733-5.15004-5

Figure 5.9 A typical medical model.

The development of the technique has been facilitated by improvements in medical imaging technology, computer hardware, 3D image processing software, and the technology transfer of engineering methods into the field of surgical medicine. A typical medical model is shown in Fig. 5.9.

The clinical application of medical models has been analyzed in a European multicenter study [16]. Results were collated from a questionnaire sent out to partners of the Phidias Network on each institution's use of MRP stereolithography models.

The 172 responses indicated the following range of applications:
- to aid production of a surgical implant
- to improve surgical planning
- to act as an orienting aid during surgery
- to enhance diagnostic quality
- useful in preoperative simulation
- to achieve patient's agreement prior to surgery
- to prepare a template for resection

Further, it was noted that the diagnoses in which a stereolithography (SL) model was employed were as follows: neoplasms (19.2%), congenital disease (20%), trauma (15%), dentofacial anomalies (28.9%), and others (16.9%). MRP is also being developed for use in dental implants. Greater accuracy was achieved with the use of rapid prototyped surgical guides for creating osteotomies in the jaw [17], and a computer-aided design/ computer-aided manufacturing (CAD/CAM) approach to the fabrication of partial dental frameworks has been developed [18].

The creation of medical models requires a number of steps: the acquisition of high-quality volumetric (3D) image data of the anatomy to be modeled; 3D image processing to extract the region of interest from surrounding tissues; mathematical surface modeling of the anatomic surfaces; formatting of data for rapid prototyping (this includes the creation of model support structures that support the model during building and are subsequently manually removed); model building; and quality assurance of model quality and dimensional accuracy. These steps require significant expertise and knowledge in medical imaging, 3D medical image processing, CAD/CAM software, and engineering processes. The production of reliable, high-quality models requires a team of specialists that may include medical imaging specialists, engineers, and surgeons.

The purpose of this report is, firstly, to describe the range of rapid prototyping technologies (including software and hardware) available for MRP, secondly, to compare their relative strengths and weaknesses, and thirdly, to illustrate the range of pitfalls that we have experienced in the production of human anatomic models. The authors have a combined experience of 17 years working in the field of MRP and have direct experience of the technologies described later. The report begins with a description of 3D image acquisition and processing and computer modeling methods required, common medical rapid prototyping techniques, followed by a discussion of model artifacts and manufacturing pitfalls. At present, there is no suitable text describing MRP or its clinical applications; however, there are two useful review papers [19,20].

5.3.2 3D image acquisition and processing for MRP

The modalities and general principles of acquiring medical scan data for rapid prototyping (RP) are described in Chapter 2, Medical imaging for RP, but some of the more important observations resulting from a large number of actual cases are discussed here. The volumetric or 3D image data required for MRP models have certain particular requirements. Specialized CT scanning protocols are required to generate a volume of data that are isotropic in nature. This means that the three physical dimensions of the voxels (image volume elements) are equal or nearly equal. This has become achievable with the introduction of multislice computed tomography (CT) scanners where in-plane pixel size is of the order of 0.5 mm with slice thickness as low as 1.0 mm [21]. Data interpolation is often required to convert the image data volume into an isotropic dataset for mathematical modeling. Further image processing steps will be required to identify and separate out the anatomy (segmentation) for modeling from surrounding structures. Segmentation may be carried out by image thresholding, manual editing, or auto-contouring to extract volumes of interest. Final delineation of the anatomy of interest may require 2D or 3D image editing to remove any unwanted details. A number of software packages are available for data conditioning and image processing for MRP and include Analyze (Lenexa, KS, www.AnalyzeDirect.com), Mimics by Materialise (Leuven, Belgium, www.materialise.com), and Anatomics (Brisbane, Australia, www.anatomics.com/). There is still a need for seamless and inexpensive software that provides a comprehensive range of data interpretation, image processing, and model building techniques to interface with RP technology.

The first models created were of bone that were easily segmented in CT image data. Bone has a CT number range from approximately 200—2000. This range is unique to bone within the human body, as it did not numerically overlap with any other tissues. In many circumstances, a simple threshold value was obtained and applied to the data volume. All soft tissues outside the threshold range were deleted, leaving only bone structures. Thresholding required the user to determine the CT number value that represented the edge of bone where it interfaced with soft tissue. Note that the choice of threshold may cause loss of information in areas where only thin bone is present.

In many circumstances, the volume of the body scanned is much larger than that actually required for model making. To reduce the model size, and therefore the cost, 3D image editing procedures may be employed. The most useful tool was a mouse-driven 3D volume editor that enabled the operator to delete or cut out sections of the data volume. The editing function deleted sections to the full depth of the data volume along the line of sight of the operator. Image editing reduced the overall model size, which also reduced RP build time. Clearer and less complex models may be generated, making structures of interest more clearly visible. Other image processing functions such as smoothing, volume data mirroring, and image addition and subtraction should be available for the production of models.

5.3.3 Rapid prototyping technologies

"Rapid prototyping" is a generic name given to a range of related technologies that may be used to fabricate physical objects directly from CAD data sources. RP enables design and manufacturing of models to be performed much more quickly than conventional manual methods of prototyping. In all aspects of manufacture, the speed of moving from concept to product is an important part of making a product commercially competitive. RP technologies enable an engineer to produce a working prototype of a CAD design for visualization and testing purposes. There are a number of texts describing the development of RP technology and its applications [22,23]. Of the many RP processes that have been applied to medical modeling, the two RP processes most extensively employed are stereolithography (SL) and fused deposition modeling (FDM).

5.3.3.1 Stereolithography

The following data provide some technical specifications of a specific type of SL machine that is in common use in medical modeling (3D systems SLA-250/40, 3D Systems, 333 Three D Systems Circle, Rock Hill, SC 29730, USA, www.3dsystems.com). However, a full description of the principles of SL is provided in Chapter 4, Section 4.2.

- laser beam diameter = 0.2−0.3 mm
- laser scanning speed = 2.54 m/s
- build platform = 250 × 250 × 250 mm
- layer build thickness = 0.05−0.2 mm
- minimum vertical platform movement = 0.0017 mm

The above specifications indicate the precision of model building that is achievable with SL. The laser focus defines the in-plane resolution, while the platform vertical increment defines the slice thickness at which the model is built. It should be noted that the imaging modality acquisition parameters are the limiting factors in model accuracy.

5.3.3.2 Fused deposition modeling

The technical specifications of a commonly used FDM machine (Stratasys FDM-3000, Eden Prairie, MN, www.stratasys.com) used for models in the cases referred to here are as follows. A full description of the working principles of FDM is provided in Chapter 4, Section 4.4.

- build envelope 254 × 254 × 254 mm

- achievable accuracy of ±0.127 mm
- road widths (extruded thermoplastic width) between 0.250 and 0.965 mm
- layer thickness (extruded thermoplastic height) from 0.178 to 0.356 mm

The above specifications indicate the precision of model building that is achievable with FDM. It can be seen that the results are broadly similar to those achieved by SL. However, SL can achieve thinner layers and more precise control over the laser position compared with the deposition of plastic material in FDM.

5.3.3.3 Computer-controlled milling

Although generally not considered one of the many rapid prototyping technologies, computerized numerically controlled (CNC) milling can successfully build some medical models [24]. This technology was applied in the construction of custom titanium implants for cranioplasty. CNC milling uses a cutting tool, which traverses a block of material, removing it on a layer-by-layer basis. Fig. 5.10 shows a model of skull defect (only half the skull has been created). The complexity of models that can be achieved using CNC milling is limited as it only cuts on one side of the model data. If the model required has any internal features or complex surfaces facing a number of directions, then CNC milling would not be suitable. An overview of the principal differences between CNC milling and RP is provided in Chapter 4.

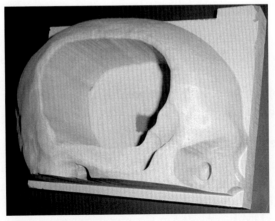

Figure 5.10 Half-skull model created by CNC milling demonstrating a large cranial defect.

5.3.3.4 Other rapid prototyping technologies

Selective laser sintering (SLS) locally heats a thermoplastic powder, which is fused by exposure to an infrared laser in a manner similar to SL. SLS models do not require support structures, and they are therefore cleaned relatively easily, thus saving labor costs. An example of the use of SLS in medical modeling is described by Berry et al. [25]. Laminated object manufacturing (LOM) builds models from layers of paper cut using a laser, which are bonded together by heating. Inexpensive sheet materials make LOM very cost effective for large volume models. However, the solid nature of the waste material means that it is not suited to models with internal voids or cavities often encountered in human anatomy. The SLS process is also described in more detail in Chapter 4.

5.3.3.5 Discussion of MRP technologies

The main factors in choosing which rapid prototyping technology are most appropriate for our clinical applications were as follows:
- dimensional accuracy of the models
- overall cost of the model
- availability of technology
- model building material

SL models are typically colorless to amber in color, transparent, and of sufficient accuracy to be suitable for MRP work. FDM models are typically made of white acrylonitrile-butadiene-styrene and are attractive both in terms of appearance and material. It has been pointed out that medical models may be dimensionally accurate to 0.62 mm ±0.35 mm [26]. It should be noted that the limiting factor in model accuracy is the imaging technique rather than the RP technology employed. In general, CT and magnetic resonance (MR) typically acquire images slices, which have slice thickness of the order of 1.0—3.0 mm, which is much greater than the limiting build resolution of any of the RP technologies.

The potential benefits of exploiting RP techniques in surgical planning have been widely acknowledged and described. The process of producing accurate physical models directly from three-dimensional scan data of an individual patient has proved particularly popular in head and neck reconstruction. In addition, most of the work done to date has concentrated on the use of three-dimensional CT data as this produces excellent images of bone. However, the process is still not conducted in the large volumes associated with industrial RP, and as such, practitioners applying these techniques to medicine often confront problems that are not encountered in industry. The small turnover associated with medical modeling also means that many manufacturers and vendors cannot justify investment in specific software, processes, and materials for this sector. These characteristics combine to make medical modeling a challenging field of work with many potential pitfalls.

The authors' many years of practical experience in medical modeling have resulted in a knowledge base that has identified the problems that may be encountered, many of which are simple or procedural in nature. This chapter aims to highlight some of these common problems, the effect they have on the resultant models, and to suggest methods that can be employed to avoid or minimize their occurrence or impact on the usefulness of the models produced.

5.3.4 Medical rapid prototyped model artifacts

Associated with all medical imaging modalities are unusual or unexpected image appearances referred to as artifacts. Some imaging modalities are prone to geometric distortion like MR [27], and this should be accounted for in soft tissue models manufactured from this source. CT does not suffer from the same distortion as MR, and models produced from this source have been proven to be dimensionally accurate [28]. In some circumstances, artifacts are easily recognizable and taken into account by the viewer, while in other circumstances, they can be problematic and difficult to explain. Artifacts present in the image data may subsequently be transferred to a medical model. In addition, due to the image processing steps and surface modeling required in the production of medical models, there is scope for the appearance of a wide range of artifacts. This section describes and illustrates some of the problems and pitfalls encountered in the production of medical models.

The procedures and potential problems associated with transferring and translating medical scan data are described in Chapters 2 and 3. However, the following observations serve to illustrate particular examples of some of the problems encountered by the authors. An example of some of the practical implications of transferring data from a hospital to an RP service provider is given in case study 5.1.2.

5.3.4.1 CT data import errors

CT data consist of a series of pixel images of slices through the human body. When importing data, the key characteristics that determine size and scale of the data are the pixel size and the slice thickness. The pixel size is calculated by dividing the field of view by the number of pixels. The field of view is a variable set by the radiographer at the time of scanning. The number of pixels in the x and y axis, respectively, is typically 512 by 512 or 1024 by 1024. If there is a numeric error in any of these parameters while data is being translated from one data format to another, the model may be inadvertently scaled to an incorrect size. The slice thickness and any interslice gap must be known, although the interslice gap is not applicable in CT where images are reconstructed contiguously or overlapping. Numeric error in the slice thickness dimension will lead to inadvertent incorrect scaling in the third dimension. This distance is typically in the order of 1.5 mm but may be as small as 0.5 mm or as high as 5 mm. Smaller scan distances result in higher quality of the three-dimensional reconstruction. The use of the internationally recognized DICOM (Digital Image Communications in Medicine, www.acrnema.org) standard for the format of medical images has largely eliminated these errors [29].

5.3.4.2 CT gantry tilt distortion

A CT scanner typically operates with the X-ray tube and detector gantry perpendicular to the long axis of the patient (z direction). The scan therefore produces the axial images that form the basis of three-dimensional CT scans. However, in some cases, the gantry may be inclined at an angle of up to 30 degrees. When a set of 2D slices is combined into an image volume for three-dimensional modeling, the gantry angle must be taken into account. With no gantry tilt, the slices are correctly aligned, and they will produce an undistorted 3D volume. Slices acquired with a gantry tilt of 15 degrees and converted into a data volume without the gantry tilt being taken into account may have a shear distortion arising from the misalignment of slices. At large angles, this is immediately visually apparent and can therefore be detected. However, at small angles, it may not be so obvious. Building a model with a small, uncorrected gantry tilt angle could be easily done and result in significant geometric inaccuracies in the resulting model. The use of the image transfer standard, DICOM, automatically provides the scan parameters including gantry tilt angle. However, the DICOM formal does not provide the direction of the angle, and it cannot therefore be relied on to automatically correct gantry tilt. It is therefore advisable to avoid gantry tilt when acquiring a three-dimensional CT image dataset; otherwise, sophisticated mathematical algorithms are required to successfully correct the data. Fig. 5.11 shows how a distorted 3D CT volume may

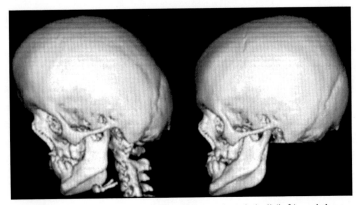

Figure 5.11 Effect of gantry tilt on a 3D surface rendered skull (left) and the same after correction (right).

be corrected using affine transformation to produce a dataset with no distortion.

5.3.4.3 Model stair-step artifact

Two elements contribute to the stepped effect seen in medical models. One contribution is from the discrete layer thickness at which the model is built. This is a characteristic of the particular RP process and material being used. Typically, these range from 0.1 to 0.3 mm. This affect can be minimized by selecting processes and parameters that minimize the build layer thickness. However, thinner layers result in longer build times and increased costs, and an economic compromise is typically found for each RP process. As the layer thickness is typically an order of magnitude smaller than the scan distance of the CT images, it does not have an overriding effect on the quality of the model.

The second effect arises from the slice thickness of the acquired CT or MR images and any potential gap between them. The stair-step artifact is a common feature on conventional and single-slice helical CT scans where the slice thickness is near to an order of magnitude greater than the in-plane pixel size [30]. The artifact is manifest as a series of concentric axial rings around the model. The depth and size of these rings depends on the CT imaging protocol, but it may be very slight where there is a thin slice used (e.g., 3-mm acquisition with 1-mm reconstruction interval). In thick-slice acquisitions (e.g., >3 mm with similar reconstruction interval to the slice thickness), the stair-step artifact will cause significant distortion to the model. Fig. 5.12 shows a stereolithography model of a full skull. The CT

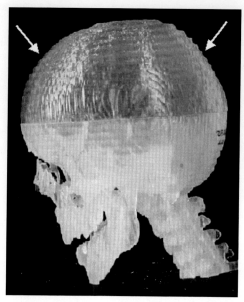

Figure 5.12 SL model with significant stair-step artifact.

scan was performed on a conventional CT scanner with 5-mm slice thickness and no interpolation of the image data to create thin slices. Note there was significant stair-step artifact around the top of the skull and on the lower edge of the mandible. The stair-step artifact was most prominent on surfaces that were inclined to the data acquisition plane as is the case for 3D surface rendered images. This model was used for surgical planning and reconstruction but was limited in the use for obtaining physical measurements.

These effects can be countered to some degree by using interpolation between the original image data. The following images illustrate the difference between using no interpolation and using a cubic (natural curve) interpolation. Due the natural nature of the cubic curve, the resulting interpolated data result in a good, smooth, and natural-appearing surface.

5.3.4.4 Irregular surface due to support structures

Both SL and FDM required support structures during the build process. These were subsequently cleaned from the model manually although generally left a rough surface. This did not affect the overall accuracy of the model but contributed to a degradation of its aesthetic appearance. Fig. 5.13 shows an SL model where surface roughness was attributed to the support

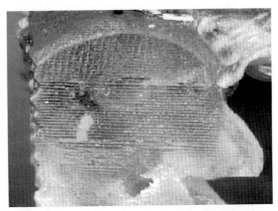

Figure 5.13 Showing surface roughness on an SL model attributed to support structures.

structures. Models were easily cleaned using light abrasive techniques, although this was felt unnecessary as the indentations were of submillimeter depth. It is unlikely that these structures would have a detrimental effect in surgical planning or implant design.

5.3.4.5 Irregular surface due to mathematical modeling

The mathematical modeling of a surface will introduce its own surface effects. The smoothness (governed by the size of the triangle mesh) of the model surface becomes poorer as the surface mesh becomes larger. A larger mesh resulted in a lower number of triangles, reduced computer file size, and quicker rendering. A smaller mesh resulted in much better surface representation, much greater computer file size, and slower rendering. Fig. 5.14 shows irregular surface structures due to the mathematical modeling process [31]. Fig. 5.14a shows a model where the mesh structure is not readily apparent and Fig. 5.14b where the model contours are more clearly observed. In both cases, the surface produced was acceptable for its own clinical application. One could imagine that the mesh resolution used in the model in Fig. 5.14b would be unacceptable for smaller models where a fine detail would be masked.

These effects can be avoided by eliminating the creation of a three-dimensional surface mesh and creating the RP build data directly from the CT image data. This essentially creates the two-and-a-half dimensional layer data for the RP machine from the CT images. Interpolation is used to create accurate intermediate layers between the CT images. This route not

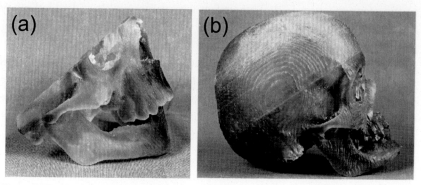

Figure 5.14 (a) Smooth surface due to high-resolution meshing algorithm and (b) meshing contours visible due to low-resolution meshing algorithm.

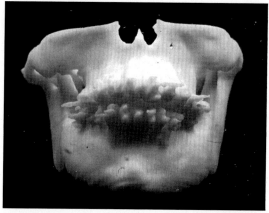

Figure 5.15 Metal artifact due to scattering of X-rays from metal within the teeth.

only eliminates surface modeling effects but also results in much smaller computer files and faster preparation.

5.3.4.6 Metal artifact

Metal artifact was present within CT scans of the maxilla and mandible due to the presence of metal within fillings of the teeth or the presence of dental implants. This was manifest as high signal intensities (in the form of scattered rays) around the upper and lower mandible. Fig. 5.15 shows an FDM model with significant metal artifact around the teeth. These ray appearances extended from a couple of millimeters to over 1 cm in length. In some circumstances, the artifact may be reduced by software during CT image reconstruction [32]. This artifact was plainly visible and added many superfluous structures to the medical models. Although no significant geometric distortion was observed on models, large spikes were visible emanating from around the teeth, which distorted the bone in the local area. The artifact may be removed by detailed slice-by-slice editing of the original CT images to produce a cleaner model. This process is very time consuming and if not performed with great care can result in anatomy of interest being removed and the subsequent model becoming unusable.

5.3.4.7 Movement artifact

CT scanning was prone to movement artifact if a patient was restless. This artifact was readily apparent in a model if the degree of movement was significantly large, i.e., greater than 1 mm. Fig. 5.16 shows a mandible with a distinct artifact present. The patient moved slightly during the acquisition of a couple of images that left a bulge of 4 mm height extending right around the mandible. In addition, present in this model were concentric axial rings of about 3 mm thickness. These corresponded to the common stair-step observed in single-slice helical CT scanning. Obviously, the degree of the movement during the scan determines the size of the movement artifact in the model. In the example shown, the artifact was felt not to be significant clinically as it did not interfere directly with the placement of a distraction device.

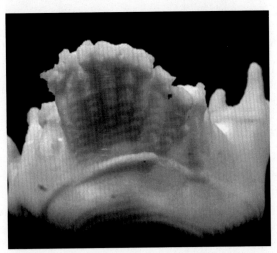

Figure 5.16 Distinct movement artifact on an FDM model of the mandible.

In another example where a model was being used for facial reconstruction, the patient moved while the scanner was acquiring data at the region where the surgery was to be performed. The movement artifact resulted in distortion of the model, so the surgeon lost confidence in its physical integrity. During the scan, around the supraorbital ridge, we believe the child rotated its head to look at a parent, which resulted in rotation of this part of the data, which was subsequently transferred to the model. In this case, the patient, a 7-year-old child, had to be rescanned under full general anesthetic. It was interesting to note that the degree of the artifact was not noted until a physical model was produced. This indicates the need for good quality assurance of the original dataset to ensure a useful model was produced.

5.3.4.8 Image threshold artifact

One of the simplest and commonest methods of tissue segmentation applied to the skull is CT number thresholding. A CT number range was identified by either ROI pixel measurements or pixel intensity profiles, which was representative of bone. If the bone was particularly thin or the threshold inappropriately measured, a continuous surface was unachievable. This left the model with a hole where the surface was not closed. In some cases, large areas of bone were removed completely, especially at the back of the orbit and around the cheekbones. Fig. 5.17 illustrates bone deletion

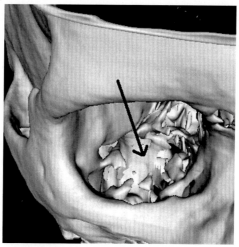

Figure 5.17 Removal of bone at the back of the orbit due to inappropriate choice of image threshold.

by data thresholding in the back of the orbit in this magnified surface shaded image. Anatomic detail is lost as the chosen threshold removed thin bone at the back of the orbit, as indicated by the black arrow. Adjusting the threshold to include bone in this case would have resulted in the inclusion of soft tissue that would have made the image more difficult to interpret. It is useful to specify what is required of a model clearly so that an appropriate threshold can be chosen to preserve tissue of interest.

5.3.5 Conclusion

Medical RP models of human anatomy may be constructed from a number of image data sources and using a range of RP technologies. They are prone to artifacts both from the imaging source, the method of manufacture, and from the model cleaning process. It is important to ensure that high-quality source data are available to assure model quality. Clinicians requesting medical models should be aware of their physical accuracy and integrity, which are generally dependent on the original imaging parameters and image processing rather than the method of manufacture, and determine that this is sufficient for the purpose. We have demonstrated a range of model artifact sources ranging from reading computer files to the removal of support structures and suggested ways to avoid or cure them. It is important that the source images be reviewed thoroughly, that robust image transfer and image processing procedures are in place, and that the model building material is fit for the purpose for which it was intended. A multidisciplinary team approach to the manufacture of medical models with rigorous quality assurance is highly recommended.

5.3.5.1 Update

Although newer and improved software and AM technologies have become available since this article was first published, the sources of the majority of the potential errors involved in data acquisition and manipulation remain the same, and the fundamental issues and problems discussed remain a concern regardless of which software and/or AM technologies are used.

References

[1] Anderl H, Zur Nedden D, Muhlbauer W, Twerdy K, Zanon E, Wicke K, et al. CT-guided stereolithography as a new tool in craniofacial surgery. British Journal of Plastic Surgery 1994;47(1):60–4.

[2] Arvier JF, Barker TM, Yau YY, D'Urso PS, Atkinson RL, McDermant GR. Maxillofacial biomodelling. British Journal of Oral and Maxillofacial Surgery 1994;32:276—83.

[3] D'Urso PS, Barker TM, Earwaker WJ, Bruce LJ, Atkinson RL, Lanigan MW, et al. Stereolithographic biomodelling in cranio-maxillofacial surgery: a prospective trial. Journal of Craniomaxillofacial Surgery 1999;27(1):30—7.

[4] Eufinger H, Wehmoller M. Individual prefabricated titanium implants in reconstructive craniofacial surgery: clinical and technical aspects of the first 22 cases. Plastic and Reconstructive Surgery 1998;102(2):300—8.

[5] Gateno J, Allen ME, Teichgraeber JF, Messersmith ML. An in vitro study of the accuracy of a new protocol for planning distraction osteogenesis of the mandible. Journal of Oral and Maxillofacial Surgery 2000;58(9):985—90.

[6] Sailer HF, Haers PE, Zollikofer CP, Warnke T, Carls FR, Stucki P. The value of stereolithographic models for preoperative diagnosis of craniofacial deformities and planning of surgical corrections. International Journal of Oral and Maxillofacial Surgery 1998;27(5):327—33.

[7] Hughes CW, Page K, Bibb R, Taylor J, Revington P. The custom made orbital floor prosthesis in reconstruction for orbital floor fractures. British Journal of Oral and Maxillofacial Surgery 2003;41:50—3.

[8] Heckmann SM, Winder W, Meyer M, Weber HP, Wichmann MG. Overdenture attachment selection and the loading of implant and denture bearing area. Part 1: in vitro verification of stereolithographic model. Clinical Oral Implants Research 2001;12(6):617—23.

[9] Winder RJ, Cooke RS, Gray J, Fannin T, Fegan T. Medical rapid prototyping and 3D CT in the manufacture of custom made cranial titanium plates. Journal of Medical Engineering & Technology 1999;23(1):26—8.

[10] Heissler E, Fischer F, Bolouri S, Lehmann T, Mathar W, Gebhardt A, et al. Aesthetic and reconstructive surgery—custom-made cast titanium implants produced with CAD/CAM for the reconstruction of cranium defects. International Journal of Oral and Maxillofacial Surgery 1998;27(5):334—8.

[11] Minns RJ, Bibb R, Banks R, Sutton RA. The use of a reconstructed three-dimensional solid model from CT to aid surgical management of a total knee arthroplasty: a case study. Medical Engineering & Physics 2003;25:523—6.

[12] Munjal S, Leopold SS, Kornreich D, Shott S, Finn FA. CT generated 3D models for complex acetabluar reconstruction. The Journal of Arthroplasty 2000;15(5):644—53.

[13] Nakajima T. Integrated life-sized solid model of bone and soft tissue: application for cleft lip and palate infants. Plastic and Reconstructive Surgery 1995;96(5):1020—5.

[14] Schwaderer E, Bode A, Budach W, Claussen CD, Danmmann F, Kaus T, et al. Soft-tissue stereolithography model as an aid to brachytherapy. Medica Mundi 2000;44(1):48—51.

[15] Mankovich NJ, Cheeseman AM, Stoker NG. The display of three-dimensional anatomy with stereolithographic models. Journal of Digital Imaging 1990;3(3):200—3.

[16] Erben C, Vitt KD, Wulf J. First statistical analysis of data collected in the Phidias validation study of stereolithography models. Phidias Newsletter 2000;(5):6—7.

[17] Sarment DP, Sukovic P, Clinthorne N. Accuracy of implant placement with a stereolithographic surgical guide. The International Journal of Oral & Maxillofacial Implants 2003;18(4):571—7.

[18] Williams RJ, Bibb R, Rafik T. A technique for fabricating patterns for removable partial denture frameworks using digitized casts and electronic surveying. The Journal of Prosthetic Dentistry 2004;91(1):85—8.

[19] Petzold R, Zeilhofer H, Kalender W. Rapid prototyping technology in medicine-basics and applications. Computerised Medical Imaging and Graphics 1999;23:277—84.

[20] Webb PA. A review of rapid prototyping (RP) techniques in the medical and biomedical sector. Journal of Medical Engineering & Technology 2000;24(4):149—53.

[21] Fuchs T, Kachelriess M, Kalender WA. Related articles, technical advances in multi-slice spiral CT. European Journal of Radiology 2000;36(2):69—73.

[22] Jacobs P. Stereolithography and other rapid prototyping and manufacturing technologies. Dearborn, MI, USA: American Association of Engineers Press; 1996.

[23] Kai CC, Fai LK. Rapid prototyping: principles & applications in manufacturing. Singapore: John Wiley & Sons Ltd; 1997.

[24] Joffe J, Harris M, Kahugu F, Nicoll S, Linney A, Richards R. A prospective study of computer-aided design and manufacture of titanium plate for cranioplasty and its clinical outcome. British Journal of Neurosurgery 1999;13(6):576—80.

[25] Berry E, Brown JM, Connell M, Craven CM, Efford ND, Radjenovic A, et al. Preliminary experience with medical applications of rapid prototyping by selective laser sintering. Medical Engineering & Physics 1997;19(1):90—6.

[26] Choi JY, Choi JH, Kim NK, Kim Y, Lee JK, Kim MK, et al. Analysis of errors in medical rapid prototyping models. International Journal of Oral and Maxillofacial Surgery 2002;31(1):23—32.

[27] Wang D, Doddrell DM, Cowin G. A novel phantom and method for comprehensive 3-dimensional measurement and correction of geometric distortion in magnetic resonance imaging. Magnetic Resonance Imaging 2004;22(4):529—42.

[28] Barker TM, Earwaker WJ, Frost N, Wakeley G. Accuracy of stereolithographic models of human anatomy. Australasian Radiology 1994;38(2):106—11.

[29] Muller H, Michoux N, Bandon D, Geissbuhler A. A review of content-based image retrieval systems in medical applications-clinical benefits and future directions. International Journal of Medical Informatics 2004;73(1):1—23.

[30] Fleischmann D, Rubin GD, Paik DS, Yen S, Hilfiker P, Beaulieu C, et al. Stair-step artefacts with single versus multiple detector-row helical CT. Radiology 2000;216:185—96.

[31] Karron D. The 'spider web' algorithm for surface construction in noisy volume data. SPIE Visualisation in Biomedical Computing 1992;1808:462—76.

[32] Mahnken AH, Raupach R, Wildberger JE, Jung B, Heussen N, Flohr TG, et al. A new algorithm for metal artefact reduction in computed tomography: in vitro and in vivo evaluation after total hip replacement. Investigative Radiology 2003;38(12):769—75.

PART 5.2

Surgical applications

CHAPTER 5.4

Surgical application case study 1—Planning osseointegrated implants using computer-aided design and additive manufacturing[*]

5.4.1 Introduction

In recent years, rapid prototyping (RP) has been used to build highly accurate anatomical models from medical scan data. These models have proved to be a valuable aid in the planning of complex reconstructive surgery, particularly in maxillofacial and craniofacial cases. Typically, RP is used to create accurate models of internal skeletal structures on which operations can be accurately planned and rehearsed [1—4]. Such models have been successfully used for positioning osseointegrated implants. Osseointegrated implants are titanium screws attached directly to a patient's bone structure and pass through the skin to provide a rigid and firm fixture

---

[*] The work described in this chapter was first reported in the references below and is reproduced here, in part or in full, with the permission of the Institute of Maxillofacial Prosthetics and Technologists and the Council of the Institute of Mechanical Engineers. Please note the term Additive Manufacturing appears in the chapter title for consistency but the original title and text uses the previously common term of Rapid Prototyping.
Bibb R, Bocca A, Sugar A, Evans P. Planning osseointegrated implant sites using computer aided design and rapid prototyping. The Journal of Maxillofacial Prosthetics & Technology 2003;6:1—4.
Bibb R, Eggbeer D, Bocca A, Evans P, Sugar A. Design and manufacture of drilling guides for osseointegrated implants using rapid prototyping techniques. Proceedings of the fourth national conference on rapid & virtual prototyping and applications. London, UK: Professional Engineering Publishing; 2003. p. 3—11, ISBN 1-86058-411-X.

Medical Modeling
ISBN 978-0-323-95733-5
https://doi.org/10.1016/B978-0-323-95733-5.00009-0

for dentures, hearing aids, and prostheses [5]. See medical explanatory note 7.2.1 for an explanation of osseointegrated implants. The accuracy of the RP models allows the depth and quality of bone to be assessed improving the selection of drilling sites before surgery. Although this process has dramatically improved the accuracy and reduced the theater time of some surgical procedures, it incurs significant time and cost to produce the anatomical model. While it does utilize RP technologies, this current route does not fully exploit the potential advantages of computer-aided design.

To address this issue, it was decided to complete as much of the planning as possible in the virtual environment and only use RP to make small templates that would guide the surgeon in theater. This route would allow the clinicians to conduct all the planning and explore many options without damaging an expensive RP model. To be successful, the approach would have to be simple to conduct and have low investment requirements.

5.4.2 The proposed approach

The approach would use three-dimensional computed tomography (CT) data to create virtual models of the elements necessary to plan the osseointegrated implants required to secure a prosthetic ear. The elements consisted of the soft tissue of the head, a copy of the remaining opposite ear, and the bone structure at the implant site. The simple and popular STL [6] format was chosen as the three-dimensional representation of the entities. This format ensures easy access to a number of software options at a reasonable cost. In this case, a popular software package used in the RP industry was used [7]. The STL file format is more fully described in Chapter 3. The entities were created as STL format files from CT data using one of a number of specialist software packages available for creating STL files from CT data [8].

The STL manipulation software was then used to mirror the copy of the ear and position it in an anatomically and aesthetically appropriate location. The software was then used to create cylinders representing the implants. These cylinders were positioned in the preferred location by the prosthetist observing a lateral view. Then, the bone quality at the implant sites could be assessed.

5.4.3 Scanning problems

Misalignment of the patient's head to one side or the other means that the optimum accuracy obtained in the axial plane during scanning is not axial to the patient. This means that entities will not be in alignment with

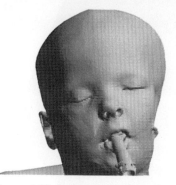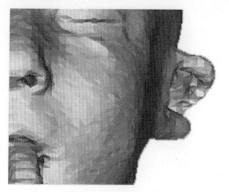

Figure 5.18 Problems resulting from poor position during computed tomography scanning.

software co-ordinate system. Although this is not a major issue, it does make control of the angles at which the entities meet more difficult to control.

Unintended displacement of the soft tissue during the CT scan results in poor representation of the anatomy. In this example, the ear of one of the patients had become folded over resulting in a deformed anatomical entity (Fig. 5.18). This made positioning the contralateral ear and the implants problematic.

5.4.4 Software problems

Initially, file size was a concern, and all of the STL files were produced at a low resolution. This resulted in small STL files that could be handled easily and rapidly by the software. In visual terms, all of the entities appeared to be well represented by the low-quality STL files. However, when attempting one of the first cases, it was found that the difference in the representation of internal air cells in the highly pneumatized bone in the mastoid was dramatically altered by the resolution at which the STL file was produced. This led to the mistaken belief that this particular case had adequate bone thickness when in fact the bone was unusually thin (illustrated in Fig. 5.19). From this experience, it was decided to produce only the small amount of bone required for the implants but at the highest possible resolution. This resulted in only one entity having a large file size, which proved to be perfectly within the capabilities of a reasonable specification computer.

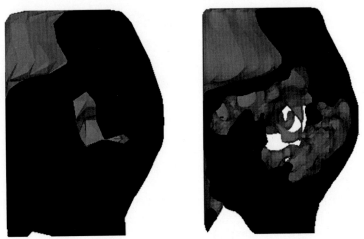

Figure 5.19 The effect of file size versus quality (low quality—left, high quality—right).

5.4.5 An illustrative case study

Three-dimensional CT data were used to create virtual models of the elements necessary to plan the osseointegrated implants required to secure a prosthetic ear. The elements consisted of the soft tissue of the head, a copy of the remaining opposite ear, and the bone structure at the implant site. The surgeon and prosthetists then used 3D software to mirror the copy of the ear and position it relative to the head in an anatomically and aesthetically appropriate location (Fig. 5.20).

Cylinders were created to represent the implants. These cylinders were positioned on the ear in the position preferred by the prosthetists observing a lateral view (Fig. 5.21). The soft tissue entities were then removed to see where the implants intersect with the bone as can be seen in Fig. 5.21.

The bone quality at the implant sites was then assessed to check that they would be suitable for implants. Sectioning the virtual model enabled the quality and thickness of the bone to be accurately measured (Fig. 5.22).

When the team were satisfied with the implant sites, a block was created that overlapped the implants and the surface of the skull (Fig. 5.23). Then,

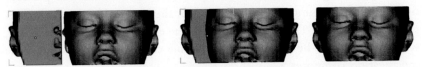

Figure 5.20 Positioning the contralateral ear.

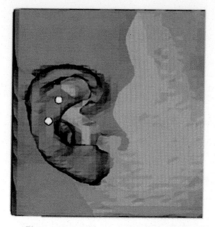 

Figure 5.21 The implant positions (*left*) and with soft tissue removed (*right*).

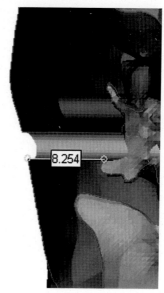

Figure 5.22 Measured section through bone at the implant site.

by using a Boolean operation, the skull and implant cylinders were sub-tracted from the block to create a template design (Fig. 5.23).

This template, shown in Fig. 5.24, was produced directly in a medically appropriate material by stereolithography [9,10]. See Chapter 4 Section 4.2 for a full description of stereolithography. The fact that at the time this was the only RP material that has been tested to a standard recognized by the

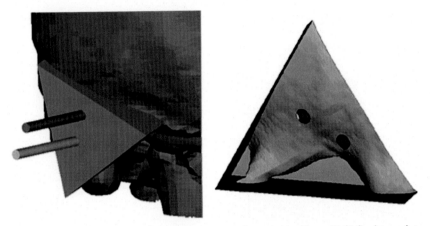

Figure 5.23 The block overlapping the bone surface and implants (*left*) final template design (*right*).

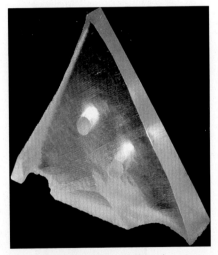

Figure 5.24 The final template produced directly using stereolithography.

FDA for patient contact in theater meant that the stereolithography template could be used directly in surgery after sterilization. Sterilization presents no practical problems and appropriate methods include ethylene oxide (55 degrees), formaldehyde, low-,temperature steam (75 degrees) and gamma irradiation. The template locates onto the anatomical features on the surface of the skull at the implant site and indicates the drilling sites to the surgeon. The mastoid and zygomatic process are exploited to provide positive anatomical features so that the template locates accurately and firmly.

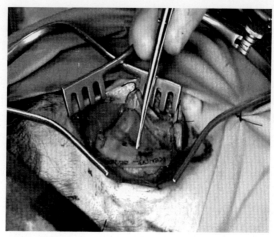

Figure 5.25 The template in position during theater.

5.4.6 Results

In surgery, the template was found to fit very accurately and securely to the area of the skull as shown in Fig. 5.25. The drilling was carried out, and the bone thickness and quality were found to be as indicated by the data. The positions indicated by the template were found to be much more accurate than those indicated by marks transferred from the soft tissue with ink and needle, by as much as 5 mm in one case. The team have now successfully carried out many similar cases using this approach with equally positive results and a considerable saving of time and money. If we consider the typical procedure for the traditional method being as follows; carve a planning ear (30 minutes), take impression of defect site (15 minutes), create template (30 minutes), planning (1 hour), and marking of template (15 minutes) the total time taken is 2.5 hours. This involves at least one technician, one surgeon, and a patient appointment, which depending on salaries and overheads could represent a cost saving of approximately £250.

5.4.7 Benefits and future development

The principle benefits resulting from this approach are reduced cost implications for planning activities. Once the entities are created from the CT data, they can be positioned and repositioned as many times as required. This allows different placement strategies to be performed and evaluated in three dimensions in a matter of minutes and with zero costs implication (other than time).

Once a plan has been agreed between the clinicians, the implant sites themselves can be assessed for bone depth and quality (within the limits of the original CT scan). If they are found to be unsatisfactory, these sites can be altered without incurring cost. When a final solution is achieved, the template model can be made in under two hours and cost dramatically less than even a localized stereolithography model of the bone structure. The CT data used in these cases were taken at 1.5 mm slice distance and proved adequate. However, reducing this slice distance would increase the quality of the three-dimensional entities.

Of course, the purchase and maintenance of the software require for this approach are significant and can be anticipated that a high volume of cases would be required to justify this investment in isolation. However, it is the experience of the authors that such software has many useful applications in head and neck reconstruction as well as other medical specialities.

5.4.7.1 Update

The use of surgical guides has developed significantly in recent years with the development of specific software and service providers, particularly in orthopedics. However, the considerations discussed in this paper are still relevant when planning cases. The next paper describes how the technique developed with more appropriate software and metal RP techniques.

References

[1] Klein HM, Schneider W, Alzen G, Voy ED, Gunther RW. Pediatric craniofacial surgery: comparison of milling and stereolithography for 3D-model manufacturing. Pediatric Radiology 1992;22:458−60.
[2] Bibb R, Brown R. The application of computer aided product development techniques in medical modelling. Biomedical Sciences Instrumentation 2000;36:319−24.
[3] Bibb R, Brown R, Williamson T, Sugar A, Evans P, Bocca A. The application of product development technologies in craniofacial reconstruction. In: Proceedings of the 9th European conference on rapid prototyping and manufacturing, 2000, Athens, Greece; 2000. p. 113−22.
[4] D'Urso PS, Redmond MJ. A method for the resection of cranial tumours and skull reconstruction. British Journal of Neurosurgery 2000;14(6):555−9.
[5] Branemark P, De Oliveira MF, editors. Craniofacial prostheses, anaplastology and osseointegration. Carol Stream, IL, USA: Quintessence Publishing Co. Inc.; 1997. p. 101−10.
[6] Manners CR. STL file format. Valencia, CA, USA: 3D Systems Inc.; 1993.
[7] Magics version 7.2. Leuven, Belgium: Materialise N. V.
[8] Mimics version 7.2. Leuven, Belgium: Materialise N. V.
[9] Jacobs PF, editor. Stereolithography and other RP&M technologies. Dearborn, MI, USA: Society of Manufacturing Engineering; 1996.
[10] RenShape H-C. 9100R stereolithography material technical specification, huntsman advanced materials. Cambridge, UK: Duxford.

CHAPTER 5.5

Surgical applications case study 2—Rapid manufacture of custom fitting surgical guides[*]

5.5.1 Introduction

Over the last decade, rapid prototyping (RP), techniques have been employed widely in maxillofacial surgery. However, this has concentrated on the reproduction of exact physical replicas of patient's skeletal anatomy that surgeons and prosthetists use to help plan reconstructive surgery and prosthetic rehabilitation [1—12].

Developments in this area are moving toward exploiting advanced design and fabrication technologies to design and produce implants, patterns, or templates that enable the fabrication of custom fitting prostheses without requiring a model of the anatomy to be made [13—17]. However, there is also growing desire from clinicians to conduct more of the surgical planning using three-dimensional computer software. While several approaches have been undertaken in the application of computer-aided surgical planning, the problem of transferring the computer-aided plan from the computer to the operating theater remains. Two solutions exist to transfer the computer plan to the operating theater, navigation systems, and surgical guides. The use of navigation systems is a specialist field in itself and will not be described here. However, research presented by Poukens, Verdonck, and de Cubber in 2005 suggests that navigation and the use of

---

[*] The work described in this chapter was first reported in the references below and is reproduced here with permission of Emerald Publishing.
Bibb R, Eggbeer D, Evans P, Bocca A, Sugar AW. Rapid manufacture of custom fitting surgical guides. Rapid Prototyping Journal 2009;15(5):346—54. ISSN: 1355—2546. https://doi.org/10.1108/13552540910993879.

Medical Modeling
ISBN 978-0-323-95733-5
https://doi.org/10.1016/B978-0-323-95733-5.00010-7

surgical guides are both accurate enough for surgical purposes [18]. RP technologies provide a potential method of producing custom-fitting surgical guides depending on the nature of the planning software used. Previous work on the application of RP technologies in the manufacture of surgical guides has concentrated on the production of drilling guides for oral and extra-oral osseointegrated implants [19–26]. This paper will also describe one drilling guide case but will also report on two cases involving the use of surgical guides for osteotomies (saw cuts through bone), which has not been previously reported.

In order to be appropriate for the manufacture of surgical guides, the RP processes have to be accurate, robust, rigid, and able to withstand sterilization. Due to these requirements, the majority of surgical guides have been produced using stereolithography (SL) and laser sintering (LS). SL and LS are described more fully in Sections 4.2.1 and 4.5.1. However, the use of SL in particular has necessitated local reinforcement of the guides using titanium or stainless steel tubes to prevent inadvertent damage from drill bits and the use of low-temperature sterilization methods such formaldehyde or ethylene oxide.

The recent availability of systems capable of directly producing fully dense solid parts in functional metals and alloys has provided an opportunity to develop surgical guides that exploit the advantages of RP while addressing the deficiencies of previous SL and LS guides. The ability to produce end-use parts in functional materials means that processes such as these may be considered rapid manufacturing (RM) processes. Surgical guides produced directly in hardwearing, corrosion-resistant metals require no local reinforcement, can be autoclaved along with other surgical instruments and are unlikely to be inadvertently damaged during surgery. The use of metals also enables surgical guides to be made much smaller or thinner while retaining sufficient rigidity. This benefits surgery as incisions can then be made smaller, and the surgeon's visibility and access are improved.

5.5.2 Methods

To date, three surgical guides have been designed and produced as described here and subsequently used in theater. The first case was a drilling guide for osseointegrated implants to secure a prosthetic ear. The remaining two cases were for osteotomy cutting guides for the correction of facial deformity. This section describes the general approach to the planning, rapid design, and manufacture of surgical guides. The following section describes an individual case where the approach has been successfully employed for an osteotomy.

5.5.2.1 Step 1: 3D CT scanning

The patients were scanned using three-dimensional computed tomography (CT) to produce three-dimensional computer models of the skull (see Chapter 2). The CT data are exported in DICOM format, which was then imported into medical data transfer software (Mimics, Materialise N.V., Technologielaan 15, 3001 Leuven, Belgium, www.materialise.com). This software was used to generate the highest possible quality STL data files of the patient's anatomy using techniques described in Chapter 4. The STL files are then imported into the computer-aided design (CAD) software.

5.5.2.2 Step 2: Computer-aided surgical planning and design of the surgical guide

The CAD package used in this study (Geomagic FreeForm, https://oqton. com/geomagic-freeform) was selected for its capability in the design of complex, arbitrary but well-defined shapes that are required when designing custom appliances and devices that must fit human anatomy. The software has tools analogous to those used in physical sculpting and enables a manner of working that mimics that of the maxillofacial prosthetist working in the laboratory. The software utilizes a haptic interface (Touch X haptic interface; https://www.3dsystems.com/haptics-devices/touch-x) that incorporates positioning in three-dimensional space and allows rotation and translation in all axes, transferring hand movements into the virtual environment. It also allows the operator to feel the object being worked on in the software. The combination of tools and force feedback sensations mimics working on a physical object and allows shapes to be designed and modified in an arbitrary manner. The software also allows the import of scan data to create reference objects or "bucks" onto which objects may be designed.

The data of the patients' anatomy were imported into the software. The surgery is then planned and simulated by using the software tools to position prostheses and implants or to cut the skeletal anatomy and move the pieces, as they would be in surgery. When the clinicians were satisfied with the surgical plan, the surgical guides were designed to interface with the local anatomy.

In general, the surgical guides were designed by selecting the anatomical surface in the region of the surgery (drilling or osteotomy) and offsetting it to create a structure 1—2 mm thick. The positions of the drilling holes or cuts are then transferred to this piece by repeating the planned surgical procedure through it. Other features may then be added, such as embossed patient names, orientation markers, or handles. When the design is completed to the clinicians' satisfaction, the human anatomy data are subtracted from the surgical guide as a Boolean operation. This leaves the surgical guide with the fitting surface as a perfect fit with the anatomical surface. Typically, the curvature and extent of the fitting surface provide accurate location when it is fitted to the patient. The final design is then exported as a high-quality STL file for rapid manufacture by selective laser melting (SLM). SLM is described in Chapter 4.

5.5.2.3 Step 3: Rapid manufacture

In order to build surgical guides successfully on the SLM machine (at the time, this was supplied by MCP Tooling Technologies Ltd. The business was subsequently taken over by Renishaw who developed a new range of machines. Today, a number of manufacturers offer laser melting machines of similar or better capabilities), adequate supports had to be created using Magics software (Version 9.5, Materialise N.V.). The purpose of the supports was to provide a firm base for the part to be built onto while separating the part from the substrate plate. In addition, the supports conduct heat away from the material as it melts and solidifies during the build process. Inadequate supports result in incomplete parts or heat-induced curl, which leads to build failure as the curled part interferes with or obstructs the powder recoating mechanism.

Recent developments in support design have resulted in supports that have very small contact points, which have improved the ease with which supports can be removed from parts. However, the parts were all oriented such that the amount of support necessary was minimized and avoided the fitting surface of the guide. This meant that the most important surfaces of the resultant part would not be affected or damaged by the supports or their removal.

The part and its support were "sliced and hatched" using the SLM Realizer software at a layer thickness of 0.050 mm. The material used was 316L Stainless Steel spherical powder with a maximum particle size of 0.045 mm (particle size range 0.005—0.045 mm) and a mean particle size of approximately 0.025 mm (Sandvik Osprey Ltd., www.smt.sandvik.com/osprey). The laser had a maximum scan speed of 300 mm/s and a beam diameter of 0.150–0.200 mm.

5.5.2.4 Step 4: Finishing

Initially, supporting structures were removed using a Dremel handheld power tool using a reinforced cutting wheel (Dremel, Reinforced Cutting Disc, ref. Number 426). However, more recently improved design of the supports has eliminated the need for cutting tools as the supports contact the part at a sharp point that can be broken away from the part easily.

The SLM parts described here were well formed with little evidence of the stair-stepping effect (resulting from the thin layers used) but showed a fine surface roughness. This roughness was easily removed by bead blasting to leave a smooth, matte finish surface. The parts were then sent to the hospital for cleaning and sterilization by autoclave.

5.5.3 Case study

Although surgical guides have been produced using RP techniques for some years, the application of surgical guides to osteotomies had not been previously attempted. This case was the first attempt at such a guide. The surgery performed in this particular case involved distraction osteogenesis to correct deformity resulting from the cleft palate. A description of distraction osteogenesis is given in medical explanatory note 7.2.3. This required a Le Fort 1 osteotomy, which is a cut across the maxilla above the roots of the teeth but under the nose in order to separate and move the upper jaw in relation to the rest of the skull. The maxilla is then gradually moved in relation to the rest of the skull, usually forwards, by mounting it on two devices that use precision screw threads to advance the position by a small increment each day. The small increment causes the bone to grow gradually so that the shape of the face can be altered. When the desired position is reached, the bone is allowed to heal completely to give strong and reshaped skeletal anatomy.

In this case, it was also the intention to include the drilling holes for the distraction devices as well as a slot for the osteotomy. The slot was then made sufficiently wide to allow the saw blade to move freely and sufficient irrigation during the cutting. The lower edge of the slot is made flat and parallel to the direction of the cut in order to provide a reference surface on which the flat saw blade rests. As the cut is in two places on either side of the maxilla, the software design tools are used to join the two parts together into one device, see Figs. 5.26 and 5.27. However, it was discovered that at this time, there was no way to simulate the bending of the distractor attachment plates using the software. While it was theoretically possible to design and manufacture custom-fitting plates and laser weld them to the distraction devices, the manufacturer of the devices would not allow the modifications.

The guide was therefore finalized, supported, and built as described above. The support structure can be seen in Fig. 5.28. The drilling positions for the distraction devices were planned on an SL model of the patient. The SLM surgical guide was then fitted to the model and the drilling sites transferred to it. The final guide is shown in Fig. 5.29.

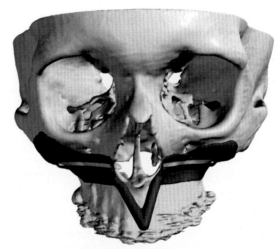

Figure 5.26 Patient data and surgical guide design.

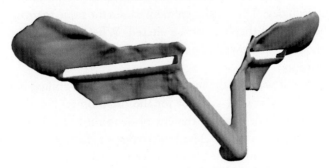

Figure 5.27 Finished surgical guide design.

Figure 5.28 The surgical guide and supports.

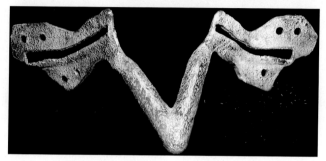

Figure 5.29 The finished surgical guide.

5.5.4 Results

The surgical guides were approved by clinicians before going to surgery, and all were deemed satisfactory for surgical use. All of the guides used in theater so far have displayed good accuracy and fitted the patients' anatomy firmly and securely, as expected. The quality of the surgical guide fit for this particular case was assessed by an experienced maxillofacial prosthetist. this was achieved by fitting it to an SL model of patient's facial skeletal anatomy where it was found to show excellent fit. Fig. 5.30 shows the guide in situ in theater.

There were no problems experienced sterilizing or using the guides. The guides all resulted in some timesaving in theater, particularly the individual case described here. The surgical outcomes were good and turned out as planned.

5.5.5 Discussion

The surgical guides described here were all deemed successful and contributed to the successful transfer of computer-aided planning to the

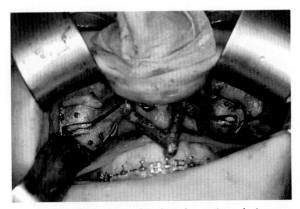

Figure 5.30 The guide being fitted to the patient during surgery.

theater. The drilling guide was very successful being thinner, more rigid, more hardwearing, and easier to sterilize than previous reported attempts that utilized SLA [21,22]. As can be seen in Fig. 5.31, the incorporation of embossed orientation markers and patient names was also beneficial and could help prevent errors in theater (the patient's name has been deliberately obscured to respect confidentiality).

However, the osteotomy guides proved challenging. There was no previous experience or publications to build on, and given the experimental nature of the two cases undertaken, the results were encouraging. The design of the guides will be significantly better in future cases based

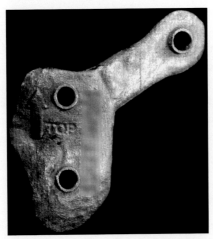

Figure 5.31 Selective laser melting (SLM) drilling guide.

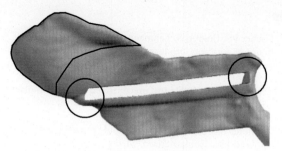

Figure 5.32 Areas for design improvement.

on the findings of these cases. The individual case described here illustrates examples of design improvement that resulted from this research. These improvements include the better positioning of handles, smaller extents of the fitting surfaces, and avoiding potential weaknesses. For example, in the case described here, the thin areas at the ends of the slots proved a potential weakness, and the fitting surface was larger than necessary and was reduced by the maxillofacial prosthetist in the laboratory as indicated in Fig. 5.32.

However, the more fundamental problem of using the approach described here to include the bending and fitting of distractor plates will be addressed in future research by exploring other software applications and techniques.

5.5.6 Conclusions

Selective laser melting has been shown to be a viable RM method for the direct manufacture of surgical guides for both drilling and cutting. Stainless steel parts produced using the SLM process in result in surgical guides that are comparable in terms of accuracy, quality of fit, and function to previous experience with surgical guides produced using other RP processes yet they display superior rigidity and very good wear resistance and are easy to sterilize.

5.5.6.1 Update

This case has been followed by many more successful cases following similar procedures and the techniques have been developed and applied to further cases following.

References

[1] Heissler E, Fischer FS, Bolouri S, Lehmann T, Mathar W, Gebhardt A, et al. Custom-made cast titanium implants produced with CAD/CAM for the reconstruction of cranium defects. International Journal of Oral and Maxillofacial Surgery 1998;27(5): 334–8.

[2] Eufinger H, Wehmoller M. Individual prefabricated titanium implants in reconstructive craniofacial surgery: clinical and technical aspects of the first 22 cases. Plastic and Reconstructive Surgery 1998;102(2):300–8.

[3] Petzold R, Zeilhofer H-F, Kalender WA. Rapid prototyping technology in medicine - basics and applications. Computerized Medical Imaging and Graphics 1999;23(5): 277–84.

[4] Joffe J, Harris M, Kahugu F, Nicoll S, Linney A, Richards R. A prospective study of computer-aided design and manufacture of titanium plate for cranioplasty and its clinical outcome. British Journal of Neurosurgery 1999;13(6):576–80.

[5] Winder J, Cooke RS, Gray J, Fannin T, Fegan T. Medical rapid prototyping and 3D CT in the manufacture of custom made cranial titanium plates. Journal of Medical Engineering & Technology 1999;23(1):26–8.

[6] D'Urso PS, Earwaker WJ, Barker TM, Redmond MJ, Thompson RG, Effeney DJ, et al. Custom cranioplasty using stereolithography and acrylic. British Journal of Plastic Surgery 2000;53(3):200–4.

[7] Bibb R, Brown R. The application of computer aided product development techniques in medical modelling. Biomedical Sciences Instrumentation 2000;36:319–24.

[8] Webb PA. A review of rapid prototyping (RP) techniques in the medical and biomedical sector. Journal of Medical Engineering & Technology 2000;24(4):149–53.

[9] Bibb R, Brown R, Williamson T, Sugar A, Evans P, Bocca A. The application of product development technologies in craniofacial reconstruction. In: Proceedings of the 9th European conference on rapid prototyping and manufacturing, Athens, Greece; 2000. p. 113–22.

[10] Sanghera B, Naique S, Papaharilaou Y, Amis A. Preliminary study of rapid prototype medical models. Rapid Prototyping Journal 2001;7(5):275–84.

[11] Hughes CW, Page K, Bibb R, Taylor J, Revington P. The custom-made titanium orbital floor prosthesis in reconstruction for orbital floor fractures. British Journal of Oral and Maxillofacial Surgery 2003;41:50–3.

[12] Knox J, Sugar AW, Bibb R, Kau CH, Evans P, Bocca A, et al. The use of 3D technology in the multidisciplinary management of facial disproportion. In: Proceedings of the 6th international symposium on computer methods in biomechanics and biomedical engineering. UK: Published on CD-ROM by First Numerics Ltd. Cardiff; 2004, ISBN 0-9549670-0-3.

[13] Vander Sloten J, Van Audekercke R, Van der Perre G. Computer aided design of prostheses. Industrial Ceramics 2000;20(2):109–12.

[14] Bibb R, Bocca A, Evans P. An appropriate approach to computer aided design and manufacture of cranioplasty plates. The Journal of Maxillofacial Prosthetics and Technology 2002;5:28–31.

[15] Eggbeer D, Evans P, Bibb R. The appropriate application of computer aided design and manufacture techniques in silicone facial prosthetics. In: Proceedings of the 5th national conference on rapid design, prototyping and manufacturing. London, UK: Professional Engineering Publishing; 2004, ISBN 1860584659. p. 45–52.

[16] Evans P, Eggbeer D, Bibb R. Orbital prosthesis wax pattern production using computer aided design and rapid prototyping techniques. The Journal of Maxillofacial Prosthetics and Technology 2004;7:11–5.

[17] Singare S, Dichen L, Bingheng L, Zhenyu G, Yaxiong L. Customized design and manufacturing of chin implant based on rapid prototyping. Rapid Prototyping Journal 2005;11(2):113—8.

[18] Poukens J, Verdonck H, de Cubber J. Stereolithographic surgical guides versus navigation assisted placement of extra-oral implants (oral presentation). In: 2nd international conference on advanced digital technology in head and neck reconstruction, Banff, Canada; 2005. p. 61. Abstract.

[19] Goffin J, Van Brussel K, Vander Sloten J, Van Audekercke R, Smet MH, Marchal G, et al. 3D-CT based, personalized drill guide for posterior transarticular screw fixation at C1-C2: technical note. Neuro-Orthopedics 1999;25(1—2):47—56.

[20] Goffin J, Van Brussel K, Vander Sloten J, Van Audekercke R, Smet MH. Three-dimensional computed tomography-based, personalized drill guide for posterior cervical stabilization at C1-C2. Spine 2001;26(12):1343—7.

[21] SurgiGuides - company information. Leuven, Belgium: Materialise N.V..

[22] Bibb R, Eggbeer D, Bocca A, Evans P, Sugar A. Design and manufacture of drilling guides for osseointegrated implants using rapid prototyping techniques. In: Proceedings of the 4th national conference on rapid and virtual prototyping and applications. London, UK: Professional Engineering Publishing; 2003, ISBN 1-86058-411-X. p. 3—12.

[23] Bibb R, Bocca A, Sugar A, Evans P. Planning osseointegrated implant sites using computer aided design and rapid prototyping. The Journal of Maxillofacial Prosthetics and Technology 2003;6(1):1—4.

[24] Sarment DP, Sukovic P, Clinthorne N. Accuracy of implant placement with a stereolithographic surgical guide. The International Journal of Oral & Maxillofacial Implants 2003;18(4):571—7.

[25] Sarment DP, Al-Shammari K, Kazor CE. Stereolithographic surgical templates for placement of dental implants in complex cases. The International Journal of Periodontics and Restorative Dentistry 2003;23(3):287—95.

[26] Van Brussel K, Haex B, Vander Sloten J, Van Audekercke R, Goffin J, Lauweryns P, et al. Personalised drill guides in orthopaedic surgery with knife-edge support technique. In: Proceedings of the 6th international symposium on computer methods in biomechanics and biomedical engineering. UK: Published on CD-ROM by First Numerics Ltd. Cardiff; 2004, ISBN 0-9549670-0-3.

CHAPTER 5.6

Surgical application case study 3—The use of a reconstructed 3D solid model from CT to aid the surgical management of a total knee arthroplasty[*]

5.6.1 Introduction

Reconstructing the knee with the aid of a prosthesis in patients with gross degenerative changes and large bone loss presents many challenges to the orthopedic surgeon; therefore, any aid in the preoperative planning would enhance the outcome of this form of surgery. Plane radiographs give little insight into the bone geometry in all three dimensions, especially the geometry of the cortex at the potential plane of resection.

The use of three-dimensional reconstructed images from CT for the assessment and planning of complex hip pathologies has been investigated and is reported in the literature [1–5] and has been shown to be helpful in the planning of surgery. More recently, the production of physical models of the bone deficient or dysplastic acetabulum using data generated from CT scans has been used in the computer-aided design and manufacture of implants [6]. The successful production of custom-made femoral

---

[*] The work described in this chapter was first reported in the reference below and is reproduced here, in part or in full, with the permission of the Institute of Engineering & Physics in Medicine.

Minns RJ, Bibb R, Banks R, Sutton RA. The use of a reconstructed three-dimensional solid model from CT to aid the surgical management of a total knee arthroplasty: a case study. Medical Engineering & Physics 2003;25(6):523–6. https://doi.org/10.1016/s1350-4533(03)00050-x

Medical Modeling
ISBN 978-0-323-95733-5
https://doi.org/10.1016/B978-0-323-95733-5.00011-9

components in total hip replacements has also been reported [7—9] as well as the use of three-dimensional models in complex cranio-facial surgery. However, their use in the reconstruction of complex bone shapes around the knee has not been reported.

5.6.2 Materials and methods

The patient was a 60-year-old lady with a long history of rheumatoid arthritis which first presented at the age of 15 with deformity of the fingers. She had a Benjamin's double osteotomy (see explanatory note in Section 7.2.4) of the left knee at 27 years of age because of the potential of sub-luxation (dislocation of the kneecap) and a synovectomy (surgery to remove inflamed joint tissue) of the right knee at 35 years of age. The left knee progressively became more varus (abnormally positioned toward the midline) and unstable, and she presented at the age of 59 years, wheelchair bound with a grossly unstable and deformed left knee. She was considered for total knee replacement and due to the gross deformity of the joint a CT was carried out. The scan was in the horizontal plane with her tibial axis aligned at right angles to the scanning plane on the machine's couch with soft firm padding (the tibia is the shin bone). Slices in the horizontal plane at 1.5 mm intervals were taken to 30 mm below the joint line, producing 20 sections. The data were stored onto a magnetic/optical disc in DICOM format for processing and converting into the appropriate file system to produce a 3-D model in the computer and consequently a solid model to scale. The whole process to generate the solid model is shown in Fig. 5.33.

The model of the knee was created using stereolithography apparatus (SLA). The SLA process is described fully in Chapter 4. The preparation from CT data to machine-build files took less than 30 minutes, and the SLA machine produced the model in less than 4 hours.

Aligning the tibia during the CT scan was an advantageous to the planning of the resection of the proximal tibia in three ways. Firstly, as the optimal surgical cut through the tibia is perpendicular to the tibial axis, this aligned approach means that the CT images could be visually inspected as sections through the tibia parallel to the plane of the intended surgical cut.

Secondly, it helps to maintain accuracy when building the physical model. CT data are captured as a series of planar images with a gap between them typically 1.5 mm, while the SLA process builds models at a layer thickness of typically 0.15 mm. Therefore, interpolation is used to create intermediate sections between the original images. A cubic interpolation

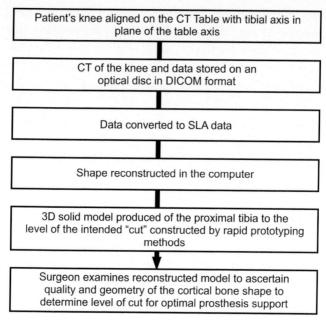

Figure 5.33 Workflow.

ensured that the intermediate sections were anatomically accurate and natural. These sections then directly drive the rapid prototyping machine that builds the physical model. Aligning the CT images ensures that the layers from which the model is built correspond exactly to the plane of the planned surgical cut. This is a significant aid when viewing the SLA model as it ensured that the layers visible in the finished model could be used to guide a perfectly level cut.

Thirdly, this two-and-half-dimensional format (called SLC) can be more accurate and less memory intensive than three-dimensional approximation formats such as the commonly used triangular faceted STL file. These formats are described in more detail in Chapter 3.

The initial assessment of the CT sections suggested that the cut should be made 15 mm below the lateral joint line. A solid model of the proximal 15 mm of the tibia was produced showing the cortical bone geometry to assess the size and shape of the supportive bone available after a cut at this level and compared with the undersurface shape and area of the tibial component. The model suggested that the bone shape and distribution would support a size small Minns meniscal bearing tibial component, the thickness of the cut bone removed by the oscillating saw was assumed to be 3 mm.

In theater, following preparation of the distal femur the tibial cut was made as planned 15 mm below the lateral joint line, orthogonal to a line from the center of the knee to the center of the ankle in the sagittal plane. The CT generated model was seen to accurately represent the clinical findings at the time of surgery confirmed when the removed bone was compared with the stereolithography model (Fig. 5.34), and a Minns meniscal bearing total knee was implanted as planned (Corin Medical Ltd., Cirencester, England) [10]. A pair of 4 mm thick meniscal bearings were found to be most appropriate in this case, and the remainder of the procedure was carried out without incident.

5.6.3 Postoperative management and follow up

Clinically, the postoperative period was uneventful; however, a postoperative X-ray the following day revealed a crack fracture of the shaft of the tibia, which was thought to have occurred at some point around the operation and was felt to be most probably due to the marked rheumatoid arthritis and disuse as no traumatic event had been noted.

This did alter the normal postoperative regime of early mobilization and was treated with 6 weeks in a full-leg-length splint prior to mobilization thereafter. The patient went home on the 10th postoperative day mobilizing nonweight bearing with a walking frame and using a wheelchair. At 6 weeks, the fracture had united, and the patient allowed to mobilize. At 12 weeks, there was a good range of knee movements the patient walking without support, pain free with much improved gait and is delighted with the result.

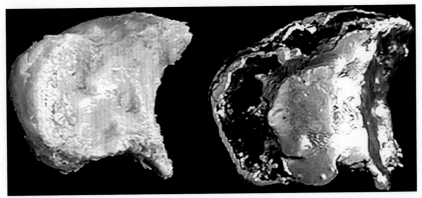

Figure 5.34 Resected bone compared to stereolithography apparatus model.

5.6.4 Discussion

The technique described above provides detailed information of the bone morphology in the region of the intended surgery and facilitated prediction of the level of transection of the deformed tibia at a level best suited to supporting the prosthesis. In addition, it provided a model on which the planned surgery could be carried out, before ever reaching the patient. This degree of preoperative information is extremely useful, and should there have been insufficient bone at the proposed level of the tibial cut, the size of any required wedge could be accurately predicted.

The fracture of the tibia may have been due to an inability of the bone to resist ordinary per-operative handling as a result of her long-standing arthritis and subsequent disuse or possibly due to the inadvertent cortical contact of one of the trephines used to prepare the proximal tibia.

The authors recognize that although it is still small, there is an increased radiation dose to the patient associated with the use of a CT scan in this technique when compared to plain radiographs normally used. A scan of the knee, being an extremity, does not put any other radiosensitive structures in the field. Although in addition to the images produced, there is also the opportunity, as was done in this case, for three-dimensional reconstruction both virtually and now as a physical model, which can be compared to the proposed prosthesis preoperatively.

The technique can be used on any tissue that can be clearly distinguished in either CT or MR images.

While we do not suggest this technique is required for any joint replacement, which is anything other than "straightforward;" however, this case has demonstrated the value of this technique in knee arthroplasty (surgical knee joint repair) in cases with complex anatomy where the bone shape and quality was difficult to predict from plain films. The creation of a three-dimensional model facilitates preoperative planning in difficult cases and provided valuable information prior to surgery.

References

[1] Bautsch TL, Johnson EE, Seeger LL. True three-dimensional stereographic display of 3-D reconstructed CT scans of the pelvis and acetabulum. Clinical Orthopaedics 1994;305:138.
[2] Barmeir E, Dubowitz B, Roffman M. Computed tomography in the assessment and planning of complicated total hip replacement. Acta Orthopaedica Scandinavica 1982;53:597.

[3] Roach JW, Hobatho MC, Baker KJ, Ashman RB. Three-dimensional computer analysis of complex acetabular insufficiency. Journal of Pediatric Orthopedics 1997;17:158.

[4] Migaud H, Corbet B, Assaker C, Kulik JF, Duquennoy A. Value of a synthetic osseus model obtained by stereo-lithography for preoperative planning: correction of a complex femoral deformity caused by fibrous dysplasia. Revue d'orthopedie et de chirurgie de l'appareil moteur 1997;83(2):156.

[5] van Dijk M, Smit TH, Jiya TU, Wuisman PI. Polyurethane real-size models used in planning complex spinal surgery. Spine 2001;26(17):1920.

[6] Munjal S, Leopold SS, Kornreich D, Shott S, Finn HA. CT-generated 3-dimensional models for complex acetabular reconstruction. The Journal of Arthroplasty 2000;15:644.

[7] McCarthy JC, Bono JV, O'Donnell PJ. Custom and modular components in primary total hip replacement. Clinical orthopaedics 1997;344:162.

[8] Bert JM. Custom total hip arthroplasty. The Journal of Arthroplasty 1996;11:905.

[9] Robinson RP, Clark JE. Uncemented press-fit total hip arthroplasty using the Identifit custom-molding technique; a prospective minimum 2-year follow-up study. The Journal of Arthroplasty 1996;11:247.

[10] Shaw NJ, Minns RJ, Epstein HP, Sutton RA. Early results of the Minns meniscal bearing total knee prosthesis. The Knee 1997;4:185.

CHAPTER 5.7

Surgical application case study 4—The custom-made titanium orbital floor prosthesis in reconstruction for orbital floor fractures*

5.7.1 Introduction

Few anatomical sites of such diminutive size have attracted so much variation in treatment as the orbital floor (the bottom of the eye socket) and its related fractures. The range of implant material in reconstruction following blow out fracture of the orbit is extensive, and the decision as to which material is used remains debated [1].

Autologous materials (those derived from human tissues) offer clear advantages with cartilage, calvarial bone, antral bone, rib, and ilium having been described [1]. These grafts offer uncertain longevity and result in tissue damage at the donor site. Artificial materials such as Silastic (Dow Corning Corporation, Auburn Plant, 5300 11 Mile Road, Auburn, MI 48611, USA) have the longest track record but a well-documented complication rate related in particular to extrusion of the graft [2]. Other artificial materials

* The work described in this chapter was first reported in the reference below and is reproduced here, in part or in full, with the permission of the British Association of Oral and Maxillofacial Surgeons.
Hughes CW, Page K, Bibb R, Taylor J, Revington P. The custom-made titanium orbital floor prosthesis in reconstruction for orbital floor fractures. British Journal of Oral and Maxillofacial Surgery 2003;41(1):50−3. https://doi.org/10.1016/S0266-4356(03)00049-4
No financial support was given. The National Center for Product Design & Development Research (PDR) supplied the stereolithography model used in making the prosthesis

Medical Modeling
ISBN 978-0-323-95733-5
https://doi.org/10.1016/B978-0-323-95733-5.00002-8

such as polyethylene sheeting (Medpor, Porex Surgical Products Group USA, Porex Surgical, Inc., 15 Dart Road, Newnan, GA 30265–1017, USA) are reported to give satisfactory results [3], and newer resorbable materials such as polydioxanone are another option [4]. The role of bioactive glass is more recently reported, but its use is limited by the size of the defect [5]. Titanium is an inert and widely used material [6,7], but in its preform, presentation can be cumbersome for use in the orbital floor, and if it has to be removed, it can present an operative challenge.

Continued development in computer-aided diagnosis and management and construction of stereolithographic models offers unparalleled reproduction of anatomical detail [8,9]. This technology is described in relation to planning in trauma surgery [10] and in planning for ablative surgery for malignancies of the head and neck (surgery to remove cancer) [11,12]. Construction of custom-made orbital floor implants is possible [13,14], although the material of choice is debated.

We describe a simple technique for construction of custom-made titanium orbital floor implants using easily available laboratory techniques combined with stereolithography models. We estimate the implant construction cost at around £300. This is largely accounted for by the cost of producing the model, which, depending on the height of orbital contour required on the model, varies between £200 and £300. The making of the implant takes about 2 hours of a maxillofacial technician's time, and the medical grade titanium sheet costs only a few pounds. This compares favorably with some of the newer alloplastic materials. This cost would drop substantially with greater use of the technique, and when reduced operating time is taken into account, the cost comparison is more favorable.

5.7.2 Technique

5.7.2.1 Imaging

The detail given here is specific to this case; a more general overview of CT scanning is given in Chapter 2 Section 2.2. Scanning protocols are observed to minimize the dose of ionizing radiation to orbital tissues [15]. Maximum detail can be obtained scanning with a 0.5 mm collimation, but the 77% increase in dosage when compared to using a 1 mm collimation may not be justified. We use a Siemens Somatom Plus4 Volume Zoom scanner with these settings: 140 kV, 120 mAs, 1 mm collimation, 3.5 feed per rotation, and 0.75 rotation time, giving a displayed dose of 45 mGy/100 mAs. Data are reconstructed using 1 mm slices with 0.5 mm increment (50% overlap) and smooth kernel. Sharp reconstruction kernels normally associated with CT imaging of bony anatomy introduce an artificial enhancement of the edge. If used as part of a three-dimensional volume based on selection of specific Hounsfield values, the enhancement artifact will be included with the bony detail, so degrading the image. The data obtained can be used to construct sharp multiplane reformats for bony detail and three-dimensional imaging for both hard-copy imaging and for stereo viewing by the surgeons.

5.7.2.2 Model construction and stereolithography apparatus

CT scans are typically taken in the axial plane at intervals exceeding 1 mm. This means that very thin bone that lies predominantly in the axial plane may fall between consecutive scans and therefore not be present in the data or three-dimensional model created from it. To overcome this, scans were taken using a smooth kernel at a slice distance of 1 mm but with a 0.5 mm overlap as described above. This improves the resolution of the data in these thin areas. The detail created is exceptionally good. The CT data were then segmented to select the desired tissue type, compact bone, using methods described in Chapter 3.

The production of models using stereolithography is described fully in Chapter 4, Section 4.2. In this case, to maintain the greatest level of accuracy, an epoxy resin was chosen (RenShape SL5220, Huntsman Advanced Materials, Everslaan 45, B-3078 Everberg, Belgium). This type of resin shows almost no shrinkage during the photo-polymerization process and therefore can produce models with excellent accuracy.

5.7.2.3 Construction of the prosthesis

From stereolithography models, the orbital defect is easily seen and assessed. The orbital defect is then filled with wax to reproduce a contour similar to the opposite side, and an impression is taken of both orbital cavities using silicone putty impression material. The orbital injury side is then reproduced by pouring a hard plaster/stone model. The defect has been filled and, therefore, appears in its proposed reconstructed form. Using pressure flasks usually used in the construction of dentures, a layer of 0.5 mm medical–grade titanium is swaged onto the stone/plaster model of the orbital floor, producing an exact replica of the proposed orbital floor and rim contour. The titanium sheet may then be trimmed to allow sufficient overlap and the positioning of a flange to fix the screws. The prosthesis is polished and sterilized for use according to local protocol for titanium implants.

5.7.3 Case report

A 54–year–old man sustained a "blow out" fracture of the left orbital floor and presented with diplopia (double vision) and restriction of upward gaze. Coronal plane CT scanning demonstrated the fracture (Fig. 5.35). A stereolithography model was constructed which shows the trap door of the fractured orbital floor well (Fig. 5.36). The model was then used for construction of a plaster cast of the orbital defect. A medical–grade titanium prosthesis was constructed from this working cast (Fig. 5.37). The prosthesis was packaged and sterilized by the hospital central sterile supplies department according to standard protocol for titanium medical implants.

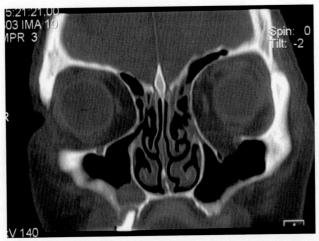

Figure 5.35 Coronal computed tomography scan demonstrating classic orbital blow-out fracture.

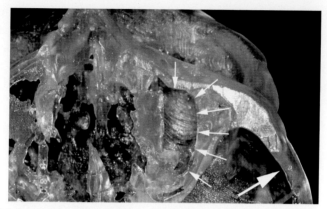

Figure 5.36 Stereolithographic model constructed from epoxy resin showing the "trap door" defect in the left orbital floor, viewed from below as if in the maxillary antrum looking up (lateral margins of the defect indicated with small arrows, under surface of zygomatic arch indicated with single large arrow).

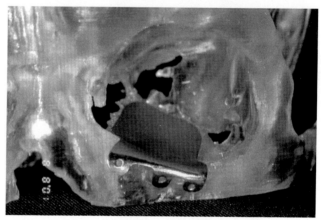

Figure 5.37 The custom titanium implant is seen on the master model.

The approach to the orbital floor was by a subciliary incision (through the lower eyelid), and the defect was exposed. Herniation and entrapment of the periglobar fat were released and the defect prepared in a standard way (this means that the damaged layer of fat that surrounds the eyeball was repaired and put back in the correct position). The prosthesis fitted perfectly and was stabilized with 1.3 mm titanium screws from a standard plating kit

(Fig. 5.38). Forced duction was confirmed as normal (this is a test to check that the eye can rotate upwards freely). Postoperative recovery was uneventful, and radiographs revealed the prosthesis to be correctly positioned (Figs. 5.39 and 5.40). At follow up, complete return to normal range of ocular movement was found with resolution of the diplopia and no evidence of complications.

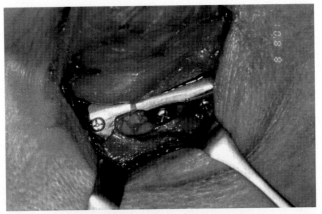

Figure 5.38 The implant inserted and fixed with 1.3 mm screws. The fit is precise.

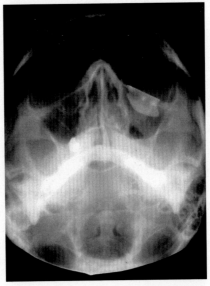

Figure 5.39 Plain radiograph in the anterior-posterior plane showing the position of the implant postoperatively.

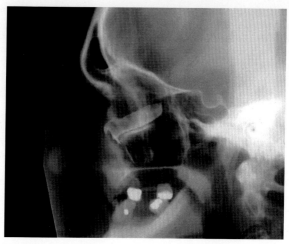

Figure 5.40 Plain radiograph in the lateral plane showing the position of the implant postoperatively.

5.7.4 Conclusion

We think that this technique has much to offer both for its simplicity and for the reliability of titanium as a prosthetic material. The laboratory techniques are simple and readily available in most maxillofacial laboratories. The models require off-site production, but their use is particularly valid in cases where defects may be complicated in three dimensions and where operating time should be reduced to a minimum. The cost of construction of models will drop substantially if numbers increase, and the technique may offer a financially viable alternative to current orbital floor prostheses.

References

[1] Courtney DJ, Thomas S, Whitfield PH. Isolated orbital blow out fractures: survey and review. British Journal of Oral and Maxillofacial Surgery 2000;38:496—503.
[2] Morriston AD, Sanderson R, Moos KF. The use of silastic as an orbital implant for reconstruction of orbital wall defects: review of 311 cases treated over 20 years. Journal of Oral and Maxillofacial Surgery 1995;53:412—7.
[3] Rubin PAD, Bilyk JR, Shore JW. Orbital reconstruction using porous polyethylene sheets. Opthalmology 1994;101:1697—708.
[4] Iizuka T, Mikkonen P, Paukku P, Lindqvist C. Reconstruction of orbital floor with polydioxanone plate. International Journal of Oral and Maxillofacial Surgery 1991;20(2):83—7.
[5] Kinnunen I, Aitasalo K, Pollonen M, Varpula. Reconstruction of orbital floor fractures using bioactive glass. Journal of Craniomaxillofacial Surgery 2000;4:229—34.

[6] Park HS, Kim YK, Yoon CH. Various applications of titanium mesh screen implant to orbital wall fractures. Journal of Craniofacial Surgery 2001;6:555—60.

[7] Dietz A, Ziegler CM, Dacho A, Althof F, Conradt C, Kolling G, et al. Effectiveness of a new perforated 0.15 mm poly-p-dioxanon-foil versus titanium dynamic mesh in reconstruction of the orbital floor. Journal of Craniomaxillofacial Surgery 2001;2:82—8.

[8] Bouyssie JF, Bouyssie S, Sharrock P, Duran D. Stereolithographic models derived from x-ray computed tomography: reproduction accuracy. Surgical and Radiologic Anatomy 1997;3:193—9.

[9] Bibb R, Brown R. The application of computer aided Product development techniques in medical modelling. Biomedical Sciences Instrumentation 2000;36:319—24.

[10] Kermer C, Lindner A, Friede I, Wagner A, Millesi W. Preoperative stereolithographic model planning for primary reconstruction in craniomaxillofacial trauma surgery. Journal of Craniomaxillofacial Surgery 1998;3:136—9.

[11] D'Urso PS, Barker TM, Earwaker WJ, Bruce LJ, Atkinson RL, Lanigan MW, et al. Stereolithographic biomodelling in cranio-maxillofacial surgery: a prospective trial. Journal of Craniomaxillofacial Surgery 1999;1:30—7.

[12] Kermer C, Rasse M, Lagogiannis G, Undt G, Wagner A, Millesi W. Colour stereolithography for planning complex maxillofacial tumour surgery. Journal of Craniomaxillofacial Surgery 1998;6:360—2.

[13] Hoffmann J, Cornelius CP, Groten M, Probster L, Pfannenberg C Schwenzer N. Orbital reconstruction with individually copy-milled ceramic implants. Plastic and Reconstructive Surgery 1998;3:604—12.

[14] Holk DE, Boyd Jr EM, Ng J, Mauffray RO. Benefits of stereolithography in orbital reconstruction. Opthalmology 1999;6:1214—8.

[15] Ionising radiation (medical exposure) regulations. UK Government Department of Health Publication; 2000.

CHAPTER 5.8

Surgical application case study 5—The use of 3D technology in the multidisciplinary management of facial disproportion*

5.8.1 Introduction

Coordinated orthodontic/surgical treatment, which allows the predictable management of dento-facial disproportion, is largely a development of the latter third of the 20th century. Traditionally, the diagnosis, treatment planning, and postoperative evaluation of patients requiring such treatment have relied heavily on the use of cephalometric analysis. This has enabled the two-dimensional quantification of dental and skeletal relationships both before and after treatment, with reference to normative data in tabulated or template form [1−3].

However, the recent development of three-dimensional measuring techniques has allowed a more clinically valid quantification of deformity and assessment of surgical outcomes [4−10]. Both computed tomography (CT), magnetic resonance imaging (MRI), and finite element (FEA) analysis have recently been employed in surgical planning and the

---

* The work described in this chapter was first reported in the reference below and is reproduced here in part or in full with the permission of First Numerics Ltd.
  Knox J, Sugar AW, Bibb R, Kau CH, Evans P, Bocca A, Hartles F. The use of 3D technology in the multidisciplinary management of facial disproportion. Proceedings of the sixth international symposium on computer methods in biomechanics & biomedical engineering, Madrid, Spain, February 2004, ISBN: 0-9549670-0-3 (Published on CD-ROM by First Numerics Ltd. Cardiff, UK).

Medical Modeling
ISBN 978-0-323-95733-5
https://doi.org/10.1016/B978-0-323-95733-5.00013-2

221

visualization of treatment objectives [11—19]. This presentation demonstrates the successful use of 3D tomography, surface laser scans, and rapid prototyping in the surgical management and postoperative evaluation of a patient presenting with maxillary hypoplasia who underwent surgical maxillary distraction.

5.8.2 Materials and method

3D virtual hard tissue images, shown in Fig. 5.41, were constructed using Mimics software, (Materialise NV, Technolgielaan 15, 3001 Leuven, Belgium), from 0.5 mm slice CT DICOM datasets, to identify tissue type. Upper and lower tissue density thresholds on the CT image were defined and the areas between the slices interpolated to improve resolution. The data was then prepared for medical modeling using stereolithography. In addition, STL files were generated so that the same data could be imported into the Geomagic FreeForm Plus software (https://oqton.com/geomagic-freeform). The preparation of data for file transfer and medical modeling is described in detail in Chapter 3.

A stereolithography model was then constructed and used to visualize skeletal discrepancy, simulate surgical movements, and adapt surgical distractors, as shown in Figs. 5.42 and 5.43. The stereolithography process is described in detail in Chapter 4, Section 4.2. The FreeForm software was

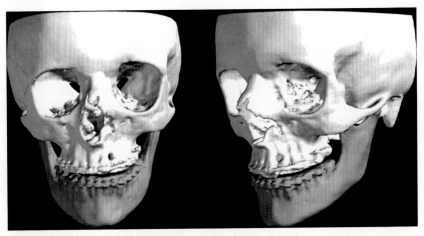

Figure 5.41 Virtual hard tissue images.

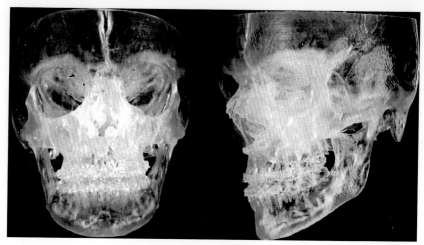

Figure 5.42 Virtual hard tissue images.

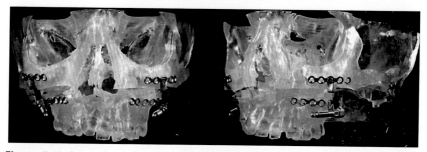

Figure 5.43 Surgical simulation and placement of distractors on stereolithography models.

used to produce a digital clay model allowing further simulation of surgical movements (see Fig. 5.44).

3D facial soft tissue images (see Fig. 5.45) were captured before and after surgery using two high-resolution Minolta Vivid VI900 3D cameras operating as a stereo-pair. The scanners were controlled with multiscan software (Cebas Computer GmbH, Lilienthalstrasse19, 69214 Eppelheim, Germany) and data coordinates were saved in the Minolta Vivid file format (called a.vvd). The scan data were transferred to a reverse modeling software package (Rapidform 2004, INUS Technology, Inc., SBC, Ludwig-Erhard-Strasse, 30-34, D-65760 Eschborn, Germany) for analysis.

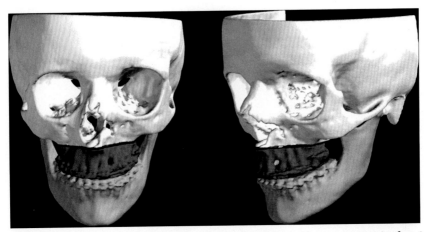

Figure 5.44 Virtual surgical simulations showing maxillary advancement at Le fort 1 level.

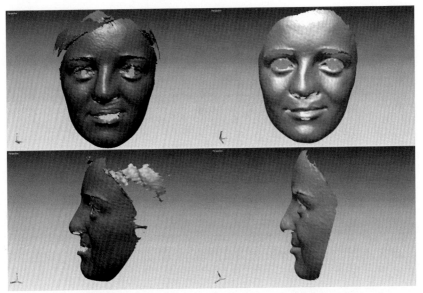

Figure 5.45 3D facial images preoperative (*left*) and postoperative (*right*).

5.8.3 Results

Surgical distraction of the maxilla at Le Fort 1 level was successfully completed, (Le Fort 1 is a cut across the maxilla above the roots of the teeth but under the nose in order to separate and move the upper jaw in relation

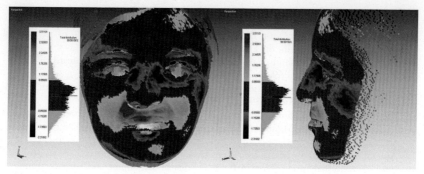

Figure 5.46 Merged pre- and postoperative facial scans demonstrating the magnitude of change in soft tissue morphology.

to the rest of the skull). The distractors were activated by 7 mm on right and left sides resulting in an equivalent advancement of the tooth-bearing portion of the maxilla. A description of distraction osteogenesis is given in a medical explanatory note in Section 7.2.3. Superimposition of surface scans allowed quantification of soft tissue changes (Fig. 5.46). Black areas indicate changes within 0.80 mm or less, which could be attributed to the error of the technique. The blue areas demonstrate negative changes of 0.85−2.30 mm. The red areas demonstrate positive changes of 0.85−3.51 mm.

5.8.4 Discussion

Changes in maxillary prominence and lip relationship can be appreciated by comparison of pre- and postoperative scans in Fig. 5.45. The magnitude of the soft tissue changes is demonstrated in Fig. 5.46. Here the primary effect of the maxillary distraction at Le Fort 1 level is an advancement of the upper lip and paranasal areas of 1.7−3.5 mm (red areas). The small advancement (pink) demonstrated in the frontal region is an artifact introduced by overlying hair. The small advancement demonstrated in the left chin is probably due to a change in lip relationship and a slight change in facial expression. The blue areas in Fig. 5.46 demonstrate a reduction in lower lip prominence is due to the change in lip relationship caused by the maxillary advancement. The change demonstrated in the sub-mandibular region and upper mid-face is suggested to be due to a reduction in body mass index.

References

[1] Broadbent BHS, Broadbent BHJ, Golden WH. Bolton Standards of dentofacial development and growth. St. Louis: Mosby; 1975.

[2] McNamara JAJ. A method of cephalometric evaluation. American Journal of Orthodontics 1984;86(6):449—69.

[3] Ackerman RJ. The Michigan school study norms expressed in template form. American Journal of Orthodontics 1975;75:282—90.

[4] Ayoub AF, Wray D, Moos KF, Siebert P, Jin J, Niblett TB, et al. Three-dimensional modeling for modern diagnosis and planning in maxillofacial surgery. The International Journal of Adult Orthodontics and Orthognathic Surgery 1996;11(3):225—33.

[5] Ayoub AF, Siebert P, Moos KF, Wray D, Urquhart C, Niblett TB. A vision-based three-dimensional capture system for maxillofacial assessment and surgical planning. British Journal of Oral and Maxillofacial Surgery 1998;36(5):353—7.

[6] Ji Y, Zhang F, Schwartz J, Stile F, Lineaweaver WC. Assessment of facial tissue expansion with three-dimensional digitizer scanning. Journal of Craniofacial Surgery 2002;13(5):687—92.

[7] Khambay B, Nebel JC, Bowman J, Walker F, Hadley DM, Ayoub A. 3D stereophotogrammetric image superimposition onto 3D CT scan images: the future of orthognathic surgery. The International Journal of Adult Orthodontics and Orthognathic Surgery 2002;17(4):331—41.

[8] McCance AM, Moss JP, Fright WR, Linney AD, James DR. Three-dimensional analysis techniques–part 1: three-dimensional soft-tissue analysis of 24 adult cleft palate patients following Le Fort I maxillary advancement: a preliminary report. The Cleft Palate-Craniofacial Journal 1997;34(1):36—45.

[9] McCance AM, Moss JP, Wright WR, Linney AD, James DR. A three-dimensional soft tissue analysis of 16 skeletal class III patients following bimaxillary surgery. British Journal of Oral and Maxillofacial Surgery 1992;30(4):221—32.

[10] Marmulla R, Hassfeld S, Luth T, Muhling J. Laser-scan-based navigation in craniomaxillofacial surgery. Journal Craniomaxillofacial Surgery 2003;31(5):267—77.

[11] Nkenke E, Langer A, Laboureux X, Benz M, Maier T, Kramer M, et al. Validation of in vivo assessment of facial soft-tissue volume changes and clinical application in midfacial distraction: a technical report. Plastic and Reconstructive Surgery 2003;112(2):367—80.

[12] Xia J, Samman N, Yeung RW, Shen SG, Wang D, Ip HH, et al. Three-dimensional virtual reality surgical planning and simulation workbench for orthognathic surgery. The International Journal of Adult Orthodontics and Orthognathic Surgery 2000;15(4):265—82.

[13] Xia J, Ip HH, Samman N, Wong HT, Gateno J, Wang D, et al. Three-dimensional virtual-reality surgical planning and soft-tissue prediction for orthognathic surgery. IEEE Transactions on Information Technology in Biomedicine 2001;5(2):97—107.

[14] Gladilin E, Zachow S, Deuflhard P, Hege H-C. A non-linear soft tissue model for craniofacial surgery simulations. In: Proceedings of the modeling & simulation for computer-aided medicine and surgery (MS4CMS), INRIA, Paris, France; 2002.

[15] Gladilin E, Zachow S, Deuflhard P, Hege H-C. Biomechanical modeling of individual facial emotion expressions. In: Proceedings of visualization, imaging, and image processing (VIIP); 2002.

[16] Gladilin E, Zachow S, Deuflhard P, Hege H-C. On constitutive modeling of soft tissue for the long term prediction of cranio-maxillofacial surgery outcome. In: Proceedings of computer assisted radiology and surgery (CARS), London; 2003. p. 343—8.

[17] Gladilin E, Zachow S, Deuflhard P, Hege HC. Realistic prediction of individual facial emotion expressions for craniofacial surgery simulations. In: Proceedings of SPIE medical imaging conference, San Diego, USA. vol. 5029; 2003. p. 520—7.

[18] Bibb R, Freeman P, Brown R, Sugar A, Evans P, Bocca A. An Investigation of three-dimensional scanning of human body surfaces and its use in the design and manufacture of prostheses. Proceedings of the Institution of Mechanical Engineers - Part H: Journal of Engineering in Medicine 2000;214(6):589—94.

[19] Bibb R, Brown R. The application of computer aided product development techniques in medical modeling. Biomedical Sciences Instrumentation 2000;36:319—24.

CHAPTER 5.9

Surgical applications case study 6—An appropriate approach to computer-aided design and manufacture of reconstructive implants*

5.9.1 Introduction

Advances in metal additive manufacturing (AM) processes have enabled complex forms designed in computer-aided design (CAD) to be manufactured in biocompatible materials such as titanium. The layer-additive nature of AM offers a greater degree of design freedom over processes such as computer numerically controlled (CNC) machining; this represents an opportunity in the production of complex implants. There are a growing number of case studies where AM technologies have been used to produce patient-specific implants in cranio-maxillofacial and orthopedic applications [1—4], but the techniques are still not yet used routinely in most hospitals.

---

* All of the cases described were undertaken as part of a multidisciplinary team.
The planning for case 1 was undertaken by Peter Evans and Adrian Sugar at Morriston Hospital Swansea, and the surgery was undertaken by Adrian Sugar. Details of this case were presented as a poster at the 2009 Institute of Maxillofacial Prosthetists and Technologists Congress.
Planning for cases 2 and 3 were undertaken by Sean Peel and Dominic Eggbeer at PDR, Cardiff Metropolitan University, and Satyajeet Bhatia at the University Hospital, Wales, Cardiff. Surgery for case studies 2 and 3 were undertaken by Saty Bhatia.
Case 4 was planned by Adrian Sugar and Peter Evans at Morriston Hospital with input from Sean Peel from PDR, Cardiff Metropolitan University. Sean Peel designed the guides and implants. Adrian Sugar undertook the surgery.

Medical Modeling
ISBN 978-0-323-95733-5
https://doi.org/10.1016/B978-0-323-95733-5.15005-7

Chapter 5.14: Prosthetic rehabilitation applications case study 3—An appropriate approach to computer-aided design and manufacture of cranioplasty plates demonstrates an evolution of techniques for cranioplasty implant production, but this section will provide an overview of other maxillofacial cases where CAD/AM technologies have been used to produce more complex patient-specific implants.

5.9.2 Case 1—Orbital rim augmentation implant

This study reports a case involving the design, manufacture, and implantation of a custom-fitting titanium implant to improve the orbital symmetry of a patient with a facial cleft. The 25-year-old female patient had a repaired Tessier number 3 facial cleft and required correction of severe right vertical orbital dystopia and enophthalmos, augmentation of the anterior zygomatico-maxillary region, revision of the facial cleft skin scar, and lip revision. The patient's right eye was nonseeing, relatively immobile, turned inward, and more than 1 cm lower than the left eye. The skin was tight over the anterior maxilla and body of the zygomatic bone where the augmentation would take place, so advancement was necessary to achieve tension-free skin cover over the augmentation when closing the facial cleft scar. The patient's major concern was about her appearance.

Evaluation of the computed tomography (CT) scan demonstrated that the whole orbit was not displaced inferiorly, but the floor was, causing the eye and orbital contents to be drawn downward by the orbital floor (Fig. 5.47). After much debate and discussion, it was decided that complete freeing of the orbital soft tissues by 360-degree periorbital dissection through a coronal flap approach followed by augmentation of the deficient orbital floor and zygomatico-maxillary region through opening up

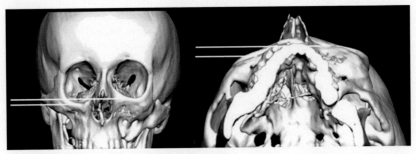

Figure 5.47 Preoperative CT showing defect on the patient's right orbit.

of the previous facial cleft scar was appropriate. It was hoped this would improve facial symmetry, eye position, and overall appearance by helping to correct the dystopia and bone deficiency. It was recognized that repositioning the medial and lateral canthal ligaments would also be required. The ophthalmologists preferred to attempt to correct the severe squint at a later stage.

Various autogenous and alloplastic augmentation materials were considered. It was decided that the ideal would be a custom-fitting solid titanium implant that would neither distort nor resorb. AM was deemed an appropriate route for the production of the implant due to its complex shape that would be impossible to achieve by swaging titanium sheet. The dimensions of the implant also made casting a titanium implant impossible to achieve at a reasonable cost, quality, and timescale. While CNC machining of custom-fitting maxillofacial implants has been demonstrated [4,5], the complexity of shape involved in this example would have presented difficulty in clamping the work piece, required multiple process steps, and involved great deal of programming time to achieve the desired result. In addition, machining this implant would have required the purchase of an appropriately oversized billet of solid titanium, which would have been expensive to purchase. The dimensions of the required billet would have been in the order of 5 × 5 cm by 5 cm (125 cm^3) to machine an implant with a volume of just 2.52 cm^3, meaning that as much as 98% of the billet would have been removed as waste.

The decision to utilize AM necessitated CAD of the implant. A variety of AM processes are available that can produce parts from a range of materials from soft thermoplastics to high-performance alloys. The principles of AM technologies are described in Chapter 4 and in a range of texts [6–9].

5.9.2.1 Materials and methods

5.9.2.1.1 Stage 1: 3D data acquisition and transfer

A CT scan was taken with 1 mm slice thickness, 0.5 mm increments, and 0.424 mm pixel size on a Toshiba Aquilion Multislice scanner. The data were imported into Mimics software (Mimics version 13, Materialise, Technolgielaan 15, Leuven 3001, Belgium), which was used to segment the bony anatomy, maxillary sinus volume, and eye globes into separate volumes. The software also allowed measurements to be taken to quantify the difference in globe height and protrusion. The 3D reconstructed data of the bony anatomy were exported as high-quality STL files utilizing the smoothing parameters available when using Mimics. The STL files were imported into CAD software (FreeForm Modeling Plus, 3D Systems, USA) to undertake the implant design. The use of this software has been reported in the design of other custom-fitting medical devices including removable partial dentures and facial prosthetics [10–15] and is described in other case studies in this book. The "fill holes" option was used to ensure the STL file was a single bound volume with no voids.

5.9.2.1.2 *Stage 2: Implant design*
The design was conducted by a maxillofacial prosthetist who had several years of experience using the FreeForm software in medical applications. On importing the patient data, the clay edge sharpness was set to 0.316 mm. This is how the quality of the model is controlled within FreeForm. The patient anatomy data was imported using the "buck" setting. This setting prevents that data from being altered in any way, thus protecting the anatomy data from accidental modification. The basic implant shape was generated by selecting and copying an area of the patient's unaffected orbit corresponding to the defect area on the contralateral side and laterally inverting (mirroring) it about the midsagittal plane. Measurement tools and reference planes were used to establish the correct contours and position of the implant to create the best possible orbital symmetry. The positioned form was then expanded to overlap the protected model of the anatomy and final shaping undertaken to create a smooth shape that effectively restored the shape of the affected anatomy.

Two 3-mm-diameter screw holes were incorporated into the design according to the ideal anchor sites, which eliminated potentially difficult drilling operations later. The model edge sharpness was refined by 0.1 mm to smooth off sharp edges. The design in FreeForm is shown in Fig. 5.48. The original anatomy was then removed using a Boolean subtraction that

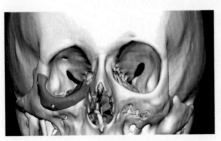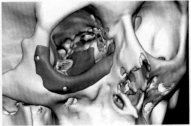

Figure 5.48 The design in FreeForm CAD.

resulted in a fitting surface that precisely matched the defect site. The final implant design was then exported as an STL file. A stereolithography model of the patient's bony anatomy was also created to enable the titanium implant fit to be validated prior to surgery.

5.9.2.1.3 Stage 3: Additive manufacture

The implant was fabricated in Ti6Al4V alloy using direct metal laser sintering [16]. This AM process uses a fiber laser to selectively melt very fine metal powder particles together to form solid layers. Once each layer is complete, the build platform is lowered by one layer thickness, and fresh powder is spread over it. The next layer is then melted and also fused to the previous layer. The process repeats until the complete metal object is created. The process also concurrently builds support structures that anchor the part being built to the build platform and support overhanging features. The machine used to produce the implant in this case was an EOSint M270 (EOS GmbH, Electro Optical Systems, Robert-Stirling-Ring 1, D-82152 Krailling, München, Germany).

The part was built using 0.03 mm layers, and it took 4 h and 47 min to build, with additional time and labor required to remove the supports and remove loose powder from the surface by bead blasting. Abrasive finishing was undertaken in the hospital maxillofacial prosthetics laboratory to achieve a smooth, polished surface. The implant was then passivated in concentrated nitric acid and autoclave sterilized.

5.9.2.1.4 Stage 4: Fitting and surgery

The finished implant was test fitted to the stereolithography model of the patient's bony anatomy to check dimensional accuracy. The implant was found to be a precise fit with secure location, as shown in Fig. 5.49.

At surgery, the orbital periosteum was raised through 360 degrees around the right orbit via a coronal incision. The previous facial scar was opened from the lower eyelid to and including the upper lip. The orbital floor was identified and freed posteriorly, and the anterior zygomatic bone and maxilla were exposed subperiosteally as required. The implant was tried in and was observed to be a perfect fit. It was fixed to the orbital rims with two 1.5-mm-diameter titanium screws, as shown in Fig. 5.50. The canthal ligaments were then identified and repositioned superiorly to a screw on the right and a mini-plate on the left. The lip scar was revised, and the skin and other soft-tissues were advanced over the implant. These tissues were then closed in layers and the scar revised.

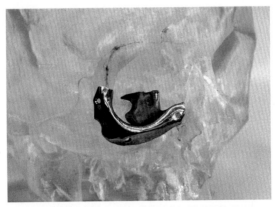

Figure 5.49 AM titanium implant test fitted to the stereolithography model.

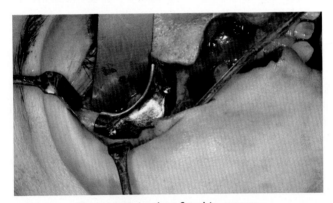

Figure 5.50 Implant fitted in surgery.

5.9.2.2 Results and conclusions from case 1

The initial postsurgical outcome was good, resulting in a successful reconstruction and much improved aesthetics. Postoperative CT scans showed the implant to be securely fixed in its intended position as shown in Fig. 5.51. There were no complications following a 3-month review and subsequent reviews.

A custom-fitting, patient-specific implant was designed using an entirely digital process and fabricated using AM in an appropriate, biocompatible titanium alloy. The prosthesis fitted both the stereolithography model and patient precisely, providing secure and positive location. However, due to the complex, organic form, no convenient datum points were available, and consequently no quantification of dimensional accuracy was undertaken to compare the implant to the CAD geometry. While this case has

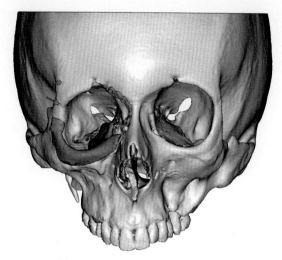

Figure 5.51 Postoperative views.

demonstrated that the digital design and AM process are capable of producing an appropriate and accurate titanium implant, further case studies and technical experiments are proposed to further evaluate the potential of the digital process, addressing design capabilities, the capability of the AM process, material physical properties, and process limitations.

5.9.3 Case 2—Orbital floor implant incorporating placement guide

The first case study demonstrated the potential of a CAD/AM approach in congenital cases involving the orbit. The second study illustrates how the same approach can be used in trauma reconstructive surgery. There has been limited discussion on the application of CAD/AM in orbital floor reconstruction in the research literature [2], but this is an area where a CAD/AM approach could improve the predictability of procedures.

The patient presented with enophthalmos (posterior displacement of the eye) of the right eye due to an orbital floor "blow out fracture." In this case, trauma to the midface had caused the floor of the orbit to fracture and displace downward in a "trap door" fashion. The goal of surgical intervention was to correct the volume of the orbit by reconstructing the orbital floor with a bespoke implant, contoured to match the unaffected side. Orbital floor reconstructions can be challenging due to the sensitive nature of the orbital anatomy, the proximity of the implant to the optic nerve, and

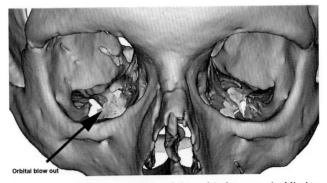

Orbital blow out

Figure 5.52 3D representation of the orbital trauma in Mimics.

the confined space in which the procedure must take place. A CAD/AM approach was considered an appropriate route to ensuring that the design was sufficiently small to minimize the necessary surgical incision and provide the best possible fit and contour.

5.9.3.1 Materials and methods

5.9.3.1.1 Stage 1: Data segmentation
The same Mimics CT segmentation techniques as described in the first case study and in Chapter 3 were undertaken to obtain an accurate 3D representation of the orbital region that showed the defect, as shown in Fig. 5.52.

The mask of the original anatomy was duplicated. The "edit mask" tool was used to manually draw in holes in the orbital floor of the unaffected side, thereby creating a thickness that could be mirrored across the midsagittal plane to the damaged side. Both the original anatomy and version with the orbital floor drawn in were exported as high-quality STL files.

5.9.3.1.2 Stage 2: Defect reconstruction
The STL files were imported into FreeForm as buck models (to protect them from accidental material removal) for implant design. An area of the left orbit, incorporating the drawn-in portion, was selected, copied, and mirrored around the midsagittal plane. This was then repositioned by eye to provide the basic reconstruction shape for the damaged right side. Once the basic position had been established, the mirrored portion was converted to clay and combined into the original anatomy buck model for shape refinement. Various tools in the "sculpt clay" palette were used to smooth and blend the clay into the buck, creating a seamless contour, upon which the implant form could be designed. Measurement tools and planes were used to check the symmetry of the final contour.

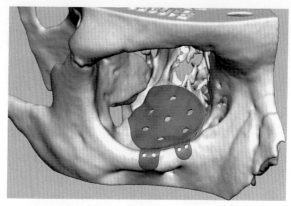

Figure 5.53 Completed orbital floor implant.

5.9.3.1.3 Stage 3: Implant and positioning guide design

The reconstructed orbital floor model was selected and converted to buck. An outline of the proposed orbital floor plate was drawn directly on to the buck surface using the "draw curve" tool. The "layer" tool was used to create an even, 0.5 mm thickness of clay that represented the implant shape, which was subsequently verified by the prescribing surgeon. The implant size was designed to be as small as possible while still providing sufficient support to the orbital content. The buck was then removed, to leave just the implant plate design. The 3-mm-diameter holes to allow fluid drainage beneath the eye globe were added to the plate using the "carve with ball tool" in random locations. Four 1.6-mm-diameter (intended for 1.5-mm-diameter screws, leaving 0.1 mm clearance) holes were then added on the tabs extending over the infraorbital rim. The completed orbital floor implant design is shown in Fig. 5.53.

To enable more accurate placement of the orbital implant, which was designed to minimize the bulk of material around the infraorbital rim, the prescribing surgeon requested the design and production of a custom-fitting guide. This was designed as a transient use device to engage with a larger area around the infraorbital rim, giving a stable and positive location. The contours were also designed to allow the implant to slot into the correct position and be secured (Fig. 5.54).

5.9.3.1.4 Stage 4: Implant and guide fabrication

AM was chosen for the same reasons as case study 1. Both the implant and guide were fabricated in Ti6Al4V alloy using laser melting technology (described in Chapter 4) by LayerWise (Leuven, Belgium). The implant

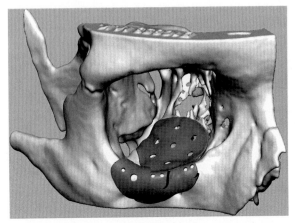

Figure 5.54 The orbital floor and placement guide.

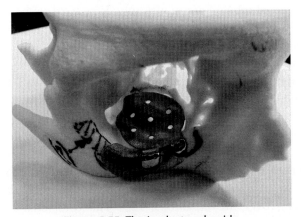

Figure 5.55 The implant and guide.

screw holes were inspected under a 40× digital microscope, which high-lighted the need to ream them to ensure the screws would fit without being forced. Both the guide and implant were then cleaned with detergent, ultrasonic cleaned, rinsed, and dried. The completed implant and guide are shown on a reference model in Fig. 5.55.

5.9.3.1.5 Stage 4: Fitting and surgery

The goal of the surgical approach was to minimize the size of the scar and insert the plate without causing surrounding tissue damage. A trans-conjunctival with lateral lid swing approach provided the access required to place the guide and locate the implant. The guide positioned adequately on the infraorbital rim without the need to use screws. The orbital plate was carefully inserted beneath the eye and located into the corresponding cut out features in the guide, providing confidence that it was correctly located on the orbital rim. Two, standard 1.5-mm Synthes Cortex screws (Synthes GmbH) were used to retain the plate.

5.9.3.2 Results and conclusions from case 2

This case demonstrated the potential of using a CAD/AM approach in orbital trauma reconstruction. The procedure was undertaken without complication, and subsequent reviews demonstrated success in achieving improved vision.

In undertaking this case, small potential improvements with the process were also identified. The nature of AM in metal results in a relatively rough surface finish compared with pressed titanium sheet and also meant that the small screw hole features were not fully round or smooth. This meant that despite clearance being left in the design, it was necessary to ream the holes, which represented a small degree of process inefficiency. In this case, it was also deemed unnecessary to countersink the screw holes, but the potential to incorporate this in future designs was noted. This would help to reduce the prominence of screw heads into overlaying soft tissue.

5.9.4 Case study 3—Multipart reconstruction

The first two case studies demonstrated relatively simple cases involving small, localized implants for a congenital and trauma case. The third case illustrates how a CAD/AM approach can be effectively used in a larger case that requires multiple components.

The patient presented with an extensive, cancerous tumor extending around the left temporal region of their head and involving the superior aspects of the orbital bones. Urgent, multidisciplinary surgery was required to remove the tumor and provide immediate reconstruction that would restore anatomic contours and enable further treatments, including radiotherapy, to be delivered effectively. This represented a significant challenge since the fast growing nature of the tumor meant that it was difficult to predict how much bone and soft tissue would need to be removed. A CAD/AM approach was considered an appropriate route to fabricating reconstructive plates, but the extra challenge of not knowing how much tissue would require removal meant the design had to incorporate a degree of flexibility by not relying on an exact fit within the defect in predetermined areas.

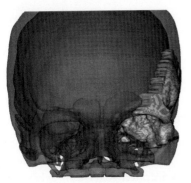

Figure 5.56 3D reconstructions of the bony anatomy and tumor extent.

5.9.4.1 Materials and methods

5.9.4.1.1 Stage 1: Data segmentation

The same Mimics segmentation techniques as previously described were used to create a 3D CAD model of the bony anatomy. In addition to this, the tumor margin was also segmented based on operator interpretation and consultation with the prescribing surgeon. Segmenting the tumor provided the best method of determining how much tissue would require removal and, therefore, the extent of the reconstruction implants. Fig. 5.56 shows the segmented 3D models of the anatomy and tumor extents. The bony anatomy and tumor were exported as high-quality STL files.

5.9.4.1.2 Stage 2: Defect reconstruction

The STL files were imported into FreeForm as clay to enable virtual resection of the tumor area and smoothing of the margin. A Boolean subtraction of the tumor from the bony anatomy was undertaken, leaving a roughly resected clay model. The resection margins were smoothed and cleaned up by manually carving, being careful not to extend too far into unaffected anatomy where the implant would locate. Once the computer-aided resection had been completed, an area of the patient's right temporal bony anatomy was selected, copied, and pasted as a new piece of clay. This was mirrored around the midsagittal plane, repositioned by eye to form the basis of the reconstruction and reduced in extent to approximately match the required shape. The original anatomy with tumor removed was converted into a buck model to protect it from carving and the reconstruction then combined into it. A variety of carving and manipulation tools were then used to blend the reconstruction into the anatomy, creating a seamless join. The completed reconstruction is shown in Fig. 5.57. This was duplicated and one version turned into a "buck" model.

Figure 5.57 The completed reconstruction.

5.9.4.1.3 Stage 3: Implant design

A three-part implant was decided as the optimum reconstruction method. This consisted of a two-part orbital implant to reconstruct the superior and lateral walls of the orbit and a temporal cranioplasty. Given that the actual extent of the tumor could not be accurately predicted, it was decided that the cranioplasty needed to sit on top of the reconstruction rather than inside the surgically created resection. The orbital plates were, however, designed to sit inside the resection since it was considered that their extent and thickness would have compromised the patient's vision in the left eye.

All design work was undertaken in FreeForm (Version 2014). Due to the detailed, low-thickness nature of the proposed design, it was necessary to refine the clay coarseness to 0.18 mm. The temporal cranioplasty plate was designed by drawing the margin directly onto the reconstruction surface of the entirely "buck" model and using the "emboss clay, raise" function to create a plate with a thickness of 0.5 mm on top of the surface. Localized fixation tabs were then added using the "layer" tool.

The orbital implants were more complicated since the intention was to set them inside the proposed surgical resection. The reconstructed model (with buck original anatomy and clay reconstruction) was duplicated again. A margin line extending beyond and enclosing the resection boundary was drawn on the surface of the model. The "emboss clay, lower" tool was used, but this time to lower the clay inside the margin line by 0.5 mm. This has the effect of removing only the clay reconstruction while leaving the "buck" anatomy unmodified. The model was then converted to "buck" and a duplication made. One version of the modified model was then used to Boolean subtract material from unmodified reconstruction model, leaving just a thin plate structure representing the contours of the orbital reconstruction. This basic form was then joined to the duplicated version of modified orbital reconstruction model for further refinement. Fixation tabs that extended onto the surface of the orbital rims were created in the same way as the cranioplasty, with the superior tab designed to interlock with corresponding contours of the cranioplasty plate to provide a location reference. Given the large extent of the orbital reconstruction, it was decided to split the plate into two separate pieces by drawing the margin on the surface of the clay and using the "profile" tool to create a 1-mm groove.

With further smoothing and refinement, the design of the orbital and temporal implants was completed while still as layers on top of the "buck" reconstructed models. Once complete, the buck models were removed to leave just the implant designs, which were then refined to 0.15-mm clay sharpness to improve edge smoothness and enable further levels of detail to be added.

CAD representations of 1.5-mm Synthes Cortex screws were positioned to sit with the countersink protruding into the surface of the design. Due to the minimal plate thickness, a small portion of the screw countersink remained protruding above the outer surface. The plate thickness was therefore locally increased around the countersinks to avoid sharp edges that could cause soft tissue damage. Boolean subtraction was then used to subtract the screws from the implants.

To allow more efficient delivery of radiotherapy, reduce the bulk of the implant, and make them less prone to distortion during the proposed AM fabrication process, it was decided to perforate the plate implant designs. A diamond pattern was drawn on a flat plane orientated approximately parallel with the plate surfaces and duplicated in a pattern. An unperforated plate margin was left, with diamond holes lying outside the margin deleted. The "extrude, cut" tool was then used to create the perforations.

A final clay refinement operation was undertaken to smooth the implant designs, which can be seen in Fig. 5.58.

The patient had combined craniofacial resection and reconstruction with neurosurgeons, maxillofacial surgeons, and an ophthalmologist.

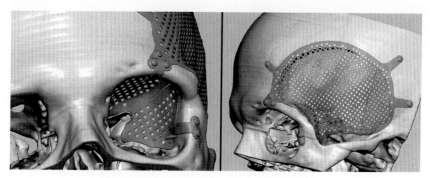

Figure 5.58 The completed implant designs.

5.9.4.2 Results and conclusions from case 3

The case incorporated lessons learned from previous experience to provide a successful outcome. The designed-in flexibility of the temporal implant reduced the reliance on the planned tumor excision, but it still enabled accurate aesthetic reconstruction of the anatomy. The interlocking features of the temporal and orbital roof plates gave extra confidence that they were located correctly. A number of areas for improvement were also noted. The tumor excision was less extensive than planned; therefore, it was only necessary to utilize the orbital roof part of the two-part orbital reconstruction plates. The orbital roof implant was also reduced in size during the operation by cutting with a bone cutter, which while possible, was not an easy task given the strength of the material. Fig. 5.59 shows a postoperative X-ray looking upward at the orbital implant, highlighting the area of reduction.

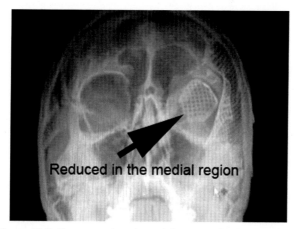

Figure 5.59 Postoperative view of the revised orbital implant.

Based on feedback from the ophthalmic surgeon, it was suggested that it is unnecessary to extend further back than the equator of the orbital globe when reconstructing the roof and lateral wall walls of the orbit. Subsequent implants of a similar nature have incorporated the feedback from the case illustrated.

5.9.5 Case 4—Posttraumatic zygomatic osteotomy and orbital floor reconstruction

As the previous case studies have illustrated, the use of computer-aided planning, CAD, and AM can help to provide a more predicable surgical outcome. Previous application of CAD/AM to zygomatic osteotomy has been limited to translating a virtual plan into theater using a polymer repositioning guide and a single moved bone piece [17]. Chapter 5.10: Surgical applications case study 7—Computer-aided planning and additive manufacture for complex, midface osteotomies notes the challenging nature of zygomatic osteotomies, especially when compounded by bone remodeling around the fracture sites. This clinical case demonstrates an evolution of the techniques described in that case study by employing anatomic models, digital reconstruction, virtual planning, and CAD/AM end-use devices: a cutting guide, repositioning guide, custom zygomatic implant, and custom orbital floor implant.

The patient's primary concern was their appearance and improving vision in their left eye. The surgical goal was to improve facial symmetry by osteotomizing bones that had been fractured, displaced, and had set in the incorrect position. A further goal was to improve the patient's vision by reconstructing the orbital floor and therefore restoring eye position during the same procedure.

5.9.5.1 Materials and methods

The complexity of this case meant that extensive, multidisciplinary planning was involved, including the development of multiple concept solutions. In the interest of brevity, only the chosen approach is described.

5.9.5.1.1 Stage 1: Data segmentation

The patient underwent a CT scan with a 0.5 mm slice thickness using a Toshiba Aquilion Multislice scanner. The DICOM format data was imported into Mimics (version 15), and the same segmentation techniques as previously described were used to create a 3D CAD model of the bony

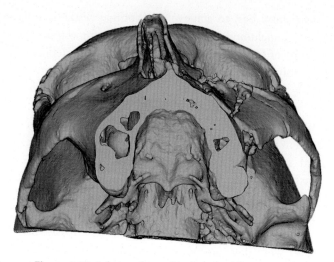

Figure 5.60 Inferior views showing the injury extents.

anatomy. Additionally, a separate "mask" of the bony anatomy was created, and the existing metal reconstruction plates from the primary reconstruction were erased manually on each CT slice. This was necessary since the surgical plan involved removing existing reconstruction plates to ensure the proposed guides and reconstructive implants would fit correctly. Fig. 5.60 shows the extent of the bone displacement.

Once completed, the original anatomy with metal removed "mask" was exported in the STL file format. This file was fabricated using stereo-lithography (SLA 250-50, 3D-Systems) to aid as a reference model to communicate the procedure to the surgical team and patient and to assist in visualizing the plan.

5.9.5.1.2 *Stage 2: Surgical planning and device design*
FreeForm Modeling Plus (version 13, 3D Systems) provided the basis for collaborative surgical planning and device design sessions between surgeon, the assisting prosthetist, and design engineer. The STL file of the bony anatomy with existing reconstruction plates removed was imported as a "buck" reference model. A portion of the patient's right facial anatomy covering the entire orbit and parts of the zygomatic, maxilla temporal bones was selected, copied, and pasted as a new "clay" piece. This was mirrored around the midsagittal plane and positioned using the surgeon's best judgment to create a target symmetrical reference.

Options for how best to achieve the optimal aesthetic outcome were then discussed, concluding that it was necessary for osteotomies of three separate sections of bone. The locations for cuts were defined by drawing representing polyline curves directly onto the anatomy. These curves would provide the basis for modeling precise grooves onto the cutting guide design. The "layer" tool was used to create a 2.5 mm thickness of "clay" on top of the anatomy over the areas required for surgical cuts. The curves were then used to create precise grooves and ridges corresponding to the proposed saw blade thickness (with 0.3 mm clearance) and angles required. Further refinement of the guide design was then undertaken using sculpting and refinement tools. Once the design was agreed, the "buck" anatomy upon which the "clay" design had been layered was subtracted, leaving a precise-fitting surface. CAD representations of standard Synthes 1.5-mm screws were then positioned at key points required to locate the guide securely. These were initially left as a visual reference to check the location in relation to the other planned positioning guide and implant screws. Fig. 5.61 shows the design.

For osteotomy planning, a duplicate of the original anatomy was made and transformed into a modifiable "clay" piece. The three separate bone sections were then selected using the "select with ball" tool and cut and pasted as new movable pieces. The pieces were then repositioned based on the surgeon's preference using the target symmetrical reference as a guide. Areas where the repositioned bone pieces overlapped the static anatomy were noted to highlight where trimming would be required. With the new positions agreed, the individual pieces were set as unmodifiable "buck"

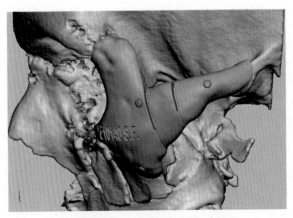

Figure 5.61 The cutting guide design.

parts to protect their condition for the remainder of the process. During the process, each bone fragment was assigned a bright, contrasting color and number to assist in communication between clinical and engineering specialities. When a model of the planned reposition was built, a color Z–Corp 510 (3D Systems, Rock Hill, USA) machine was selected for its ability to maintain the virtual coloring on the physical model. Similarly, a model was built with the separated bone fragments as loose, individual pieces to facilitate procedure rehearsal (Fig. 5.62).

To translate the planned osteotomy into the operating theater, a positioning guide, similar in concept, but larger than those described in Chapter 5.10: Surgical applications case study 7—Computer-aided planning and additive manufacture for complex, midface osteotomies was proposed. It was also agreed that this would also provide an appropriate mechanism to accurately locate custom implants that would remain once the guide had been removed, which dictated a complex design with locating features. Two implants were required: one to fix the zygomatic complex in the correct position and the second to reconstruct the orbital floor based on the new bone positions. The goal was to ensure the implants were as low profile as possible while providing the necessary fixation. FreeForm was used to model a 0.7-mm-thick and 5-mm-wide plate that spanned from just anterior of the auditory meatus to the zygomaticofrontal suture. Orbital floor reconstruction and implant design was undertaken using techniques described and adapted from case 2. CAD representations of 1.5-mm Synthes Cortex screws were located along the zygomatic implant and the locating tabs of the orbital floor implant. The thickness of the implants was increased locally around the screws to increase strength and maintain a

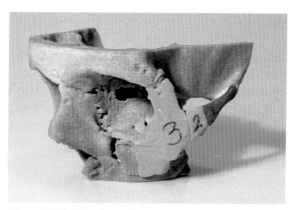

Figure 5.62 The planned osteotomy illustrated in color.

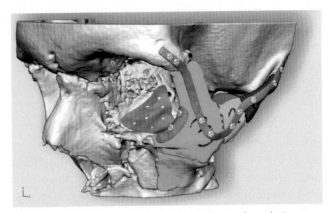

Figure 5.63 The orbital and zygomatic implant designs.

smooth contour around the countersunk screw heads. The completed implant and guide designs are shown in Fig. 5.63.

The positioning guide was agreed as a crucial component in ensuring that the bone pieces were located correctly and that the implants were fixed correctly. It needed to provide a secure method of engaging the fixed portion of anatomy in two locations: at the posterior border of the zygomatic arch and on the supraorbital rim. The positioning guide was designed using the "layer" and multiple sculpting tools in FreeForm, and the implants profiles were used to Boolean subtract material from inside the guide shape (taking care to include clearance and remove undercuts that would prevent them slotting in). CAD representations of 1.5-mm Synthes Cortex screws were located on the positioning guide. These were required to hold the guide firmly in place while the osteotomy pieces were located and the implants fixed. The design is shown in Fig. 5.64.

5.9.5.1.3 Stage 3: Device fabrication

The implant designs were signed-off by the prescribing surgeon for fabrication. The agreed STL files were exported and sent to external ISO13485 accredited manufacturers. The zygomatic implant was built using the selective laser melting process (LayerWise, Leuven, Belgium). A basic support-removal and bead-blasted finish was requested. Grade 23 Ti6Al4 ELI was specified as the build material on the basis of its biocompatibility and favorable mechanical properties. The orbital floor implant (0.5 mm nominal thickness) was built using the same process and with the same finish on the inferior surface. The superior surface was

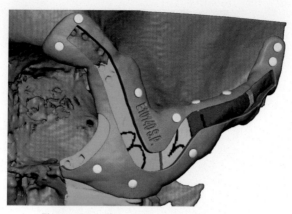

Figure 5.64 The positioning guide design.

Figure 5.65 The fabricated implants and guides.

polished. Postreaming of the fixation holes was undertaken as a precautionary measure in the hospital lab upon delivery. The guides were built using the laser sintering process (Renishaw, Wotton-under-Edge, UK). A dental grade cobalt—chrome was specified as the build material owing to its compatibility for in vivo use as a transient device, high stiffness, and a lower cost relative to titanium. The implants and guides are shown in Fig. 5.65.

5.9.5.1.4 Stage 4: Surgery
The parts were checked and sterilized prior to surgery. In surgery the positioning guides and implants fitted as intended, enabling the accurate reproduction of the surgical planning. One of the surgical guides and its associated implant is shown in Fig. 5.66.

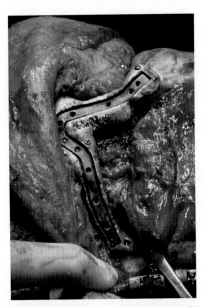

Figure 5.66 The positioning guide and implants in position during surgery.

5.9.5.2 Results and conclusions from case 4

The design process combined with the use of custom, AM-produced guides and implants resulted in an excellent clinical outcome. Both of the guides fulfilled their functional requirements. The cutting guide fitted onto the existing anatomy precisely and securely. Each bone segment was osteotomized in the location planned virtually prior to the surgery. Extra in-theater clarification and discussion were required to properly define the appropriate saw angle for the medial cut at the edge of the guide. A cutting ledge of greater thickness than the main body of the guide could mitigate this delay in future cases.

The repositioning guide fitted securely, accommodating each bone segment accurately in its preplanned position. During the process of fitting the smaller segments into the repositioning guide, however, the locations of the pieces anteriorly had to be cross-referenced with the planning imagery; they retained some freedom to move even with the guide in place. This could not be solved mechanically without endangering the guide's ability to interface with the pieces without undercut complications. A solution for future cases may be to indicate the intended end point of each bone piece using embossed markings on the surface of the guide itself.

Each implant was positioned successfully and fixed without complication. However, despite the successful clinical outcome, the design process was highly involved (required numerous interactions between the clinical and design engineering team), and it required multiple concept iterations, and therefore, it could be argued as being cost inefficient. Although time was not a significant factor (since it was an elective procedure), and this was a particularly complex case, it helped to highlight the need for process efficiency improvements that would help to make the use of computer-aided planning, design, and AM techniques more economically viable.

References

[1] Mazzoli A, Germani M, Raffaeli R. Direct fabrication through electron beam melting technology of custom cranial implants designed in a PHANToM-based haptic environment. Materials & Design 2009;30:3186−92.
[2] Salmi M, Tuomi J, Paloheimo KS, Björkstrand R, Paloheimo M, Salo J, et al. Patient-specific reconstruction with 3D modeling and DMLS additive manufacturing. Rapid Prototyping Journal 2012;18(3):209−14.
[3] Murr LE, Gaytan SM, Martinez E, Medina F, Wicker RB. Next generation orthopaedic implants by additive manufacturing using electron beam melting. International Journal of Biomaterials 2012;2012:245727. http://www.hindawi.com/journals/ijbm/2012/245727/abs/.
[4] Wehmöller M, Warnke PH, Zilian C, Eufinger H. Implant design and production: a new approach by selective laser melting. In: Lemke HU, Inamura K, Doi K, Vannier MW, Farman AG, editors. CARS 2005: computer assisted radiology and surgery, vol 1281; 2005. p. 690−5. ISBN: 0-444-51872-X.
[5] Weihe S, Wehmöller M, Schliephake H, Hassfeld S, Tschakaloff A, Raczkowsky J, et al. Synthesis of CAD/CAM, robotics and biomaterial implant fabrication: single-step reconstruction in computer-aided frontotemporal bone resection. International Journal of Oral and Maxillofacial Surgery 2000;29(5):384−8.
[6] Chua CK, Leong KF, Lim CS. Rapid prototyping: principles and applications. 3rd ed. WSPC; 2010, ISBN 978-9812778987.
[7] Gibson I, Rosen DW, Stucker B. Additive manufacturing technologies: Rapid Prototyping to Direct digital manufacturing. Springer; 2009, ISBN 978-1441911193.
[8] Noorani RI. Rapid Prototyping: principles and applications. John Wiley & Sons; 2005, ISBN 978-0471730019.
[9] Hopkinson N, Hague R, Dickens P, editors. Rapid manufacturing: an industrial revolution for a digital age: an industrial revolution for the digital age. Wiley Blackwell; 2005, ISBN 978-0470016138.
[10] Bibb R, Eggbeer D, Evans P. Rapid prototyping technologies in soft tissue facial prosthetics: current state of the art. Rapid Prototyping Journal 2010;16(2):130−7.
[11] Bibb R, Eggbeer D, Evans P, Bocca A, Sugar AW. Rapid manufacture of custom fitting surgical guides. Rapid Prototyping Journal 2009;15(5):346−54.
[12] Bibb R, Eggbeer D, Williams R. Rapid manufacture of removable partial denture frameworks. Rapid Prototyping Journal 2006;12(2):95−9.
[13] Eggbeer D, Evans P, Bibb R. A pilot study in the application of texture relief for digitally designed facial prostheses. Proceedings of the Institute of Mechanical Engineers Part H: Journal of Engineering in Medicine 2006;220(6):705−14.
[14] Eggbeer D, Bibb R, Williams R. The computer aided design and rapid prototyping of removable partial denture frameworks. Proceedings of the Institute of Mechanical Engineers Part H: Journal of Engineering in Medicine 2005;219(3):195−202.

[15] Williams RJ, Bibb R, Eggbeer D, Collis J. Use of CAD/CAM technology to fabricate a removable partial denture framework. The Journal of Prosthetic Dentistry 2006;96(2):96—9.

[16] EOS Materials data sheet—EOS Titanium Ti64 and EOS Titanium Ti64 ELI, EOSINT M 270 Systems (Titanium Version), EOS GmbH—Electro Optical Systems, Robert-Stirling-Ring 1, D-82152 Krailling, München, Germany.

[17] Herlin C, Koppe M, Béziat JL, Gleizal A. Rapid prototyping in craniofacial surgery: using a positioning guide after zygomatic osteotomy—a case report. Journal of Cranio-Maxillofacial Surgery 2011;39(5):376—9.

CHAPTER 5.10

Surgical application case study 7—Computer-aided planning and additive manufacture for complex, mid-face osteotomies[*]

5.10.1 Introduction

The concept of using additive manufacturing (AM) to transfer a computer-aided surgical plan to the operating theater has been discussed in the literature [1–9]. Since these earlier case studies, the application of advanced computer-aided design (CAD) and fabrication technologies has grown to encompass more complex surgeries, including the cutting and repositioning of bones (osteotomy) in multiple vectors. The accurate osteotomy of bone in multiple vectors presents a significant challenge, especially in the mid-face region, where small movements can have a large impact on the surgical result. The type of surgery described in this case study would not be possible to any sufficient degree of accuracy without the use of computer-aided techniques. The application of CAD techniques and RP to enable complex, multiaxis osteotomies of the mid-face has not been previously reported.

To date, three surgical guides have been designed and produced as described here and subsequently used in theater. These were for very similar cases but with each involving slightly different, case-specific challenges. All patients had blunt trauma to the face, which had caused the bones to fracture and become displaced posteriorly and laterally. The patients all had

[*] The processes described in this chapter were undertaken in collaboration with Mr. Adrian Farrow and Steven Hutchison at Raigmore Hospital, Inverness, Scotland

Medical Modeling
ISBN 978-0-323-95733-5
https://doi.org/10.1016/B978-0-323-95733-5.15001-X

facial asymmetry and problems with their vision and therefore required surgery to restore symmetry and function.

The cases reported here describe the use of stereolithography (SL) as an AM method of producing a patient-specific, transient use guide device. Since the studies reported in Chapter 5.5 Surgical Application Case Study 2—Rapid Manufacture of Custom Fitting Surgical Guides, where metal guides were used, further work has been undertaken to demonstrate the ability of the chosen SL material to withstand autoclave sterilization and measure the baseline bio burden of the material having gone through the cleaning and sterilization process. The SL material chosen, ClearVue (3D-Systems, USA), had also undergone prerequisite testing by the manufacturer to USP Class 23/6, which demonstrated low toxicity and tissue reaction. The resin also exhibits a high degree of dimensional stability and accuracy in the presence of moisture. SL was also deemed appropriate since the guides produced spanned a relatively small area, was not used to drill or cut along, and could be made sufficiently robust through material thickness.

5.10.2 Methods

Over-the-Internet screen sharing was used to undertake the plan in real time with the surgeon and prosthetist at the hospital available to guide the process, which was driven by a design engineer. The approach will be illustrated through one of the cases.

5.10.2.1 Step 1: 3D CT scanning and virtual model creation

The patients were scanned using computed tomography (CT) to capture the facial bony anatomy (see Chapter 2, Section 2.2). Techniques described in Chapter 3 were then used to produce a 3D rendering of the bony anatomy and export the data as an STL file. The extent of the asymmetry and displacement of the zygomatic arch can be seen in Fig. 5.67.

5.10.2.2 Step 2: Computer-aided surgical planning

The STL files were then imported into the CAD software, FreeForm (Geomagic, 3D Systems, USA) using the "buck" setting and duplicated. FreeForm was selected for its capability in the design of complex, arbitrary but well-defined shapes that are required when designing custom appliances and devices that must fit human anatomy. FreeForm is described in Chapter 3 Section 3.6.6. In order to create a symmetry-based target position for the

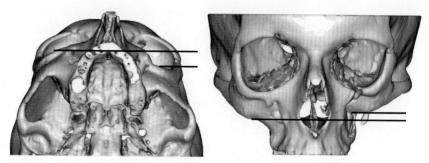

Figure 5.67 The 3D reconstruction of the existing bony anatomy, highlighting the relative displacement of the left zygomatic bone.

osteotomized bone, a portion of the patient's right, unaffected side was selected using the "select clay with box" tool, then mirrored across the mid sagittal (midline) plane to the left side. Manual re-positioning of the mirrored piece was used to create the ideal target contour, to which the repositioned bones should match. The zygomaticofrontal suture was used as a key point of rotation in positioning the mirrored section since it was around this point that the fracture was hinged. The mirror was agreed by the prescribing surgeon, and it is shown in Fig. 5.68.

In each of the cases, it was deemed unnecessary to use a cutting guide to free the zygomatic bone, but that it was important to plan the approximate cutting locations in order to ensure the subsequent guide design would fit correctly. In the illustrated case, four primary cuts were required: two on the infraorbital rim to free a small section of rim and the anterior portion of

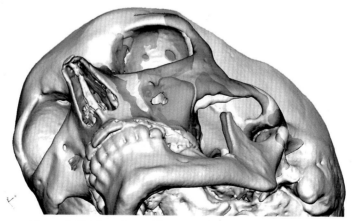

Figure 5.68 The mirrored bone overlaid on the defect area.

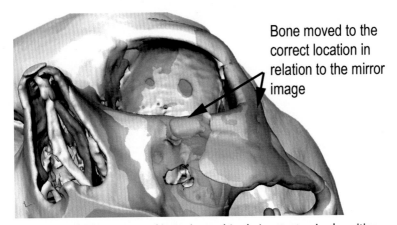

Bone moved to the correct location in relation to the mirror image

Figure 5.69 The sections of bone located in their osteotomized position.

the zygomatic bone; a cut along the zygomaticofrontal suture; and a cut on the posterior portions of the zygomatic arch. These cuts were undertaken on the duplicate model in order to preserve a version of the original anatomy and were created using the "select with ball" tool, cutting the selected areas, and pasting them as new pieces of clay. The two freed sections of bone for osteotomizing are shown in Fig. 5.69. This also left a version of the bony anatomy with the cut sections removed. The free sections of bone were then positioned by eye, using the superficial area of the zygomaticofrontal suture as a point of rotation and by making minor movements. The final bone positions that were agreed by the surgeon can be seen in Fig. 5.69.

The digitally osteotomized bone pieces were then joined into the model of the cut away anatomy, thereby creating the intended surgical result, onto which a guide could be designed. The model was selected and converted to "buck" to protect it from further modification by material removal.

A small section of the reconstructed left skull was then exported as an STL file for fabrication using SL. The SL model provided a physical jig, around which mini plates that would provide permanent fixation of the bone sections were bent in the prosthetics laboratory before being cleaned and sterilized before surgery.

5.10.2.3 Step 3: Design of the surgical positioning guide

The aim of the guide was to provide contours that would firmly locate using screws on the fixed areas of bone and enable the osteotomy pieces to

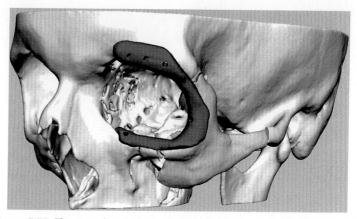

Figure 5.70 The completed guide design in computed-aided design (CAD).

locate and be fixed according to the planned position. It was agreed that the device needed to avoid being fixed where prebent plates would provide permanent fixation, low enough profile to minimize surgical incisions, but still provide adequate overlapping contours to locate on the fixed bone and engage the osteotomy pieces. The basic shape was created using the "layer" tool, then refined using various carving and shaping tools. Holes that would enable temporary fixation of the device using standard screws were also created. The completed design shown in Fig. 5.70 was validated by the prescribing surgeon before being exported as an STL file ready for SL fabrication.

5.10.2.4 Step 4: Guide additive manufacture

The guide STL file was imported into build preparation software (Magics Version 16, Materialise N.V.) and orientated to reduce the number of supports on the fitting surfaces and minimize the effects of stair stepping (as described in Chapter 4) on the hole features. Proprietary software (Lightyear V1.1, 3D-Systems, USA) was used to set SL build parameters and prepare the guide for fabrication in 0.1 mm layers using an SLA 250−50 machine (3D-Systems). ClearVue resin (3D-Systems) was used due to it having undergone testing to demonstrate suitable levels of biocompatibility. In the illustrated case, two guides were produced, one as a lab reference and the other for use in surgery. Clean up and postprocessing involved removing support structures, wiping off excess resin from the guide surface, scrubbing then soaking in fresh isopropanol, drying, and postcuring in a UV oven. The SL-produced guide is shown in Fig. 5.71.

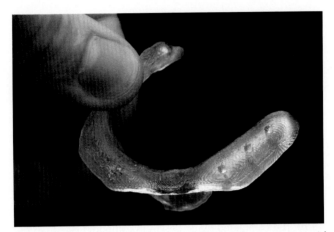

Figure 5.71 The stereolithography (SL)-produced osteotomy guide.

5.10.3 Results

The surgical guide used to illustrate the process (and in all cases where they have been used) was assessed by the clinicians before going to surgery and were deemed satisfactory for use. Surgery took approximately 1 hour (see Figs. 5.72 and 5.73).

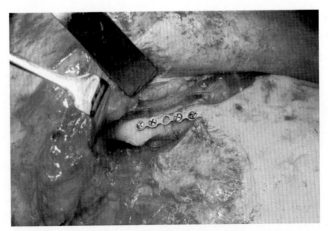

Figure 5.72 The guide in situ in theater.

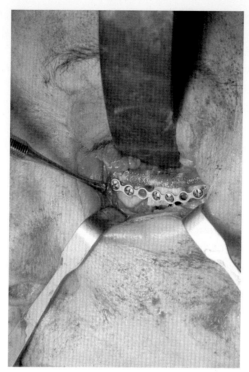

Figure 5.73 The guide being fitted to the patient during surgery.

5.10.4 Discussion

The application of surgical guides described here all aided in the successful transfer of computer-aided planning to the theater. Although improvements in accuracy and efficiency cannot be quantified, postsurgery analysis of each case provided the opportunity to evaluate the approach, understand the limitations and inform the development of revised design and manufacture guidelines. The first case (used to illustrate the process) demonstrated the value of a collaborative approach to osteotomy guide design. The profile was sufficiently small to enable easy insertion, yet it enabled the guide to be located securely to the fixed sections of anatomy and the osteotomy pieces to locate with minimal play (which would have been manifested as a rocking motion around the guide).

Potential improvements were also noted by the surgical team. The value of using a physical guide to accurately reproduce the planned vectors and locations of the bone cuts in theater would have been useful to ensure accuracy of the subsequent osteotomy. This would be of increasing value in more complex cases. Directional arrows or other features that ensured the surgeon could easily identify anatomical landmarks in relation to the guide would also provide increased confidence that it was correctly located.

5.10.5 Conclusions

The approach illustrated represents a clinically viable and appropriate method of undertaking complex, mid-face osteotomies. While the case illustrated did not require a cutting guide, similar techniques to those discussed in Chapter 5.5 Surgical Application Case Study 2—Rapid Manufacture of Custom Fitting Surgical Guides, could be incorporated if cut precision was more critical. The use of SLA polymer was deemed appropriate in the cases undertaken to date, but the use of metal could assist where a guides need to span a large area with sufficient stiffness or provide a harder edge to cut against.

References

[1] Poukens J, Verdonck H, de Cubber J. Stereolithographic surgical guides versus navigation assisted placement of extra-oral implants (oral presentation). In: 2nd international conference on advanced digital technology in head and neck reconstruction, Banff, Canada; 2005. p. 61. Abstract.

[2] Goffin J, Van Brussel K, Vander Sloten J, Van Audekercke R, Smet MH, Marchal G, et al. 3D-CT based, personalized drill guide for posterior transarticular screw fixation at C1-C2: technical note. Neuro-Orthopedics 1999;25(1—2):47—56.

[3] Goffin J, Van Brussel K, Vander Sloten J, Van Audekercke R, Smet MH. Three-dimensional computed tomography-based, personalized drill guide for posterior cervical stabilization at C1-C2. Spine 2001;26(12):1343—7.

[4] SurgiGuides - company information. Leuven, Belgium: Materialise N.V.

[5] Bibb R, Eggbeer D, Bocca A, Evans P, Sugar A. Design and manufacture of drilling guides for osseointegrated implants using Rapid prototyping techniques. In: Proceedings of the 4th national conference on Rapid and virtual prototyping and applications. London, UK: Professional Engineering Publishing; 2003, ISBN 1-86058-411-X. p. 3—12.

[6] Bibb R, Bocca A, Sugar A, Evans P. Planning osseointegrated implant sites using computer aided design and rapid prototyping. The Journal of Maxillofacial Prosthetics and Technology 2003;6(1):1—4.

[7] Bibb R, Eggbeer D. Rapid manufacture of custom fitting surgical guides. Rapid Prototyping Journal 2009;15(5):346—54.

[8] Sarment DP, Sukovic P, Clinthorne N. Accuracy of implant placement with a stereolithographic surgical guide. The International Journal of Oral & Maxillofacial Implants 2003;18(4):571—7.

[9] Sarment DP, Al-Shammari K, Kazor CE. Stereolithographic surgical templates for placement of dental implants in complex cases. The International Journal of Periodontics and Restorative Dentistry 2003;23(3):287—95.

CHAPTER 5.11

Surgical applications case study 8—Virtual surgical planning and development of a patient-specific resection guide for treatment of distal radius osteosarcoma[*]

5.11.1 Introduction

The necessity to derive solutions for the enhancement and streamlining of surgical procedures is a vitally important goal toward achieving improved patient outcomes. From the surgeon's perspective, such solutions would comprise devices to assist in the surgical resection process to minimize the removal of healthy bone from a patient, alongside the use of patient-tailored, custom solutions to address an underlying condition (e.g., bone replacement implants). In response to these demands and coinciding with innovations in the fields of material science, advance design, and additive manufacturing, devices are now emerging that address these criteria through the so-called patient-specific paradigm. Using such technologies, many innovative medical device innovations have been demonstrated with

---

[*] The work described in this chapter was first reported in the references below and is reproduced here with permission of the organising committee of the Solid Freeform Fabrication Symposium. This research was conducted by the Australian Research Council Industrial Transformation Training Centre in Additive Biomanufacturing (IC160100026). All resources for the study were provided courtesy of the School of Engineering at Deakin University.

Mohammed MI, Ridgway MG, Gibson I. Development of virtual surgical planning models and a patient specific surgical resection guide for treatment of a distal radius osteosarcoma using medical 3D modeling and additive manufacturing process. 2017 International Solid Freeform Fabrication Symposium. University of Texas at Austin; 2017.

Medical Modeling
ISBN 978-0-323-95733-5
https://doi.org/10.1016/B978-0-323-95733-5.15002-1

applications ranging from medical modeling [1—3], implantable devices [4,5], prosthetics [6,7], and orthotics [8,9].

With respect to surgical interventions, patient-specific technologies are a rapidly emerging area in preclinical planning, bone resection, and implantable devices with demonstrated applications in osteotomies [10], cancer resection [11,12], and bone replacement [4,5,13], among other procedures. Osteosarcoma surgeries (removal of bone tumors) can be a somewhat involved process requiring the use of chemotherapeutic action [14,15], alongside the resection of the offending bone tissue [15,16] in an attempt to completely remove the compromised tissues and halt any form of metastatic progression. Once performed, the residual uncompromised bone also requires fixation into its original orientation using either donor bone tissue or a load-bearing implant alongside a fixation plate in order for the patient to regain normal use of the anatomy in question. Given the necessity to remove sections of corrupt bone from a patient, such procedures could significantly benefit from the use of patient-specific technologies to streamline resection and piloting of drill holes for the fixation plate. Additionally, given the unique topography of a patient's anatomy, it would be desirable to move beyond the typical, generically sized fixation plates, to ones that more accurately mirror the patient's bone topography, thereby realizing a better fit and improving the ability to retain the original orientation of the resected bone. Indeed, there is significant potential for the fixation plate to be used in conjunction with a custom resection guide as part of a more comprehensive treatment option.

In this study, we aim to develop a platform by which advanced design and 3D printing concepts can be used as a methodology to examine complex osteosarcoma in a simulated patient case study. Custom resection technology for such procedures has been demonstrated previously [11,17] but has been relatively crude in its design and could significantly benefit from further innovations. Initially, we replicate a patient case study where the individual presented with a periosteal osteosarcoma of the radius, and we replicate the tumor morphology on a publicly available patient representative model, derived from CT scans. The advantage of this approach is that we can simulate the particular case study in a platform that is free from typical ethical constraints when using patient-specific data yet still be reflective of a working example for examining the planning of device development. We believe our demonstration to be the first example of such an approach. Using the devised patient representative model, we develop a custom resection guide that integrates functionality to avoid collision with the osteosarcoma, makes use of natural surface contours to provide optimal placement, and integrates resection/drill/pin openings and a handle to easily maneuver the guide. We believe our guide to be the most sophisticated example of a patient-specific guide found in the literature to address surgical intervention of distal radius osteosarcoma resection. The guide fundamentally realizes all features required to perform all surgical intervention elements prior to fixation plate and implant/bone donor bone placement. To take the work a stage further, we also develop a patient-specific fixation plate and mock scaffold implant intended for use with the guide, providing a complete system for surgical intervention. In the present study, we focus our attention on the design and product prototyping elements to demonstrate the potential feasibility of our approach. Ultimately, our methodology provides a relatively low cost means of examining the planning of the device development under simulated circumstances, whereby the anatomic features of a patient are reproduced accurately, at low cost and with no impact on the patient's wellbeing. We believe there to be significant potential for the devised devices to be 3D printed in biocompatible material for use in actual clinical circumstances.

5.11.2 Methods

5.11.2.1 Anatomic data modeling

Models of the patient data were constructed using anatomic data sets derived from CT scan data, based on the standardized Digital Imaging and Communications in Medicine (DICOM) file type with an image slice thickness of approximately 0.6 mm. To convert the data into a working 3D digital model, Mimics 19.0 (Materialise, Belgium) was used to isolate and extract the bone data based on a Hounsfield unit (HU) threshold of 226−3071. Within Mimics bone models for the hand, wrist, ulna, and radius were segmented from the wider data set and exported as STL files for further postprocessing.

5.11.2.2 Cancer design

The segmented bone STL models were exported into 3-Matic Research 11.0 (Materialise, Belgium) for additional postprocessing to remove errors and major defects from the Mimics conversion process. To avoid ethical issues relating to the use of clinical patient data, we appealed to a published case study of a periosteal osteosarcoma of the radius [18], which we digitally

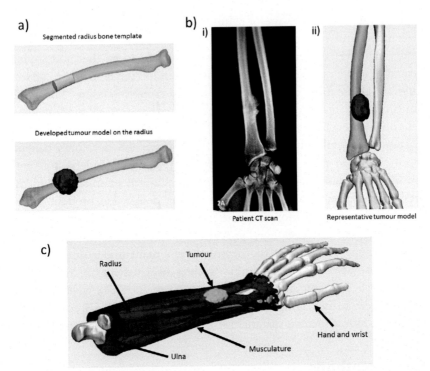

Figure 5.74 (a) Illustration of the surface osteosarcoma model formation process, (b) a comparison of the CT scan image of the actual patient case study [18] and the digital representative model of the same case study, and (c) a concept view of the osteosarcoma on the patient.

reproduced as our example case for design of the resection guide (Fig. 5.74). In this patient case, the tumor was defined as being attached to the surface and wrapped around the bone. CT scan images revealed the outer geometry of the tumor from several reference viewpoints. The written and visual evidence of the osteosarcoma in the patient study provided sufficient information to approximate the geometry of the tumor and to form our hypothetical digital tumor model for examination.

5.11.2.3 Resection guide design

The combined model of the patient-specific anatomy of the arm and the tumor model were used as the template for the resection guide design, which was developed using 3-Matic. To construct the primary features of the resection guide, we manually designed elements in CAD software (Solidworks, Dassault Systèmes, France). These features were then integrated onto a surface projection of the patient bone model, adjacent to the tumor site. Using this approach, the patient-specific bone projection segments would allow for the ideal placement of the guide and self-alignment using the natural contours of the bone as a locator guide. The use of traditional CAD allows for the precise design of the bulk portions of the resection guide, such as the cutting slots, drill holes, Kirschner wire fixation ports, and an ergonomic handle for maneuvering/placement.

When designing the resection guide, consideration needs to be made as to the thickness of the part to ensure sufficient rigidity, but without enlarging the device to a point that results in complications during placement. Rigidity is important to ensure the device does not break during use and that flex is minimized to retain geometric accuracy. Keeping the device footprint to a minimum is important to improve the patient's recovery, and to avoid complications resulting from infection, general practice is to minimize the area exposed during a surgical intervention. Therefore, the smaller the device, the smaller is the area that needs to potentially be exposed during surgery. As it is unknown a priori what the ideal geometric attributes of the device should be, we examined various design iterations until a satisfactory candidate design was achieved, which was then manufactured for virtual surgery.

5.11.2.4 Implant design

In this study, we aimed to construct a concept of a patient-specific bone fixation plate that could be used in conjunction with the resection guide. To achieve this, we built a plate based on the typical design of a standard distal radius surgical plate [19]. The plate is designed in 3-matic based on the surface topography of the mock patient's bone model and around the planned drill holes on the resection guide. Using this methodology, we hoped to demonstrate a precise fit of the final bone fixation plate and alignment of the planned pilot holes. Additionally, as the system has been designed around the original positioning of the bone, once the plate is in position, we anticipate the remaining bone segments to return to their original position. For simplicity in the initial design, screw holes were designed to be nonthreaded and with an upper tapering section to allow the screws to sit flush with the mock plate.

Alongside the fixation plate, we design a conceptual scaffold-type implant to be used in place of donor bone material. The scaffolds would potentially be representative of titanium implant that could allow for osseointegration or some form of bioprinted device. Currently, our intent for the implant is purely cosmetic and for illustrative purposes and is not intended as a developed treatment component. In future work, we aim to develop this component further.

5.11.2.5 Prototype manufacturing

This study focused on the design and surgical device developments facet. As they will not be used for direct clinical use, the various design iterations of the anatomic features, the resection guide, and implant were realized using fused filament fabrication (FFF) printing in commercially available ABS plastic. This material provides a good representation of the final medical devices, alongside the patient's representative anatomy. Evaluation prints were performed using a commercially available desktop printer (Flash Forge Dreamer, USA), which allow for a rapid turnaround time for part production. It is anticipated that the final resection guide will likely be printed using either FFF or PolyJet printing methodology in biocompatible materials (PC–ISO, MED 610, etc.). However, for evaluation purposes, the current ABS models would be comparable to parts made in such materials.

5.11.3 Results

5.11.3.1 Cancer anatomic modeling

The cancerous model was constructed by directly manipulating the patient-specific radius bone model to shape the tumor. Initially, a template portion of the distal region of the radius in the approximate location of the tumor was segmented from the wider radius model of the patient and enlarged in the x-y-z direction by a scale factor of 2, yielding an enlargement of $\Delta X = 25.6$ mm, $\Delta Y = 12.5$ mm, and $\Delta Z = 32.0$ mm. The resulting section was hollowed performing a Boolean subtraction of the original radius model and trimmed to match the length of the template segmented region. The model was then segmented into two equal halves along the coronal plane, before being manually extruded to approximate the size of the tumor in all spatial dimensions. As we were not privy to the original patient's medical imaging scans, we based the overall shape of the tumor model based on the relative sizes of the X-ray images of the presented case study [18] and by appeal to the typical gross specimen of a resected periosteal osteosarcoma [20]. Fig. 5.74a illustrates the template segment approach used to form the tumor, alongside the final model on the radius bone. Fig. 5.74b shows a direct comparison of the patient CT scan and the developed 3D model. Despite the differences in the patient anatomy, the general shape and location of the osteosarcoma have been accurately replicated and would be representative of an actual patient case. Fig. 5.74c finally shows a representative model of the bone, musculature, and tumor as it is likely to be observed.

5.11.3.2 Resection guide development

Resection of the compromised sections of bone can require the removal of between 6 and 20 mm of "normal" soft tissue, to ensure the complete removal of a tumor, thereby avoiding recurrence [21]. In the process of limb salvage surgical strategies, it is important to minimize the removal of healthy bone tissue, to allow the best platform for functionality recovery. However, a defined consensus has yet to be determined as to the minimal amount of healthy bone removal to ensure local nonrecurrence of tumors. This lack of consensus has implications upon the design choices of the given resection guide design, and for this study, we optimized our design around a 10-mm tolerance, which would be a good compromise between ensuring tumor removal while minimizing the removal of healthy bone.

For the present case study, the resection guide will need to be designed to allow for easy fit to the patient's bone but with sufficient clearances to avoid issues with placement over the surface osteosarcoma, which protrudes outward from the bone. We therefore examined an approach of using a surface projection of the bone surface, which will form the basis of the surface locator portion of the resection guide. Fig. 5.75a illustrates the design process by which the template bone surface data are constructed and manipulated to form the guide. The projection process comprises the initial isolation of a template region of bone, which is approximately 45—50 mm from each extremity of the tumor. This region is then enlarged in all directions by 3 mm, before a Boolean subtraction of the original bone structure, to create a hollow model that encompasses the whole topography of the template bone section. This is then digitally trimmed, firstly to open the hollow model at each extremity before being divided into two halves at the upper and lower section of the model. The lower section is then discarded, and a segment approximately 4—5 mm either side of the tumor extremities is removed, resulting in the two halves of the resection guide surface locator model. The model is then trimmed at the distal region of the model to avoid placement issues at the interface with the wrist. Developing

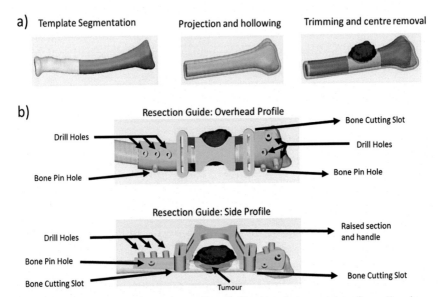

a) Template Segmentation   Projection and hollowing   Trimming and centre removal

b)

Resection Guide: Overhead Profile

Bone Cutting Slot

Drill Holes

Drill Holes

Bone Pin Hole

Bone Pin Hole

Resection Guide: Side Profile

Raised section and handle

Drill Holes

Bone Pin Hole

Bone Cutting Slot

Bone Cutting Slot

Tumour

Figure 5.75 (a) Illustration of patient-specific bone surface template formation (purple) and (b) annotated diagrams of the final resection guide concept.

the guide to use the patient's natural bone surface topography over the selected regions of the radius introduces natural markers for the positioning of the final resection guide, which is based on both the thickness of the bone at each half of the projected bone surface and the vertical contours of the distal region of the bone. Therefore, the guide would only have one ideal placement orientation whereby there was no movement in the guide once in position. This methodology ensures the geometric accuracy of the pinning, drilling, and resection features of the guide. It is also envisaged that to a minor degree, the protrusion of the tumor would provide further insight as to the placement of the guide based on the designed raised section of the guide.

Following projection of the patient's bone surfaces, a classical CAD approach was used to construct the drill/Kirschner wires holes and the resection slots. To aid with the design parameters, we appealed to standard instruments that would be used in an orthopedic setting and that would be suitable given the size of the radius bone. The first constraint of the resection slot was to be placed 10 mm away from each side on the tumor. The maximum width of the radius was digitally measured to be approximately 23 mm; therefore, we opted to design slots suitable for a standard

20-mm sagittal blade, with a blade thickness of 1.27 mm (Stryker, USA). The slots comprised a rounded rectangular configuration, with a total internal length of 28 mm and internal width of 1.8 mm, as can be seen in Fig. 5.75. This geometry was selected to ensure that there were sufficient clearances to encompass the full width of the radius bone, while also allowing enough space for the excursion width of the blade (22 mm). The slots were also placed to be orthogonal to the bone at their location and made with a thickness of 4 mm. To complete the structural components, a bridge section over the osteosarcoma was developed, which raised the guide a height of 25 mm above the height of the cutting slots. The bridge section also integrated two arcs, for finger handling of the guide by a user. The bridge section was initially made to a thickness of 3 mm; however, it was found to have considerable flex, to a point where aggressive handling led to breakages. We therefore opted to increase the thickness to 6 mm, which considerably increased the rigidity of the final model, to a point where flex was completely eliminated. In future development, we hope to characterize the tensile strength of the model and for now focus on the design for additive manufacturing elements of the guide.

From direct discussion with orthopedic surgeons, it was discovered that for a typical radius, the fixation plate ideally requires a minimum of three fixation points at either side of the resected bone to provide adequate support. In light of this, we incorporated this as part of our design constraints for both the resection guide and implant plate. For a typical surgical procedure on the distal radius, 2.4-mm surgical locking screws can be used for fixation [10]. We therefore designed the drill holes to have an internal diameter of 2.5 mm and outer diameter of 5.5 mm, which allowed for easy access of a standard surgical drill guide and the use of a 1.8-mm drill bit to form a pilot hole for the screws. The requirement of designing the drill holes to be used with a standard metallic surgical drill guide as opposed to simply using the 3D printed guide eliminates any complications that may occur due to the drill damaging the guide and releasing debris in the surrounding tissue. The drill holes were located long the center line of the radius, relative to the resection slots and at a regular spacing of approximately 5 mm from the outer diameters of either the resection slots or drill holes. This spacing allowed for adequate clearances so that the screws would not intersect each other when placed in the bone. To confirm this, a "digital screw" was created and used to guide the placement of the drill holes. Finally, Kirschner wires holes were created at either extremity of the guide to ensure the guide could be temporarily pinned in position during resection. Kirschner wires (also known simply as K wires) can be used as an effective methodology to hold resection templates in place during surgery [10,12] and are an important feature to retain the geometric accuracy of the guide during use. Wire holes with an inner diameter of 1.5 mm and outer diameter of 4.5 mm were designed based on standard Kirschner wire sizes and were angled such that when the pins are in position, the tension acts to lock the guide in place. The overall final guide concept can be seen in Fig. 5.75b.

5.11.3.3 Fixation plate and implant concept

To complement the patient-specific guide, a concept implant was developed to both replace the resected bone and to provide support to the segmented bone sections. As with the resection guide, these components were designed to be patient specific and match the original anatomic features of the patient. Fig. 5.76a illustrates the design process for construction of the fixation plate. Initially, using the bone projection that was used to form the base of the guide, the basic form of the plate is realized. As the resection guide had already incorporated the drill holes in the design, we could use the center point of the holes as a template for the screw holes on the plate. As the final design would likely be manufactured using Ti-64, it is important to reduce the overall mass of the part to a minimum. Therefore, the template systematically underwent topology optimization to reduce the overall mass of the plate to realize the final design, as seen in Fig. 5.76a.

The implant to replace the resected bone will comprise an open scaffold structure. We have previously reported the rapid development of scaffold structures that can be applied in titanium implantable devices for light weighting [4]. However, in this study, the scaffold design is primarily for conceptualization purposes, but potentially, it could be applied to next-generation bioprinted scaffolds or a titanium implant as the technologies mature. Briefly, the process for scaffold formation involves the segmentation of the bone template, before applying a fixed geometry unit cell structure throughout the digital model, as described previously [22]. In this instance, we opted for a Gyroid lattice unit cell structure, with a unit cell size of $3 \times 3 \times 3$ mm, and the final scaffold structure can be seen in Fig. 5.76b. Finally, a virtual representation of the implant, fixation plate, and patient anatomy was constructed to better visualize the parts in situ and can be seen in Fig. 5.76c.

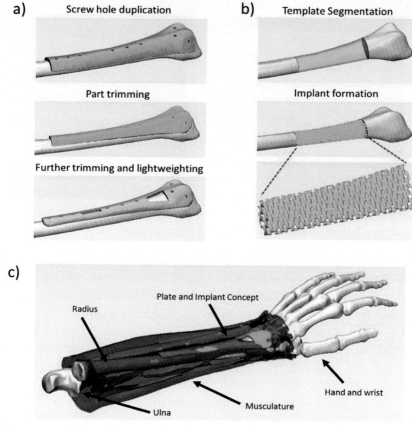

Figure 5.76 (a) Design process for fixation plate development, (b) design process for implant scaffold development, and (c) final digital patient model with fixation plate and implant in place.

5.11.3.4 Mock surgery

To qualitatively test the efficacy of the devised resection guide, the implant, and fixation plate, a mock surgery was performed, whereby the resection guide was pinned onto a 3D printed model of the patient's arm and osteosarcoma, as can be seen in Fig. 5.77(a)(i). As we did not have access to a sagittal saw, we opted to mark the region of the cut using a thin blade, and the model was cut using a circulating saw. Several repeat cuts were performed, and the variance in the cutting position from the midpoint of the cutting guide varied by approximately ±0.6 mm. In addition to the cuts, a standard 1.8-mm drill bit was used to drill pilot holes in the model for attachment of the screws into the fixation plate concept. It was found that

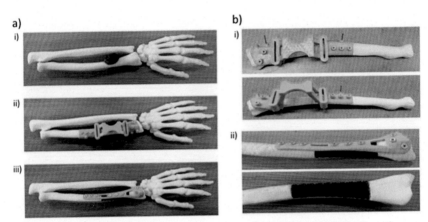

Figure 5.77 (a) 3D printed hand, wrist, ulna, and radius models with (i) surface osteosarcoma (black), (ii) the osteosarcoma and resection guide concept in situ, and (iii) the resected radius with fixation plate and scaffold implant. (b) Close-up views of the radius with (i) the resection guide pinned in place and (ii) the fixation plate and scaffold concept.

the drill holes were achieved to an accuracy of ±0.3 mm of the intended center point. Once all mock resection and drill piloting had been performed, the pins and guide were removed, and the implant and plate were screwed into position. The entire procedure from start to finish took approximately 2–3 min to perform, reflecting the potential time it would take the surgeon for this portion of the procedure within theater.

Fig. 5.77a shows the stages of the model, before and after mock resection. It was found that the implant fixes accurately into the bone model, along with the placement of the screwed fixation plate. Equally, the resection guide had sufficient rigidity to allow for aggressive handling when simulating the placement of Kirschner wires, marking cuts, and drilling of the pilot holes. It was noted that there was some ability for the screws to deviate from their desired directional placement, which we believe is due to the constraints of the model being fabricated in ABS plastic as opposed to real bone. It is believed that if the procedure is performed on actual bone that the rigidity of the bone would allow for greater directional conformity of fixation screws. Overall, we are confident that there is considerable potential of our approach for development as a clinical treatment option, should the resection guide be manufactured in biocompatible materials and the implant/fixation plate in Ti64, which is the current benchmark for titanium implants.

5.11.4 Conclusion

In this study, we present our initial findings in the development of a resection guide and implantable treatment concepts, which could serve as methodologies for the treatment of a surface osteosarcoma on the distal radius. The guide was found to accurately match the contours of the mock patient's bone anatomy and provided a high degree of efficacy to minimize the removal of potential healthy bone tissues, with resection tolerances being approximately ±0.6 mm from the desired location. We also demonstrate a patient-specific fixation plate and mock implant, which conform very well to the bone topography, providing efficacious fixation and reproduction of the patient's original bone anatomy. It is our hope that the methodology outlined in this preliminary study would provide guidance to professionals seeking to develop a robust design methodology for resection guide and fixation plate development and that could be applied to a range of different clinical case studies.

References

[1] Salmi M, et al. Accuracy of medical models made by additive manufacturing (rapid manufacturing). Journal of Cranio-Maxillofacial Surgery 2013;41(7):603–9.
[2] Mitsouras D, et al. Medical 3D printing for the radiologist. RadioGraphics 2015;35(7):1965–88.
[3] Cohen A, et al. Mandibular reconstruction using stereolithographic 3-dimensional printing modeling technology. Oral Surgery, Oral Medicine, Oral Pathology, Oral Radiology & Endodontics 2009;108(5):661–6.
[4] Mohammed MI, Fitzpatrick AP, Gibson I. Customised design of a patient specific 3D printed whole mandible implant. KnE Engineering 2017;2(2):104–11.
[5] Peel S, Eggbeer D. Additively manufactured maxillofacial implants and guides—achieving routine use. Rapid Prototyping Journal 2016;22(1):189–99.
[6] Bibb R, Eggbeer D, Evans P. Rapid prototyping technologies in soft tissue facial prosthetics: current state of the art. Rapid Prototyping Journal 2010;16(2):130–7.
[7] Mohammed MI, et al. Advanced auricular prosthesis development by 3D modelling and multi-material printing. KnE Engineering 2017;2(2):37–43.
[8] Fitzpatrick A, Mohammed MI, Collins PK, Gibson I. Design of a patient specific, 3D printed arm cast. KnE Engineering 2017;2(2):135–42.
[9] Kelly S, Paterson A, Bibb RJ. A review of wrist splint designs for Additive Manufacture. 2015.
[10] Dobbe JGG, et al. Patient-specific distal radius locking plate for fixation and accurate 3D positioning in corrective osteotomy. Strategies in Trauma and Limb Reconstruction 2014;9(3):179–83.
[11] Ma L, et al. 3D-printed guiding templates for improved osteosarcoma resection. Scientific Reports 2016;6:23335.
[12] Bellanova L, Paul L, Docquier P-L. Surgical guides (patient-specific instruments) for pediatric tibial bone sarcoma resection and allograft reconstruction. Sarcoma 2013;2013:7.

[13] Arabnejad S, et al. Fully porous 3D printed titanium femoral stem to reduce stress-shielding following total hip arthroplasty. Journal of Orthopaedic Research 2016:n/a.

[14] Errani C, et al. Palliative therapy for osteosarcoma. Expert Review of Anticancer Therapy 2011;11(2):217—27.

[15] Luetke A, et al. Osteosarcoma treatment — where do we stand? A state of the art review. Cancer Treatment Reviews 2014;40(4):523—32.

[16] Ta HT, et al. Osteosarcoma treatment: state of the art. Cancer and Metastasis Reviews 2009;28(1):247—63.

[17] Victor J, Premanathan A. Virtual 3D planning and patient specific surgical guides for osteotomies around the knee. A Feasibility and Proof-Of-Concept Study 2013;95-B(11 Suppl. A):153—8.

[18] Abdelwahab IF, et al. Dedifferentiated periosteal osteosarcoma of the radius. Skeletal Radiology 1997;26(4):242—5.

[19] Stryker. The new comprehensive stryker VariAx distal radius locking plate system: operative technique. Stryker; 2009.

[20] Nouri H, et al. Surface osteosarcoma: clinical features and therapeutic implications. Journal of Bone Oncology 2015;4(4):115—23.

[21] Oryan A, Alidadi S, Moshiri A. Osteosarcoma: current concepts, challenges and future directions. Current Orthopaedic Practice 2015;26(2):181—98.

[22] Mohammed MI, Badwal PS, Gibson I. Design and fabrication considerations for three dimensional scaffold structures. KnE Engineering 2017;2(2):120—6.

PART 5.3

Prosthetic applications

CHAPTER 5.12

Prosthetic rehabilitation applications case study 1—An investigation of three-dimensional scanning of human body surfaces and its use in the design and manufacture of prostheses*

5.12.1 Introduction

Three-dimensional surface scanning, or reverse engineering, has been used in industry for many years as a method of integrating the surfaces of complex forms with computer-generated design data [1]. Noncontact scanners operate by using light and camera technology to capture the exact position in space of points on the surface of objects. Computer software is then used to create surfaces from these points. These surfaces can then be analyzed in their own right or integrated with computer-aided design (CAD) models. The general principles of noncontact surface scanning are described more fully in Chapter 2. This section includes a description of the

---

* The work described in this chapter was first reported in the reference below and is reproduced here in, part or in full, with the permission of the Council of the Institute of Mechanical Engineers.
Bibb R, Freeman P, Brown R, Sugar A, Evans P, Bocca A. An investigation of three-dimensional scanning of human body parts and its use in the design and manufacture of prostheses. Proceeding of the Institute of Mechanical Engineers Part H: Journal of Engineering in Medicine 2000;214(H6):589—94.

Medical Modeling
ISBN 978-0-323-95733-5
https://doi.org/10.1016/B978-0-323-95733-5.15003-3

potential difficulties that may be encountered when employing the technique and includes suggested methods to overcome them.

The scanner used in the work described in this section was a structured white light system that uses a projected fringe pattern of white light and digital camera technology to capture approximately 140,000 points on the surface of an object (Steinbichler USA, 40,000 Grand River, Suite 101, Novi, MI 48375). Scanners using this type of Moiré fringe pattern have been used in the past in the assessment of spinal deformity [2]. In this case, the area to be scanned is distinguished from its surroundings by altering the contrast. For example, a white object may be placed on a dark background and vice versa.

Due to the high accuracy of this type of scanner, nominally accurate to within 0.05 mm, movement of the object was avoided during scanning. Even small movements result in noise and affect the quality of the data captured. In this case, the patient must remain motionless for approximately 40 seconds. Other systems that have been investigated for use in measuring and recording changes in a patient's topography employ multiple cameras and a fast capture time to eliminate the problems associated with motion. However, a smaller number of data points are captured at a slightly lower accuracy [3]. Other systems based on scanning have been used to manufacture custom orthotics for podiatric patients [4,5]. At the time of this work, the application of captured surface data in the manufacture of facial prostheses had not been fully investigated.

While prosthetic rehabilitation of the human face offers many potential applications that could exploit this technology, scanning human faces presents particular problems. A primary difficulty is presented by the presence of hair. Hair does not form a coherent surface, and the scanner will not pick up data from areas such as the eyebrows and lashes. This problem can be overcome to a certain degree by dusting a fine white powder over the hair. When considering the scanning of faces, the area around the eyes may also be particularly difficult. As described above, movement leads to the capture of inaccurate data, so to minimize problems caused by blinking during the scan, it is more comfortable for the subject to keep their eyes closed. This also alleviates discomfort caused by the bright light emitted by the scanner. If the eye is held open during the scan, watering of the eye may cause problems. In addition, the surface of the eyeball is highly reflective, making data capture difficult.

Line of sight issues are also encountered when scanning faces. For example, a single scan will not acquire data where the nose casts a shadow.

However, this is overcome by taking several overlapping scans (as illustrated in Chapter 2).

This case study describes how these issues were approached by an investigation into the scanning of human faces and describes the application of these techniques in the manufacture of a facial prosthesis to restore the appearance of a patient recovering from the excision of a rare form of tumor called an olfactory neuroblastoma. Removal of this tumor had necessitated the surgical removal of the patient's left eye.

5.12.2 Methods

5.12.2.1 Preliminary trial of facial scanning

As a preliminary investigation of the practicality of scanning human faces, a male subject was scanned using the system described above. Initial attempts at scanning the face of the seated subject were poor due to slight involuntary movement of the head despite being seated in a comfortable position. Therefore, additional support was fashioned from a block of polystyrene foam to locate the back of the head and minimize movement. With the subject thus supported in a semireclining position, a series of three scans were taken, each from a different viewpoint. Scans were taken from the patient's left side, from directly in front, and from the right side. A fourth scan was taken from a central position below the second scan to allow the acquisition of data from the area below the eyebrow ridge. Each scan took approximately 40 seconds, during which the subject remained motionless. The whole process of arranging the subject and taking these four scans took approximately 10 minutes.

Once completed, the scans were aligned using the proprietary scanner software. To achieve this alignment, four notable points or landmarks were manually selected in an overlapping area in each of two separate scans. The software then aligns the landmarks and calculates the best fit between the two data sets. Consequent scans were aligned in a similar fashion.

In this experiment, four scans resulted in the capture of accurate data describing the whole face with the exception of areas obscured by hair, such as eyebrows, eyelashes, and facial hair. Other areas lacking data not immediately obvious in the figure include areas beyond the line of sight, such as the nostrils. It was therefore concluded that the approach could be applied to the scanning of patients provided they could be kept still during the scan.

5.12.2.2 Scanning a surgical subject

In this case, the subject was a patient recovering from reconstructive craniofacial surgery. The surgery to excise the tumor necessitated the removal of bone and soft tissue including the left eye. After successful operations to replace the orbital rim, an osseointegrated (bone anchored) prosthetic was planned for the missing eye and surrounding tissue [6]. To aid in the construction of this prosthesis, the right (unaffected) side of the patient was scanned, and the data were used to create a laterally inverted ("mirrored") model that would be used as a guide when creating the prosthesis.

Four scans were taken of the patient's face with the chin supported on a polystyrene block to minimize movement. As the data were intended to aid in the construction of a prosthetic of an open eye, a scan of the open eye was attempted. The scans were taken from angles similar to those described previously; however, care was taken to ensure that the bright light from the scanner did not shine directly into the patient's eye. As before, data were not captured from areas obscured by the eyebrows and lashes. Small un-avoidable movements of the eye and eyelids and the reflective nature of the surface of the eye itself affected the accuracy of the captured data. However, as this inaccuracy was extremely small, it did not affect the overall quality of the data. The resulting scan data is shown in Fig. 5.78. The whole process of scanning the patient took approximately 10 minutes.

The next step required the building of a model of the area around the unaffected eye. To further aid the creation of the prosthesis, the data would be laterally inverted ("mirrored" left to right) before building the model. This model could then be used to guide the production of a prosthetic with good size, fit, and aesthetic symmetry.

To create a model from the scan data, it was translated into an STL file format [7]. This is a triangular, faceted surface normally used in rapid prototyping systems. The STL file format is more fully described in Chapter 3. However, before the data could be used to produce a model, gaps in the data, such as the area at the eyebrows, were filled. This was achieved by using surface creation software to create a patch that continues the shape of the captured data surface. The patch was created to follow the natural curves of the surrounding data and replicate the surface as best as possible. This required a certain amount of judgment on the part of the operator. However, this case did not present great difficulty in this respect, as the missing areas were relatively small.

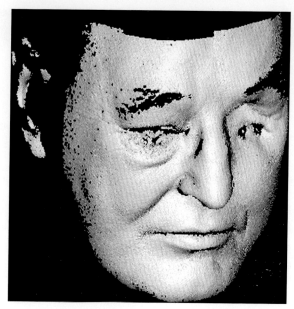

Figure 5.78 The aligned scan data.

The file size was reduced at this stage by removing unnecessary points. This was achieved without sacrificing accuracy because there were a vast number of points in the captured data coupled with the fact that the accuracy of the scan data is greater than can be achieved by subsequent rapid prototyping processes. An offset surface was created from the captured data and the gap between them closed to create a finite bound volume. To minimize file size and model cost, only the specific area of interest was selected. The resulting data were stored as an STL file. The STL file size was reduced to 3.6 megabytes. This was then laterally inverted ("mirrored"), as shown in Fig. 5.79. In this case, the process of creating a valid STL file from the scan data took approximately 3 hours, but this will vary from case to case.

The model in this case was produced using laminated object manufacturing (LOM) [8]. The model is shown in Fig. 5.80. The scan data and STL file were archived in case it is necessary to reproduce the model or for future reference.

5.12.2.3 Prosthesis manufacture

This case involved the manufacture of an osseointegrated implant-retained silicone prosthetic of the left eye (see the explanatory note in Chapter 7 for

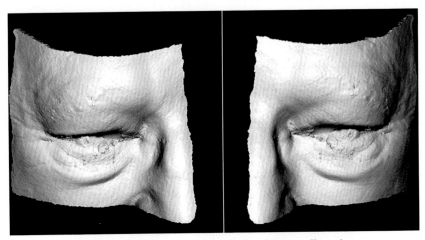

Figure 5.79 STL file and "mirrored" file of the unaffected eye.

Figure 5.80 An LOM model of the mirrored unaffected eye.

an explanation of osseointegrated implants). Previously, two titanium fixtures had been attached to the zygomatic bone (cheekbone) and allowed to integrate to the bone. Six months later, they were exposed, and percutaneous titanium abutments were attached to them (this means that the abutments passed through the skin to form an anchor for the subsequent prosthesis). The eventual prosthesis would attach to these abutments with magnets. To aid the construction of the prosthesis, the LOM model was used to cast a wax replica. The prosthetist then removed excess material from the wax until it approximated the required shape. The traditional procedure would have required the prosthetist to carve this piece from wax. This would have taken the prosthetist approximately half a day in this particular case, during which the patient would have been required to sit with the prosthetist for visual reference. The use of the model, therefore, not only saved approximately half a day of work for the prosthetist (from a total of three) but, importantly, also reduced the time required for the patient to attend the clinic.

The aperture for the eye was opened to allow the positioning of the artificial eye. This is considered crucial to the overall success of the prosthesis [6]. It was noted that, compared with the traditional methods, the mirrored nature of the model allowed far greater accuracy when locating the artificial eye, especially concerning anterior—posterior positioning. Once the eye position was fixed, the fine details were built up in wax. The areas immediately around the eye were dealt with in particular, as this is where the original scan data, and therefore the LOM model, had lost some detail.

An impression was taken from the patient and used to shape the rear surface of the prosthesis. A small acrylic base plate that would hold the magnets used to locate the prosthesis was also made. When the prosthetist was satisfied with the visual appearance of the fine details and the fit of the prosthesis, it was cast in color-matched silicone in the usual manner.

5.12.3 Results

Accuracy

The accuracy of the scan data is nominally within 0.05 mm. From the data, the theoretical height of the model was 76.76 mm. LOM models are nominally accurate to within 0.2 mm. However, when measured, the height of the completed LOM model was found to measure 76.7 mm. Therefore, the accuracy of the model can be estimated in the order of ± 0.1 mm. As all human faces are somewhat asymmetric and the surface of the skin is pliable, the wax replica was manually manipulated and adjusted to fit the desired area. Therefore, an accuracy of around 0.1 mm is more than adequate for facial prosthesis manufacture. The model also proved to be a good match in terms of reproducing a realistic visual appearance for the prosthesis.

Outcome analysis

The success of this experiment proves the feasibility of three-dimensional scanning of human body surfaces. The ease and relative speed of the scanning allow the complex forms of human features to be permanently captured without hindrance or discomfort to the patient. The accuracy of the data was found to be more than adequate for prosthesis construction. This method would compare favorably with the current practice of taking impressions, proving to be quicker, more accurate, and aiding the reproduction of a realistic visual appearance. In particular, the use of "mirrored" medical models was felt to be of great help to the prosthetist when positioning artificial eyes in orbital prostheses.

The cost of the scanning equipment is considerable, and it may be difficult for hospitals to justify the initial investment. For this reason, it may prove more feasible for hospitals to use external service providers for the cases where the approach is expected to produce superior results. The cost of the scanning described in this paper would probably amount to several hundred pounds with the LOM model costing approximately £120. The costs incurred by this approach should be balanced against the improved results and crucially the time saved over traditional methods. The reduction in time taken allows more patients to be treated, reducing waiting lists (a major goal of the British National Health Service).

The noncontact nature of the scanning means there is less discomfort for the patient and no distortion of soft tissues caused by the pressure applied when taking impressions. This advantage in combination with the ability to "mirror" data may have many applications in rehabilitation. It is difficult, for example, to take a satisfactory impression of a breast; therefore a similar technique may be used in the creation of symmetrical prostheses for mastectomy patients. From the results of this case study, it can be concluded that 3D scanning and medical modeling can save a significant amount of time for both the patient and the prosthetist. Lateral inversion and high accuracy can be a significant aid in prosthesis manufacture, especially for large or complex cases. These techniques may be a valuable aid to shaping and positioning the prosthesis, but the skill and knowledge of the clinicians will determine the best method of creating, color matching, and attaching the prosthesis to the patient.

Update

Scanning and processing software has developed considerably since this paper was first published, greatly speeding up the process. However, the fundamental approach and advantages identified remain true, and many of the issues encountered with line of sight, hair, and involuntary movement are still present and still need to be carefully considered when scanning patients. LOM has become largely redundant despite its advantages for applications such as this. Any other additive manufacture (AM) processes could be used to produce acceptable models from this method.

References

[1] Motavalli S. Review of reverse engineering approaches. Computers & Industrial Engineering 1998;35(1—2):25—8.

[2] Wong HK, Balasubramaniam P, Rajan U, Chang SY. Direct spinal curvature digitization in scoliosis screening: a comparative study with Moiré contourography. Journal of Spinal Disorders 1997;10(3):185—92.

[3] Tricorder Technology plc. Tricorder ships new measurement software. In: December 8th Press Release, Tricorder Technology Ltd. The Long Room, Copperhill Lock, Summerhouse Lane, Harefield, Middlesex, UB9 6JA, UK; 1998.

[4] Bergman JN. The Bergman foot scanner for automated orthotic fabrication. Clinics in Podiatry Medicine and Surgery (Treatment biomechanical assessment using computers) 1993;10(3):363—75.

[5] Bao HP, Soundar P, Yang T. Integrated approach to design and manufacture of shoe lasts for orthopaedic use: reverse engineering in industry: research issues and applications. Computers & Industrial Engineering 1994;26(2):411—21.

[6] Branemark P, De Oliveira MF, editors. Craniofacial prostheses, anaplastology and osseointegration. Carol Stream, IL, USA: Quintessence Publishing; 1997. p. 101—10.

[7] Manners CR. STL file format. Valencia, California, USA: 3D Systems Inc.; 1993.

[8] Jacobs PF. Stereolithography and other RP&M Technologies. Dearborn, MI, USA: Society of Manufacturing Engineering; 1996.

CHAPTER 5.13

Prosthetic rehabilitation applications case study 2—Producing burns therapy conformers using noncontact scanning and additive manufacturing[*]

5.13.1 Introduction

This case study describes the use of three-dimensional (3D) noncontact scanning, computer-aided design (CAD) software and rapid prototyping (RP) techniques in the production of burn therapy masks, also known as conformers. Such masks are used in the management of hypertrophic scars on the face resulting from burn injuries (see the explanatory note in Chapter 7).

Two case studies were undertaken where noncontact laser scanning techniques were used to capture accurate data of burn patients' faces. The surface data were then manipulated using two different CAD techniques to achieve a reduction in prominence of the scarring. This reduction in height of the scarring on the vacuum-forming mold results in a conforming

---

[*] The work described in this chapter was first reported in the reference below and is reproduced here, in part or in full, with the permission of First Numerics Ltd.
Bibb R, Bocca A, Hartles F. Producing burns therapy conformers using noncontact scanning and rapid prototyping. Proceedings of the sixth international symposium on computer methods in biomechanics and biomedical engineering, Madrid, Spain, February 2004, ISBN: 0-9549670-0-3 (Published on CD-ROM by First Numerics Ltd. Cardiff, UK).

Medical Modeling
ISBN 978-0-323-95733-5
https://doi.org/10.1016/B978-0-323-95733-5.00018-1

facemask that fits the face precisely while applying localized pressure to the scars. This pressure on the scars produces the beneficial effect from such masks. Once manipulated to achieve this effect, the data were then used to create vacuum-forming molds via a selection of RP methods.

The effectiveness of the CAD techniques and RP processes for this application is evaluated. The case studies illustrate the benefits of the approach in comparison to traditional practices while indicating operational and technical difficulties that may be encountered. Finally, the cost effectiveness, patient benefits, and opportunities for further research are discussed.

Closely fitting masks have been shown to provide a beneficial effect on the reduction of scarring resulting from burns, particularly to the face and neck [1–4]. These masks are typically vacuum formed from the strong, clear plastic material, polyethylene terephthalate glycol (PETG). Traditionally, the vacuum-forming mold is made from a plaster cast of the patient, which itself is made from an alginate impression. Taking a facial impression is uncomfortable, time consuming for the patient, and may be particularly disturbing following the physical and psychological trauma of burns.

Published work has indicated that optical scanning and computer-aided manufacturing techniques can be used for various clinical applications [5–7] including the fabrication of burn masks [8–10]. The potential benefit of this approach is the noncontact nature of the data capture, which has been shown to be more accurate, quicker, more comfortable, and less distressing for burn patients compared with the traditional impression. The aim of this research was to explore the practical implications of employing such an approach to the treatment of facial burns and to assess various methods of adapting and physically reproducing the data to create a vacuum-forming mold.

5.13.2 Methods

3D surface scanning has been used in industry for many years to integrate surfaces of objects with computer-generated designs. Noncontact scanners operate by using structured light or lasers and digital camera technology to capture the exact position in space of a large number of points on the surface of objects. Computer software is then used to create surfaces based on these points. These surfaces can then be analyzed or integrated with

CAD models. The general principles of noncontact surface scanning are described more fully in Chapter 2.

The optical scanner used in this work uses a laser and digital camera technology to capture the surface of an object (Vivid 900, Konica Minolta Photo Imaging UK Ltd., Milton Keynes, UK). This scanner was selected because the specifications suggested that the accuracy, resolution, and range of capture were more than adequate for capturing the human face. It also benefited from ready availability, manufacturer after sales support, comparatively low price, and compact size compared with other systems that have been reported, which have been specialized and expensive or locally made prototypes [9,10].

Although the acquisition time for this type of scanner is only a fraction of a second, movement would still lead to inaccuracy in the captured data. Therefore, the patients remained motionless in a comfortable position during the acquisition. All light-based scanners are limited by line of sight during each acquisition, and this is typically overcome by taking several overlapping scans (as illustrated in Chapter 2). However, in these cases, a pair of scanners was used to capture both sides of the patient's face. The scanners were positioned low down to ensure data were captured from the areas under the chin and eyebrow ridge [9].

Once the data points are acquired, software is used to create surfaces based on them. The simplest method of creating a surface from point data is polygonization. Neighboring data points are joined together to form triangular facets, which form the computerized surface model.

5.13.2.1 Case 1

This patient was recovering from burns to the head, neck, and arms. Scans were taken as described above, taking approximately 5 min. Eyebrows, eyelashes, and blinking affected the accuracy of the data around the eyes, but as the mask is intended to avoid the eyes, this did not present problems.

The point data were then polygonized to create a triangular faceted surface model of the patient's face. This was created in the STL file format (the STL file format is more fully described in Chapter 3), which is commonly used in RP. However, RP requires STL files that represent a single fully enclosed volume. Therefore, before the data can be used to produce a physical model, gaps in the data have to be filled, and the surface needs to be given a thickness to produce a finite bound volume. This was achieved by using CAD software that extruded the perimeter of the of the

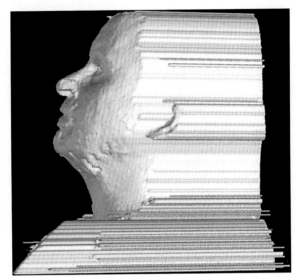

Figure 5.81 Extruded data to form a solid computer model.

captured data surface toward an arbitrary plane to create a solid model, as shown in Fig. 5.81 (FreeForm, 3D Systems, USA). The resulting STL file was thus created in less than 5 min.

In traditional practice, the plaster replica of the patient's face would be ground back in areas of scarring to produce localized pressure on scar sites while conforming comfortably to the rest of the face. For this research, the reduction in the height of the scarring was undertaken on the computer before producing the physical RP model. Although software has been reported that has been designed for this purpose, it is not widely available [9]. Therefore, this research specifically applied readily available software.

To achieve the desired affect, software that is commonly used to prepare STL files for RP was used (Magics, Materialise NV, Technologielaan 15, 3001 Leuven, Belgium). A "smoothing" function in this software averages out the STL surface. The effect is to reduce the height of raised features on the surface and produce a simpler, smoother surface. However, the affect is applied to the whole surface of the object.

The second approach utilized CAD software (FreeForm) that enables the user to conduct a virtual sculpture on 3D computer models using a touch–feedback stylus. The software tools mimic those of traditional handcrafting. This makes the software easy to learn and intuitive to use, particularly for prosthetists, and it has been successfully used by the authors in other maxillofacial laboratory applications.

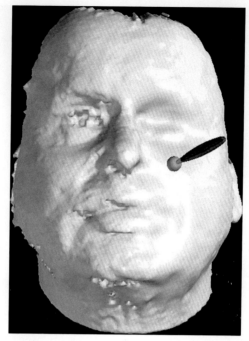

Figure 5.82 Smoothing the data (exaggerated for clarity).

Initially, a smoothing operation was carried out, which produced an effect similar to that obtained using the RP software. Secondly, the smoothing function was used only over areas selected by the user. Finally, a carving tool was used to carve away small areas of scarring locally. The effect is illustrated in an exaggerated manner in Fig. 5.82. In practice, a combination of these functions enabled the prosthetist to produce a surface that met the needs of the individual case rapidly.

The data was then cropped so the physical model would form a good vacuum-forming mold. Then laminated object manufacture (LOM) was used to produce the vacuum-forming mold shown in Fig. 5.83.

5.13.2.2 Case 2

The scan for the second case was carried out as before. In the interests of comparison, a mold was manufactured using stereolithography (SL) (3D Systems Inc.). SL materials and process time are more expensive than LOM; therefore, to reduce cost, the data were reduced to a thin shell, as shown in Fig. 5.84. As the glass transition temperature of SL materials may be

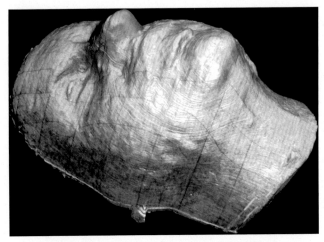

Figure 5.83 LOM vacuum-forming mold.

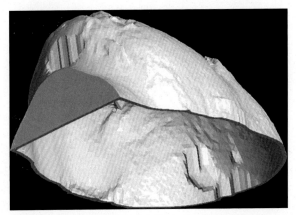

Figure 5.84 Shelled solid computer model.

exceeded during vacuum forming, the shell was filled with plaster, as shown in Fig. 5.85. This prevented the SL shell from distorting under the load and heat encountered in vacuum forming.

5.13.2.3 Experimental molds

The ThermoJet (3D Systems Inc.) RP process prints three–dimensional models using a wax material in a layered manner (the ThermoJet process is no longer available). The layers are very thin, leading to models with excellent surface finish. However, the low melting point of the wax

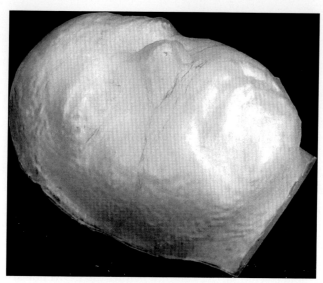

Figure 5.85 Plaster-filled SL mold.

precluded its use as a vacuum-forming mold. Instead, reversing the shelling operation used previously resulted in a negative pattern. The intention was to cast plaster a vacuum-forming mold from the wax pattern. However, in practice, the pattern was so fragile that it was destroyed during transport to the laboratory.

Other centers have successfully employed computer numerically controlled (CNC) machining in this and similar applications [8,9]. In comparison to RP processes, CNC is a viable option for this application as it is unlikely to encounter undercuts or reentrant features. To investigate the approach, a trial mold (case 1) was machined from a medium-density board, typically used in industrial model making. Unlike reported techniques that utilize soft foams for ease and speed of machining, this mold proved to be perfectly adequate for direct use as a vacuum-forming mold, requiring no surface treatment or modification [9].

5.13.2.4 Mask manufacture

The masks were manufactured in the usual manner by vacuum forming sheet PETG over the RP mold. It is common practice to drill holes through molds to provide even spread of the vacuum. However, holes were not drilled in these cases, and the masks were formed perfectly well without them.

5.13.3 Results

5.13.3.1 Treatment outcome

Both cases responded extremely well to treatment and showed considerable reduction in scarring. The patients' masks fitted well, performed as intended, and proved equally as effective as those produced by traditional methods.

5.13.3.2 Vacuum molding

The LOM model proved to be an excellent mold and was unaffected by vacuum forming. The plaster-filled SL mold also proved to be a satisfactory mold. Both types of mold can be worked with grinding tools should they need to be altered to produce subsequent masks as treatment progresses. However, it was noted that the transparent nature of the SL material combined with the white plaster infill made the surface features difficult to see clearly. The layered surface finish of neither the LOM nor SL mold was transferred to the vacuum-formed PETG mask.

5.13.3.3 Accuracy

The accuracy of the scan data is nominally 0.1 mm. As the skin is somewhat mobile and pliable, the accuracy of the captured data proved to be more than adequate. SL models are typically accurate to within 0.1 mm of the data from which they are built, and LOM models are nominally accurate to within 0.2 mm. These accuracies proved to be well within the requirements of this application. For this application, relatively inexpensive CNC machines are capable of producing satisfactory molds.

5.13.4 Discussion

The success of this study illustrates the efficacy of 3D scanning, CAD, and RP for this application. The accuracy of the scan data was found more than adequate. The noncontact nature of the scanning imposes less discomfort on patients and eliminates the potential risk of distortion of soft tissues, which may occur when taking impressions. The method therefore compares favorably with current practice. In addition, the reduction in anxiety or claustrophobia when taking facial impressions should be seen as an advantage.

Although this research deliberately utilized readily available, manufacturer supported equipment and software, the cost of the scanners, associated software, and RP equipment remains considerable, and it may be difficult for hospitals to justify the investment. As with all treatments, the costs must be balanced against the benefits.

5.13.5 Conclusions

As reported by other researchers, the principal benefits observed in this research are increased comfort and speed of the process compared with the traditional impression [9,10]. The scans took no more than a few minutes of the patient's time and presented no discomfort or distress. The timesaving also benefits the prosthetist.

The vacuum-forming molds produced by LOM and CNC require no special treatment. The production of the molds from the scan data presented no inconvenience to the prosthetist, and they could be delivered in a matter of days in most circumstances.

5.13.5.1 Update

Although the LOM process is now obsolete, the methods described here are still entirely useable with other techniques including backfilled shelled 3D printed molds and CNC milling.

References

[1] Rivers EA, Strate RG, Solem LD. The transparent facemask. American Journal of Occupational Therapy 1979;33:108.
[2] Shons AR, Rivers EA, Solem LD. A rigid transparent face mask for control of scar hypertrophy. Annals of Plastic Surgery 1981;6:245—8.
[3] Powell EM, Haylock C, Clarke JA. A Semi-rigid transparent facemask in the treatment of post burn hypertrophic scars. British Journal of Plastic Surgery 1985;38:561—6.
[4] Leach VSE. A method of producing a tissue compressive impression for use in fabricating Conformers in hypertrophic scar management. The Journal of Maxillofacial Prosthetics and Technology 2002;5:21—3.
[5] Sanghera B, Amis A, McGurk M. Preliminary study of potential for rapid prototype and surface scanned radiotherapy facemask production technique. Journal of Medical Engineering & Technology 2002;26:16—21.
[6] Bibb R, Brown R. The application of computer aided product development techniques in medical modelling. Biomedical Sciences Instrumentation 2000;36:319—24.
[7] Bibb R, Freeman P, Brown R, Sugar A, Evans P, Bocca A. An investigation of three-dimensional scanning of human body surfaces and its use in the design and manufacture

of prostheses. Proceedings of the Institution of Mechanical Engineers - Part H: Journal of Engineering in Medicine 2000;214(6):589—94.

[8] Whitestone JJ, Richard RL, Slemker TC, Ause-Ellias KL. Fabrication of total-contact burn masks by use of human body topography and computer-aided design and manufacturing. Journal of Burn Care & Rehabilitation 1995;16:543.

[9] Rogers B, Chapman T, Rettele J, Gatica J, Darm T, Beebe M, et al. Computerized manufacturing of transparent facemasks for the treatment of facial scarring. Journal of Burn Care & Rehabilitation 2003;24:91—6.

[10] Lin JT, Nagler W. Use of surface scanning for creation of transparent facial orthoses: a report of two cases. Burns 2003;29:599—602.

CHAPTER 5.14

Prosthetic rehabilitation applications case study 3—An appropriate approach to computer-aided design and manufacture of cranioplasty plates[*]

5.14.1 Introduction

It has long been recognized in product design and engineering that computer-aided design and rapid prototyping (CAD/RP, also called additive manufacturing, AM) can have significant advantages over traditional

[*] Some of the work described in this chapter was first reported in the reference below and is reproduced here, in part or in full, with the permission of the Institute of Maxillofacial Prosthetics and Technologists.
Bibb R, Bocca A, Evans P. An appropriate approach to computer aided design and manufacture of cranioplasty plates. The Journal of Maxillofacial Prosthetics & Technology 2002;5(1):28−31.
The authors would like to gratefully acknowledge Sarah Orlamuender, Greta Green, and James Mason from (at the time of original case study development) SensAble Technologies Inc. for their assistance in this project. The authors would also like to thank Brendan McPhillips, who at the time of writing was Principal Maxillofacial Prosthetist and Technologist/Laboratory Manager, Maxillofacial Laboratory, Royal Preston Hospital, for the images of cranioplasty plate pressing in case study 3.
Further thanks extend to Stefan Leonhardt and Richard Evans at 3D Systems and Dr. Brent Golden, Dr. Ramon Ruiz at Orlando Health (Florida, USA), and Fluvio L. Lobo Fenoglietto at DASH (Digital Anatomy Simulations for Healthcare, Florida, USA) for the details described in case study 5.

Medical Modeling
ISBN 978-0-323-95733-5
https://doi.org/10.1016/B978-0-323-95733-5.00019-3

techniques, particularly in terms of speed and accuracy. These advantages can be realized at all stages from concept through to mass production.

As these processes have become more widespread in industry, attempts have been made to transfer the technology to medical procedures. For example, computer-aided production methods such as RP have been used to build highly accurate anatomic models from medical scan data. These models have proved to be a valuable aid in the production of reconstructive implants such as cranioplasty plates (a cranioplasty plate is an artificial plate that is fitted to the skull to restore the shape of the head and protect the brain). Typically, RP is used to create accurate models of internal skeletal structures, such as skull defects [1]. The cranioplasty plate is then hand-crafted in wax on the model by the prosthetist [2–4]. Alternatively, the anatomic model can be used to create molds or formers [5]. Although this process has dramatically improved the accuracy of cranioplasty plate manufacture, it incurs significant time and cost to produce the anatomic model. This current route does not fully exploit the potential advantages of an integrated and optimized CAD/RP process. A major impediment to the application of CAD in cranioplasty design is the fact that it requires the creation of complex, naturally occurring free-form shapes that are necessary to accurately reconstruct the defect. Although efforts have been made to investigate the use of CAD in cranioplasty plate design, they have proved to be time consuming and only served to highlight these limitations [6]. Advances in voxel-based CAD software described in Chapter 3, Section 6, such as FreeForm, have enabled these challenging reconstructions to be undertaken more efficiently [7,8].

In addition to advances in CAD, metal and polymer AM have enabled implants, such as cranioplasty plates to be fabricated directly from CAD models [9]. This section will describe the development of CAD techniques from an initial concept through to current best practice that employs AM. Although the tools available within FreeForm have developed since early applications, the essential functions described in the initial study remain relevant to current versions.

5.14.2 First case

To investigate the application of voxel CAD technologies for reconstruction, a clinical case requiring a cranioplasty plate was attempted using FreeForm. The case began with importing a three-dimensional model of the cranial defect. These data were derived from a 3D CT scan and imported into the software as a "buck." This means that the user can feel the surface of the skull but not alter it with any subsequent tools. The next step was to create a piece of "virtual clay" that could then be worked into the correct shape. In prosthetics terms, it may be more appropriate to refer to this as virtual or digital wax-up.

First, a sketch plane is positioned by eye over the defect and a two-dimensional perimeter drawn around it. The planes are then offset either side of the defect and the perimeters joined to form a three-dimensional piece of digital wax approximately the right size, as shown in Fig. 5.86. Material removal tools were then used to "grind" away material. The tools used are similes of physical sculpting tools such as scrapers, grinding wheels, etc. However, the size and shape of the tools can be arbitrarily altered to suit the job in hand. A wide "grinding" tool was used to work the surface back, as shown in Fig. 5.87. In addition, the simulated physical properties of the virtual clay can be altered to represent differing hardness and tack strength. When using the system, the user can feel tactile resistance when removing material, and when they reach the surface of the skull, resistance is total. The approach is therefore a digital equivalent of waxing up on a medical model.

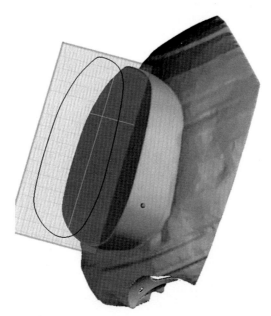

Figure 5.86 Extrusion of working material.

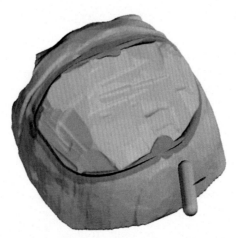

Figure 5.87 Working the material back.

However, operating in the digital domain enables some useful techniques to be exploited. In this case, control curves were created on the surface of the plate. These curves can then be moved using control points on the surface or tangents to the surface blue arrow. A single control curve applied to the sagittal plane is shown in Fig. 5.88. Many control curves can be added in various planes to achieve a very smooth and well-defined natural curvature that matches the surrounding tissue. In addition, whole areas can be selected by painting on the model with the stylus. Then various

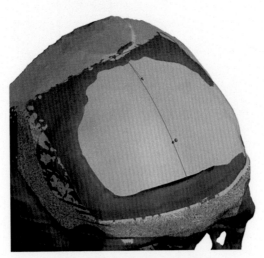

Figure 5.88 Controlling curvature.

operations can be performed on the entire selected area in a single operation. For example, the area could be pulled out to create a bulge or pushed in to create a depression. Fig. 5.89.

The material removal and smoothing process was then repeated for the inner surface illustrated in . Then the buck (skull) is subtracted from the wax to leave a free-form shape that would repair the defect, as shown in Fig. 5.90. To allow the implant to be inserted from the outside of the skull,

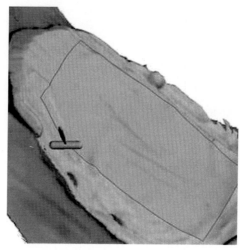

Figure 5.89 Working back the inner surface.

Figure 5.90 The result of the Boolean subtraction.

the inner edges of the plate were worked back, as shown in Fig. 5.91. The final shape can then be accurately fabricated using RP techniques. This can then be used as a pattern from which an implant can be made using an appropriate material.

In this case, a highly accurate stereolithography (SL) model was made of both the implant and the defect to test the design for fit and accuracy; see Figs. 5.92 and 5.93. In this case, the fit was excellent and comparable to the

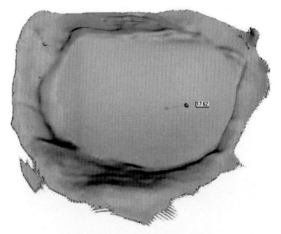

Figure 5.91 The final design.

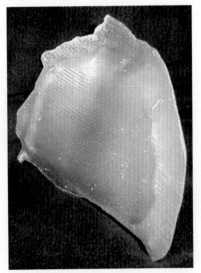

Figure 5.92 The SL model of the plate.

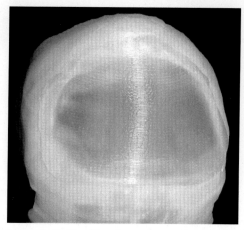

Figure 5.93 Checking fit on SL model of defect.

results achieved when using SL models and traditional wax-up methods. The SL resin used is approved for patient contact under theater conditions, which will allow the plate to be test fitted to the actual defect at the time of surgery. However, the material is not approved for implantation, and therefore the actual implant used will be cast from the SL master into an approved acrylate.

5.14.3 Second case

This case followed similar techniques to the initial case yet proved to be more challenging due to the complex nature of the implant required. The added complexity further illustrated the potential benefits of utilizing this approach. The unaffected side of the patient's skull was copied and laterally inverted to provide a base part from which a well-matched implant could be designed (Fig. 5.94). As in all cases attempted in this manner thus far, the laterally inverted copy does not precisely fit the defect; however, the FreeForm software enables the form to be modified and adapted to produce a prosthetic design that shows good aesthetic appearance and a precise fit at the margins. The software also enabled the implant design to be produced in a very smooth manner, which will prove suitable for a double processed and polished acrylic implant. As in the previous case, the implant design was manufactured using SL and test fitted to an existing model of the patient defect, as shown in Fig. 5.95.

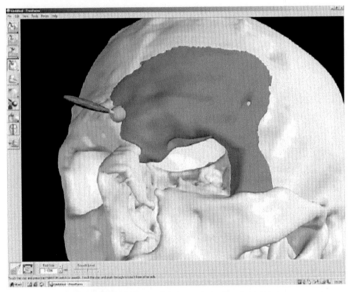

Figure 5.94 Designing a complex implant.

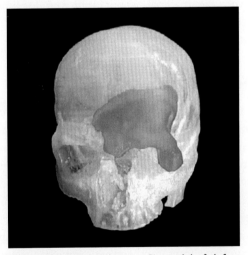

Figure 5.95 SLA implant on SL model of defect.

5.14.4 Third case—Press tool design

The first case studies demonstrated the feasibility of a CAD approach. With minor refinement of the techniques, a clinically viable method of producing mold tools that enabled pressed titanium plates was developed. This method has subsequently become widely adopted and utilized by the UK National Health Service and can be illustrated through a case study.

Techniques described in the first two studies were used to reconstruct the defect. At the point of completed reconstruction, the model consisting of buck anatomy and clay reconstruction was duplicated. The "remove buck" option was used to leave just the clay reconstruction, which showed jagged edges around the area where it had previously blended into the anatomy (Fig. 5.96). The clay coarseness was refined by 0.1 mm, which had the effect of smoothing the jagged edges and creating a small gap between the original anatomy and reconstruction (Fig. 5.97). This gap defined the boundary between anatomy and CAD reconstruction, thus creating a margin, beyond which a pressed plate should extend. The smoothed reconstruction was then combined with a copy of the original anatomy,

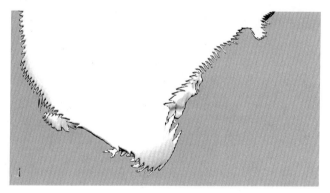

Figure 5.96 Jagged edges of the unrefined model.

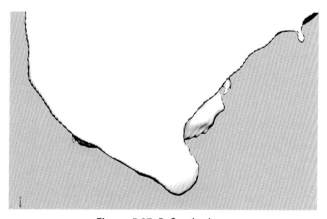

Figure 5.97 Refined edges.

creating a reconstructed CAD model, with a physical groove representing the defect margin. This was cropped according to a reasonable margin around the defect that would allow the plate to be pressed.

The model was then exported as an STL file, and it was fabricated using SL. The SL model was then backfilled with plaster and used as the male section of a steel-bolstered press tool (Fig. 5.98). Conventional, lab-based methods of pressing sheet titanium similar to those described in research literature [10] were then used to complete the implant fabrication process (Fig. 5.99).

Figure 5.98 Plaster-back, steel-bolstered press tool.

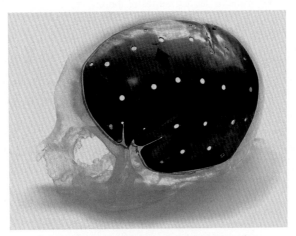

Figure 5.99 Completed pressed titanium plate.

5.14.5 Fourth case—Implant design for metal AM fabrication

With further CAD modeling, it is also possible to use these techniques to design the definitive implant that incorporates a wide range of design features that can be produced using metal AM technologies. This has the potential to improve functional and aesthetic outcomes. This can be illustrated through the same case study as previous but going a stage further to design an implant that sits just inside the defect.

At the stage where reconstruction has been completed and a duplication of the model has been made, the process is different. A boundary line enclosing the defect was drawn just outside the defect margin on the buck anatomy. The "emboss clay" function was used to lower the clay of the reconstruction (note that the buck remains unmodified) by the desired thickness of the plate, in this case, 1 mm. The entire model was then converted to buck. In this case, a hexagonal pattern was desired to create a perforated plate, which would be lighter than a comparable solid structure. A boundary line, inside which the pattern would be created, was drawn on the buck model just inside the defect margin. A repeating, uniform, hexagonal mesh pattern was created using Photoshop (Adobe Systems Software, USA), and the "emboss with wrapped image" tool was used to create a 1-mm raised thickness of clay in the shape of the pattern. The "layer" tool was then used to manually draw in a 1-mm-thick plate around the periphery of the plate to the point where it met the defect margin, also filling in incomplete perforations. The same tool was used to design 0.6-mm-thick tabs that extended over the defect margin to provide fixation points for the plate and prevent it recessing too far into the defect during surgery. The "remove buck" tool was then used to remove the reference buck model, leaving just the implant design. "Select lump of clay" was used to select the implant design, with "invert selection" and delete used to remove any unwanted, unattached pieces of clay. CAD designs of the intended 1.5-mm screws that incorporated countersink head features were then imported and positioned according to the implant tab locations, ensuring they did not protrude beneath the fitting surface of the implant. Boolean subtraction was undertaken to remove the screws from the plate, leaving the holes with countersink. The completed plate design clay coarseness was then refined by a factor of 0.1 mm to smooth jagged features and create a tolerance to the fixation holes. "Select lump of clay" was used again to select the implant design, with "invert selection" and delete used to remove any

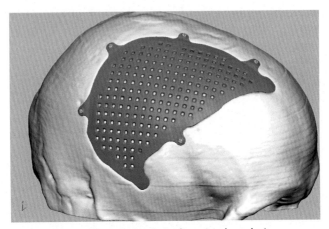

Figure 5.100 CAD cranioplasty implant design.

unwanted, unattached pieces of clay. The implant design, shown in Fig. 5.100, was then exported as an STL file.

For subsequent fabrication, it is also necessary to undertake further checks to validate both the design and integrity of the STL file.

For the illustrated case study, electron beam melting (Q10, Arcam AB) (described in Chapter 4) was the intended fabrication method. The plate was not fabricated due to cost limitation.

5.14.6 Fifth case—Implant design for PEEK AM fabrication

Titanium cranioplasty implants have notable clinical and economic limitations, including heightened patient sensitivity in extreme temperatures, difficulty to modify postproduction or during surgery, interference with radiotherapy, and, in the case of metal AM, high production costs. Polyetheretherketone (PEEK) is a semicrystalline engineering polymer with biocompatible grades that make it suitable for cranioplasty plate production. Being a polymer, it offers reported clinical advantages, including being radiolucent, chemically inert, strong, elastic, not creating artifacts on imaging, comfortable, and not conducting temperature (resulting in fewer issues with patient sensitivity in cold/hot climates) [11]. Research has also noted a trend toward lower complication rates and failure rates with PEEK implants versus autologous or titanium [12]. Currently, the vast majority of PEEK cranioplasty plates are produced using computer numerically controlled machining from a solid block of material, which is expensive, wasteful, and leads to some levels of geometric design constraints. Although PEEK is a thermo-polymer, AM production is challenging, particularly in a regulatory compliant environment. This means that despite potential clinical and cost advantages, custom PEEK AM—produced cranioplasty plates are only just becoming available.

This case study illustrates the development of a manufacturing process using the EXT 220 MED (developed by Kumovis GmbH, a 3D Systems company, Germany), a dedicated PEEK AM machine intended to produce cranioplasty plates and other custom implants.

The manufacturing process contains different developments:

- technical description of manufacturing-related requirements on a printed PEEK cranial implant
- description of a validated design envelope
- evidence documents on mechanical performance
- evidence documents on the biocompatibility of the PEEK printing process
- manufacturing-related SOPs

This development was done to allow medical device manufacturers and healthcare providers to leverage existing data points to support their regulatory roadmap and aim for medical device regulation (MDR) and Food and Drug Administration (FDA)-compliant manufacturing.

Since all cranioplasty plates are patient specific and therefore have the potential for inconsistency, part of the pathway toward regulatory approval requires the development of a safe design envelope. This is also necessary since the mechanical properties of parts produced using an extrusion-based 3D-printing process are influenced by the orientation as well as the cross-section of each layer.

A critical quantity of implant designs was analyzed based on literature research and clinical data to understand variations and to define worst-case geometries, which represented the boundaries of a design envelope (see Fig. 5.101) that was used to generate evidence data. Multiple criteria such as curvature, thickness, surface, cross-section, and diameter of the implant influenced the printing parameters.

In this illustrated preclinical study example, cranioplasty design was undertaken using methods similar to those described in case study 1, using Geomagic FreeForm Plus. Mirroring the healthy side of the skull was used as a template for the implant design, as shown in Fig. 5.102.

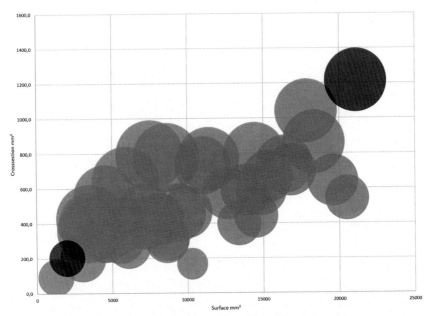

Figure 5.101 Cranioplasty plate patient population data.

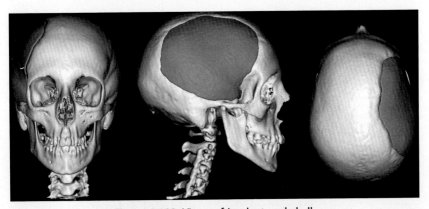

Figure 5.102 Views of implant and skull.

This preclinical study was used to ensure the design made by biomedical engineers in the hospital was suitable for the new manufacturing process, followed the technical design requirements, and fitted into the design envelope backed up with mechanical and biologic data. In addition to the engineering aspect associated with the workflow, the study helped to answer clinical areas of interest, including the following. How would the additive-manufactured implant behave intraoperatively compared with best practice reported in the research literature? How could the likelihood of warpage and shrinkage of the material be controlled when using AM? These questions resulted in a fitting test and a comparison between an additive-manufactured PEEK implant and an implant milled by an external service provider and medical device manufacturer.

The additive-manufactured design being checked for geometric accuracy intraoperatively by a physician is shown in Fig. 5.103. Furthermore, more studies have been performed by multiple physicians to verify the comparability with implants produced using the here described manufacturing process with multiple different designs.

This evidence study covered the clinical perspective to compare established milled implants and PEEK additive-manufactured implants using the EXT 220 MED and the developed manufacturing process from Kumovis, a 3D-Systems company (Fig. 5.104). To release a 3D-printed patient-specific implant to the market, different regulatory roads can be followed:

- FDA market: traditional 510k
- FDA market: medical device production systems (process still under development within FDA)

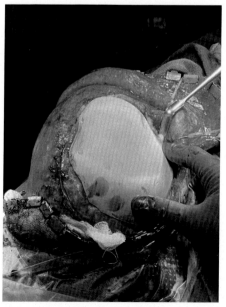

Figure 5.103 The printed design checked for geometric accuracy intraoperatively.

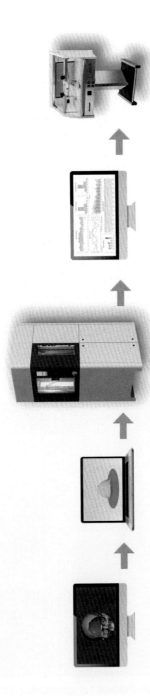

Figure 5.104 The Kumovis medical device production system.

- MDR market: CE marking
- MDR market: manufacturing at the point of care according to MDR2017/745 Article 5/5

However, each approach requires a validated manufacturing workflow starting with segmentation of anatomic data and implant design. It continues with material handling and manufacturing, including optimum support structure designs while embedding documentation of the critical process parameters and postprocessing with qualified hardware. It concludes with the sterilization process. 3D-Systems developed solutions for each step and can support with evidence-based data to achieve regulatory compliance.

Preparation for fabrication required the consideration of support structures. As with most extrusion AM processes, reducing the amount of support structures helps to reduce the amount of postprocessing required. Furthermore, appropriate support design will help to prevent warpage and ensure a homogenous surface quality. In this example, orientation and support strategy were developed to ensure an excellent surface quality as well as a contact area of the implant to the bone that does not require postprocessing. The final implant is shown in Figs. 5.105 and 5.106.

One of the major challenges of PEEK AM production is achieving consistency in mechanical properties and reducing internal part stresses that

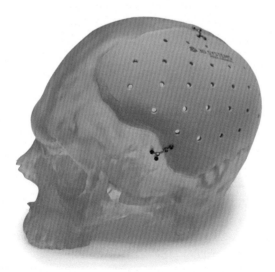

Figure 5.105 Cranial implant on skull model.

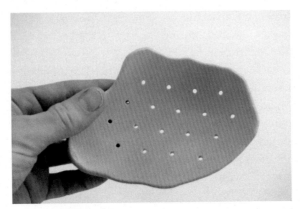

Figure 5.106 Final cranial implant.

can cause warping. In this case, a tight temperature-controlled build chamber combined with the optimized build orientation of geometries helped to ensure consistent part properties. Given each implant is custom, mechanical testing was undertaken on worst-case designs of cranioplasty implants, proving that properties equivalent to accepted criteria reported in the literature could be achieved. To further ensure the absence of any foreign particles between the layers, control the mechanical properties of the part, and ensure medical grade manufacturing, the build chamber of the machine was developed to ISO class 7 equivalence. Furthermore, a dedicated tool set was used to remove support structures. Support witness marks were smoothed using hand tools and manual finishing with dedicated high-speed rotary tools.

The next variable to consider was material. Material manufacturers have recognized the opportunity to make custom implant fabrication more affordable and closer to the point of care. There are a growing number of companies producing PEEK filament with the necessary preapprovals that make it suitable for long-term implant production. Evonik VESTAKEEP i4 3DF PEEK (Evonik Industries AG, Germany) filament was used in the workflow described in this case. Manufacturers of medical devices, such as cranioplasty implants, must work with qualified machines and validated processes and fulfill regulatory requirements. Biologic validation of the PEEK printing process was undertaken on coupons that represent a final finished device. The regulatory requirement of biologic validation involved tests according to ISO 10993 [13], included:

- reverse mutation assay using bacteria (ISO 10993-3)
- in vitro mammalian cell gene mutation assay (ISO 10993-3)
- cytotoxicity (ISO 10993-5)
- implantation effects (ISO 10993-6)
- sensitization (ISO 10993-10)
- acute systemic toxicity (ISO 10993-11)
- material-mediated pyrogenicity (ISO 10993-11)
- investigation of SVOC—GC/MS fingerprint investigation after exhaustive extraction (ISO 10993-18)
- investigation of NVOC—LC/MS fingerprint investigation after exhaustive extraction (ISO 10993-18)
- investigation of inorganic ions—ICP/MS fingerprint investigation after exhaustive extraction (ISO 10993-18)
- irritation test (intracutaneous reactivity) (ISO 10993-23)

These data can be leveraged from medical device manufacturers or healthcare institutions to support their own regulatory compliance.

These types of implants are normally delivered unsterile to the representative hospitals. The responsibility for cleaning and sterilization lies within the sterilization department. To support them, cleaning and sterilization (hot-steam sterilization) tests have been performed on printed parts. Furthermore, the influence on the mechanical properties after cleaning and sterilization has been investigated. It could be shown that worst-case geometries still fulfill all acceptance criteria after multiple sterilization cycles.

Fixing the PEEK cranioplasty plate during surgery would require the use of conventional screws and miniplates. Using this process, surgeons have the possibility to use a range of self-tapping screws and miniplates, and they intraoperatively decide screw position and can intraoperatively drill holes for fixation, as shown in Fig. 5.107.

This workflow was tested with commercially available screw and plate systems to support the user with evidence data.

This case study has illustrated the complexity of regulatory, design, and manufacturing considerations required for custom cranioplasty production using PEEK AM. The manufacturing-related process developments from 3D Systems can support healthcare providers and industrial implant manufacturers to fulfill regulatory requirements and bring their products quicker to market.

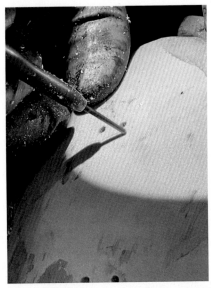

Figure 5.107 Intraoperatively deciding screw position and drilling holes for fixation.

Figure 5.108 Different shapes and sizes of plate.

Examples of different shapes and sizes are shown in Fig. 5.108. AM with PEEK has the potential to be material efficient, faster, and, overall, more cost-effective compared with alternative manufacturing technologies.

5.14.7 Future development and benefits

The evolution of cranioplasty plate production is interwoven with access to CAD software, advances in manufacturing processes, and developments in regulatory requirements. Research has explored the relationship between these factors, with economics also playing a significant role on which technique is used by hospitals [14].

Benefits of using CAD are apparent. The CAD model of the implant can be altered and refabricated any number of times if it is required. Data can be archived easily. There is also reduced dependence on lab-based materials during the design process (the process as a whole is clean and requires minimal floor space, equipment, and consumables). Research has also explored the potential of cranioplasty design automation and compared the accuracy of results with alternative techniques [15]. Reducing the time associated with design is a potentially significant step in reducing the cost further. This combination of advantages could lead to significant improvements in treatment times and cost effectiveness [15]. An estimate of potential cost and time saving is indicated in Table 5.1.

Notwithstanding the benefits, further work is still required to better understand the economic implications of using AM for production. Current lab-based techniques used in the production of pressed sheet titanium custom cranioplasty plates have developed over many years, particularly in the United Kingdom. Although there are design limitations, sheet materials can be modified easily even at the point of surgery and are low cost. AM methods used in the production of custom implants, such as cranioplasty plates, are still expensive and require extensive industrial facilities. In-hospital lab production is designed around low-volume production where the high capital and infrastructure costs make the investment in AM production unfeasible. Polymer AM production of cranioplasty plates in PEEK could offer a more cost-effective option for hospitals producing larger numbers of implants, but further work is required to fully assess the economics implications.

Recent changes in MDRs may, however, accelerate the trend toward CAD/AM production of cranioplasty plates. The high degree of variability associated with in-hospital lab-based production makes it challenging to verify accuracy and repeatability. Global regulations place great emphasis on using controlled, accurate, and repeatable manufacturing processes. With increased emphasis on quality management within both hospitals and industry, combined with increasing accessibility of CAD/AM technology, it is reasonable to anticipate the use of pressed sheet titanium cranioplasty methods will be phased out.

Table 5.1 Comparison of current technique and potential future technique.

Current technique			Future technique		
3D CT scan			**3D CT scan**		
Send data to RP service provider	1 day	£5	Derive 3D CAD data from CT data	5 min	Minimal
Derive 3D CAD data from CT data	5 min	Minimal	Import into FreeForm	5 min	Minimal
Derive RP files from CAD data	1 h	£22	Design implant	1 h	£22
Build RP model of anatomy	1 day	£1000	Direct manufacture of implant	2 h	£200
Post model to prosthetist	1 day	£18	Sterilize for surgery		
Prosthetist wax up on model	1 day	£250			
Make plaster mold					
Cast implant					
Sterilize for surgery					
Total	4 days	£1295	Total	3 h	£222

References

[1] Klein HM, Schneider W, Alzen G, Voy ED, Gunther RW. Pediatric craniofacial surgery: comparison of milling and stereolithography for 3D-model manufacturing. Pediatric Radiology 1992;22:458—60.

[2] Bibb R, Brown R. The application of computer aided product development techniques in medical modelling. Biomedical Sciences Instrumentation 2000;36:319—24.

[3] Bibb R, Brown R, Williamson T, Sugar A, Evans P, Bocca A. The application of product development technologies in craniofacial reconstruction. In: Proceedings of the 9th European conference on rapid prototyping and manufacturing; 2000. p. 113—22. Athens, Greece.

[4] D'Urso PS, Redmond MJ. A method for the resection of cranial tumours and skull reconstruction. British Journal of Neurosurgery 2000;14(6):555—9.

[5] Joffe J, Harris M, Kahugu F, Nicoll S, Linney A, Richards R. A prospective study of computer-aided design and manufacture of titanium plate for cranioplasty and its clinical outcome. British Journal of Neurosurgery 1999;13(6):576—80.

[6] Taha F, Lengele B, Boscherini D, Testelin S. In: Moos N, editor. "Phidias report" No. 6 June 2001. Danish Technological Institute Teknologiparken, 8000 Aarhus C-Denmark; 2001.

[7] Massie TH, Salisbury KJ. PHANToM haptic interface: a device for probing virtual objects. In: Proceedings of the 1994 international mechanical engineering congress and exposition (code 42353), American Society of Mechanical Engineers, Dynamic Systems and Control Division (publication) DSC, v 55-1; 1994. p. 295—9.

[8] FreeForm modelling plus software, geomagic, 3D-systems, 333 three D systems circle. SC 29730, USA: Rock Hill.

[9] Lethaus B, Poort ter Laak M, Laeven P, Beerens M, Koper B, Poukens J, et al. A treatment algorithm for patients with large skull bone defects and first results. Journal of Cranio-Maxillo-Facial Surgery 2011;39(6):435—40.

[10] Bartlett P, Carter LM, Russell JL. The Leeds method for titanium cranioplasty construction. British Journal of Oral and Maxillofacial Surgery 2009;47:238—40.

[11] Shah AM, Jung H, Skirboll S. Materials used in cranioplasty: a history and analysis. Neurosurgical Focus 2014;36(4):E19. https://doi.org/10.3171/2014.2. FOCUS13561.

[12] Punchak M, et al. Outcomes following polyetheretherketone (PEEK) cranioplasty: systematic review and meta-analysis. Journal of Clinical Neuroscience 2017;41:30—5. https://doi.org/10.1016/j.jocn.2017.03.028.

[13] International Organization for Standardization. ISO19003-1:2018 Biological evaluation of medical devices — Part 1: evaluation and testing within a risk management process. Geneva: ISO; 2018.

[14] Peel S, Eggbeer D. Additively manufactured Maxillofacial implants & guides - achieving routine use. Rapid Prototyping Journal 2016;22(1):189—99.

[15] Peel S, Eggbeer D, Burton H, Hanson H, Evans P. Additively manufactured versus conventionally pressed cranioplasty implants: an accuracy comparison. Proceedings of the Institution of Mechanical Engineers - Part H: Journal of Engineering in Medicine 2018;232(9):949—61.

CHAPTER 5.15

Prosthetic rehabilitation applications case study 4—Evaluation of advanced technologies in the design and manufacture of an implant-retained facial prosthesis*

5.15.1 Introduction

Despite the widespread application of computer-aided design and rapid prototyping (CAD/RP) technologies in the production of medical models to assist maxillofacial surgery, advanced technologies remain underdeveloped in the design and fabrication of facial prosthetics. Research studies to date have achieved some limited success in the application CAD/RP technologies, but very few have addressed the whole design and manufacture process or incorporated all of the necessary components [1–11]. In particular, the components concerned with retention have been neglected. Given that implant retention is now widely considered state of the art, the incorporation of this into digital facial prosthesis techniques must be addressed. Implant-retained prostheses are described in an explanatory text in Chapter 7.

---

* This paper was written by Dominic Eggbeer, and it is based on Ph.D. research he conducted at the National Centre for Product Design & Development Research (PDR) under the supervision of Richard Bibb and in collaboration with Morriston Hospital, Swansea. The authors would like to thank Frank Hartles, Head of the Dental Illustration Unit, Media Resources Center, Wales College of Medicine, Biology, Life and Health Sciences, Cardiff University for his help in using the Konica-Minolta scanners.

Medical Modeling
ISBN 978-0-323-95733-5
https://doi.org/10.1016/B978-0-323-95733-5.00020-X

Being able to undertake all of the prosthesis design and construction stages without the patient present has the potential to dramatically reduce clinic time and make the entire process more flexible. Reducing the number of clinic visits and the time involved in them would help to reduce patient inconvenience and improve efficiency and flexibility by allowing the prosthetist to work on any given design at any period.

This paper reports on part of ongoing doctoral research that ultimately aims to identify the target specification requirements for advanced digital technologies that may be used to drive development of advanced technologies, so they will be suitable for the design and manufacture of complex, soft tissue facial prostheses. The study reported here tested the capability of currently available technologies in the design and manufacture of an implant-retained prosthesis. Evaluation of the results will clarify the current position and, where they fall short, direct further research that will identify the direction and magnitude of the developments required.

5.15.2 Existing facial prosthetics technique

Facial prosthesis design and construction techniques have changed little in 40 years, and they are described well in textbooks [12,13] and papers [14,15]. By their nature, prostheses are one-off, patient-specific devices that cannot benefit from batch or mass manufacture. Handcrafting techniques are therefore used to fabricate the prosthesis form and retentive components and, in some cases, join them to prefabricated components that enable the prosthesis to be attached to the implants.

Various retention methods may be used to secure a facial prosthesis such as magnets, bar and clip, adhesives, or engaging anatomic undercuts. However, in many cases, implant-retained prostheses are now considered the optimum solution. In implant-retained cases, the prosthesis typically consists of three components; the soft tissue prosthesis itself, a rigid substructure incorporating the retention parts, and the corresponding retention parts that remain attached to the patient. The attachment between the two retention components can be by bar and clip or by magnets. Bar and clip gives the highest retention force, and the strength may be altered by crimping the metal clips. Magnets can provide a range or retentive forces (around 500−1000 g) depending on the number and type used. Magnets may either be screwed directly on to the abutments or located on a framework. The prosthesis-mounted components may be bonded directly into the silicone if the prosthesis is small or a substructure is not necessary.

Prosthesis design is typically undertaken by shaping wax on a plaster replica of the patient's anatomy. Realism is predominantly achieved though the prosthetist's ability to interpret the correct location and physically recreate the anatomic shape and detail. Color matching of the silicone also helps to complete a good blend into the surrounding anatomy.

Although these existing techniques are time-consuming, they can be applied to a wide range of situations. Previous studies have shown that to be effective, digital technologies must be sympathetically integrated into these existing techniques so that the skills and flexibility of the prosthetist are not hampered [9—11].

5.15.3 Review of advanced technologies in facial prosthetics

A review of previous research highlights a range of advanced technologies that may be used to design and manufacture a facial prosthesis [1—11].

Data capture	Noncontact surface scanning to digitize the surface of the affected anatomy
	Various structured white light scanners, laser scanners, computerized photogrammetry
Design	Flexible CAD software
	FreeForm (FreeForm Plus, 3D-Systems, USA, http://geomagic.com/en/products/freeform-plus/overview); Magics (Materialise N.V., Technologielaan 15, 3001 Leuven, Belgium, www.materialise.com); Rhino (Robert McNeel & Associates, 3670 Woodland Park Avenue North, Seattle, WA 98103, USA, www.rhino3d.com)
Manufacture	Rapid Prototyping Processes
	ThermoJet wax printing (3D Systems. The ThermoJet process is now obsolete with the most analogous replacement process being the 3D-Systems ProJet 3510CP and CPX wax printers); selective laser melting (at time of original publication—MCP Tooling Technologies Ltd., UK. The technology has now been acquired by Renishaw Plc, New Mills, UK); Stereolithography (3D Systems), selective laser sintering (3D Systems) and various computer numerically controlled machining processes

The review of previous work has shown that these advanced techniques and technologies demonstrate a number of limitations, and it identifies a range of technical challenges. Specifically, there are three notable areas: the capture of data that describes the anatomy and implant abutment features, the design and alignment of the prosthesis components, and the manufacture of components in appropriate materials.

5.15.3.1 Data capture

Although noncontact surface scanning technologies have been used to capture anatomic forms, limitations have also been identified. Areas of hair, undercut surfaces, highly reflective surfaces, and patient movement give poor results [3—6]. Insufficient data resolution and errors in the form of "noise" also limit the ability of scanning technologies to capture sharp edges and small geometric features at the scale required [16]. The ideal scanning technology must therefore be capable of capturing both anatomic surfaces and implant components with sufficient accuracy, resolution, and speed to overcome these limitations.

5.15.3.2 Design

To design the various components of the prosthesis so they accurately fit together using CAD, the operator must be able to import, manipulate, create, and align both anatomic and geometric forms. Engineering CAD software packages typically work with geometric shapes and provide methods of aligning components. However, engineering software is poorly suited to handing complex and individual anatomic forms. CAD software such as FreeForm (3D-Systems, USA) provides a more intuitive solution to handling anatomic forms (more akin to a digital sculpting package) yet does not provide suitable tools for aligning the various components. A suitable CAD software package must provide tools for precisely aligning geometric shapes as well as the manipulation of complex anatomic forms.

5.15.3.3 Manufacture

Material requirements for maxillofacial prostheses are varied according to the separate components. For the soft tissue elements that are currently made from color-matched silicone, no technology exists that is able to build the final prosthesis form directly from CAD data. Therefore, a pattern must be produced instead. The review of previous research and experiments carried out at PDR and Morriston Hospital has shown that producing the pattern in a material compatible with conventional sculpting techniques is highly desirable. Building the pattern in wax allows the prosthetist to easily adjust the pattern using their existing techniques and skills, particularly during test fitting on the patient [9—11]. The substructures need to be accurate and rigid enough to contain the retentive elements and the forces experienced during attachment and removal. The retentive components that remain attached to the patient via the implants must be noncorrosive and unreactive (similar to jewelry) and rigid enough to withstand the retentive forces. The materials for all of the components must also not react with each other and must resist the effects of being included in the manufacture of the final prosthesis from the pattern, such as mold heating. Finally, they must also provide adequate wear resistance, resist permanent distortion, and provide adequate retention for the service life of the prosthesis.

5.15.4 Case 1

To assess the capabilities of current advanced technologies in the design and manufacture of an entire implant-retained prosthesis accurately, an exploratory study was undertaken. The study would not only evaluate the ability of current technologies, but measurements and observations made would inform future research. A bar and clip, implant-retained auricular prosthesis case was selected. A 3D CT scan had already been undertaken and the data used to plan the placement of two implants in a single-stage operation. A healing period of 6 weeks was allowed before prosthesis construction.

5.15.4.1 Data capture

An impression and dental stone replica that recorded the implant abutment locations and the surrounding anatomy were made using conventional methods. In addition, the patient was digitally scanned using a pair of laser scanners (Konica-Minolta Vivid 900 laser scanners, Osaka, Japan) to allow for subsequent digital prosthesis design. Previous work has shown that these scanners had a relatively fast capture time and an accuracy level appropriate to the scanning of faces [17,18]. The actual number of points captured per mm^2 is determined by a scanner's field of view. At a distance of 1.35 m, the scanners each captured an area of 445×333 mm, resulting in a point density of one point per 0.69 mm^2. A paired setup was used to capture a wider field of view without having to move the patient. In this configuration, the scanners are triggered consecutively (simultaneous capture would lead to the scanners interfering with each other). The patient was seated with their head positioned 1.35 m away from the scanners, and a 14-mm focal length lens was used. Although the specified capture time is 0.6 s for each camera, a short pause between scans meant that the patient had to remain motionless and with the same facial expression for approximately 8 s. The point cloud scan data was aligned and converted to an STL file using Rapidform software (INUS Technology Inc., Seoul, South Korea). Shadow areas behind the ears were not captured. However, the three-dimensional CT data that had been acquired for the implant planning was also available to be used as the basis for the prosthesis design.

5.15.4.2 Design

The scan data was imported into the sculpting CAD package, FreeForm, using the "thickness" option to make a solid model. Although the scanners have been shown to be able to capture anatomic detail well, this study demonstrated that the data resolution was insufficient to describe the implant abutments accurately (see Fig. 5.109). However, the data were good enough to allow the abutment locations to be identified, which allowed the overall prosthesis form to be designed around them with sufficient accuracy. The patient's opposite healthy ear was obtained from the CT data, imported into FreeForm, and mirrored to the defect site (see

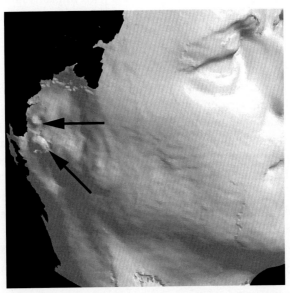

Figure 5.109 Scan data of the defect site (arrows indicating the implant abutments).

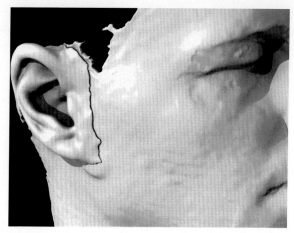

Figure 5.110 The mirrored ear positioned at the defect site (before blending).

Fig. 5.110). The tools in FreeForm were then used to blend this ear into the surrounding anatomy and then subtracted from it to leave an accurate fitting surface using a technique that has also been reported in the digital design of other prostheses [9—11].

5.15.4.3 Manufacture

The final design of the prosthesis pattern was physically manufactured using the ThermoJet printing process in a wax material. The use of wax allowed modifications to include the retentive components to be made using conventional methods. A color-matched silicone prosthetic ear was then fabricated for the patient using conventional methods.

5.15.4.4 Initial findings from case 1

This initial trial highlighted the limitations of noncontact scanning to capture anatomy and finely detailed abutments with sufficient resolution.

5.15.5 Case 2

5.15.5.1 Data capture

The first case showed that the data captured was insufficient to be used in the design of the retentive components. Therefore, the same case was repeated using a higher-resolution structured white light scanner (Steinbichler Optotechnik GmbH, Germany). This type of scanner is typically used for engineering and has a much longer capture time. Therefore, it was used to digitize the dental stone replica of the patient produced using conventional impression methods. This scanner captures approximately nine points per mm^2 (three per mm in the x, y plane) and around 140,000 points per scan over an area of approximately 250×250 mm and a working range of approximately 180 mm. To make the abutment locations easier to scan, magnetic keepers (Technovent Ltd., Leeds, UK) were screwed on to the abutments to provide a flat surface, and the model was coated in a fine matt white powder to reduce reflectivity. Six overlapping scans covering the entire model were taken and the data aligned using Polyworks software (InnovMetric Software Inc., Quebec, Canada). STL file data were created from the point cloud information using Spider (Alias-Wavefront, Toronto, Canada) and the data imported into Magics. Magics provides alignment and modification tools for STL file data and was used to digitally remove the abutment caps. The flat surfaces of each abutment cap in turn were aligned by selecting a triangle on the top surface and using a Magics function to make it downfacing to the x-y plane. The "sectioning" and "cut" tools were then used to remove the exact depth of the cap. This effectively left a perfectly flat surface representing the top surface of the abutments. The modified STL data were then imported into FreeForm.

5.15.5.2 Design

As in the previous study, FreeForm was used to manually align the ear taken from the CT data to the digital cast based upon the estimated aesthetic requirements and possible substructure location. However, unlike the first case, the data quality in this case study enabled the design of all of the components to be attempted. Digital versions of the screws used to attach frameworks to the abutments and cylinder components were designed using the drawing and rotation tools in FreeForm. A circular-section framework linking the two cylinders was created using the "add clay" tool. As in conventional methods, this followed the thickest section of the ear to ensure all of the components would fit within the ear profile.

Clip designs were created using the two-dimensional drawing and "extrude" tools. These were copied three times and manually located along the bar structure at key points of maximum prosthesis thickness. To secure the clips into the prosthesis body and to assist application, a substructure shell that would be bonded to the silicone was required. This had to provide enough clearance for the clips to spring open and closed but provide firm anchorage for bonding to the silicone. Digital "clay" that enclosed the framework and clips, leaving just their top features as a point of attachment, was built up from the cast model. The "paint on selection" tool was used to select and copy then paste the raised section surrounding the framework. The "create offset piece" tool was then used to create a shell 1.5 mm thick surrounding the clips and bar. Boolean subtraction operations and hand carving were used to finalize the shell before joining it to the clip components. The bar, clip, and shell design is shown in Figs. 5.111 and 5.112.

The prosthesis profile was thickened around the clip areas to accommodate the shell component, and a Boolean subtraction operation was used

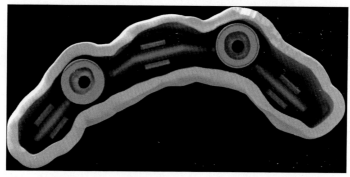

Figure 5.111 The bar located in the clips inside the substructure shell (in FreeForm).

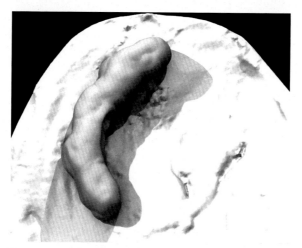

Figure 5.112 The located substructure shell (in FreeForm).

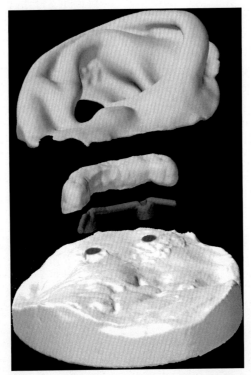

Figure 5.113 An exploded view of the components in FreeForm.

to create a fitting recess. The bar design was finalized with Boolean sub-traction operations to provide a location for the screws, and smoothing operations were applied around the joints with the cylinder features. Hemispherical dimples were also created where cylinders were located on the abutments. Fig. 5.113 shows an exploded view of the components in the FreeForm environment. Each of the components was then exported as high-quality STL files ready for RP fabrication (see Chapter 3 for more information about the STL file format).

5.15.5.3 Manufacture

A range of RP technologies were selected to produce the components in the most suitable materials available. Selective laser melting (SLM) was selected to produce the bar component due to its ability to produce parts in corrosion-resistant and rigid metals and alloys (see Chapter 4 for a description of the SLM process). The bar was created using SLM in 0.05–

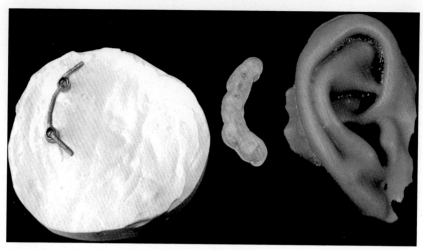

Figure 5.114 The manufactured components: SLM bar (*left*), stereolithography substructure (*middle*), and ThermoJet pattern (*right*).

mm-thick layers using 316L stainless steel. Grit blasting and light polishing were used to remove the rough finish produced by the process.

Stereolithography was used to produce the shell component in DSM Somos 10110-epoxy resin (see Chapter 4 for more information on stereolithography). ThermoJet wax printing was used to produce the prosthesis pattern. The physical components are shown in Fig. 5.114.

5.15.6 Results

The parts designed and produced in the second case study did produce a whole and complete prosthesis. However, the fit between the components required a small amount of adjustment. Therefore, although the finished components would not have been suitable for use in an actual prosthesis, it was possible to evaluate them when fitted to the dental stone model.

For the fit between the implant abutments and the bar, a passive fit as described by Henry [19] was not achieved, and although the bar did screw on to the stone model securely, visible gaps remained. In addition, the surface finish of the SLM bar was slightly pitted when compared with soldered gold bars.

For the fit between bar and clips/substructure, clip retention was initially good, but repeated application caused the clips to wear. This suggested that the relatively soft epoxy resin is not suitable for use in a clip that undergoes repeated applications.

For the fit between the substructure and prosthesis pattern, the fit was tight due to the rough finish left after removing supports from the ThermoJet part's downfacing surfaces. This was easily corrected using a heated scalpel, enabling a secure location and close fit.

Regarding detail, the fragile nature of the ThermoJet wax prevented the pattern's edges from being made any thinner. Thin edges allow the prosthesis margin to blend naturally into the skin when made in silicone. These thin edges were added by the prosthetist using sculpting wax. Heated sculpting metal tools were also used to blend the join between the different waxes and add further anatomic details that helped to achieve a more realistic appearance.

5.15.7 Discussion

This study has highlighted the potential of digital technologies to assist facial prosthesis design, but it also demonstrated that there are many limitations that must be addressed to improve their effectiveness. The limitations encountered are discussed in three categories: data capture, design, and manufacture.

5.15.7.1 Data capture

Scanning small, detailed abutments proved particularly difficult. The distributed nature of point cloud data captured by optical scanning technologies and the effects of noise meant that even with a high point density, small features were subject to loss of edge definition. Inadvertent patient movement during data capture exasperates this problem. With current technologies, a compromise must be made between detail and speed of capture. However, as digital camera technology and computer processor power increases, it can be foreseen that the desired capability to capture rich, high-quality data sufficiently quickly to enable the scanning of patients may be achieved in the near future.

5.15.7.2 Design

This study has shown that digital techniques can be used to design all of the components of a prosthesis. However, as the major aim of embracing digital technologies is the gain in efficiency, it is clear that more work is needed to address software capability. This study has shown that although digital design is possible, it requires the use of multiple software packages to achieve specific tasks. This reduces efficiency, increases costs, and introduces more opportunities for error as data is translated or transferred from one source to another.

Future studies will explore alternative software solutions and identify practical methods of overcoming the issues identified in this research. The authors intend to evaluate other potentially suitable technologies in future studies once an initial specification has been developed.

5.15.7.3 Manufacture

The technologies used in this study have shown that they are capable of producing a complete prosthesis. However, the processes all require improvement to match or improve upon existing techniques. The ThermoJet wax process produced a good quality pattern in an appropriate and useful material that integrated with existing skills and techniques. However, it requires the ability to generate thinner edges if it is to produce a complete pattern without modification. The ThermoJet process is now also obsolete. The closest analogous product currently available that prints in wax is the ProJet CP range of 3D printers (3D-Systems, USA).

The SLM bar was sufficiently strong and rigid enough for the application, and the material should prove corrosion resistant enough for most patients. The bar, however, did not fit as precisely as would be expected of a bar made by existing techniques, and the surface was slightly pitted. As the overall shape and accuracy appeared adequate, finer control of the process may yield parts with better detail and surface finish.

The stereolithography shell component was accurate and rigid enough for the application. The retention strength of the clips was not very high, although it may have been high enough for the purpose. However, the clips did not withstand repeated use and quickly wore down, severely degrading retention strength. The process therefore could prove adequate for the purpose if a harder-wearing material was available.

5.15.8 Conclusions

Literature to date and the findings of this study have demonstrated that although advanced technologies enable the digital design and RP fabrication of complete facial prostheses, further work is needed before they produce results comparable to existing techniques. Without a specification against which potential technologies may be measured and toward which they may be developed, quantifying success is based on subjective assessment and expert opinion. The authors intend to use the findings of this study to direct further research that will develop a specification that will provide quantifiable and objective measures against which advanced technologies may be assessed.

References

[1] Wolfaardt J, Sugar A, Wilkes G. Advanced technology and the future of facial prosthetics in head and neck reconstruction. International Journal of Oral and Maxillofacial Surgery 2003;32(2):121—3.
[2] Coward TJ, Watson RM, Wilkinson IC. Fabrication of a wax ear by rapid-process modelling using Stereolithography. International Journal of Prosthodontics 1999;12(1):20—7.
[3] Bibb R, Freeman P, Brown R, Sugar A, Evans P, Bocca A. An investigation of three-dimensional scanning of human body surfaces and its use in the design and manufacture of prostheses. Proceedings of the Institution of Mechanical Engineers - Part H: Journal of Engineering in Medicine 2000;214(6):589—94.
[4] Cheah CM, Chua CK, Tan KH, Teo CK. Integration of laser surface digitizing with CAD/CAM techniques for developing facial prostheses Part 1: design and fabrication of prosthesis replicas. International Journal of Prosthodontics 2003;16(4):435—41.
[5] Cheah CM, Chua CK, Tan KH. Integration of laser surface digitizing with CAD/CAM techniques for developing facial prostheses Part 2: development of molding techniques for casting prosthetic parts. International Journal of Prosthodontics 2003;16(5):543—8.
[6] Reitemeier B, Notni G, Heinze M, Schöne C, Schmidt A, Fichtner D. Optical modeling of extraoral defects. The Journal of Prosthetic Dentistry 2004;91(1):80—4.
[7] Tsuji M, Noguchi N, Ihara K, Yamashita Y, Shikimori M, Goto M. Fabrication of a maxillofacial prosthesis using a computer-aided design and manufacturing system. Journal of Prosthodontics 2004;13(3):179—83.
[8] Verdonck HWD, Poukens J, Overveld HV, Riediger D. Computer-assisted maxillo-facial prosthodontics: a new treatment protocol. International Journal of Prosthodontics 2003;16(3):326—8.
[9] Eggbeer D, Bibb R, Evans P. The appropriate application of computer aided design and manufacture techniques in silicone facial prosthetics. In: Proceedings of the 5th national conference on Rapid design, prototyping, and manufacture; 2004. p. 45—52.
[10] Evans P, Eggbeer D, Bibb R. Orbital prosthesis wax pattern production using computer aided design and rapid prototyping techniques. The Journal of Maxillofacial Prosthetics and Technology 2004;7:11—5.
[11] Sykes LM, Parrott AM, Owen CP, Snaddon DR. Application of rapid prototyping technology in maxillofacial prosthetics. International Journal of Prosthodontics 2004;17(4):454—9.
[12] Thomas KF. Prosthetic rehabilitation. Chicago: Quintessence Publishing; 1994, ISBN 1-85097-032-7.
[13] McKinstry RL. Fundamentals of facial prosthetics. Arlington: ABI Professional; 1995, ISBN 1-886236-00-3.
[14] Seals RR, Cortes AL, Parel S. Fabrication of facial prostheses by applying the osseointegration concept for retention. The Journal of Prosthetic Dentistry 1989;61(6):712—6.
[15] Postema N, van Waas MA, van Lokven J. Procedure for fabrication of an implant-supported auricular prosthesis. Journal of Investigative Surgery 1994;7(4):305—20.

[16] Chen LC, Lin GC. An integrated reverse engineering approach to reconstructing free-form surface. Computer Integrated Manufacturing Systems 1997;10:49—60.

[17] Kau CH, Zhurov A, Scheer R, Bouwman S, Richmond S. The feasibility of measuring three-dimensional facial morphology in children. Orthodontics and Craniofacial Research 2004;7:198—204.

[18] Kau CH, Knox J, Richmond S. Validity and Reliability of a portable 3D optical scanning device for field studies. In: Proceedings of the 7th European craniofacial congress, bologna: monduzzi editore-international proceedings division; 2004.

[19] Henry PJ. An alternative method for the production of accurate casts and occlusal records in the osseointegrated implant rehabilitation. The Journal of Prosthetic Dentistry 1987;58(6):677—94.

CHAPTER 5.16

Prosthetic rehabilitation applications case study 5—Additive manufacturing technologies in soft tissue facial prosthetics: Current state of the art*

5.16.1 Introduction

Patients who suffer from facial deformity, either congenital, traumatic, or from ablative surgery, are treated by maxillofacial units using a variety of surgical and prosthetic techniques. Maxillofacial prosthetics and technology covers the treatment and rehabilitation of these patients by producing facial prostheses using artificial materials. Improvements in medicine, surgical techniques, and in particular cancer survival rates are resulting in ever increasing patient numbers. However, these same drivers are also leading to higher costs, which is putting pressure on healthcare providers to improve efficiency during a period when the number of newly qualified maxillofacial prosthetists and technologists is only matching the number retiring [1].

These pressures have led researchers to explore whether the cost and time savings associated with advanced design and product development

* The work described in this chapter was first reported in the reference below and is reproduced here with the permission of Emerald Publishing Ltd.
Bibb, R., Eggbeer, D., Evans, P., 2010. Rapid prototyping technologies in soft tissue facial prosthetics: current state of the art. Rapid Prototyping Journal 16 (2), 130–137, ISSN: 1355-2546, https://doi.org/10.1108/13552541011025852.

Medical Modeling
ISBN 978-0-323-95733-5
https://doi.org/10.1016/B978-0-323-95733-5.00021-1

technologies can be realized in maxillofacial prosthetics. Technologies such as three-dimensional surface capture (3D scanning), three-dimensional computer-aided design (3D CAD), and layer additive manufacturing processes (or rapid prototyping (RP) and manufacturing, AM) have been investigated in maxillofacial prosthetic applications. However, the literature is mostly comprised of reports of single-case studies that describe a given technology or application. Much early work involved making an anatomic form using AM processes such as stereolithography or laminated object manufacture [2—5]. This research however did not attempt to integrate the AM technologies into existing prosthetic practice or was limited to the production of an anatomic form that was used as a pattern for replication into more appropriate materials via secondary processes such as silicone molds or vacuum casting. Later research attempted to use AM methods to produce molds from which prosthesis forms could be molded, but these required extensive time and CAD facilities and did not attempt to exploit the advantages of RP processes that could be better integrated with existing prosthetic practice [6,7]. Other researchers attempted to exploit the ability of the ThermoJet (3D Systems Inc., 333 Three D Systems Circle, Rock Hill, SC 29730, USA) process to produce prosthesis forms in a wax material that was comparable to the waxes used in the typical prosthetic laboratory [8—11]. This enabled a more integrated approach that incorporated the advantages of AM into the existing workflow. However, these studies lacked critical evaluation of the physical properties of the prostheses produced or the technical capabilities and appropriateness of the AM technologies used for developing an integrated and efficient digital prosthetic process.

A wider investigation was therefore required to explore how these technologies may be applied in combination to a variety of applications. The research described here was part of a wider project to investigate whether available design technologies could be successfully applied to produce gains in efficacy and efficiency in a busy maxillofacial unit. The work resulted in a PhD thesis that covers all of the work in greater detail than is possible in this paper [12]. Four case studies (with three previously reported) were explored over a 4-year period in collaboration with a regional maxillofacial unit and complemented by two technical experiments (with one previously reported). This chapter focuses on the aspects of the research that addressed the current capabilities and limitations of layer AM technologies, more commonly referred to as rapid prototyping and manufacturing technologies, in maxillofacial prosthetics. The paper summarizes 4 years of study and draws conclusions on the current state of the art of AM in maxillofacial prosthetics and includes recommendations and target technical specifications toward which future AM developments should be made to meet the needs of patients and clinicians in this field.

5.16.2 Methodology

The case studies were planned and carried out using an action research (AR) approach. AR methods utilize an iterative process of research design, research implementation, and evaluation. AR approaches and case studies are typically applied in the social sciences for studying "real-life" situations where the researcher cannot control all of the variables or the research environment [13]. Typically, these involve complex, changing situations and small samples. The nature of maxillofacial treatment means that each case is unique, and therefore repetitive or series studies are not possible in a clinical setting, making the AR approach appropriate for this research. Therefore, the research was undertaken through iterative case studies, the findings from each case study informing the design and implementation of the subsequent case study or experiment. The practical methods utilized in each case study or experiment varied according to the needs of the individual case study. A number of the cases required flexibility and adaptation to unforeseen challenges. Throughout the research, a small number of complementary experiments were carried out to establish technical capabilities that informed subsequent case studies and did not require clinical investigation.

5.16.3 Summary of case studies

This section summarizes the methods and results from the cases that directly employed AM technologies. Additional, complementary case studies and experiments were carried out into related digital technologies such as 3D scanning but are not reported here.

5.16.3.1 Case 1—Orbital prosthesis

This study was intended to explore the use of CAD and AM methods that had been identified in the literature. The case utilized computed tomography (CT), FreeForm Modeling Plus Computer-Aided Design software (SensAble Technologies Inc., 15 Constitution Way, Woburn, MA 01801, USA), and ThermoJet RP (3D Systems Inc., 333 Three D Systems Circle,

Rock Hill, SC 29730, USA) combined with conventional fitting and finishing techniques to produce the final prosthesis.

This case explored the idea that clinical three-dimensional CT data acquired for maxillofacial surgery purposes could also be subsequently applied to aid in prosthetic design and manufacture. As it would be inappropriate to undertake unnecessary CT scans without full clinical justification, this work was based on exploiting existing CT scans that the patient had previously undergone for diagnostic or surgical planning reasons. This case demonstrated that such CT data was sufficient to capture the gross facial anatomy, but it could not capture fine details such as skin texture and wrinkles, for example at the corners of the eyes. This is because a typical CT scan taken for maxillofacial surgery will require a field of view wide enough to acquire images of the whole head; this would be typically 25 cm. A typical CT scanner in clinical use will have a pixel array of 512×512 leading to a pixel size of 0.488 mm. As skin textures typically have depths in the range $0.1-0.8$ mm, it is clear that the pixel size is too great to be able to adequately describe skin texture [14]. Although it is technically feasible that a CT scan could be optimized to enable the capture of skin texture, the change in CT protocol required would increase scan time and X-ray dosage for the patient, which would be difficult to justify.

This case also indicated that FreeForm was an appropriate CAD tool for the design of the basic shape of a prosthesis based on the surface data captured from the patient. The ThermoJet process successfully manufactured the prosthesis form. The CAD design and final ThermoJet model are shown in Fig. 5.115. It was noted by the prosthetist undertaking the case that the wax ThermoJet material was compatible with his conventional maxillofacial laboratory techniques. The wax prosthesis form was therefore finished and fitted to the patient in the normal manner. It was then used as a sacrificial pattern in the production of the final silicone rubber prosthesis.

Figure 5.115 Orbital prosthesis designs in CAD and ThermoJet wax pattern.

This case proved that the CAD and AM technologies described in the literature were appropriate for further development in facial prosthetics and was subsequently published [15].

5.16.3.2 Case 2—Texture experiment

The first case highlighted the importance of capturing and reproducing fine skin features such as wrinkles and texture. The experiment involved developing three-dimensional skin texture relief on an anatomical shape using FreeForm CAD and reproducing the relief at a series of depths identified from dermatology literature ranging from approximately 0.1 to 0.8 mm in depth [14]. The parts were produced using the ThermoJet process. The ThermoJet process was found to be capable of reproducing a convincing skin texture on a prosthesis, and the study was published in 2006 [16]. One of the CAD models and the resulting ThermoJet model are shown in Fig. 5.116.

5.16.3.3 Case 3—Auricular prosthesis A

A review of the literature revealed that although AM had been used in maxillofacial prosthetics, the cases had all addressed only the creation of the overall anatomic form of the prosthesis. The accepted "gold standard" for maxillofacial prostheses is the osseointegrated implant-retained silicone prosthesis. Implant-retained maxillofacial prostheses typically consist of multiple components, each having different physical requirements. A typical maxillofacial prosthesis will consist of the following items.

Figure 5.116 Skin texture sample in CAD and ThermoJet wax pattern.

- main body of the prosthesis to provide the anatomical form and appearance in a soft silicone material
- rigid substructure to support the main body and provide a firm location for retention mechanisms
- retention mechanisms in the prosthesis (clips or magnets)
- retention mechanisms attached to the patient (bar for clips or magnets)

This case investigated the use of 3D scanning, CAD, and AM techniques in the design and manufacture of a magnet-retained ear prosthesis. 3D scanning was used to capture a plaster replica of the patient's defect site and contralateral (unaffected) ear, and then FreeForm CAD and ThermoJet technologies were used combined with standard fitting and finishing techniques. The resulting prosthesis showed that advanced technologies and AM could be used to develop a magnet-retained ear prosthesis design, and the case was reported in 2006 [17].

5.16.3.4 Case 4—Auricular prosthesis B

This was built on the experience gained in the previous case but addresses the bar and clip retention method. As previously, the defect site and contralateral ear were scanned, and all of the prosthesis components were designed using FreeForm CAD. This case explored the application of a number of AM processes in the manufacture of the various prosthesis components. As it had been shown to be successful in previous cases, the ThermoJet process was used to produce a wax pattern of the main body of the prosthesis. The rigid substructure of the prosthesis was attempted using stereolithography (SLA, 3D Systems Inc.). The design of the substructure incorporated the retention clips. The bar (which remains attached to the patient by the implants) is usually made from a suitable metal such as titanium or gold. In this case selective laser melting (SLM, then MTT Technologies Ltd., now Renishaw, UK) was utilized to produce the bar in 316L stainless steel. This study identified significant limitations in capturing sufficient detail of the implant abutments in order to design a precisely fitting bar. The stereolithography substructure proved adequate in design, rigidity, and strength and was found to be very comparable to the physical properties of the light-cured acrylic materials that are typically used in current maxillofacial laboratories. However, it was readily apparent upon attaching and removing the substructure to the bar that the integral clips provided insufficient retention strength as they could be easily removed using very slight finger pressure. As the retention strength was so apparently low, it was decided that accurate measurement of the retention force was

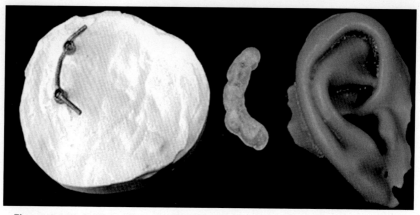

Figure 5.117 SLM retention bar, SLA substructure, and ThermoJet wax pattern.

redundant. In addition, it was also readily apparent that the service life would be insufficient for clinical use. Visible wear was apparent on the clips after fewer than 100 cycles of attachment and removal, at which point the retention strength reduced to the point where the clips no longer functioned. The SLM bar demonstrated that the process had the potential to produce a clinically acceptable bar if the quality of the detail and finish could be enhanced. The SLM bar, SLA substructure, and ThermoJet ear are shown in Fig. 5.117. These findings were reported in 2006 [18].

5.16.3.5 Case 5—Nasal prosthesis

This case was undertaken to apply the findings of the previous cases and compare them to entirely conventional methods. The case involved the design and manufacture of a nasal prosthesis incorporating magnetic retention. To overcome the limitations of 3D scanning the implant abutments described in previous cases, a 3D scan of a plaster replica of the patient's defect site was used rather than scanning the patient directly. The design was undertaken using FreeForm, and stereolithography was used to produce a rigid substructure, and the ThermoJet process was used to produce the main prosthesis body in wax, as shown in Fig. 5.118. This case demonstrated that digital methods (combined with conventional fitting and finishing) were capable of producing a clinically acceptable facial prosthesis. It was shown that there were no significant differences in the positional accuracy and reproduction of anatomic shape achieved by both techniques. The only significant difference was in the quality of the margin of the prosthesis. The margins of facial prostheses are made to be as thin as

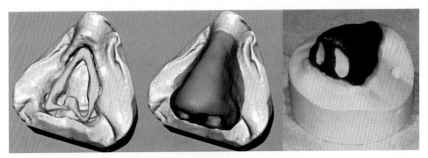

Figure 5.118 CAD designs for nasal prosthesis and ThermoJet wax pattern.

possible. This causes the silicone to become transparent and extremely flexible, which allows the edge of the prosthesis to be closely adapted to the patients' skin and follow their facial contours without leaving a conspicuous gap. The digitally produced, designed, and manufactured prosthesis was not able to reproduce a sufficiently thin margin, which led to a noticeable gap.

In addition to investigating the technical capabilities of advanced technologies, this case was also used to compare time and cost effectiveness. This case suggested that reductions in the overall design and construction time were possible when utilizing digital techniques. The application of digital technologies in this case gave a 1-hour reduction in the time spent by the prosthetist and reduced the patient time spent in the clinic by 2 hours and 35 minutes. Although these reductions in time do not appear hugely significant, the ability to go straight to color matching effectively removes a much longer period where the patient must be either in clinic or waiting nearby. The waiting and in-clinic period may effectively be reduced from a day to a single morning of work.

The reduction in consultation and laboratory time was, however, offset by the additional stages of scanning and AM fabrication. Although this extra time does not require an operator, clinician, or patient input, it does add to the overall delivery time. AM fabrication is typically an overnight process with an additional delay for postage if built by a service provider. The 40-minute period required to set up scans and process the data was represented by four block periods: three 5-minute sessions to set up the individual scans and a final 20-minute period to process the data. This effectively meant that the operator had to attend to the scanner despite being able to carry on with other work. The same situation is reflected for the patient and prosthetist during periods between construction stages, such as material curing and boiling out of molds.

5.16.3.6 Case 6—Direct manufacture of retention bar experiment

This experiment addressed only the design and direct manufacture of a prosthesis retention bar. To overcome the accuracy issues when capturing the implant abutments, this experiment utilized a touch probe scan of a plaster cast of a patient's defect site (Roland Pix-30, Roland ASD, 25691 Atlantic Ocean Drive, B-7, Lake Forest, CA 92630, USA). The design of the bar was undertaken using FreeForm, and the bar was produced directly using SLM. Unlike the previous attempts to produce a bar using SLM, this bar was produced on an SLM machine specifically designed to produce small, accurate parts (SLM-100, then MTT Technologies Ltd., now Renishaw). The resulting bar proved to be a good fit when it was screwed on to the abutments located in the replica plaster cast, as shown in Fig. 5.119. Accuracy and surface finish of the bar were deemed acceptable for clinical use by the prosthetists.

5.16.4 Discussion

The cases studies were analyzed in three areas: quality, economic impact, and clinical implications. Quality refers to characteristics such as fit (marginal integrity of the prosthesis against the skin and fit between components), accuracy (anatomic shape and position), resolution (reproduction of folds, wrinkles, texture, and substructure components), and materials (mechanical and chemical properties).

Figure 5.119 SLM bar fitted to patient cast.

5.16.4.1 Fit

The margins produced using traditional techniques were measured to range from 40 to 130 μm using a dial test indicator. Although some AM technologies are capable of producing very thin layers, most could not produce parts in a material that could be readily incorporated into the workflow of the maxillofacial prosthetist. The vast majority of maxillofacial prosthetic sculpting is carried out using wax. Currently available AM processes that are capable of sufficiently thin layers include Objet 3D printing (Objet Geometries Ltd.), Perfactory (EnvisionTEC), Solidscape (Solidscape Inc.), ThermoJet (3D Systems), and ProJet (3D Systems). Of these, only the Solidscape, ThermoJet, and ProJet systems are capable of building in a wax-based material that could be incorporated into conventional lab techniques. The Solidscape machine was deemed too slow for the size and mass of the typical facial prosthesis. Although the ThermoJet process proved capable of producing patterns in an appropriate wax material and produced parts in 40-μm-thin layers, limitations with the CAD process and fragility of the wax meant that patterns were fragile, and very thin edges were prone to breaking when support structures were removed. In order to produce sufficiently thin margins, the edges of the wax AM pattern were heated and blended into the lower section of the dental stone mold using flexible metal sculpting tools.

Despite some reports in the literature, there is no recognized standard method of assessing the fit between maxillofacial prosthesis retention components, e.g., between bar and implant abutments [19–21]. Clinical methods often rely on identifying gaps by visual inspection and finger pressure tests for movement. The cases reported here suggested that the SLM process was capable of producing a satisfactory bar but that the optical scanning methods were not able to capture data of sufficient quality directly from the patient to enable the design of the bar.

5.16.4.2 Accuracy

The accuracy of a maxillofacial prosthesis is essentially a subjective visual assessment. The prosthesis should be convincing in restoring the appearance of the particular individual. Therefore, it is very difficult to assess the accuracy using a quantifiable system. The approach taken in these studies has been to assess the accuracy of the prostheses created using AM techniques to those produced using entirely conventional techniques. In case 5, a five-point scale was used to rate the digital and conventionally produced prostheses in terms of quality of edge, positional accuracy, and overall shape. Thirteen clinical staff from the Maxillofacial Unit at Morriston Hospital were asked to rate each of the prostheses based upon photographs provided of each prosthesis. All were blinded to the production methods used (none were involved in providing these prostheses), and the images were provided sequentially. Table 5.4 shows the results. The results were analyzed using a paired, Student's t-test ($P = .05$) to identify the significance between the results.

In terms of edge quality, there was statistical significance in opinions between the two prostheses ($P = .003695$) in favor of the conventional prosthesis. The average score for the conventional prosthesis was 3 (st dev = 1.35). The average score for the digitally produced prosthesis was 1.8 (st dev = 0.93). For positional accuracy, there was no statistical significance in opinions between the two prostheses ($P = .179533$). The average score for the conventionally produced prosthesis was 4.1 (st dev = 0.76). The average score for the digitally produced prosthesis was 3.5 (st dev = 1.2). For shape, there was no statistical significance in opinions between the two prostheses ($P = .064649$). The average score for the conventionally produced prosthesis was 4 (st dev = 1.15). The average score for the digitally produced prosthesis was 3.2 (st dev = 1.1). In this respect the AM methods used were deemed capable of producing an accurately formed and located maxillofacial prosthesis (Table 5.2).

5.16.4.3 Resolution and texture

Although the 3D scanners investigated in these cases proved able to accurately capture overall anatomical form very well, the case studies demonstrated that 3D scanning technologies are not yet able to capture data of a sufficiently detailed nature to enable the reproduction of delicate skin folds, wrinkles, and texture. This limitation was also apparent when attempting to capture the precise shape and location of implant abutments. The use of plaster cast replicas of patient anatomy showed that some touch-probe scanners are capable of scanning to a high enough resolution, but these involve unwanted extra process steps resulting in higher costs and longer delivery times.

Case 2, the texture experiment, demonstrated that CAD packages are able to create three-dimensional relief at an appropriate scale to produce convincing skin textures. It also demonstrated that the ThermoJet process was able to physically reproduce these textures over an anatomically shaped

Table 5.2 Responses to aspects of prosthesis quality.

	Poor		Fair		Average		Good		Excellent	
	A	B	A	B	A	B	A	B	A	B
Feature	3	5	1	5	3	2	5	1	1	0
Edge quality	0	1	0	1	3	4	5	4	5	3
Positional accuracy	0	1	2	1	3	6	3	3	6	2
Shape	0	1	2	1	2	6	3	3	6	2

A = conventionally produced prosthesis, B = digitally produced prosthesis

surface. However, although the experiment was successful on small samples, the associated computer file sizes were large (in the order of 100 Mb). To apply texture in this manner to the surface of a large facial prosthesis may present difficulties due to large data file sizes. A large contributor to this problem is the fact that most AM technologies rely on the STL file format, which is very inefficient for describing highly detailed surface relief.

5.16.4.4 Physical properties

Attempts to incorporate retention clips into substructures manufactured using AM techniques proved unsuccessful. Although AM technologies were shown to be capable of producing a functioning retention incorporated into the rigid substructure to a sufficient accuracy, the retention strength was poor, and they wore rapidly, reducing retention strength still further. As a prosthesis is likely to be applied and removed twice a day over a typical 12-month service life, retentive components should be expected to provide strong retention over 1460 cycles. The AM retention clips proved to suffer an unacceptable loss of retention strength after fewer than 100 cycles.

Retention bars need to be small, stiff, and have good surface finish to enable easy cleaning. Due to their proximity to the skin, nonreactive metals are required, such as gold, cobalt—chrome, stainless steel, tantalum, nitinol, or titanium. The bars produced in this study using 316L stainless steel and cobalt—chrome showed great potential and indicated that with optimized build parameters and minimal finishing, a clinically acceptable bar can be produced using AM techniques.

To date, silicone elastomer rubber has the most suitable physical properties for maxillofacial prostheses and can be color matched to a produce a highly convincing appearance [22]. Currently, no AM technology is capable of producing a prosthesis in silicone rubber or with physical properties similar to those required. This remains a great challenge to the AM industry. In the cases reported here, the optimum AM process available proved to be the ThermoJet process. The wax parts produced were successfully incorporated into the conventional prosthetic production process with adaptation to produce finer edges, add fine details, or make minor adjustments to the shape.

5.16.4.5 Economics

The economic impact of changes in maxillofacial prosthetics is difficult to quantify because each case is unique, and costs vary regionally. In the

United Kingdom, the situation is further complicated as some costs are direct, such as materials, while many of the greater costs are hidden in overheads and fixed costs, such as clinical staff salaries.

Digital techniques can potentially have a significant impact on the cost effectiveness of maxillofacial prosthesis delivery. The research undertaken and illustrated in the time savings measured in the nasal prosthesis case study explored the impact of the digital workflow compared to the traditional practice. As such, it is difficult to identify the individual impact that the AM technologies can have unless they are incorporated as part of a well-resolved digital technology—based workflow.

The time savings indicated in Tables 5.3 and 5.4 identified that there was potential for savings in direct costs and opportunity costs. Although the productivity of rapid prototyping technologies cannot be objectively assessed from a single case and there would be a learning curve associated with moving to a new procedure, the fundamental differences between a digital workflow and traditional techniques enable a more flexible approach to workflow management and reduce the necessity of patient attendance at clinics. Data acquisition can be undertaken rapidly and efficiently in a clinical setting. This enables the design stages to be scheduled and undertaken at the discretion of the prosthetist without requiring the patient to be present. The reduction in both the duration of patient attendance and the number of occasions they are required to attend would result in significant savings in travel, accommodation, and missed appointments (referred to as

Table 5.3 The time taken to construct the nasal prosthesis using conventional methods.

Stage	Prosthetist time (minutes)	Patient time (minutes)	Setting/curing time (minutes)
Initial consultation and impression taking	50	50	Included
Production of stone replica	50	0	15
Base-plate design	40	15	0
Pattern design	195	140	0
Mold production	95	0	40
Color match	60	60	0
Curing	0	0	75
Finishing	80	60	0
Total	9 h 30 min	5 h 25 min	2 h 10 min

Table 5.4 The time taken to construct the nasal prosthesis using AM technologies.

Stage	Prosthetist/ operator time (minutes)	Patient time (minutes)	Setting/ curing/ fabrication time (minutes)
Initial consultation and impression taking	50	50	Included
Production of stone replica	55	0	15
Scanning and conversion to STL	40	0	120
Base-plate design	15	0	0
Pattern design	76	0	0
Pattern fabrication	15	0	180
Sub-structure fabrication	20	0	90
Mold production (assume same as conventional)	95	0	40
Color match (assume same as conventional)	60	60	0
Curing (assume same as conventional)	0	0	75
Finishing (assume same as conventional)	80	60	0
Total	8 h, 30 min	2 h, 50 min	8 h, 40 min.

"did not attends"). In addition, the savings to the patient in terms of time away from home or work is also beneficial.

In a digital workflow, the physical production of prosthetic components can be achieved in a small batch basis, which would enhance the cost effectiveness of running AM machines in a clinical setting. However, current technologies would require a level of investment that would be difficult to justify for all but the largest and busiest prosthetic units.

Clearly, future developments in this area depend on the identification of a market opportunity. Maxillofacial prosthetics is relatively small when compared with other medical sectors, which combined with the varying models of healthcare delivery across different nations makes estimating the size of the facial prosthetics market extremely difficult. However, the demand for facial prosthetics is increasing with the improved detection and surgical intervention of head and neck cancer. A 2006 survey conducted by Watson et al. [23] found that maxillofacial prosthetists in 50 hospitals in the UK produced 4259 prostheses annually. This includes other work typically undertaken by maxillofacial prosthetists such as breast, nipple, hand, and finger prostheses.

Although the UK enjoys a comprehensive maxillofacial prosthetics service through the National Health Service, other nations that rely on healthcare insurance have lower levels of provision. However, even taking that into account, it would be reasonable to anticipate similar levels of activity throughout countries in Western Europe, North America, Japan, Australia, and New Zealand. Using UK figures, a crude estimate based on population sizes would suggest that more than 64,000 facial prostheses are made each year in the wealthiest nations. However, there is potentially enormous demand for facial prostheses throughout the developing world that is currently unmet. The developing world could benefit greatly from rapid, low-cost methods of providing many thousands of facial prostheses. It is therefore reasonable to suggest that a potentially valuable global market could be developed for AM processes dedicated to the manufacture of facial prostheses.

5.16.5 AM specification

This research analyzed current best practice in prosthetic design and manufacture to identify key characteristics of prostheses. These characteristics were used to identify target specifications for digital technologies that would meet the particular needs of maxillofacial prosthetics. It is anticipated that dissemination of these target specifications will enable the AM industry to adapt or develop processes and machines specifically for the rapid and cost-effective production of maxillofacial prosthetics. Table 5.5 contains the target specifications for AM technologies.

Table 5.5 Target specifications for AM technologies.

Bar structure	
Material	Stiffness approximately equal to or greater than 18 carat (75% gold) > 75 GPa Suitable to polish Bio-compatible
Resolution	Sufficient to build 1.3 mm diameter holes with sharp detail

(Continued)

Table 5.5 Target specifications for AM technologies.—cont'd

Pattern	
Wax material	Softening temperature in the order of 35–43°C; Melt point approximately 60–63°C; 0% flow at 23°C as per ISO standard 15854:2005; In the order of 25%–30% flow at 37°C (specification of Anutex wax by Kemdent)
Resolution	Equal to or better than a ThermoJet printer $300 \times 400 \times 600$ dpi at 40 μm layer thickness

Sub-structure/clips	
Resolution	Equal to or better than an Objet or Perfactory; Objet = $600 \times 300 \times 1600$ dpi, Perfactory = 90 μm minimum pixel size; 15 μm minimum layer thickness
Material	Wear/fatigue resistant (approximately 1460 cycles to represent 1 year of use). Clips approximately 150 kg/mm^2 hardness equivalent 18 carat gold Able to bond to the prosthesis body Resist hot water at 90°C during mold release Hydrophobic (will not soften in the sustained presence of moisture/body fluids) Water sorption equal to or less than 0.6 mg/cm^2 (that of heat-processed acrylic)
Other	Clip strength should be adjustable

Prosthesis body	
Color	Production process capable of creating millions of colors from digital color matches of a patient's skin wrapped around a CAD model
Resolution	Equal to or better than a ThermoJet printer $300 \times 400 \times 600$ dpi at 40 μm layer thickness
Material	A20-30 Shore hardness, >500% elongation at break, >16 kN/m tear strength, 4.8 N/mm^2 tensile strength
Environment	Degradation resistant to UV light, dirt and body secretions

5.16.6 Conclusions

This research has highlighted the fact that AM technologies have not been developed specifically toward the needs of maxillofacial prosthetics. However, although much research has demonstrated the potential effectiveness that AM technologies have in maxillofacial prosthetics, the research described here has indicated that this potential cannot be fully exploited by currently available technologies. The full benefits of digital technologies will only be achieved through the adoption of an appropriately devised, implemented, and evaluated workflow. In addition, AM technologies need to be developed to address the specific materials and process requirements of the field of maxillofacial prosthetics.

References

[1] Wolfaardt J, Sugar A, Wilkes G. Advanced technology and the future of facial prosthetics in head and neck reconstruction. International Journal of Oral and Maxillofacial Surgery 2003;32(2):121—3.

[2] Chen LH, Tsutsumi S, Iizuka T. A CAD/CAM technique for fabricating facial prostheses: a preliminary report. International Journal of Prosthodontics 1997;10(5): 467—72.

[3] Coward TJ, Watson RM, Wilkinson IC. Fabrication of a wax ear by rapid-process modelling using stereolithography. International Journal of prosthodontics 1999;12(1): 20—7.

[4] Bibb R, Freeman P, Brown R, Sugar A, Evans P, Bocca A. An investigation of three-dimensional scanning of human body surfaces and its use in the design and manufacture of prostheses. Proceedings - Institution of Mechanical Engineers Part H, Journal of Engineering in Medicine 2000;214(6):589—94.

[5] Chua CK, Chou SM, Lin SC, Lee ST, Saw CA. Facial prosthetic model fabrication using rapid prototyping tools. Integrated Manufacturing Systems 2000;11(1):42—53.

[6] Cheah CM, Chua CK, Tan KH, Teo CK. Integration of laser surface digitizing with CAD/CAM techniques for developing facial prostheses. Part 1: design and fabrication of prosthesis replicas. International Journal of Prosthodontics 2003a;16(4):435—41.

[7] Cheah CM, Chua CK, Tan KH. Integration of laser surface digitizing with CAD/CAM techniques for developing facial prostheses. Part 2: development of molding techniques for casting prosthetic parts. International Journal of Prosthodontics 2003b;16(5):543—8.

[8] Verdonck HWD, Poukens J, Overveld HV, Riediger D. Computer-Assisted Maxillofacial Prosthodontics: a new treatment protocol. International Journal of Prosthodontics 2003;16(3):326—8.

[9] Reitemeier B, Notni G, Heinze M. Optical modeling of extraoral defects. Journal of Prosthetic Dentistry 2004;91(1):80—4.

[10] Sykes LM, Parrott AM, Owen P, Snaddon R. Applications of rapid prototyping technology in maxillofacial prosthetics. International Journal of Prosthodontics 2004;17(4):454—9.

[11] Chandra A, Watson J, Rowson JE, Holland J, Harris RA, Williams DJ. Application of rapid manufacturing techniques in support of maxillofacial treatment: evidence of the requirements of clinical application. Proceedings - Institution of Mechanical Engineers, Part B: Journal of Engineering Manufacture 2005;219(6):469—76.

[12] Eggbeer D. The computer aided design and fabrication of facial prostheses. PhD Thesis. Cardiff, UK: University of Wales Institute Cardiff; 2008.

[13] Yin RK. Case study research: design and methods. 3rd ed. London, UK: Sage Publishing; 2003.

[14] Lemperle G, Holmes RE, Cohen SR, Lemperle SM. A Classification of facial wrinkles. Plastic Reconstructive Surgery 2001;108(6):1735—50.

[15] Evans P, Eggbeer D, Bibb R. Orbital prosthesis wax pattern production using computer aided design and rapid prototyping techniques. Journal of Maxillofacial Prosthetics & Technology 2004;7:11−5.

[16] Eggbeer D, Evans P, Bibb R. A pilot study in the application of texture relief for digitally designed facial prostheses. Proceedings - Institution of Mechanical Engineers Part H, Journal of Engineering in Medicine 2006;220(6):705−14.

[17] Eggbeer D, Bibb R, Evans P. Assessment of digital technologies in the design of a magnetic retained auricular prosthesis. Journal of Institute of Maxillofacial Prosthetists and Technologists 2006;9:1−4.

[18] Eggbeer D, Bibb R, Evans P. Towards Identifying specification requirements for digital bone anchored prosthesis design incorporating substructure fabrication: a pilot study. International Journal of Prosthodontics 2006;19(3):258−63.

[19] Brånemark PI. Osseointegration and its experimental background. Journal of Prosthetic Dentistry 1983;50:399−410.

[20] Jemt T. Failures and complications in 391 consecutively inserted fixed prostheses supported by Brånemark implant in the edentulous jaw: a study of treatment from the time of prosthesis placement to the first annual check-up. The International Journal of Oral & Maxillofacial Implants 1991;6(3):270−6.

[21] Kan JY, Rungcharassaeng K, Bohsali K, Goodacre CJ, Lang BR. Clinical methods for evaluating implant framework fit. The Journal of Prosthetic Dentistry 1999;81(1):7−13.

[22] Aziz T, Waters M, Jagger R. Analysis of the properties of silicone rubber maxillofacial prosthetic materials. Journal of Dentistry 2003;31:67−74.

[23] Watson J, Cannavina G, Stokes CW, Kent G. A survey of the UK maxillofacial laboratory service: profiles of staff and work. British Journal of Oral and Maxillofacial Surgery 2006;44(5):406−10.

CHAPTER 5.17

Prosthetic rehabilitation applications case study 6—Evaluation of direct and indirect additive manufacture of maxillofacial prostheses using additive manufacturing*

5.17.1 Introduction

Increasing patient numbers, the need to improve process efficiency, the desire to add value to the profession, and the lack of access to facial prostheses provision in some areas of the world have led researchers to investigate the potential benefits of computer-aided technologies. Technologies such as three-dimensional surface scanning, computer-aided design (CAD), and rapid prototyping/additive manufacturing (RP/AM) have been applied in a number of research cases, yet they are not in widespread clinical application. Within the published literature, computer-aided technologies have been employed in different ways, with the most common method being to digitize pattern design and incorporate this into a conventional mold and final prosthesis production [1–12]. These methods rely on lab-based methods to produce molds or time-consuming techniques such as

---

*This work was first reported in the reference below and is reproduced here with kind permission of Sage Publishing.

Evaluation of direct and indirect additive manufacture of maxillofacial prostheses, Proceedings of the Institution of Mechanical Engineers, Part H, Journal of Engineering in Medicine 2012;226(9):718–728. ISSN: 0954-4119, https://doi.org/10.1177/0954411912451826.

Medical Modeling
ISBN 978-0-323-95733-5
https://doi.org/10.1016/B978-0-323-95733-5.00022-3

vacuum casting. Computer-aided mold tool production has also been attempted [13]; however, the techniques presented were not sympathetic to the skills of maxillofacial prosthetics, prosthodontics, or anaplastologist professions and were not able to address the subtlety of design that makes a facial prosthesis realistic.

Previous research has predominantly been reports of individual cases and has not attempted critical evaluation. Consequently, they do not provide robust evidence to support or dismiss either the clinical efficacy or cost effectiveness of computer-aided technologies. Such evidence is recognized as critical to the adoption of a high-value, technology-based approach [14]. A more recent, comprehensive review concluded that "the full benefits of digital technologies will only be achieved through the adoption of an appropriately devised, implemented and evaluated workflow" [15].

Furthermore, silicone elastomer is proven in clinical application, and material characteristics have been identified [16], but how this compares to currently available AM materials has not been explored.

This paper addresses the limitations of previous research and considers three aspects of evaluating the effectiveness of computer-aided technologies in facial prosthesis production: workflow, aesthetic outcome, and material characteristics.

Computer-aided methods were evaluated through a case study and controlled experiments to ISO standards. Through consultation with the prosthetist undertaking the case, criteria for aesthetic evaluation were established, and barriers encountered in previous research were considered. This helped to ensure an appropriate and intuitive process with outcomes that could be evaluated. This paper also compares the physical properties of an AM material that could be used in the fabrication of facial prosthetics with those of a benchmark silicone commonly used in facial prosthesis production.

5.17.2 Methods

A magnet-retained nasal prosthesis case was chosen as a case study to compare computer-aided with conventional methods. The patient had undergone a rhinectomy (total nose removal) following cancer and had been wearing a prosthesis for 2 years prior to this revision. Informed patient consent was obtained, and the study was undertaken as part of routine treatment to provide a new prosthesis. This ensured that minimal additional procedures were required. The processes are illustrated in Fig. 5.120.

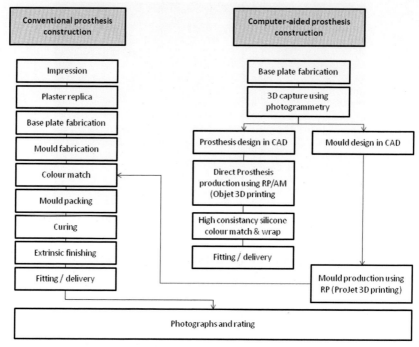

Figure 5.120 Prosthesis construction methods.

On reviewing available AM technologies, it was apparent that no existing technology is capable of producing a realistic, detailed, colored facial prosthesis directly. Therefore, an analysis of existing workflows revealed two possible applications of AM that could potentially improve the efficiency of prosthesis construction while maintaining a viable outcome:

(i) Direct AM prosthesis production of the body of the prosthesis from a digital design, which could then be wrapped in a very thin layer of color-matched, detailed silicone;

(ii) Indirect production of the prosthesis body in a color-matched silicone by molding in a mold produced from a digital design and made using AM.

Both approaches relied on 3D photogrammetry to capture patient anatomy data and FreeForm CAD for the initial design of the prosthesis form. A benchmark prosthesis was also fabricated by a highly skilled chief maxillofacial prosthetist with 22 years' experience using conventional methods based on recognized best practice and published literature [17]. This is shown in Fig. 5.121.

Figure 5.121 The completed prosthesis using conventional methods.

5.17.2.1 Common stage patient scanning and design

A base plate was fabricated using light-cured acrylic material (TranSheet, Dentsply, York, USA) on the original replica cast. This was also coated in a gray paint to aid three-dimensional topography capture.

Photogrammetry (3DMD, Face Capture system, Atlanta, USA) was used to capture the facial topography with the base plate in position and with the patient's original prosthesis (since they were happy with the shape). The proprietary 3D Patient (3DMD) software was used to create a stereolithography (STL) file of the two data sets. Fig. 5.122 shows the resulting mesh structure around the nasal area.

The STL data were imported into the CAD software, FreeForm Modeling Plus (Version 11, SensAble Technologies, Boston, USA) as "digital clay" models with 0.2-mm edge sharpness using the "hole filling" option. FreeForm has previously been shown to be an appropriate software application for the CAD of prostheses [6,7,18].

Areas of the face where the prosthesis margins required positive pressure to form a blended seal with the skin were adjusted using the "smudge" tool to press and reduce the thickness of skin by 1−2 mm. The "tolerance map" tool was used to gauge the depth of the modification (Fig. 5.123).

The digital face was then converted to a "buck" model, which protects it from further modification. Areas of "clay" representing air voids were then built up and the model turned into a "buck," which prevents further unwanted modification (Fig. 5.124).

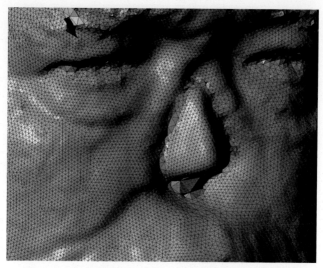

Figure 5.122 The STL file polygon mesh data of the baseplate and surrounding anatomy.

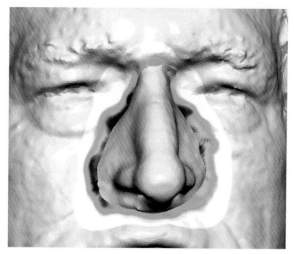

Figure 5.123 Creating relief around the prosthesis margin (see color section).

The original prosthesis form was used as the basis for the new version. This was modified to reintroduce nostrils and blend into the surrounding anatomy and had texture details added using techniques previously described [19]. The final result is shown in Fig. 5.125.

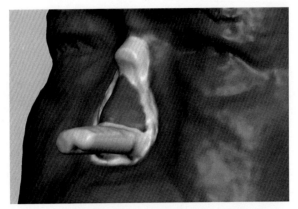

Figure 5.124 Area of cavity inside the prosthesis.

Figure 5.125 The complete digital design with texture.

5.17.2.2 Indirect approach—Mold design

A copy of the completed design was made to assist in mold design. The "buck" model of the face was subtracted in a Boolean operation, leaving the prosthesis form. The main volume of the form was selected, and then "inverse selection" and "delete" were used to remove any stray unattached pieces of "clay." The edge sharpness was refined and smoothed to 0.18 mm, which improved the smoothness of the margins. The digital prosthesis form is shown in Fig. 5.126. The "reduce for export" option was used reduce the STL file to 50 MB to make it manageable by a modern desktop computer.

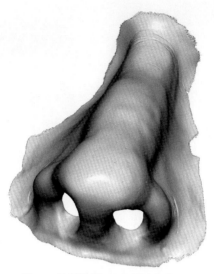

Figure 5.126 The prosthesis form.

Mold design was then undertaken in FreeForm. The copy of the prosthesis joined to the face was used to create the outer mold surface. The nostrils were blocked off with "clay" to avoid major undercuts. A copy of the completed design was then made. A line representing the mold edge was drawn on the surface of the model around 15 mm offset from the prosthesis margin. The "emboss with curve" tool was used to create an overlaying shell with a thickness of 2.5 mm on the model copy. The original version of the design was then used to perform a Boolean subtraction, leaving just the outer shell representing the top section of the mold. A flat section around the center of the mold was created to allow the mold to be clamped together.

The lower mold section was created on the face model by building up "clay" material to represent the internal cavity and leaving a cavity into which the silicone prosthesis would be formed. The design was "shut off" with the outer mold at the nostril openings, and the area around the base plate was kept clear to allow silicone to encase them. Care was taken to avoid large undercuts that would prevent the prosthesis from being released from the mold. An area of the face around the nose was selected to form the rear section of the mold tool. This was shelled to 2.5 mm thick, and the rear section was removed to reduce material and therefore cost. A flat area corresponding to that on the outer shell was also created to allow the mold to be clamped together. The final mold design is shown in Fig. 5.127.

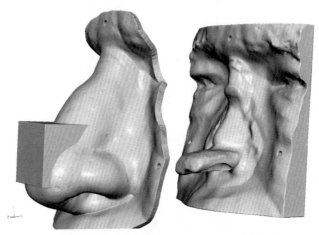

Figure 5.127 Completed two-part nose mold design.

5.17.2.3 Indirect approach—Additive manufacture

The two mold sections were fabricated using 3D printing (ProJet HD 3000 Plus, 3D-Systems, Rock Hill, USA) in Xtreme High Definition mode (the resolution of the machine in this mode is given as $750 \times 750 \times 1600$ DPI (x–y–z), and the layer thickness is 16 µm). This provided sufficient detail to reproduce the textures created in the computer model and a smooth surface finish. The build took 16 h, 40 min. Once complete, the wax supporting material was removed from the mold halves (90 min at 80°C in a temperature controlled oven), and they were cleaned and grit blasted to create a smooth, matt surface finish (approximately 30 min of manual labor).

The 3D printed mold was used to complete the prosthesis body by molding silicone in a manner similar to conventional methods. A base-shade color-matched silicone (Reality Series, Spectromatch Ltd., UK) was first mixed and used as the basis for creating different tones and shades that matched the surrounding skin. The inner surface of the front mold was painted with a mixture of base shade and variations to match the local surrounding anatomy colors and flocking to mimic capillaries before being packed out with the base shade. The mold was then closed and clamped ready for curing (Fig. 5.128). Fig. 5.129 shows the final result.

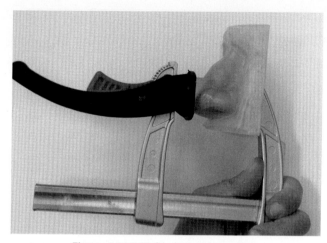

Figure 5.128 Mold clamped together.

Figure 5.129 Completed prosthesis from the AM-fabricated mold.

5.17.2.4 Direct approach

The direct prosthesis body production method used the "PolyJet modeling" 3D printing process (Objet Connex 500, Objet Geometries, Rehovot, Israel) in a soft, transparent, acrylate-based material (TangoPlus—an Objet trade name) with a specified Shore hardness of 26-A. This was the only available AM process capable of producing objects in a

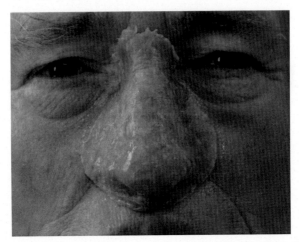

Figure 5.130 Completed prosthesis from the silicone-wrapped direct AM-fabricated pattern.

soft material with similar physical properties to the silicone rubber typically used in prosthesis production. However, at the time of the study, the material was not approved for clinical application or undergone skin sensitivity trials.

The directly manufactured prosthesis body resulted in a clear transparent form. Therefore, a novel method was required to produce a realistic color-matched surface. A high-consistency HC20 silicone (Technovent Ltd., Bridgend, UK) was mixed with base shade and flocking. This was mill rolled to create a thin, pliable sheet approximately 0.4 mm thick. This was wrapped around the prosthesis pattern that had been precoated with G604 Primer (Technovent Ltd.) used to form a strong adhesive bond. Another layer of silicone sheet was then wrapped over to create a stronger color, and the edges were blended out by pressing against a hard surface with a metal sculpting tool. This was cured at 60°C for 3 h. The final result is shown in Fig. 5.130.

5.17.2.5 Qualitative rating of aesthetic outcome

Of the three production methods, only two were judged clinically viable and worth rating in terms of aesthetic quality: the conventionally produced prosthesis and CAD/AM mold version.

Photos looking straight on, at a 45-degree angle, and side view were taken to record the aesthetic result of each prosthesis. The set of three images for each prosthesis was printed on separate sheets for the reviewers to rate the results. 19 people who worked within the hospital unit in other dental specialties but were blinded to the design and construction methods used were asked to evaluate the prostheses. A Likert five-point rating scale was used to evaluate four aspects of prosthesis appearance: positional accuracy, shape, color, and quality of edge. A Student's t-test (two tails, type 2, $P = .05$) was also used to identify the significance between the results.

5.17.2.6 Material testing

Since no specific standard yet exists for evaluating the performance of AM-produced samples, mechanical testing was undertaken as a pilot study to provide a benchmark. Three aspects of mechanical performance were tested: tensile testing, elongation at break (ISO 37:2005), and tear strength (ISO 34−1:2010). Each test was repeated five times for each material at room temperature (approximately 21°C). Three-dimensional computer models were made of the test specimens using CAD software (ProEngineer Wildfire 4, PTC, Needham, MA, USA) according to the dimensions specified in the ISO standards and shown in Tables 5.6 and 5.7. The CAD files were exported as STL files (Figs. 5.131 and 5.132) suitable for AM. All of the samples were produced in a single build using an Objet Connex 500 in TangoPlus material (Objet Ltd., Rehovot, Israel).

For the benchmark comparison, test specimens were produced from a noncolored silicone rubber (M511, Technovent Ltd., Newport, UK). This is the same chemistry of silicone that was used for the traditionally manufactured silicone prosthesis but sold by a different company and was not precolored. Test specimens were cut from a cast sheet of the silicone by mechanical die cutter to the same ISO specifications.

Tensile testing and elongation at break were ascertained using a Lloyds LR50KPlus testing machine set with a 1-kN load cell and an elongation speed of 500 mm/min. The same Lloyds testing machine was used to establish tear strengths; a 1-kN load cell was used, and the specimens were fixed into two crossheads and torn at the speed of 100 mm/min.

5.17.3 Results

5.17.3.1 Aesthetic outcomes

The results of the Likert scale ratings are shown in Tables 5.8 and 5.9. The Sutdent's t-test results are shown in Table 5.10.

Table 5.6 Dimensions of the tensile test bar.

Overall length A/mm	Width of ends B/mm	Length of narrow portion C/mm	Width of narrow portion D/mm	Transition radius outside E/mm	Transition radius inside F/mm
115	25.0 ± 1	33 ± 2	6 ± 0.4	14 ± 1	25 ± 2

Table 5.7 Dimensions of the tear test strip.

Overall length A/mm	Width of ends B/mm	Length of cut C/mm
≥ 100	15 ± 1	40 ± 5

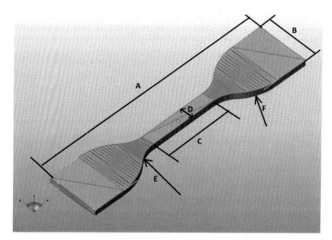

Figure 5.131 The image of tensile testing sample (ISO 34-1:2010).

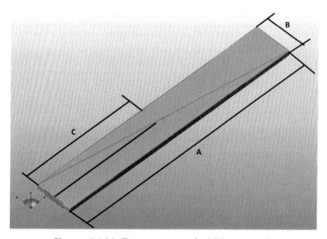

Figure 5.132 Tear test sample (ISO 37:2005).

Table 5.8 Mean average ratings and standard deviations of aesthetic quality for the conventionally produced prosthesis.

Conventional	Mean	Std. Dev.
Position	3.158	1.068
Shape	2.474	1.219
Color	3.000	1.202
Edge	1.947	1.129

Table 5.9 Mean average ratings and standard deviations of aesthetic quality for the AM mold-produced prosthesis.

RP mold	Mean	Std. Dev.
Position	3.842	0.834
Shape	4.000	0.745
Color	3.842	0.688
Edge	3.526	0.772

Table 5.10 Student's t-test results identifying significance between the rated aesthetic outcomes of each prosthesis.

Aesthetic factor	Significance
Position	0.034241
Shape	0.000043
Color	0.011873
Edge	0.000014

5.17.3.2 Mechanical properties

The calculation of tensile strength is as follows:

$$Ts = \frac{Fm}{Wt}$$

Where, Ts is tensile strength (MPa), Fm is force (N), W is width of the gauge section (mm), and t is thickness of the test length (mm). The average dimensions, maximum force, and tensile strength values of selected materials are shown in Table 5.11.

Elongation is defined as the increase of the length of narrow portion (ΔL) subjected to a tension force, divided by the original length of the test sample (L). It also can be called strain (ε). The calculation of strain is shown as follows:

Table 5.11 Average dimensions, maximum force and tensile strength values of TangoPlus and M511 silicone.

Testing sample	Average maximum force (Fm)/N	Average width (W)/mm	Average thickness (t)/mm	Average tensile strength (Ts)/MPa
TangoPlus	19.05	6.02	2.99	1.1
M511 Maxillofacial silicone rubber	61.46	6.01	3.00	3.4

$$\varepsilon = \frac{\Delta L}{L} \times 100\%$$

The average values of elongation at break were 365.36% for TangoPlus and 1181.87% for M511 maxillofacial silicone rubber.

The calculation of tear strength is shown as follows:

$$T_s = \frac{F}{d}$$

Where F is the maximum force/N, and d is the median thickness of the test piece/mm. The curves of tear load versus extension are shown in Fig. 5.133. The average tear strength of TangoPlus was 0.897 N/mm, and M511 Maxillofacial silicone was 5.471 N/mm.

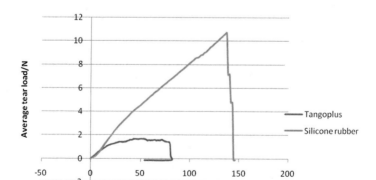

Tear load versus extension

Figure 5.133 Tear load versus extension for M511 silicone and TangoPlus.

5.17.4 Discussion

The qualitative rating results indicated a high degree of confidence that the digital mold design and production process created an aesthetically acceptable prosthesis. The results demonstrate a significant difference in favor of the AM mold prosthesis over the conventional version in all aspects, but especially in edge quality. Given that the final prosthesis was also produced in a suitably biocompatible material, with demonstrated clinical acceptance, it provided a prosthesis that was fit for purpose and viable for use. The shape was also rated significantly better on the digital version, perhaps due to the ability to analyze and adjust it from viewing angles that are more difficult to achieve when observing a patient sitting in a chair (Fig. 5.134).

From a process perspective, computer-aided technologies improved flexibility of working for the prosthetist. Design work was undertaken independently of the patient, and mold production was semiautomated. The computer-aided mold technique described also has the potential to remove a full day of clinic and waiting between processes for the patient. Conventional stages of taking a physical impression and hand carving were condensed into two short consultation periods, the first to 3D capture the facial topography and the second to color match the silicone and final fit the prosthesis.

The research presented also highlights specific limitations of using a computer-aided approach. These can be classified as process, cost, material, and technical limitations. Despite demonstrated improvements in process flexibility, further, in-depth measurement of the resources and time taken

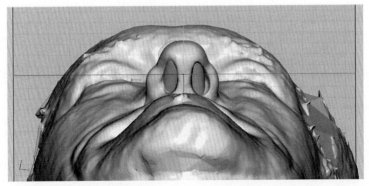

Figure 5.134 Viewing the prosthesis design from below with a measurement plane used to evaluate positioning.

to fabricate prostheses is required to accurately determine the actual cost and overall viability of computer-aided technologies. This is challenging when dealing with low case numbers and with every case being unique, but essential if a computer-based approach is to be widely adopted in clinical practice. Even with process efficiency savings, there is a significant offset in the cost of technology investment and machine time that should be considered if critically comparing the economics of each method.

Although the concept of direct AM body production was demonstrated, the mechanical testing results highlight the limitations that prevent it being used in the manufacture of a definitive prosthesis. Although the TangoPlus material currently represents the closest match to the physical properties of the benchmark silicone, it is not sufficiently robust. TangoPlus has a tensile strength of just 1.06 MPa and a tear strength of less than 1 N/mm, which when subjected to daily wear and tear would likely result in premature breakdown of thin wall sections. Since the AM material used is an acrylate-based, ultraviolet-curing photopolymer, prolonged exposure to ultraviolet light and other weathering may cause further degradation of mechanical properties, and this should be investigated in future work.

To make the direct AM fabrication process viable, it would be necessary to additively manufacture prostheses in a suitably biocompatible material with mechanical properties closer to the benchmark silicones currently used. The ability to selectively print multiple materials and transparent materials means that, in principle, the Objet process could be expanded to also selectively print color to produce a color-matched prosthesis body in a soft, pliable material. However, this would require a great deal of materials and process research and development.

Another aspect that requires further technical refinement is digital base plate design. Due to technology limitations of photogrammetry or any other currently available 3D surface scanning technique, it is not possible to capture the relatively large volume of the gross facial topography and the very small abutment or magnet details in the same scan with sufficient resolution and accuracy. Refinement of computer-aided techniques in base plate and retention mechanism design would enable an entirely digital prosthesis creation route. Experiments to refine suitable techniques are on-going.

5.17.5 Conclusions

Two alternative methods of using computer-aided technologies were used to produce a facial prosthesis. The method utilizing the AM-produced mold resulted in a prosthesis that was judged by experts to be clinically acceptable and rated superior to a benchmark prosthesis produced using conventional methods. Despite the potential for using AM to produce a prosthesis body directly, poor mechanical properties and untested biologic responses of the chosen material currently prevent it from being used in clinical application. Further research is required to test the biologic response of AM materials, improve the mechanical properties, and optimize the digital design process around direct fabrication. Further work is also necessary to incorporate base plate design to create an entirely digital process.

Both computer-aided methods enabled the prosthetist to work in a more flexible manner without relying on long patient consultations. They also reduced the length of consultation time for the patient, who only had to attend for a surface scan and then a color match on a separate day.

This research contributes toward the understanding of how computer-aided technologies may most effectively be used in clinical extraoral prosthesis cases and provides direction to future research efforts.

Acknowledgements

This work was first reported in the reference below and is reproduced here with kind permission of Sage Publishing. Evaluation of direct and indirect additive manufacture of maxillofacial prostheses, Proceedings of the Institution of Mechanical Engineers, Part H, Journal of Engineering in Medicine, 2012; 226(9): 718-728, ISSN: 0954-4119, https://doi.org/10.1177/0954411912451826.

References

[1] Chen LH, Tsutsumi S, Iizuka T. A CAD/CAM technique for fabricating facial prostheses: a preliminary report. International Journal of Prosthodontics 1997;10(5): 467–72.
[2] Coward TJ, Watson RM, Wilkinson IC. Fabrication of a wax ear by rapid-process modelling using stereolithography. International Journal of Prosthodontics 1999;12(1): 20–7.
[3] Chua CK, Chou SM, Lin SC, Lee ST, Saw CA. Facial prosthetic model fabrication using rapid prototyping tools. Integrated Manufacturing Systems 2000;11(1):42–53.
[4] Runte C, Dirksen D, Delere H, Thomas C, Runte B, Meyer U, et al. Optical data acquisition for computer-assisted design of facial prostheses. International Journal of Prosthodontics 2002;15(2):129–32.

[5] Cheah CM, Chua CK, Tan KH, Teo CK. Integration of laser surface digitizing with CAD/CAM techniques for developing facial prostheses. Part 1: design and fabrication of prosthesis replicas. International Journal of Prosthodontics 2003;16(4):435—41.

[6] Verdonck HWD, Poukens J, Overveld HV, Riediger D. Computer-Assisted maxillofacial prosthodontics: a new treatment protocol. International Journal of Prosthodontics 2003;16(3):326—8.

[7] Sykes LM, Parrott AM, Owen P, Snaddon DR. Applications of rapid prototyping technology in maxillofacial prosthetics. International Journal of Prosthodontics 2004;17(4):454—9.

[8] Evans P, Eggbeer D, Bibb R. Orbital prosthesis wax pattern production using computer aided design and rapid prototyping techniques. The Journal of Maxillofacial Prosthetics & Technology 2004;7:11—5.

[9] Reitemeier B, Notni G, Heinze M. Optical modeling of extraoral defects. The Journal of Prosthetic Dentistry 2004;91(1):80—4.

[10] Chandra A, Watson J, Rowson JE, Holland J, Harris RA, Williams DJ. Application of rapid manufacturing techniques in support of maxillofacial treatment: evidence of the requirements of clinical application. Proceedings of the Institution of Mechanical Engineers - Part B: Journal of Engineering Manufacture 2005;219(6):469—76.

[11] Eggbeer D, Bibb R, Evans P. Assessment of digital technologies in the design of a magnetic retained auricular prosthesis. Journal of The Institute of Maxillofacial Prosthetists & Technologists 2006a;9:1—4.

[12] Eggbeer D, Bibb R, Evans P. Specifications for non-contact scanning, computer aided design and rapid prototyping technologies in the production of soft tissue, facial prostheses. In: Proceedings of the 7th national conference on Rapid design, prototyping & manufacturing. Beaconsfield: MJA print; 2006. p. 67—77.

[13] Cheah CM, Chua CK, Tan KH. Integration of laser surface digitizing with CAD/CAM techniques for developing facial prostheses. Part 2: development of molding techniques for casting prosthetic parts. International Journal of Prosthodontics 2003;16(5):543—8.

[14] Wolfaardt J, Sugar A, Wilkes G. Advanced technology and the future of facial prosthetics in head and neck reconstruction. International Journal of Oral and Maxillofacial Surgery 2003;32(2):121—3.

[15] Bibb R, Eggbeer D, Evans P. Rapid prototyping technologies in soft tissue facial prosthetics: current state of the art. Rapid Prototyping Journal 2010;16(2):130—7.

[16] Aziz T, Waters M, Jagger R. Analysis of the properties of silicone rubber maxillofacial prosthetic materials. Journal of Dentistry 2003;31:67—74.

[17] Thomas K. In: Thomas S, editor. The art of clinical anaplastology. UK: 4Edge Ltd; 2006 (Chapter 4).

[18] Eggbeer D, Bibb R, Evans P. Digital technologies in extra-oral, soft tissue facial prosthetics: current state of the art. Journal of The Institute of Maxillofacial Prosthetists & Technologists 2007;10:9—18.

[19] Eggbeer D, Evans P, Bibb R. A pilot study in the application of texture relief for digitally designed facial prostheses. Proceedings of the Institution of Mechanical Engineers, Part H: Journal of Engineering in Medicin 2006c;220(6):705—14.

CHAPTER 5.18

Prosthetic rehabilitation applications case study 7—Computer-aided methods in bespoke breast prosthesis design and fabrication*

5.18.1 Introduction

Around 38,000 women were diagnosed with breast cancer in 2006 in England, Wales, and Scotland [1]. In the 2007−08 period in England and Wales, this resulted in 19,334 breast removals [2,3]. In England, there were 4209 breast reconstructions and 7786 breast prostheses in 2007−08 [2]. There are various options available to reconstruct the breast including autologous reconstruction and prosthetic implant [4]. These may require additional surgery, which may not be suitable or desirable. In the majority of these cases, brassiere-retained, external prostheses are provided.

The goal of an external prosthesis is to restore the aesthetic contours of the chest region. It is essential for the prosthesis to be comfortable to wear for long periods during the day [5]. Questions as to whether a prosthesis weight should help to maintain good posture are contested; comfort, closeness of fit, and aesthetics are the primary considerations reported [6,7].

---

* The work described in this chapter was first reported in the references below and is reproduced here, in part or in full, with the permission of Sage Publishing.

Eggbeer D, Evans P. Computer-aided methods in bespoke breast prosthesis design and fabrication. Proceedings of the Institution of Mechanical Engineers, Part H, Journal of Engineering in Medicine 2011;225(1):94−9. https://doi.org/10.1243/09544119JEIM755. Minor updates have been undertaken throughout this case study to reflect current knowledge. Further updates are described in the Discussion section.

Medical Modeling
ISBN 978-0-323-95733-5
https://doi.org/10.1016/B978-0-323-95733-5.00023-5

Studies have shown that satisfaction is significantly associated with how well the prosthesis fits, the weight, and movement [8,9]. Traditionally, prostheses are available commercially as "off the shelf" in a range of sizes or are custom made, typically by maxillofacial prosthetists. Both are usually fabricated in a soft, skin-like silicone. The bespoke route offers an improved fit, better contour, improved color match, and therefore a more lifelike appearance. However, techniques used in the production of bespoke prostheses are time consuming, complicated, and material intensive. There is also limited literature describing the typical, lab-based stages involved with producing bespoke breast prostheses [10].

Lab-based techniques also do not account for the shape of the breast when it is supported by a brassiere. There remains a need to develop an efficient method of producing well-fitted, patient-specific breast prostheses.

The introduction of computer-aided technologies such as three-dimensional topographic scanning, photogrammetry, computer-aided design (CAD), and rapid prototyping/manufacturing (RP&M) into patient-specific medical applications has provided new opportunities to improve the delivery of prostheses and other patient-specific medical devices [11–15]. However, an efficient production chain for the delivery of bespoke prostheses that match the contralateral, brassiere-supported breast has not been reported. This paper introduces a technique that utilizes three-dimensional photogrammetry, haptic computer sculpting, and RP&M methods to produce bespoke breast prostheses.

5.18.2 Methods

Four patients to date have had a prosthesis fabricated using the technique described; however the methods will be illustrated through a single patient case study. An overview of the digital procedure is show in Fig. 5.135.

A 3dMDtorso, four-pod photogrammetry system (3dMD, USA) was used to capture the chest topography of the patient without a brassiere and with a plain, unpatterned white brassiere. The anatomy was captured pointing slightly upward and toward the breast fold (Fig. 5.136). Data acquisition took ~1.5 ms and was undertaken as part of a multipatient clinic.

Data processing took approximately 3 minutes and created a mesh of 363,346 and 190,441 triangles for the brassiere- and nonbrassiere-wearing anatomy respectively. The meshes represented surface topography of the chest area and also included a color map, which provided sufficient detail to

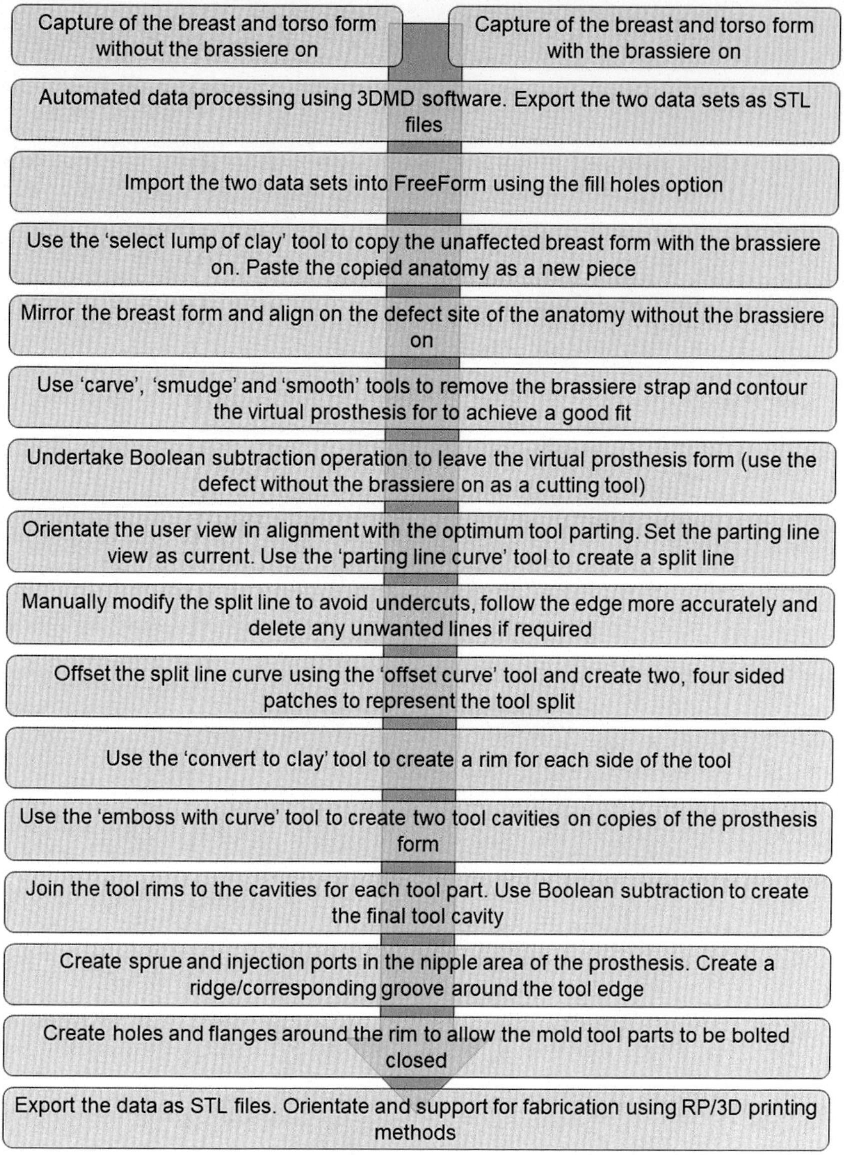

Capture of the breast and torso form without the brassiere on	Capture of the breast and torso form with the brassiere on

Automated data processing using 3DMD software. Export the two data sets as STL files

Import the two data sets into FreeForm using the fill holes option

Use the 'select lump of clay' tool to copy the unaffected breast form with the brassiere on. Paste the copied anatomy as a new piece

Mirror the breast form and align on the defect site of the anatomy without the brassiere on

Use 'carve', 'smudge' and 'smooth' tools to remove the brassiere strap and contour the virtual prosthesis for to achieve a good fit

Undertake Boolean subtraction operation to leave the virtual prosthesis form (use the defect without the brassiere on as a cutting tool)

Orientate the user view in alignment with the optimum tool parting. Set the parting line view as current. Use the 'parting line curve' tool to create a split line

Manually modify the split line to avoid undercuts, follow the edge more accurately and delete any unwanted lines if required

Offset the split line curve using the 'offset curve' tool and create two, four sided patches to represent the tool split

Use the 'convert to clay' tool to create a rim for each side of the tool

Use the 'emboss with curve' tool to create two tool cavities on copies of the prosthesis form

Join the tool rims to the cavities for each tool part. Use Boolean subtraction to create the final tool cavity

Create sprue and injection ports in the nipple area of the prosthesis. Create a ridge/corresponding groove around the tool edge

Create holes and flanges around the rim to allow the mold tool parts to be bolted closed

Export the data as STL files. Orientate and support for fabrication using RP/3D printing methods

Figure 5.135 Illustration of the digital process tools.

create the final prosthesis. Fig. 5.137 shows the surface topography captured of the patient's chest with the defect.

The mesh data were exported as the de facto industry standard STL file (stereolithography) and imported into CAD software for prosthesis design

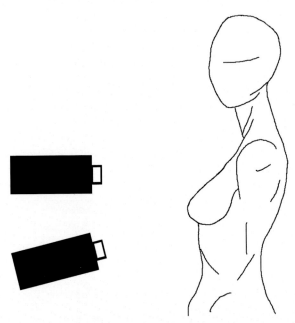

Figure 5.136 Illustration of the angle used to capture the breast contour and defect.

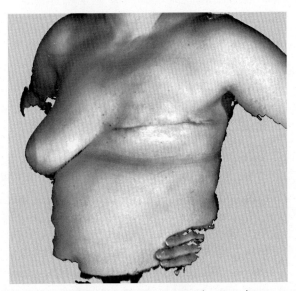

Figure 5.137 The surface topography captured using photogrammetry.

(Geomagic FreeForm Modeling Plus, version 10). FreeForm CAD was chosen due to its suitability when working with anatomic data and tools analogous to conventional lab sculpting methods [11]. The fill holes option was used to create a solid model in the CAD software, and an edge sharpness of 0.4 mm was chosen to provide sufficient detail. The unaffected side of the chest with the brassiere on was copied, pasted, and mirrored to the defect side on the second scan (Fig. 5.138).

Software carving and shaping tools were used to further shape the digital prosthesis design to fit the defect side with the brassiere off and to remove the brassiere strap. Particular attention was given to the fit of the prosthesis under the arm, which is an area of particular discomfort when wearing a stock-sized prosthesis. Once satisfied with the contours, a Boolean subtraction operation was undertaken to form the fitting contours and leave just the digital prosthesis design. This could be checked for correct size and position using the with-brassiere data (Fig. 5.139).

At this point, there were two options: (1) fabricate the prosthesis pattern using RP&M tools, and then use lab techniques to create a mold, or (2)

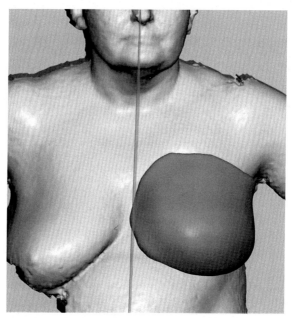

Figure 5.138 The mirrored breast form in CAD.

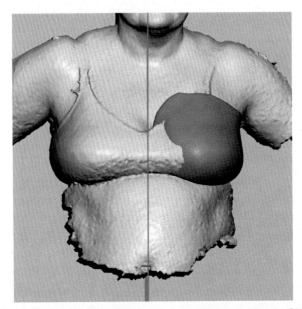

Figure 5.139 Checking the size and prosthesis position in CAD.

design a tool using the CAD software and produce that using RP&M technologies. Option two was chosen since this further reduced the dependence on lengthy lab-based techniques.

The parting line for the mold was first chosen and highlighted in FreeForm. This represented the most bulbous line around the rim of the prosthesis where the tool would split. This line was offset by 15 mm using the "offset curve" tool. Two, four-sided patches were created using the split line and offset split line. These represented the splitting surface of the tool. A new empty piece of clay with an edge sharpness of 0.4 mm was created for each tool part. The patches were given a thickness of 3 mm using the "convert to clay" tool to form the rim of each tool side. The "emboss with curve tool" was then used to create the cavities of each tool side with a thickness of 1.8 mm on a copy of the prosthesis form. The rim sections were joined to the cavities for each tool side, and holes with flanges were created around the rim to allow the tool to be bolted closed. A Boolean subtraction operation was then undertaken with the prosthesis form used to remove material from the tool. A groove and corresponding ridge were also created around the rim to prevent silicone from leaking around the split

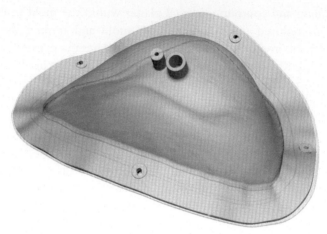

Figure 5.140 The completed prosthesis tool design in CAD.

line. Sprue injection and vent holes were then created on the top section of the mold to allow the final prosthesis material to be injected and air to escape. Optional holes and flanges can then be created, which allow the tool to be bolted closed. The completed mold design is shown in Fig. 5.140. The sections were exported as STL files.

Magics (Materialise, Belgium) was used to support the two mold sections. The sections were fabricated in 0.15-mm layers using stereolithography (SLA 250/50, 3D-Systems Corporation, USA), in WaterShed XC resin (DSM Somos, USA). This resin was chosen due to its ease of hand finishing and translucency, which would make it easier to see when the mold was fully filled with silicone. The tool sections were built on edge to allow both to fit within the build volume. The build took overnight to complete. Once completed, the sections were cleaned and hand finished with glass paper to remove the layer steps and to achieve a smooth, matt surface. The mold was then passed to the maxillofacial prosthetics lab.

Research has suggested that the weight of the prosthesis should match the remaining breast to prevent back and neck problems for the patient [16]; however, this has been contested more recently [6,7]. All previous patients treated complained about the prosthesis weight. It was therefore decided to reduce the weight of the prosthesis from a theoretical 1100 g (if the prosthesis was entirely gel silicone-based) to a final weight of 845 g using a low-density, open-cell foam polyurethane core. This was fabricated by part filling and contouring dental plaster within the mold to create an insert. This reduced the volume of the mold and formed the desired contour for the insert. A two–part, foaming polyurethane material (Technovent Ltd, UK) was injected into the mold and left to cure.

The SLA mold was first coated with a 1–mm layer of silicone elastomer (M511, Technovent Ltd, UK), which was applied incorporating Cosmesil base color pigments to achieve a "skin–like" coloration (Technovent Ltd, UK). The polyurethane core was placed on the lower half of the mold and the mold closed and bolted together. The catalyzation of the elastomer was accelerated by heating the coated mold at 80°C. The silicone elastomer gel (M512, Technovent Ltd, UK) was pigmented to match the patient's skin color and then injected through the injection port with a syringe (Fig. 5.141).

The mold was left for 24 hours to allow the gel to set. Once set, the prosthesis was removed from the mold and trimmed to remove any flash from the molding process. The injection port area was sealed over with a silicone nipple to complete the prosthesis. The completed prosthesis is shown in Fig. 5.142.

From data capture to final prosthesis delivery took approximately 5.5 prosthetist hours, plus additional automated fabrication time. Table 5.12 compares conventional, lab-based methods with the computer-aided method.

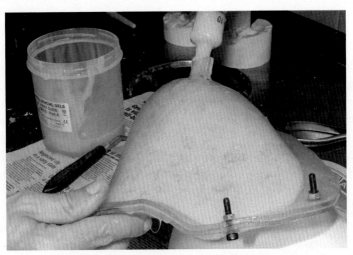

Figure 5.141 Injecting the color-matched silicone.

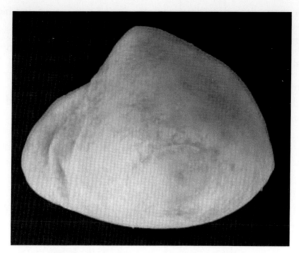

Figure 5.142 The completed prosthesis.

5.18.3 Discussion

The techniques described provided the patient with a satisfactory cosmetic result, which established symmetry with the healthy, brassier-supported breast and that fitted well against the chest wall. Follow-up reports by the breast care nurse team suggested that this helped to ensure stability and comfort of the prosthesis when the patient was mobile. The application of computer-aided technologies also provided other process benefits. Using photogrammetry removed the need for impression taking, which is labor and material intensive, time consuming, intrusive for the patient, and messy. The advantages and limitations of photogrammetry and other three-dimensional topographic scanning methods compared with direct impression techniques have been discussed and, in this case study, provided an ideal solution [16,17]. The design stages did not require the patient to be present, and FreeForm provided the ideal tool for shaping the prosthesis. Mold designing in FreeForm could be more intuitive and efficient; however it is estimated that it was still faster than conventional methods.

Alternative mold tool production methods such as selective laser sintering (EOS, GmbH, Germany) and fused deposition modeling/fused filament fabrication could also be viable for mold production but would be more difficult to achieve the desired level of surface finish.

The high cost of computer-aided technologies has been reported as a limitation in patient-specific medical applications [18]. However, given the

Table 5.12 A comparison of traditional methods and computer-aided highlighting the resource reduction and opportunities identified.

Conventional	Computer-aided method	Resource reduction/ opportunity
Take an impression. Create a replica model of the anatomy.	Photogrammetry with bra and without bra.	Time saving. Reduced material usage. Improved accuracy. Captures the defect and breast form in a brassiere.
	Import data into FreeForm.	
Manually carve wax prosthesis form.	Design prosthesis form using CAD.	Time saving. More flexible working No need for patient to be present. Reduced material usage.
Mold the prosthesis In a two-part plaster mold.	Identify and mark the tool split line.	Automated split line identification.
Boil out wax residue. Apply separator.	Design the tool sections.	Time saving. Material saving. Mistakes easy to correct.
	Export as STL data. Orientate and support for RP&M process. Fabrication. Hand finishing.	Can be duplicated easily. Automated fabrication, but additional cost over conventional methods.
Create foam mold insert. Apply color-matched silicone to the mold cavity. Inject with skin-tone color silicone. Cure. Trim and deliver.	Same as conventional.	

process and prosthesis improvements and the related indirect cost savings associated with patients undergoing shorter consultations, the computer-aided method has clear advantages. The equipment also has application in many other patient-specific device design and surgical scenarios [1–15].

Since this original work was published, the Maxillofacial Lab at Morriston Hospital, Swansea, UK, has continued to refine the methods used to provide custom breast prostheses in response to patient feedback. Demand has also increased dramatically due to the Covid 19 global pandemic, which has placed great strain on surgery to reconstruct breasts using autologous tissue or implants postmastectomy. Breast prosthesis design now incorporates a flap beneath the prosthesis that helps it to secure under the brassier. Prostheses have also been made lighter by creating a foamed silicone core structure. Lower-cost extrusion-based 3D printing methods in polylactic acid have also replaced stereolithography mold production. 3D printed versions of the prosthesis form are duplicated in plaster material, which is hand finished by wet sanding. Plaster molds are more resilient to making multiple prostheses and resist breaking or warping during the curing process. The downside to this approach is increased material use, heavy mold weights, and issues with mold growth over prolonged periods of storage.

5.18.4 Conclusions

The technique presented provided a prosthesis that accurately fitted the defect chest wall of the postmastectomy patient, providing a comfortable fit and improved retention over an "off-the-shelf" alternative. By designing the prosthesis digitally and using the mirrored contour of the breast within the brassier, a better symmetry was achieved for the wearer.

Manufacturing a custom prosthetic appliance is more cost and labor intensive than an off-the-shelf solution. Further research is required to quantify the resources and time taken. This technique has, however, illustrated how computer-aided methods are able to offer a cost-effective alternative to the traditional labor-intensive techniques involving impression taking by reducing the length of patient consultation, reducing the number of patient visits, reducing the quantity of materials used, and providing a more flexible and repeatable method of working.

References

[1] Available online from UK National Statistics. http://www.statistics.gov.uk/downloads/theme_health/MB1-37/MB1_37_2006.pdf.
[2] Available online from NHS Hospital Episode Statistics. http://www.hesonline.nhs.uk/Ease/servlet/ContentServer?siteID=1937&categoryID=210.

[3] Available online from Health Solutions Wales. http://www.wales.nhs.uk/sites3/docmetadata.cfm?orgId=527&id=105899.

[4] Rozen WM, Rajkomer AKS, Anavekar NS, Ashton MW. Post-mastectomy breast reconstruction: a history in evolution. Clinical Breast Cancer 2009;9(3):145–54.

[5] Glaus SW, Carlson GW. Long-term role of external breast prostheses after total mastectomy. Breast Journal 2009;15(4):385–93.

[6] Hojan K. Does the weight of an external breast prosthesis play an important role for women who undergone mastectomy? Reports of Practical Oncology and Radiotherapy 2020 ;25(4):574–8. https://doi.org/10.1016/j.rpor.2020.04.015.

[7] Manikowska F, Ozga-Majchrzak O, Hojan K. The weight of an external breast prosthesis as a factor for body balance in women who have undergone mastectomy. Homo November 29, 2019;70(4):269–76. https://doi.org/10.1127/homo/2019/1114.

[8] Livingston PM, White VM, Roberts SB, Pritchard E, Hayman J, Gibbs A, et al. Women's satisfaction with their breast prosthesis: what determines a quality prosthesis? Evaluation Review 2005;29(1):65–83.

[9] Gallagher P, Buckmaster A, O'Carroll S, Kiernan G, Geraghty J. Experiences in the provision, fitting and supply of external breast prostheses: findings from a national survey. European Journal of Cancer Care, June 01 2009;18.

[10] Thomas KF. Prosthetic rehabilitation. London: Quintessence Publishing Co. Ltd.; 1994.

[11] Evans PL, Eggbeer D, Bibb R. Orbital prosthesis wax pattern production using computer aided design and rapid prototyping techniques. The Journal of Maxillofacial Prosthetics and Technology 2004;7:11–5.

[12] Eggbeer D, Bibb R, Williams R. The computer aided design and rapid prototyping of removable partial denture frameworks. Proceedings of the Institution of Mechanical Engineers, Part H: Journal of Engineering in Medicine 2005;219(3):195–202.

[13] Eggbeer D, Bibb R, Evans PL. Towards Identifying specification requirements for digital bone anchored prosthesis design incorporating substructure fabrication: a pilot study. International Journal of Prosthodontics 2006;19(3):258–63.

[14] Verdonck HWD, Poukens J, Overveld HV, Riediger D. Computer-assisted maxillofacial prosthodontics: a new treatment protocol. International Journal of Prosthodontics 2003;16(3):326–8.

[15] Sykes LM, Parrott AM, Owen P, Snaddon DR. Applications of rapid prototyping technology in maxillofacial prosthetics. International Journal of Prosthodontics 2004;17(4):454–9.

[16] Kovacs L, Eder M, Hollweckb R, Zimmermanna A, Settlesc M, Schneiderd A, et al. Comparison between breast volume measurement using 3D surface imaging and classical techniques. Breast 2007;16(2):137–45.

[17] Weinberg SM, Naidoo S, Govier DP, Martin RA, Kane AA, Marazita ML. Anthropometric precision and accuracy of digital three-dimensional photogrammetry: comparing the Genex and 3dMD imaging systems with one another and with direct anthropometry. Journal of Craniofacial Surgery 2006;17(3):477–83.

[18] Sanghera B, Naique S, Papaharilaou Y, Amis A. Preliminary study of rapid prototype medical models. Rapid Prototyping Journal 2001;7(5):275–84.

CHAPTER 5.19

Prosthetic rehabilitation applications case study 8—Immediate nasal prosthesis following rhinectomy*

5.19.1 Introduction

The pursuit of computer-aided methods is particularly valuable where it offers improved patient experiences and increases service provision to more people. The impact of losing a portion of the face can be enormous [1]. One study indicates a statistically significant improvement in the domains of body image, positive emotions, and negative emotions with facial prostheses [2]. Maxillofacial prosthetists/anaplastologists and surgeons witness the anguish and impact to patients, which is especially acute immediately after surgery. Although patients are typically consulted and prepared as well as possible for waking up with anatomic features removed, the potential to offer an immediate aesthetic solution could be of value in softening the impact to both the patient and relatives/friends.

A rhinectomy involves surgical removal of all or part of the nose, usually as a result of cancer. Nasal prostheses restore contours and are designed to blend seamlessly into the surrounding anatomy using thin edges that flex with facial expressions. Previous research presented in literature and in Chapter 5.16 of this book describe how computer-aided technologies can be used in nasal prosthesis design and manufacturing. Creating a realistic nasal prosthesis takes a blend of artistic skill and knowledge of materials and

---

*This technique was developed by the Maxillofacial Laboratory at Morriston Hospital, Swansea, UK, and had been presented at regional workshops and seminars prior to publication here.

Medical Modeling
ISBN 978-0-323-95733-5
https://doi.org/10.1016/B978-0-323-95733-5.00024-7

manufacture processes. Definitive prostheses are typically provided around 6 weeks following surgery once swelling and anatomy changes have settled and implants have sufficiently osseointegrated. During this period, the patient is typically provided with a gauze covering for their nasal region. Even with computer-aided technologies, the process requires chairside consultation, usually spanning 2 days for impression taking, carving, molding, and color matching for the same procedure over many outpatient visits. Prostheses are usually expected to last around 18 months. One previous study has explored the idea of providing an immediate, temporary prosthesis following rhinectomy [3], but limitations associated with the fragility of healing tissue postsurgery have limited greater use of this approach. Notwithstanding this limitation, providing a temporary, but immediate nasal prosthesis could help to ease the impact of losing a nose for the patient, thus could be considered a valuable use of computer-aided technologies.

This case study describes the process of using computer-aided technologies to provide an immediate nasal prosthesis for patients who have had a rhinectomy due to a cancerous tumor in the nasal region. This includes methods to overcome previously noted limitations associated with healing tissue after surgery. Although not a permanent solution, this type of prosthesis was developed to ease the impact of losing a nose and to create a transition to the definitive prosthesis fabricated further along in the recovery process.

5.19.2 Methods

5.19.2.1 Planning (general protocols)

Suitable patients are identified during multidisciplinary team meetings, during which an initial treatment plan is developed in collaboration between the surgical and Maxillofacial Laboratory team. A computer tomography (CT) scan will have been undertaken to determine the tumor extent and to plan zygomatic implants, which will be used to retain the definitive prosthesis. This will typically be in the order of 1 mm slice thickness with 0.5 mm overlap, which is required to capture the thin bony areas of the maxilla.

5.19.2.2 Case study technique

The CT data were imported into Mimics V24 (Materialise, Belgium), and separate "masks" were created for bone (206–3071 HU [Hounsfield Units]) and soft tissue (−92 to 3071 HU). Operations that are described in more detail in Chapter 5.16 were used to crop the data and create a three-dimensional (3D) reconstruction of the soft tissue area of the midface. The soft-tissue 3D reconstruction was then exported as an STL file using the high-quality setting, before being imported into Geomagic FreeForm Plus (3D-Systems, USA). The "fill holes" option was used, and a clay coarseness of 0.7 mm was selected to provide a sufficient balance between resolution and computing power needed for modeling. The same process was repeated with the bony anatomy, since the guides for placing two zygomatic implants were designed in parallel within the same file.

In this case, as with most cases, the nose was swollen around the area of the tumor and due to infection edema. The nose was therefore reshaped in FreeForm Plus using a mirror of the less-affected side as a guide, followed by various modeling techniques, which are described in Chapter 5.16. Once modeling was completed, the model was cropped with a margin of approximately 25 mm around the nose, and the model was offset to the inside by 1.5 mm. This represented the intended nominal thickness of the prosthesis. The model was then optimized for 3D printing in a polylactic acid (PLA) material by shelling to approximately 2 mm thick and cropping the rear face off to reduce the material. This is not an essential process; the model could be left unshelled to leave a flat base and eliminate the need for support structures. The completed model, shown in Fig. 5.143, was then exported as an STL file.

The file was prepared for 3D printing fabrication using Idea Maker software for a Raise3D Pro2 material extrusion printer in 0.1 mm layer thickness (https://www.raise3d.com/products/pro2-3d-printer/). Once completed, the model was duplicated in plaster using alginate as an impression. This technique was used due to previous experiences where silicone curing had been inhibited by some 3D printed materials, including PLA and stereolithography resins. PLA is also not suitable for oven curing due to its low glass transition temperature.

Silicone color swatches were used to determine the most suitable base prosthesis shade at a presurgery consultation with the patient. The patient then underwent rhinectomy surgery with placement of zygomatic implants using custom surgical drill guides.

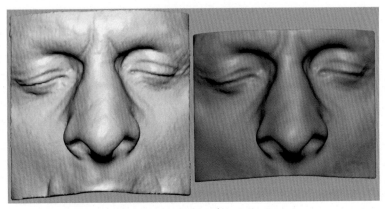

Figure 5.143 The completed design in FreeForm.

High-consistency rubber (HCR) silicone was used for the prosthesis (Spectromatch, UK). This has a gum-like consistency and is available in precolored options. The HCR silicone was weighed according to the necessary 10:2 ratio of base to catalyst, folded together, and then run multiple times through a roll mill to mix. Small amounts of colored flocking were added, and the thickness of the roll mill was adjusted to create a sheet of silicone approximately 1.5 mm thick. The sheet was draped over the plaster mold of the nose (Fig. 5.144) and manipulated into the nostril areas, trimmed, and blended around the margins (Fig. 5.145). The tip of the nose was reinforced with extra silicone to prevent it from collapsing. This was blended in by hand.

Further color flocking and textures were added using hand tools and cotton buds. A thin layer of translucent roll-milled silicone sheet was then

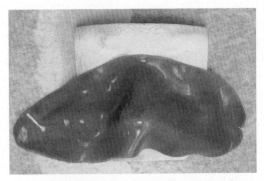

Figure 5.144 HCR silicone sheet draped over the plaster mold.

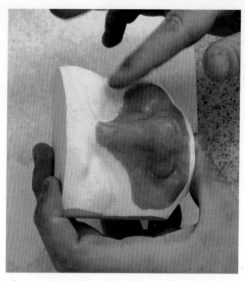

Figure 5.145 HCR silicone manipulated into the nostril areas, trimmed, and blended around the margins.

added to the top surface, blended in, and textured using a wire brush (Fig. 5.146). Finally, the margin was trimmed again, before being blended into the surrounding anatomy, breaking up the edge to hide the obvious line. The process of silicone fabrication took approximately 15 minutes. The final result before being cured is show in Fig. 5.147.

The completed prosthesis was then oven cured at 50°C for at least 8 hours before being removed from the plaster mold and trimmed.

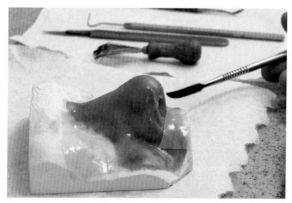

Figure 5.146 Translucent roll-milled silicone sheet added to the top surface, blended in, and textured.

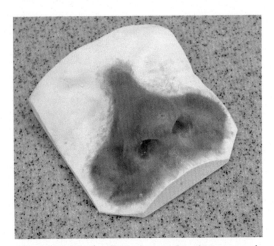

Figure 5.147 The final result before being cured.

Chlorohexidine was used to cold sterilize the prosthesis in theater prior to it being attached to the patient once the primary surgery was complete. A small number of sutures can also be used for a firmer attachment. Ribbon gauze packing is typically used around two plastic airways that allow the patient to breathe more freely through the prosthetic nose. See Fig. 5.148.

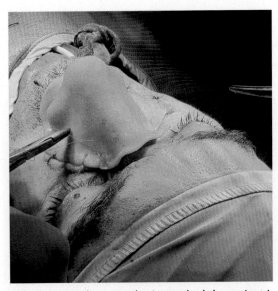

Figure 5.148 The immediate prosthesis attached the patient in surgery.

5.19.2.3 Follow-up (general protocols)

Patients who have undergone tumor removal often receive postoperative radiotherapy. The temporary prosthesis is typically removed prior to this, is cleaned, and attached to glasses. This helps to reduce the problems associated with tissue friability during radiotherapy.

After a period of around 6 weeks, once postoperative radiotherapy has been completed and implants have sufficiently osseointegrated, patients are typically referred for a definitive prosthesis.

5.19.3 Discussion

The method presented represents a relatively cost-effective way to provide patients with a temporary prosthesis following rhinectomy. There are, however, opportunities to improve upon the workflow and to provide further information on the healthcare economic considerations of this method. Case studies with no benchmark or control are difficult to assign statistical significance to. Healthcare economic considerations must be undertaken based on quantitative assessment of the resources used, which are in addition to standard treatment with no immediate prosthesis, and qualitative assessment of patient benefit. This would, however, be challenging given the small number of patients. A multicenter study could be used to compare patient feedback without immediate prosthesis reconstruction and those who have received one.

This technique also makes use of relatively expensive software, Geomagic FreeForm/FreeForm Plus, for prosthesis design, although the 3D printing hardware was low cost. Since high fidelity is not the primary goal, there are opportunities to use lower cost options, such as Meshmixer (Autodesk Inc, USA), which have been used in surgical applications and dental device design [3–6]. Where CT scanning is not used, further research could also evaluate the application of mobile phone— or tablet-based 3D scanners, such as the Structure Sensor (Occipital, Inc, USA), which are likely to offer sufficient resolution for this type of application.

Further research is also required to establish why silicone curing is sometimes inhibited by 3D printed materials. In the future, it is also foreseeable that direct 3D printing fabrication of temporary prostheses may be feasible [7], although a significant amount of research and development is required to achieve an accurate color match and sufficiently thin prosthesis margins.

5.19.4 Conclusions

This research contributes toward the understanding how computer-aided technologies may be used to provide a temporary nasal prosthesis for patients who have undergone rhinectomy. The technique described has been made possible using a combination of computer-aided technologies and silicone molding to create a sterilizable prosthesis suitable for short-term use. Future work will analyze the resources used in this method and consider them in the context of patient satisfaction.

References

[1] Jablonski RY, et al. Outcome measures in facial prosthesis research: a systematic review. The Journal of Prosthetic Dentistry 2021;126(6):805—15. https://doi.org/10.1016/j.prosdent.2020.09.010.

[2] Dholam KP, et al. Development of a psychosocial perception scale and comparison of psychosocial perception of patients with extra oral defects before and after facial prosthesis. The Journal of Prosthetic Dentistry 2021;128. https://doi.org/10.1016/j.prosdent.2021.03.017.

[3] Yoshioka F, et al. Innovative approach for interim facial prosthesis using digital technology. Journal of Prosthodontics 2016;25(6):498—502. https://doi.org/10.1111/jopr.12338.

[4] Abo Sharkh H, Makhoul N. In-house surgeon-led virtual surgical planning for maxillofacial reconstruction. Journal of Oral and Maxillofacial Surgery April 2020;78(4):651—60. https://doi.org/10.1016/j.joms.2019.11.013.

[5] Buzayan MM, Etajuri EA, Seong LG, Abidin ZBZ, Sulaiman EB, Ahmed HMA. First steps of a digital workflow to build up a virtual articulator using open-source Autodesk Meshmixer software. International Journal of Computerized Dentistry March 24, 2022;25(1):71—81.

[6] Farook TH, et al. Designing 3D prosthetic templates for maxillofacial defect rehabilitation: a comparative analysis of different virtual workflows. Computers in Biology and Medicine 2020;118:103646. https://doi.org/10.1016/j.compbiomed.2020.103646.

[7] Unkovskiy A, et al. Direct 3D printing of silicone facial prostheses: a preliminary experience in digital workflow. The Journal of Prosthetic Dentistry 2018;120(2):303—8. https://doi.org/10.1016/j.prosdent.2017.11.007.

CHAPTER 5.20

Prosthetic rehabilitation applications case study 9—Applications of 3D topography scanning and multimaterial additive manufacturing for facial prosthesis development and production*

5.20.1 Introduction

Facial defects can arise because of congenital deformities, disease infiltration, and trauma. Given the prominence of the face and how it influences human interactions, such disfigurements can have a profoundly negative impact on the quality of life of a patient. With respect to rehabilitation, there are primarily two treatment options, which comprise surgical

---

* The work described in this chapter was first reported in the references below and is reproduced here with permission of the organizing committee of the Solid Freeform Fabrication Symposium. We would like to thank the Royal Melbourne hospital for their input on the clinical aspects of this project. We would also like to thank the School of Engineering at Deakin University who provided funds and resources for this pilot project and their technical staff who assisted in the 3D printing of the models. Mohammed MI, Tatineni J, Cadd B, Peart G, Gibson I. Applications of 3D topography scanning and multi-material additive manufacturing for facial prosthesis development and production. Proceedings of 27th Solid Freeform Fabrication Symposium. Austin TX, USA; 2016. 1695−707.

Medical Modeling
ISBN 978-0-323-95733-5
https://doi.org/10.1016/B978-0-323-95733-5.00025-9

intervention or the use of a prosthesis. The decision-making process as to which option is the most suitable is not so clearly defined and is dependent on several factors, including the size/severity of the condition, age, etiology, and critically, the patient's own personal preference [1,2]. Examining prosthesis-based rehabilitation, there are several immediate advantages when compared with surgical intervention such as the immediate aesthetic improvement, its simplicity over surgery, the ability to explore several design iterations without impact on the patient, and its low cost. More recently, there has been a surge of interest in a tissue engineering approach to replace missing or compromised organs [3,4]; however, despite the promise of this technology, there are still many issues to resolve before this is likely to become a reality. Therefore, prosthetic treatment provides a more robust, tried, and tested approach that has a quick turnaround time for part production and does not have complications associated with surgical intervention, such as tissue rejection.

Traditional prosthesis production is a long, labor-intensive process, requiring the use of several invasive and subjective techniques over the entire fabrication process. A summary of the fabrication stages can be found in Fig. 5.149. Typically, the process begins with some form of casting approach, using plaster, to ascertain the topography of the defective area, or

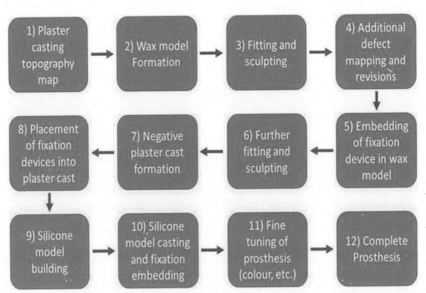

Figure 5.149 Traditional process chain for prosthesis fabrication.

uncompromised anatomy that could be used as a template for the prosthesis [5,6]. In some instances, the plaster is placed over the entirety of a patient's face, requiring breathing to be performed through a straw until the plaster sets. Consequently, due to the discomfort of this process, the patients can often move during the casting process, resulting in an inaccurate topography map. Following the formation of plaster cast, a wax model is formed and manipulated to realize the finished prosthesis model. At this stage, fixation device alignment is also performed, embedding either mechanical or magnetic abutments into the wax model. The process of alignment and finishing of the wax model are all performed manually, with the result being dependent on the artistic skill of the clinician, which is arguably subjective in nature [7]. Once the wax model is finalized, it is formed into a plaster negative, which in turn is used to form the final silicone prosthesis. Any additional touches to the model, such as the addition of colors and hair, are again performed manually. This makes the complete process of prosthesis production a very arduous and labor-intensive process.

Traditional prosthesis development is at a turning point, where several disruptive 3D technologies are likely to transform traditional prosthesis production. Modern optical scanning techniques readily allow for the rapid, high-resolution reproduction of surface topographies, to a precision of <100 μm while capturing useful data such as texture maps of a patient's skin [8−10]. Additionally, data is obtained noninvasively and can be performed using laser-free methodologies, greatly improving the potential uptake of this technology in a clinical capacity. Modern computer-aided design (CAD) software readily allows for the conditioning of 3D scan data to create a complete model, with the added advantage of allowing for the digital storage of design iterations and the final model. Such digital data conditioning and storage, when applied to prosthesis production, compares very favorably to traditional handcrafting techniques. Finally, modern 3D printing technologies offer the ability to reproduce a high-resolution digital model easily and rapidly into a physical part [10−16]. Fabrication can also be achieved using a vast array of flexible and biocompatible materials, allowing for augmentation of existing practices and, potentially, direct prosthesis fabrication.

In this study, we have investigated the use of optical scanning, CAD, and 3D printing in the direct production of various facial prosthesis (ear and nose). Previous studies have primarily focused on the use of such technologies to produce a cast to augment traditional techniques or models which are made in rigid plastics. By contrast, the novelty in this work is to use high-resolution 3D printing of flexible materials for the direct production of a prosthesis that not only is of a higher quality surface finish compared with previous studies but also mimics the tactile feel and pigmentation of human soft tissues. We also investigate the production of advanced prosthesis models, comprising multimodel and multimaterial designs, which more closely reproduce human tissue through mimicry of both skin and cartilage. Our technique offers several advantages over traditional techniques as the use of optical scanning for topography mapping is noninvasive, can acquire data within minutes, and realizes anatomically precise, high-resolution data, ideal for prosthesis production. A digital CAD approach for prosthesis design is superior when compared with traditional casting and handcrafting approaches as design iterations can be easily digitally stored, do not require any fabrication/material consumption, and allow for operations such as mirroring of a model with relative ease. Finally, the use of high-precision 3D printing allows for rapid digital part realization, can produce a sophisticated prosthesis comprising complex multimaterial/models, and can readily allow for precise reproduction of a given model should duplicates be required. We believe the techniques presented in this work realize a patient specific, low-cost, and high-resolution approach to streamlining prosthesis optimization and production.

5.20.2 Experimental

In this study we have investigated the use of several 3D digitizing, rendering, and printing technologies to directly create prosthesis replica from a person's anatomy. The complete process chain can be seen in Fig. 5.150, where surface topography maps are made using an optical scanner, designs are postprocessed using CAD, and the final model is realized using high-resolution, multimaterial 3D printing.

5.20.2.1 Topography mapping and model construction

In this study, a laser-free, optical scanning system (Spider, Artec, Luxembourg) was employed to obtain the surface topography of a nose and ear of a volunteer. Laser-free scanning technology alleviated any health concerns resulting from laser exposure to the eyes. The scanner used had image scan resolution of approximately 50—100 μm, which is more than adequate to resolve all the major and minor details of the anatomic part rendered. The scanner operates alongside a proprietary software (Artec

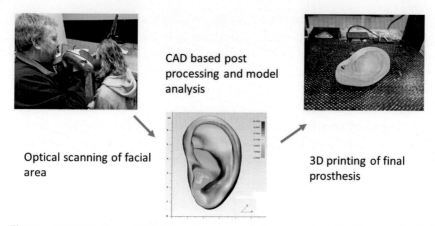

CAD based post processing and model analysis

Optical scanning of facial area

3D printing of final prosthesis

Figure 5.150 Diagram illustrating the proposed process chain for prosthesis production.

studio 10, Artec, Luxembourg), which allow for the real-time visualization of the scan data during acquisition. It was found that several translations of the scanner were required to obtain the full surface map, comprising translations at approximately 10 cm/s in a lateral and vertical arcing motion. The scanner allowed for a part to be rendered rapidly within approximately 3—5 minutes.

The scanner software processes the input data as a point cloud, which is converted into a full contour map. Rudimentary operations can also be performed to condition the data, removing spurious noise, cropping of unrequired data, filling of holes/gaps in the model, and smoothing of contours. The resulting output is as an enclosed model, so further post-processing is required to hollow a part and realize features such as nasal cavities and ear canals.

5.20.2.2 Model construction and postprocessing

Data from the scanner was conditioned using more sophisticated CAD software to process the surface topography data and to construct a more advanced design for composite model printing. In this study, all additional postprocessing was performed using the 3-Matic software package (Materialise, Belgium), which can allow for direct STL manipulation, error checking, and part thickness analysis. Data conditioning comprised smoothing of rough surfaces, removal of features that could lead to print failure, and performing procedures such as part hollowing.

In this study, we aimed to realize an advanced prosthesis that more closely mimics human physiology in terms of tactile feel and pigmentation. As a demonstrator, we took a constructed model of the ear and reconstructed it to build separate models of the cartilage and a composite of the softer tissues. Facets of the cartilage model were constructed by a purely design-based approach, so to obtain anatomic accuracy, the model was constructed by cross-referencing anatomic drawings of the ear cartilage and by direct feel of the test subject's actual ear. Approximations were then used to determine the layer thickness of the skin relative to the cartilage in the model.

5.20.2.3 3D printing

Once rendering and postprocessing had been completed, the final models were directly 3D printed to produce the final prosthesis. To obtain a high accuracy in digital reproduction, a high-resolution 3D printer was used (Connex 3 500, Stratasys, USA), which can print models in up to three individual materials or blends thereof to an accuracy of 16−30 μm. This level of accuracy was more than adequate to reproduce all major and minor surface contours while also realizing a high-quality surface finish to the prosthesis, like that obtained by traditional techniques. The printer operates exclusively using a set of materials that are developed by Stratasys for use in this printer, with no option to use nonproprietary materials. This system is generally designed for product prototyping and design evaluation purposes, so facets relating to exact composition and biocompatibility of the materials used are not known. Exploration of the material properties will be the subject of our future work.

Initially, a model is loaded into the printer's software as an STL file and is allocated a specific material combination before being sliced into the individual printing layers. Simultaneously a water-soluble support material is allocated as required to ensure the build integrity. The printer operates using a PolyJet technology, whereby a proprietary liquid photo-curable polymer is delivered by the print head and subsequently flattened and cured by a UV lamp. Once a model is complete, a final cleaning phase is required to remove the support material before the part is ready for use. The printing process can take on average 2−3 hours for a part to be printed, depending on its size and orientation.

The printer used in this study was capable of printing in several different materials, which have a variety of different color and mechanical properties. For instance, the TangoPlus material range is a flexible material, while the

Vero materials are rigid, colored materials. For the final prosthesis parts, we examined a combination of both TangoPlus and Vero materials, so we could obtain a final model with a flexible tactile feel but also with adjustable colors. It was our hope in this study to realize combinations that mimic the feel and pigmentation of a person's actual tissue. Further, the advanced multimodel approach allows for further control of the tactile feel of a given anatomic part.

5.20.3 Results

5.20.3.1 Scanning and model creation

Various scans were performed on a test subject to reproduce the surface topography of their nose and left ear. In this instance, the subject was a healthy individual who suffered no facial defects, allowing for comparison of the printed prosthesis with the original anatomy. When performing scans, the fast fusion mode of the scanner was used, which allowed for rapid, real-time visualization of the scans as they were being performed. Scans were performed translating the scanner through the various orientations previously described. Following completion of the data acquisition, several additional data-conditioning phases were performed using Artec studio to remove spurious nose and to reconstruct gaps in the data. Following completion of the scans, the software allowed for several automated procedures to crop the part of interest from the wider facial data and to convert the scan from a surface to an enclosed mesh.

It was found that there were limitations to the scanning process, and areas of the skin that were shiny/reflective were difficult to render during scanning. This limitation could be overcome using a mat finish power to dampen optical reflections; however, that was not required in this study. Additionally, regions that had low levels of light exposure were equally difficult to render, such as the nasal cavities and ear canal. As such features are critical to the final prosthesis, they had to be rendered independently using 3-matic.

5.20.3.2 Basic model postprocessing

Following initial formation, the models were checked to remove errors such as inverted normals, multiple shells, noisy shells, etc. This procedure ensures the best quality of digital data to be used for the subsequent design phases. In general, the rendered models from the scanner looked reasonably close to the final prosthesis models, so only relatively simple postprocessing was required. With respect to the nose data, the nasal cavities were completely enclosed, so the first procedure was to remove excess digital material to open these areas and to provide access to the reverse side of the model. The necessity for this is that generally a recipient of such a prosthesis will have use of their nasal ducts, so unrestricted access here would allow for a potential patient to retain the ability to smell, breathe, etc. With respect to the ear data, by cross-referencing the major contours against the original test subject, it was found that several of the contours formed by the cartilage had been lost during the postprocessing within the Artec software. Therefore, these areas were manually reconstructed using the "push pull" and extrude functions within 3-Matic. Fig. 5.151aillustrates the final model of the ear prosthesis.

With respect to the ear, when a thickness analysis was performed, it was found that there were regions that were <350 μm thick. While this is not an issue to the aesthetics of the digital mode, should this be printed, this region would be extremely fragile. From initial test prints at this thickness in the rubber-like TangoPlus material, it was found that the parts ruptured during the support material removal phases. Several design iterations were examined increasing the minimum thickness in steps of approximately 200 μm, where it was found that for thicknesses greater than 1 mm, the models could be cleaned without rupturing. Therefore, the thickness analysis of the part is critical to the integrity of the final prosthesis using the TangoPlus material, and the parts were thoroughly inspected to ensure no regions were of a thickness less than 1 mm. Fig. 5.151b shows a model of the ear before and after the thickness analysis and subsequent model region growing.

5.20.3.3 Multi-component ear modeling

The anatomy of the human ear comprises a single piece of cartilage tissue in the ear, which makes the design process relatively simple in comparison to more complex cartilage structures such as the nose. Initially, attempts were made to render the cartilage of the ear as a stand-alone model. To achieve this, the original ear model was duplicated and reduced in size by 96% to create a replica structure that was positioned approximately 1.5 mm into the original ear model. This new model was then reformed by referencing anatomic drawings of a generic human ear cartilage, the contours captured within the scanned ear model and by direct feel of the human subject. The

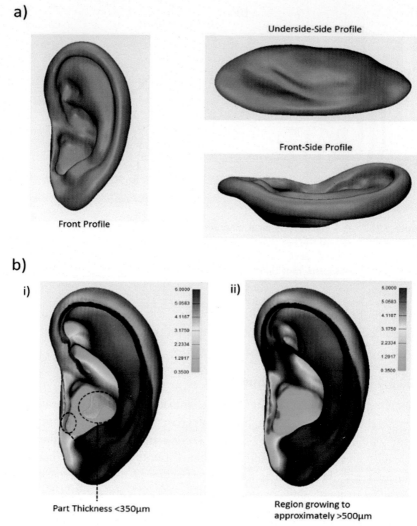

Figure 5.151 (a) Various orientations of the ear prosthesis model. (b) Color map thickness analysis of the ear model for (i) the raw data model input from the 3D scanner software and (ii) postprocessed model to grow the thickness of the inner ear section to approximately ≥ 1 mm.

cartilage reconstruction process can be performed to a high degree of precision using digital thickness analysis to augment the design process; however, arguably, the use of feel to determine model topography is a subjective metric. It is noted that this part of the process is merely for qualitative assessment of the model and forms a very minor part in the

a)

b)

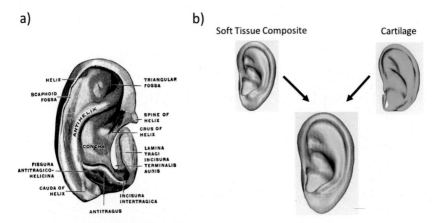

Soft Tissue Composite   Cartilage

Figure 5.152 (a) A classical medical diagram of the cartilage within the ear, (b) The multi-model design of the ear prosthesis comprising the soft tissue composite and the cartilage. For visualisation purposes both segments are also represented individually. *(a) Adapted from Griffin et al. [17].)*

design process and does not compromise the precision of the final rendering. The final model that was achieved can be seen in Fig. 5.152, which closely matched the medical diagrams.

5.20.3.4 Prosthesis 3D printing

Following completion of the models, initial tests were performed to ascertain the printing precision, model integrity, mechanical conformity, and pigmentations that could be achieved using a multimaterial combinatory approach. The Connex printer is capable of printing with three materials simultaneously. However, as flexibility is achieved using the transparent TangoPlus material, only two additional materials (colors) could be used. Given the availability of colors by the manufacturers, it was decided that blends of magenta and yellow would provide the best options for skin pigmentation mimicry. An effective color map was produced of the different material combinations when using TangoPlus and Vero materials, as can be seen in Fig. 5.153a. Another issue that arose was the percentage blend of the rigid Vero with the flexible TangoPlus material, where a threshold of >50%−60% Vero material compromised the tactile feel beyond the desired flexibility found in a typical prosthesis. The TangoPlus material used in this study was a translucent variant, which on its own provided the softest tactile feel but was not a suitable color for a prosthesis. It was found that a

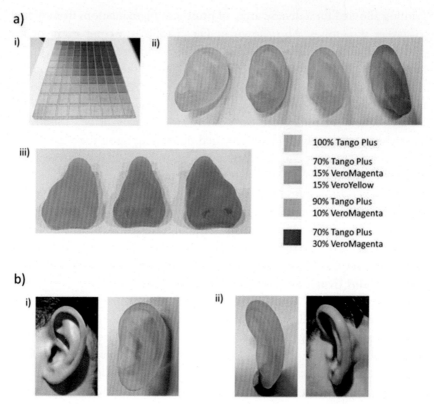

Figure 5.153 (a) (i) an image of the material map that shows the various color combinations possible with the 3D printer using two Vero colors and the flexible TangoPlus material and images of various printed prosthesis of (ii) an ear and (iii) a nose model. Material combinations are also highlighted in the legend. (b) Comparative images of the subject's ear and the 3D printed model from (i) side and (ii) reverse profiles.

minimum of 10% Vero material was required to provide any visually noticeable pigmentation into a given prosthesis. Various combinations of Tango and Vero material were explored to print the final prosthesis, and the various material combinations examined are illustrated in Fig. 5.153a.

The high resolution of the Connex printer and its effective use of dissolvable support material allows for the reproduction of both the ear and nose models and with a smooth surface finish. The reproduced models were found to be a very good match to the original anatomy, both in terms of major surface contours and overall size. Various nose and ear models can be seen in Fig. 5.153a(iii), where it can be clearly seen how the multimaterial printing allowed for a diverse array of prosthesis pigmentations from lighter to darker skin tones. Also shown in Fig. 5.153b is a comparison of the printed ear prosthesis relative to the volunteer subject's original ear. It can be seen that the final prosthesis very closely matches the original contours of the ear, validating the efficacy of this technique. All 3D printed models were realized using a single material combination across the entirety of the model. When examining the ear models more closely, it was found that the richness of the visible color changed depending upon the thickness of the material. By comparison, the nose models did not exhibit such features. This is perhaps due to the thickness tolerances of these models, which were generally between 1 and 6 mm allowing for varying degrees of translucency. In contrast the minimum thickness of the nose was approximately 15 mm; therefore at these thicknesses, all translucency properties are negligible. This facet is similar to human physiology, where the thickness of the soft tissues overlaying the cartilage of the ear has a variety of different pigmentations owing to its relative thickness. We hope to examine such facets more closely in future work.

The printed models were qualitatively assessed for their tactile feel and mechanical compliance to that or the original human subject. Tests comprised the ability to flex the lobe area, a vertical compression test from the lobe to the upper portion of the ear, and a lateral compression test pinching across the center of the ear. It was found that the printed models all exhibited similar characteristic to the original anatomy, such that upon relaxation of the compressive force, the ears would return to the neutral position of the model, as can be seen in Fig. 5.154a. It was noted that the printed material was more rigid, despite its elasticity, than human anatomy. This resulted in a greater force having to be applied to achieve the various modes of compression, highlighting limitations of current printable, flexible materials. The purpose of this work was to present our methodology regarding the scanning, design, and printing capabilities, and we hope to measure mechanical properties of the prosthesis more quantitatively in future studies. It is also noted that as the percentage combination of the Vero material was increased in the model, the necessary applied force became larger. At a percentage of 60% Vero, the elasticity of the model was compromised beyond an acceptable level and became noticeably rigid. This therefore limits the wider color combinations possible while still retaining acceptable levels of elasticity to mimic human mechanical properties. We hope in future work to quantify such forces both for real anatomy and printed prosthesis.

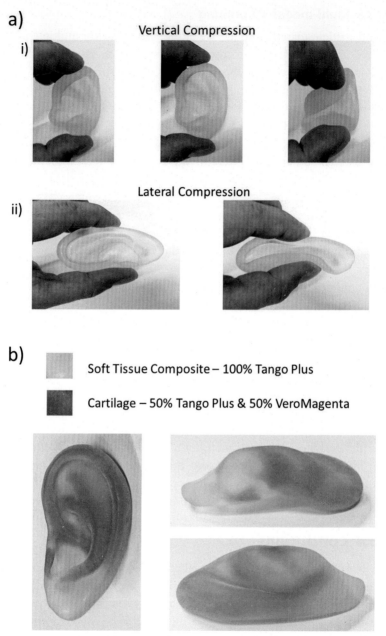

Figure 5.154 (a) Qualitative compression of the model to demonstrate the realism in the tactile feel for both (i) vertical and (ii) lateral compression. (b) Multimodel printing of the ear, with independent material combinations for the cartilage and residual soft tissue composite.

5.20.3.5 Multi-model 3D printing

Beyond a single material model, we attempted to realize multiple models, rendered using independent material combinations, thereby mimicking the softer and harder tissues of the ear. The Connex printer allows for the overlay of multiple models and for independent material allocation to each when printing. We therefore processed the advanced ear model such that the soft tissue composite was printed with 100% TangoPlus and the cartilage model was printed with 50% TangoPlus and 50% VeroMagenta. These material combinations were used primarily for visualization purposes such that each element could be visually differentiated, and the final printed model can be seen in Fig. 5.154b. The Connex printer was found to provide excellent multimaterial printability, with the two materials seamlessly blending into a single structure without any impact on the final surface finish of the model. Therefore, the PolyJet printing process of the Connex is considered ideal for rendering of blended materials as well as multiple models for direct prosthesis printing.

Once again, the mechanical properties of the model were assessed by qualitative compression tests. On this occasion it was found that there was much greater rigidity to the ear model, as expected; however the ability to be compressed vertically and laterally was not compromised, and the model would also return to the neutral position upon relaxation of the compressive force. It was noted that the movement of the ear lobe was identical to the single material/model prints. Outcomes of the multimodel printing illustrate that the mechanical properties of the printed models can be further modified to reach increasing levels of complexity, as found in actual human anatomy. We hope in the future to assess the mechanical properties more quantitatively and to explore the ability to blend the color combinations more seamlessly into a more realistic final model.

5.20.4 Conclusions

This study has demonstrated the potential of the 3D design and multi-model/material printing, augmented with the use of optical surface scanning, to produce realistic prosthetic models of both the ear and nose. The fabricated prosthesis was realized to a high degree of accuracy and surface finish, and we believe the technique to be suitable to render additional prosthetic parts, such as orbital prosthesis. We realized advanced prosthesis models beyond the traditional single model variants using novel design techniques to render components such as cartilage alongside other soft tissues (skin, etc.). Using the multimaterial printing approach, we could tailor the skin pigmentation of the prosthesis to a variety of skin tones while also mimicking the mechanical properties of the original anatomy. The mechanical properties can be further tailored using a multimodel approach. Currently, there are limitations in the complexity of the skin tones that can be mimicked without compromising the mechanical properties, due to the percentage of material combinations. Additionally, the overall printed prosthesis' tactile feel is more rigid than actual anatomy. These limitations are primarily due to the materials used in the 3D printer, but we believe as the technology matures over the coming years, these limitations will be resolved. Ultimately, the findings in the work validate our approach for direct prosthesis production that overcome limitations relating to the subjective nature of current prosthesis fabrication and allow for production within a single day. This compares favorably to traditional techniques where typically a prosthesis if fabricated over several weeks/months. Ultimately, this technique holds considerable potential for implementation within a clinical setting, streamlining to overall process for prosthesis production, and could see applications in other niche areas such as soft robotics or anatomic modeling.

References

[1] Kretlow JD, McKnight AJ, Izaddoost SA. Facial soft tissue trauma. Seminars in Plastic Surgery 2010;24(4):348–56.
[2] Ranganath K, Hemanth Kumar HR. The correction of post-traumatic pan facial residual deformity. Journal of Maxillofacial and Oral Surgery 2011;10(1):20–4.
[3] Mannoor MS, et al. 3D printed bionic ears. Nano Letters 2013;13(6):26342639.
[4] Jung-Seob L, et al. 3D printing of composite tissue with complex shape applied to ear regeneration. Biofabrication 2014;6(2):024103.
[5] Mantri SS, Thombre RU, Pallavi D. Prosthodontic rehabilitation of a patient with bilateral auricular deformity. The Journal of Advanced Prosthodontics 2011;3(2):101–5.
[6] Ozturk AN, Usumez A, Tosun Z. Implant-retained auricular prosthesis: a case report. European Journal of Dentistry 2010;4(1):71–4.
[7] Jani RM, Schaaf NG. An evaluation of facial prostheses. The Journal of Prosthetic Dentistry 1978;39(5):546–50.
[8] Ciocca L, et al. CAD/CAM ear model and virtual construction of the mold. The Journal of Prosthetic Dentistry 2007;98(5):339–43.
[9] Sansoni G, et al. 3D imaging acquisition, modeling and prototyping for facial defects reconstruction. SPIE Proceedings 2009;7239.
[10] Palousek D, Rosicky J, Koutny D. Use of digital technologies for nasal prosthesis manufacturing. Prosthetics and Orthotics International; 2013.
[11] Al Mardini M, Ercoli C, Graser GN. A technique to produce a mirror-image wax pattern of an ear using rapid prototyping technology. The Journal of Prosthetic Dentistry 2005;94(2):195–8.

[12] Bos EJ, et al. Developing a parametric ear model for auricular reconstruction: a new step towards patient-specific implants. Journal of Cranio-Maxillofacial Surgery 2015;43(3):390−5.

[13] He Y, Xue GH, Fu JZ. Fabrication of low-cost soft tissue prostheses with the desktop 3D printer. Scientific Reports 2014;4:6973.

[14] Kuru I, et al. A 3D-printed functioning anatomical human middle ear model. Hearing Research 2016:206−13.

[15] Liacouras P, et al. Designing and manufacturing an auricular prosthesis using computed tomography, 3-dimensional photographic imaging, and additive manufacturing: a clinical report. The Journal of Prosthetic Dentistry 2011;105(2):78−82.

[16] Subburaj K, et al. Rapid development of auricular prosthesis using CAD and rapid prototyping technologies. International Journal of Oral and Maxillofacial Surgery 2007;36(10):938−43.

[17] Griffin MF, Premakumar Y, Seifalian AM, Szarko M, Butler PEM. Biomechanical characterisation of the human auricular cartilages; implications for tissue engineering. Annals of Biomedical Engineering 2016;44:3460−7.

CHAPTER 5.21

Prosthetic rehabilitation applications case study 10—Advanced auricular prosthesis development by 3D modeling and multimaterial printing*

5.21.1 Introduction

Traditional methods of maxillofacial prosthetic design have proven to be highly effective means of producing a given facial prosthesis. However, the process largely remains qualitative in nature, with the artistic skill of the clinician being the most pivotal factor in the fit and final aesthetic appearance of the end prosthesis [1]. Traditional processes are also highly labor intensive, as parts are handmade, and the procedure to take an impression of the defective surface can be highly invasive and uncomfortable for the patient [2,3]. To eliminate these disadvantages, and to potentially reduce the time and work needed to develop a prosthesis, an alternative to the traditional method is required.

---

* The work described in this chapter was first reported in the references below and is reproduced here with permission of the organizing committee of the Solid Freeform Fabrication Symposium.
  Mohammed MI, Tatineni J, Cadd B, Peart G, Gibson I. Advanced auricular prosthesis development by 3D modeling and multi-material printing, In: DesTech 2016: Proceedings of the International Conference on Design and Technology. Knowledge E; 2016. pp. 37−43.

Medical Modeling
ISBN 978-0-323-95733-5
https://doi.org/10.1016/B978-0-323-95733-5.00026-0

Modern computer-aided design (CAD), patient scanning, and 3D printing technology are poised to transform the way in which prosthesis devices are developed and realized, replacing more traditional, manual approaches [4−6]. Medical imaging data can now feasibly be used to construct patient-representative models and could potentially supersede impression/cast-based approaches when attempting to reconstruct a defective or uncompromised area of anatomy. The use of medical imaging data offers several advantages both in terms of resolution of topography reproduction, in addition to being minimally invasive to the patient and part of routine clinical practice, and also capturing useful information of the wider patient anatomy for model construction and virtual placement analysis [7,8]. Modern digital CAD software offers multiple advantages to the traditional design process, as no model fabricating is required, and designs are digitally evaluated and stored, reducing costs and readily allowing for future reproduction. Additionally, digital reproduction of surrounding anatomy can allow for virtual placement and aesthetic evaluation without the necessity for the patient to be present. Finally, 3D printing is the ideal technology for translation of digital models to working devices, owing to the high precision of modern devices and the diverse possibility of materials combination that could be employed during fabrication. Indeed, PolyJet multimaterial printing technology allows for not only near perfect digital reproduction, but it can realize complex color combinations, mimicking skin pigmentation, in addition to adjustable mechanical properties, to realize the tactile feel of human tissue. In this work, we propose a process using medical imaging data to construct a virtual model of a patient's uncompromised anatomy/defective area before the captured data are conditioned to obtain a final model of the proposed prosthesis. Finally, the finished prosthesis is manufactured through high-resolution, multimaterial 3D printing. As a proof of principal, we examined the reproduction of a volunteer's ear and into a 3D model before printing in a flexible material to mimicking the tactile feel of the volunteer's original anatomy. We also examine design possibilities that would allow for complex rendering of multiple color combinations on the single model, thereby mimicking more closely the complex color attributes found on human skin tissue. Ultimately, the presented approach realizes a rapid, cost-effective, and anatomically precise methodology for prosthesis development, with potential for increased realism resulting from multimaterial printing. We believe that such techniques would complement or potentially surpass traditional techniques and practices for prosthesis development and production.

5.21.2 Experimental

5.21.2.1 CT scan data and model construction

Models of the ear were constructed directly from publicly available CT scan data of anonymized patient data (http://dicom.nema.org/) in the standard Digital Imaging and Communications in Medicine (DICOM) format. The data set comprised individual slices, with an approximate thickness of 0.625 mm, of a patient's head, which includes the ears, for prosthesis production. All DICOM data were analyzed using Mimics (Materialise, Belgium). Fig. 5.155 illustrates the workflow from medical imaging interrogation, model creation/postprocessing, and 3D printing.

5.21.2.2 Model construction and postprocessing

Models from the scanner software were further processed using the 3-Matic software package (Materialise, Belgium), which can allow for direct STL manipulation, error checking, and part thickness analysis. Data conditioning comprised smoothing of rough surfaces, removal of features that could lead to print failure, and performing procedures such as part hollowing to obtain the final ear model. To realize the multicolored prosthesis design, we examined an approach of independent material allocation to multiple segments of the complete model. To achieve this, the final ear model was segmented into seven individual STL files, which when combined form the prosthesis. To segment the ear model, we once again used 3-Matic to perform operations such as thickness/depth analysis and slicing.

5.21.2.3 3D printing

Once all postprocessing had been completed, the final model was 3D printed to produce the final prosthesis. In this study, a high-resolution 3D

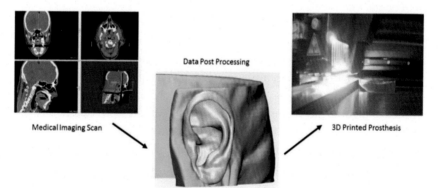

Medical Imaging Scan   Data Post Processing   3D Printed Prosthesis

Figure 5.155 A summary of the process chain for the production of the ear prosthesis.

printer was used (Connex 3 500, Stratasys, USA), which can print models in up to three individual materials or blends thereof to an accuracy of 16–30 µm. Materials range from colored, hard, and elastic materials, so we are capable of reproducing a prosthesis model with a tactile feel and pigmentation more closely representing that of human tissues. The high resolution of the printer also ensures allowing the major and minor surface contours to be readily reproduced while also realizing a high quality.

Initially, a model is loaded into the printer's software as an STL file and is allocated a specific material combination before being sliced into the individual printing layers. Simultaneously a water-soluble support material is allocated as required to ensure the build integrity. When printing the multisegment models, as each is created as an individual STL file, the printer software allows for independent material allocation to each of the segments. In this work, the segments comprised varying combinations of Tango plus, a clear, rubber-like material, with VeroMagenta, a rigid, magenta–colored material. Therefore, different depths of potential skin pigmentation could be achieved in the model by either increasing or decreasing the percentage content of VeroMagenta with Tango Plus.

5.21.3 Results

5.21.3.1 Scanning and model creation

The DICOM data were input into the software Mimics for 3D model construction. The software is capable of isolating specific tissue data across the three typical geometric planes, before rendering the selected data into a digital 3D model. The software operates using a series of threshold functions that highlight regions of a particular Hounsfield unit (HU), allowing for the rapid differentiation of different tissue types such as bone, soft tissue, etc. For the examined dataset, a threshold function of −275 to 93 HU was found to be optimal to fully encompass all the soft tissue information from the patient data. This threshold encompasses the entirety of the CT scan data, so cropping was required to isolate the ears, from which to make the prosthesis, and the rear circumference of the head for virtual placement analysis. Once the regions of interest had been isolated across the three geometric planes, the Mimics software allows for the conversion of these data into a 3D representative model of the patient. Fig. 5.156a shows the final model, which was then exported as an STL file for postprocessing using more sophisticated software.

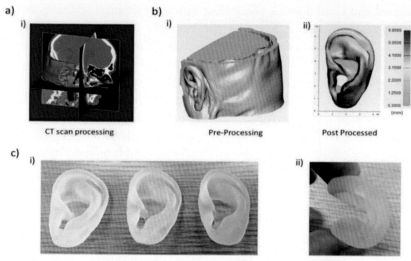

Figure 5.156 (a) Construction of patient representative model from CT scan, (b) (i) data extraction prior to CAD processing and (ii) thickness analysis postprocessing of the CT scan data, and (c) (i) various 3D prints of the ear model in varying material combinations and (ii) demonstration of the elastic deformation of the ear model.

5.21.3.2 Prosthesis modeling

The output STL model from Mimics was loaded into 3-Matic and checked to remove any errors during the rendering process that would otherwise cause issues during 3D printing, such as multiple shells, inverted normal, etc. Generally, the models produced by Mimics were very close to the final printable model and only required minor postprocessing. To form the auricular prosthesis, the left ear was segmented from the wider patient data and the reverse side smoothed to yield a standalone ear model. A part thickness analysis of the isolated ear uncovered regions of the model that were too thin (<300 μm) to be produce a robust 3D printed model using the employed system, Fig. 5.156b(ii). To remedy this, compromised areas were reconstructed digitally, using the push–pull, extrude, and smoothing functions within 3-Matic to ensure the model had a minimum thickness of 1 mm and that all major contours were reproduced. Fig. 5.156b illustrates the final model of the ear prosthesis.

5.21.3.3 Prosthesis 3D printing

Upon completion of the ear prosthesis model, 3D printing was performed using a varied material combination, comprising entirely the Tango plus material and combinations comprising various percentage additions of the VeroMagenta material. Some of the resulting models can be seen in Fig. 5.156c(i) where ear one comprises 100% Tango Plus, ear two comprises 20% VeroYellow and 80% Tango Plus, and ear three is the complex multisegment/material combination print, which will be discuss in the next section. The Tango plus material used in this study was a translucent variant, which on its own provided the softest tactile feel but was not a suitable color for a prosthesis. It was found that a minimum of 10% Vero material was required to provide any visually noticeable pigmentation into a given prosthesis. Ultimately, it was found there was significant scope to adjust the pigmentation of the models to encompass lighter and darker skin tones, and we hope to investigate further material combinations in future work. In addition to pigmentation mimicry, the tactile feel of a typical prosthesis was also reproduced, whereby the printed prosthesis could be elastically deformed. Fig. 5.156c shows a demonstration of the elastic deformation of an ear model, whereby upon relation of the compressive force, the ear would spring back to its original shape. It is noted that as the percentage blend of the VeroMagenta with the Tango plus increased beyond the threshold of 50%−60% Vero material, this significantly affected flexibility of the printed prosthesis. We hope to examine and quantify the mechanical properties in future studies more closely.

5.21.3.4 Multi-segment model

Following the creation of the final digital prosthesis, the model was segmented into subsections based on the height of the ear, resulting in the ear being partitioned into six primary segments, over four layers of the model (Fig. 5.157a). Each of these layers was allocated with individual material assignment, with varying percentage content of VeroMagenta and Tango Plus, leading to a deepening of the color during the transition from layer one to three. Additionally, a fourth layer was implemented to balance the color tones in layers two and three. A complete breakdown of the spatial orientation of the segments can be seen in Fig. 5.157a. Preliminary findings reveal the potential of this approach to increase the color complexity of the prosthesis, thereby increasing the potential mimicry of human skin pigmentation. We hope to develop on these findings in future work.

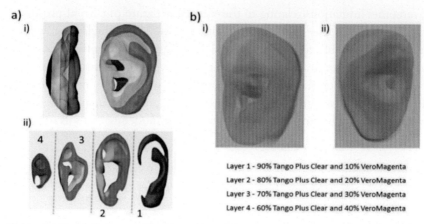

Layer 1 - 90% Tango Plus Clear and 10% VeroMagenta
Layer 2 - 80% Tango Plus Clear and 20% VeroMagenta
Layer 3 - 70% Tango Plus Clear and 30% VeroMagenta
Layer 4 - 60% Tango Plus Clear and 40% VeroMagenta

Figure 5.157 (a) (i) Front and side profiles of the segmented ear model, highlighting the orientation of the subsection in the overall model and (ii) an exploded view of the separate section of the prosthesis and how they are orientated over four sublayers of the model. (b) A 3D print of the final prosthesis.

5.21.4 Conclusions

This study has demonstrated the potential of the 3D design and multi-segment/material printing, alongside the use of medical imaging data, to produce realistic prosthetic models of the ear. The fabricated prosthesis was realized to a high degree of accuracy and surface finish and could easily be applied to realize alternative prosthetic parts, such as the nose or orbital prosthesis. Using the multimaterial printing approach, we could tailor the skin pigmentation of the prosthesis to a variety of skin tones while also mimicking the mechanical properties of the original anatomy. Beyond the use of a material combination, we discovered that additional color complexity can be created by segmentation of a digital model and the assigning of unique material combination to each segment. By creating darker variations of a base color for the prosthesis, realism such as the depth of color perceived in shaded regions can be realized, further increasing the realism of the designed models. Currently, there are limitations in the complexity of the skin tones that can be mimicked without compromising the mechanical properties, due to the percentage material combinations, but we believe as the technology matures over the coming years, these limitations will be resolved. It was found that direct prosthesis production overcomes limitations relating to the subjective nature of current prosthesis fabrication, and this allows for production within a single day, which compares favorably against traditional techniques where prosthesis turnaround time takes several weeks/months. Ultimately, this technique holds considerable potential for implementation within a clinical setting, streamlining to overall process for prosthesis production and could see usefulness in other niche areas such as anatomic modeling and soft robotics.

References

[1] Jani RM, Schaaf NG. An evaluation of facial prostheses. The Journal of Prosthetic Dentistry 1978;39:546−50. https://doi.org/10.1016/S0022-3913(78)80191-7, 1978/05/01.

[2] Mantri SS, Thombre RU, Pallavi D. Prosthodontic rehabilitation of a patient with bilateral auricular deformity. The Journal of Advanced Prosthodontics 2011;3:101−5. https://doi.org/10.4047/jap.2011.3.2.101.

[3] Ozturk AN, Usumez A, Tosun Z. Implant-retained auricular prosthesis: a case report. European Journal of Dentistry 2010;4:71−4.

[4] Subburaj K, Nair C, Rajesh S, Meshram SM, Ravi B. Rapid development of auricular prosthesis using CAD and rapid prototyping technologies. International Journal of Oral and Maxillofacial Surgery 2007;36:938−43. https://doi.org/10.1016/j.ijom.2007.07.013.

[5] Eggbeer D, Bibb R, Evans P. Assessment of digital technologies in the design of a magnetic retained auricular prosthesis. 2006.

[6] Mohammed MI, Tatineni J, Cadd B, Peart P, Gibson I. Applications of 3D topography scanning and multi-material additive manufacturing for facial prosthesis development and production. In: Proceedings of the 27th annual international solid Freeform fabrication symposium; 2016.

[7] Ciocca L, Mingucci R, Gassino G, Scotti R. CAD/CAM ear model and virtual construction of the mold. The Journal of Prosthetic Dentistry 2007;98:339−43. https://doi.org/10.1016/S0022-3913(07)60116-4.

[8] Sansoni G, Trebeschi M, Cavagnini G, Gastaldi G. 3D imaging acquisitiion, modeling and prototyping for facial defects reconstruction. SPIE Proceedings 2009;7239.

CHAPTER 5.22

Prosthetic rehabilitation applications case study 11—Augmented patient-specific facial prosthesis production using medical imaging modeling and 3D printing technologies for improved patient outcomes*

5.22.1 Introduction

The face is a vital part of the human anatomy and is pivotal to our daily human interactions, communication, and sense of identity [1]. As such, complications relating to disfigurements and defects of the face can have a profoundly negative impact on a person's quality of life. Craniofacial defects can occur either from birth or during the course of a person's life and manifest primarily as a result of congenital deformities, disease infiltration, severe trauma, or following surgical intervention to treat a more critical condition such as cancers. To remedy such complications, surgical

* The work described in this chapter was first reported in the reference below and is reproduced here with permission of Taylor & Francis publishing. We would like to thank the Royal Melbourne Hospital for their input on the clinical aspects of this project. We would also like to thank the School of Engineering at Deakin University who provided funds and resources for this pilot project and their technical staff who assisted in the 3D printing of the models.
Mohammed MI, Cadd B, Peart G, Gibson I. Augmented patient specific facial prosthesis production using medical imaging modeling and 3D printing technologies for improved patient outcomes. Virtual and Physical Prototyping 2018;13(3):164—76.

Medical Modeling
ISBN 978-0-323-95733-5
https://doi.org/10.1016/B978-0-323-95733-5.00027-2

interventions can correct the majority of situations, but they can often fall short of the patient's expectations, either due to the residual scarring during healing or when the anatomy is beyond any form of surgical recovery [2]. In such circumstances, prosthetic-based treatment becomes the most desirable, if not the only, option for the patient [3–5]. Inherently, surgical intervention is highly invasive, not without its potential complications, and can result in irreversible alterations to the face following implementation. By contrast, prosthetic-based treatment is typically minimally invasive, provides an immediate aesthetic improvement, and is nonpermanent, allowing the ability to explore several design iterations without impact on the patient [3,6,7]. However, the decision-making process as to which option is the most suitable is not so clearly defined and is dependent on a number of factors, ranging across the size/severity of the condition, age, etiology, and critically, the patient's own personal preference [3,8,9]. Beyond these typical techniques, bioprinting is emerging as a potential third treatment option, with the promise of reproducing fully functioning living tissues for craniofacial abnormalities [10–12]. However, despite the promise of this technology, there are still many issues to resolve before the potential of this field can be realized as a clinical treatment option. Ultimately, facial prosthetics treatment is arguably the best option for patients as it is a robust and proven technology, with fewer associated complications compared with surgical intervention.

Examining prosthesis production more closely, there are many associated production phases, many of which are relatively subjective and qualitative in nature. Primarily, these phases can be divided into wax modeling, silicone casting, and postprocessing. A summary of each phase can be found in Fig. 5.158. The start of the process is to obtain an accurate topology of the patient's compromised area as well as perhaps the uncompromised anatomy that will form the basis of the prosthesis template [4,13–16]. Depending on the preferred fabrication methodology of the prosthetist, the skin topology mapping phases can be relatively invasive and uncomfortable for the patient. Various methodologies can be employed, which traditionally comprised techniques such as plaster casting [3] or the use of a fast-setting elastomer [17]. More recently, minimally invasive techniques have been employed such as the use of medical imaging scan data [5,15] and optical surface scans [14,16,18]. Once the topology is obtained, it is converted into a wax model for ease of reworking into a representation of the final prosthesis. This wax model is also a versatile platform for assessing the alignment and placement of the fixation device relative to the patient's

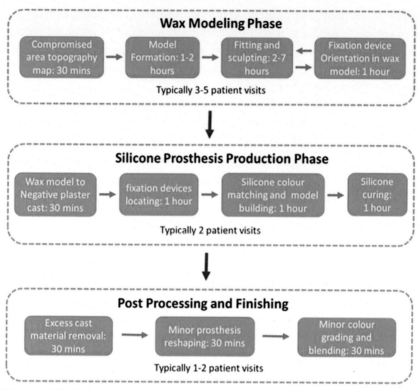

Figure 5.158 Typical process for prosthesis fabrication indicating for each phase the time and approximate number of patient visits.

anatomy. All manipulations of the model during this phase are performed manually, so the efficacy of the end result can be somewhat subjective and dependent on the skill of the prosthetist, with results varying from clinic to clinic [19]. The basic shape of the prosthesis is devised based upon either reference to the uncompromised anatomic topology, images of the patient, and in instances where there is no template anatomy for reference, the prosthetist makes a "best guess" on the shape of the prosthesis. This element of the design and fabrication process is arguably the most time consuming for the prosthetist and can take many weeks to complete depending on the complexity and nature of the prosthesis being developed. Once the wax model is finalized, it is formed into a plaster negative, which in turn is used to form the final silicone prosthesis, often using multiple pigmentations within the silicone to reproduce the patient's skin tone. Typically, the cast silicone prosthesis then requires some minor postprocessing to remove residual silicone material left behind from the casting process, in addition to minor adjustments and possibly additional coloration or gluing of hair, etc. The complete production of a prosthesis can be quite an involved process requiring several weeks to months to complete alongside multiple patient visitations but can be significantly longer if the prosthesis shape has to be created with no reference anatomy.

Traditional prosthesis development is at a turning point, where several disruptive 3D technologies are likely to transform production, potentially leading to reduced part turnaround time, improved quality of the end products, and lowering of production costs. In particular, modern computer-aided design (CAD) and 3D printing technologies have realized a range of patient-specific medical applications, from surgical planning models [20,21], bone replacement implants [15−22], orthotics [23,24], and facial prosthetics [7,14,25−30]. Despite the promise of CAD and 3D printing, no single technology has the capacity to produce a prosthesis model in either an acceptable biocompatible material or with sufficient reproduction of realistic skin pigmentation and tone. Given the advancements in the field of 3D technologies over the last decade, it is suspected that this may not be the case in the near future. However, for now the greatest added value of such technologies is in the streamlining of current fabrication processes, particularly with regard to anatomic reproduction.

In this study, we investigated the use of medical imaging−based anatomic modeling, CAD, and 3D printing in streamlining current prosthesis production methodologies. We focus our attention on a patient case study who presents us with the complete loss of the left ear and an uncompromised right ear. The patient has been making use of a prosthesis for several decades, however remaining deeply unsatisfied due to the inferior nature of the device. The patient was previously using medical glue with respect to fixation, which only provided temporary bonding and was prone to regular failure or only partial bonding contact. Critically, the patient was unhappy with this fixation and the irregularity in size/shape of the prosthesis when compared with their remaining ear. The patient therefore agreed to the implantation of titanium fixation abutments in place of the use of adhesives. We therefore focus our attention on the use of 3D technologies to produce an accurate reproduction of the patient's uncompromised ear from their medical CT scans derived for the abutment implantation process. A patient-specific model was produced and digitally mirrored to form the basis of a template model. Once created, we will use high−resolution 3D printing to create a part with a smooth surface finish to minimize the necessity for any postprocessing of the template, thereby further streamlining the process. We then will undergo the standard fabrication process outlined in Fig. 5.158, before assessing the applicability of the technique for clinical implementation. Previous studies have applied the use of optical scanning and 3D printing methodologies for prosthesis production [30,31]; however, we aim to focus on the use of medical imaging data for model preproduction. Additionally, the majority of studies have been demonstrated to only show how production could be achieved [14,31] and have focused less on the complete methodology from start to finish to achieve satisfactory device fixation in addition to fit and coloration. Therefore, we aim to address this and to demonstrate a robust methodology for prosthesis production as applied to both an actual patient case study, but which could equally be implemented using the existing infrastructure, and protocols within a clinical setting. Ultimately, we believe the techniques presented in this work surpass current production protocols both in terms of superiority of the end prosthesis, satisfaction of the patient, and reduced cost and production time.

5.22.2 Experimental

5.22.2.1 Patient-specific model construction

In this study, we use medical CT scan data from which to form our patient-specific auricular template model. As part of the patient treatment, they had undergone surgical implantation of titanium fixation abutments, making available their medical imaging CT scans for use in the study. This is important as we will be using these historical scans to construct the 3D prosthesis template, meaning no unnecessary radiation exposure to the patient, alongside making use of current imaging protocols within the hospital. CT scan images were obtained from the radiology department from the Royal Melbourne Hospital (Melbourne, Australia). Scans were received in the standardized Digital Imaging and Communication in Medicine (DICOM) file format captured on a Somatom Definition CT scanner (Siemens, Germany). The scan data comprised 129 individual slices with a slice thickness of 0.75 mm and pixel size of 0.4 mm. The patient's dataset was analyzed using Mimics 18.0 (Materialise, Belgium), which compiled the dataset into the axial sagittal and coronal planes for further processing. Using Mimics software, the patient's various tissue types could be isolated using a Hounsfield thresholding method, whereby the effective levels of radiation attenuation can be grouped and extrapolated to form the final 3D models. Once the basic model had been constructed, the file was exported in STL format for further postprocessing to remove additional digital artifacts and condition the data to be 3D printed.

5.22.2.2 Model construction and postprocessing

The model of the patient's soft tissue was exported from Mimics into 3-Matic Research 10 (Materialise, Belgium) for further postprocessing to isolate the healthy ear from the bulk of the data. Prior to any data processing, because the model is constructed from a nonideal dataset, preprocessing is required to remove any digital errors (inverted normal, overlapping triangles, small subshells, etc.), to improve the uniformity of the mesh and reduce the overall triangle count. Such measures ensure the data integrity toward the final printable model and reduce the computational processing cost for data operations. The software is also used to smooth the digital model to create a more natural and continuous surface for the derived anatomy. Finally, the software is used to artificially produce a flat base to the model, so the final template can be easily handled and used in the casting equipment by the prosthetist.

5.22.2.3 3D printing

Once rendering and postprocessing had been completed, several test models were printed using a variety of material combinations to produce the final prosthesis template. To obtain geometric accuracy in digital reproduction, a high-resolution 3D printer was used (Connex 3 500, Stratasys, USA), which can print models in up to three individual materials, or blends thereof, to an accuracy of 16–30 μm. This level of manufacturing precision was more than adequate to reproduce all major and minor surface contours while also realizing a low surface roughness to the prosthesis template.

Initially, a model is loaded into the printer's software as an STL file and is allocated a specified material combination before being sliced into the individual printing layers. Simultaneously a water-soluble support material is allocated as required to ensure the build integrity in regions containing overhanging features. The printer operates using PolyJet technology, whereby a liquid photo curable polymer is delivered by the print head onto the build areas, and subsequently flattened with a precision roller, before being cured by a UV lamp. Once a model is complete, cleaning is required to remove the support material before the part is ready for use and was achieved using a high-pressure water jet station system. The printing process can take on average 2–3 hours for the part to be printed, depending on its size and orientation.

The materials used by the printer are available in a range of shore hardnesses, ranging from highly rigid to "rubber-like" consistencies. For instance, the TangoPlus material range is a flexible material with a shore hardness of between 73—77 ASTM D-2240 scale A units, while the Vero materials are more rigid with a shore hardness of between 83—86 ASTM scale D units [32]. Therefore, the mechanical properties of the printed part can be adjusted to examine the influence this would have on the casting process. The final prosthesis casts we fabricated used three material combinations, comprising TangoPlus only, VeroMagenta only, and a combination of 50% TangoPlus and 50% VeroMagenta, to establish which mechanical variant was optimal for release from the alginate gel cast formation.

5.22.2.4 Silicone auricular prosthesis production

Once the 3D printed cast of the patient's ear was formed, the silicone prosthesis was formed using the standard manufacturing techniques, as described in Fig. 5.158. Briefly, this comprised the formation of an alginate impression of the 3D printed model, which is used to form a wax model of the ear. The wax model was then hand formed into the desired shape of final prosthesis through several iterative stages of minor modification and inspection upon the patient. During this phase, the magnetic fixation cups are integrated into the wax model using a silicone impression holding the wax ear model relative to the fixation abutments. Once this phase is complete, the wax model is converted into a plaster cast, comprising a negative impression of the ear, the relative orientation of the abutments, and the topology of the prosthesis contact area on the patient's head. The cast is then used to form the final silicone prosthesis, where during the curing phase the magnetic fixation cups are embed into the device. Once cured, final postprocessing is done to remove excess material and to correct coloration before the prosthesis is ready for use by the patient.

5.22.3 Results

5.22.3.1 Scanning and model creation

Initially the patient's CT scan was uploaded into the Mimics software to create the patient-specific model of the head soft tissue. From this, the uncompromised ear was isolated and modeled to construct the template. Additionally, the wider soft tissue, including the defective area, the bone, and fixation implants were reconstructed to perform a virtual, digital placement analysis of the templated ear to assist the prosthetists with visualization of the auricular prosthesis final form. The Mimics software is programmed with several predefined thresholds for isolation of either the soft tissue (CT) at −700 to 225 Hounsfield units (HU) or skin tissue (CT, adult) at −718 to −177 HU. However, both thresholds were found to be unsuitable for this particular dataset, with considerable levels of random noise still present. Therefore, a custom threshold of −298 to 531 was employed as a good compromise of noise reduction without compromising detail. It was found that with this threshold, much of the internal soft tissue was also highlighted, so cropping of the internal regions was performed to remove unwanted elements. Cropping was conducted across the three spatial planes to remove the majority of the unrequired internal soft tissue information while retaining the outer head template data. Minimization of the model size is also necessary as it ultimately reduces the computational processing requirements for postprocessing and digital rendering during design. As a final stage of data improvement, a mask smoothing operation was performed that allowed for the final patient model to be free of the majority of minor surface defects. Fig. 5.159 illustrates the thresholding/cropping process and the formation of the initial template model.

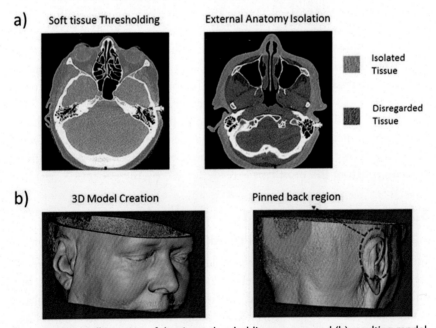

Figure 5.159 (a) Illustration of the tissue thresholding process and (b) resulting model from the optimized threshold highlighting the region of the ear that deviates from the natural position.

When constructing the model, it was noticed that the patient's ear did not sit in a natural position and was rather pressed against the patient's head. Upon further examination of the CT scan data, it was discovered that the headrest used to provide comfort to the patient during the scan, was deflecting the ear from its natural position, as can be seen in Fig. 5.159b. As there is no way to alter the data following image capture, this irregularity could not be removed and highlights the limitations of this modeling process. Feasibly the model could be digitally altered to regain the natural position of the ear; however, it was not attempted in this study over the clinician's preference to perform alterations during the physical wax cast-forming phase, as adjustments can be realized relatively quickly and easily. In future studies or use of the technology for this type of process, the radiographer should investigate alternative orientations of the patients to allow for a natural orientation of the target tissue that is to be modeled.

It was found that the constructed model was still in an unsatisfactory form for printing, with many minor surface defects that required resolving to build a smooth surface for the template. The model constructed in Mimics was input into 3-Matic, and various operations were performed to fill in gaps in the data and smooth the skin surface. At this point, the uncompromised ear was trimmed to isolate the model from the wider anatomic data. The model was then reworked to sit on a flat base, allowing for both the ease of printing but was also desirable for the subsequent casting phases, providing a flush and level base when pressed into the alginate gel. To create the flat base, the portion of the head to which the ear was originally attached was extruded to a distance of 5 mm, and the surrounding sections were filled, creating a "watertight" model. The base of the resulting model was then trimmed to create the flat base. Finally, the model was digitally mirrored to create a model suitable for the opposite side of the patient's head in the defective area.Fig. 5.160a illustrates the various patient-specific models during the design phase. At this point the model was complete and checked for any further digital errors before being in a state ready for 3D printing. Fig. 5.160b shows various images of the 3D printer during the manufacturing phase alongside the final models.

During the design phase, the wider anatomy of the patient was also postprocessed to remove digital errors resulting from the conversion process and to smooth the anatomy for a more natural look. This was performed on the soft tissue and titanium implants so that a virtual mock-up of the patient's anatomy could be created and used for a qualitative placement analysis of the resulting mirrored ear model. Fig. 5.160c(i) illustrates the

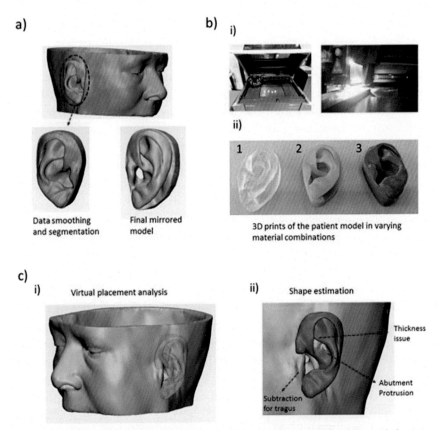

Figure 5.160 (a) Illustration of the optimized final patient-specific data and the isolated ear model before and after mirroring, (b) (i) photographs of the 3D printer during operation, and (ii) the final 3D printed models using a combination of 1%—100% TangoPlus, 2%—50% TangoPlus and 50% VeroMagenta, and 3%—100% VeroMagenta.

digital representation of the ear model on the wider anatomy of the patient. The analysis uncovered that the model would potentially require further adjustment, specifically to increase the thickness of the model at the abutment fixation points and within the inner portion of the ear, in addition to the modifications to regain the natural look of the ear. It was also predicted by the analysis that the final prosthesis model should be adjusted to accommodate the residual skin tissue on the patient that mimicked the tragus, as can be seen in Fig. 5.160c(ii).

5.22.3.2 Prosthesis cast production

Once manufactured the 3D printed model was implemented in the standard prosthesis production process. Initially an alginate gel was created and poured into a casting cylinder with the 100% TangoPlus material 3D printed model. This model was initially selected as it was the most flexible of the models and therefore was believed to allow for the easiest release from the gel. The alginate gel sets within approximately 5 minutes and the 3D printed model was easily removed to leave a negative impression of the model. Fig. 5.161a shows the alginate gel and the 3D printed template. It was found that the 3D printed model could be easily removed from the set alginate gel without causing any damage to the impression, despite its complex folds and overhanging features typically found on the ear. Given the success of this 3D printed model, none of the other 3D printed models were tested.

At this point the impression is filled with a medical-grade molten wax to create the base model of the prosthesis, as seen in Fig. 5.161a. The wax is allowed to set before the gel is split into two parts to release the wax model. Comparing the wax model to the 3D printed cast, Fig. 5.161b, it was seen that there was excellent reproduction of all features of the patient's ear. It was however noted that due to the inherent shrinkage of the wax during setting, the model shrunk by approximately 10% of its original size when compared against the existing anatomy of the patient. Despite this limitation, for the purposes of this study, we proceeded with the existing wax model to form the prosthesis to limit the inconvenience to the patient in terms of visitations. At a later date, we did however examine the use of 3D printed models that were enlarged by 5%, 10%, and 15% to form the wax model and found that a 10% enlargement resulting in adequate correction of shrinkage effects.

Once formed, the wax model was reworked by hand to remove any excess material and to form the opening for the tragus. The model is then offered up to the defective area on the patient, in the relative orientation at which the prosthesis will be located, and a quick-setting silicone compound (Affinis Perfect Impression, Coltene, Switzerland) is used to form an impression of the defective area, as can be seen in Fig. 5.161b. Once set, the silicone is easily peeled off the patient and is used to form a plaster cast reference of the patient's anatomy, alongside the location of the fixation abutments (Fig. 5.161c). This approach allows for further sculpting of the wax model without the necessity for the patient to remain within the hospital as the model required further modifications to return the ear to a more "natural" position. Additionally, as discovered from the digital placement analysis, the thickness of the wax model had to be increased to adequately house the fixation abutments and magnetic retainers. The model

a) Alginate gel casting and wax model forming

b) Wax model development and fixation referencing

c) Defective area casting and wax model finalisation

  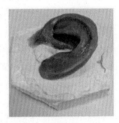

d) Final plaster cast of the patient specific prosthesis

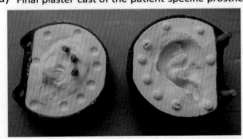 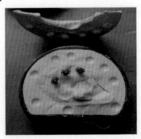

Figure 5.161 Various stages from 3D printed model to the final prosthesis plaster cast. (a) Alginate gel casting using the 3D printed patient model before preparation of the wax model, (b) comparison of the adjusted wax model next to the 3D printed model and phases of silicone surface mapping and model positioning, (c) formation of fixation template cast and final phases of wax model production, and (d) conversion of plaster cast and fixation locations into the final plaster cast for silicone production.

was once again manipulated by hand to incorporate these corrections and to embed the magnetic retainers before the model was once again placed on the patient to examine the size, fit, and position relative to the patient's uncompromised ear, as can be seen in Fig. 5.161c. Further minor modifications were made before the wax model was coated with plaster and set to form a negative mold into which the final silicone prosthesis would be formed. The final mold for the auricular prosthesis can be seen in Fig. 5.161d and comprised three primary subcomponents.

5.22.3.3 Silicone prosthesis production and fitting

The final prosthesis is formed by initial preparation of various shades of silicone to match the patient's unique skin pigmentation. As skin color can vary depending on the overall skin tone, sun exposure, skin thickness, and natural shadows, mimicry of a person's skin tone can require the use of several different colors of silicone to achieve an acceptable level of realism. Coloration is based on the judgment of the clinician and obtained by mixing a matching color or combination of pigmentations into transparent silicone before comparing against the skin of the patient until an acceptable match is obtained. It was also noted that when building up a particular skin tone, the coloration can alter due to the thickness of the translucent silicone, as has been demonstrated previously [15]. Therefore, an approach of forming a color that is marginally lighter than the perceived skin tone of thicker areas is adopted. Ultimately, this stage of the prosthesis formation process is highly qualitative in nature and highly dependent on the artistic skill of the clinician.

In this study, seven individual silicone colors were used to create the final prosthesis and meticulously painted into the cast to build up the final prosthesis, as can be seen in Fig. 5.162a. Once complete, the upper portion of the mold, which contains both the abutments and magnetic retainers, is pressed into the lower section and clamped together before being placed into an oven at a temperature of approximately 150°C for curing. During the curing process, the magnetic retainers are permanently fixed into the prosthesis, and the use of the abutments in the upper cast section ensures the correct orientation. The curing process can take approximately 0.5−1 hour, and the postcured silicone prosthesis can be seen in Fig. 5.162a. At this stage, there is typically some residual silicone material that has cured in the interface of the two cast halves and is removed using a miniature rotating

a) Silicone painting and curing

b) Final Prosthesis and antomical comparison

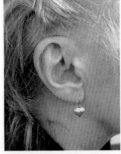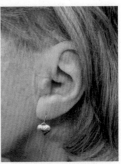

Final Prosthesis   Uncompromised Ear   Prosthetic Ear

c) Auricular prosthesis and model comparison

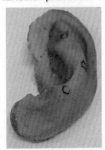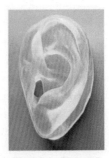

Old prosthesis   New prosthesis   3D Printed Model

Figure 5.162 Various stages of the final silicone prosthesis production. (a) Silicone painting stages to build the final model and the postcured model, (b) the final prosthesis compared to the uncompromised anatomy of the patient, and (c) a comparison of the previous and new prosthesis alongside the 3D printed template model.

sanding tool, leaving a smooth, continuous finish. The formed prosthesis is then placed onto the patient to test the fit before some further minor modifications are made to both ensure the prosthesis sits comfortably on the patient's skin and to make minor coloration corrections. Ultimately the prosthesis was a very good fit and very closely resembled the look of the uncompromised ear, both in shape and color, as can be seen in Fig. 5.162b. A qualitative comparison was also made with the patient's previous prosthesis, Fig. 5.162c made by the same clinician. It can be seen that the new prosthesis realizes a far superior end result with respect to the size and reproduction of the major features, confirming the efficacy of the presented approach.

5.22.3.4 Cost-benefit analysis

Ultimately, it was found that the use of 3D printing technology improved not only the end symmetry and shape of the final prosthesis but provided a cost-effective alternative to traditional practises. Using data from the Royal Melbourne Hospital, a typical facial prosthesis can cost on average $6000, not including costs for the surgical implantation or fixation abutments. Many of these costs are for the prosthesis consumables (magnets, fixation caps, silicone, pigments, finishing burs, etc.), which can amount to approximately $750—$1000 of the overall cost, placing $5000—$5250 in labor and overhead costs for production. The typical production time for an auricular prosthesis is approximately 15 hours for a prosthetist with 5 or more years' experience. Therefore, reduction in the labor time is the most influential factor to reducing manufacturing cost. During this study, we estimate that the use of the 3D printed model reduced the production time by approximately 5—6 hours, in addition to a reduction of two to three patient visits. Factoring in operator cost for design work and the cost of the 3D printed model, we estimate the production cost at approximately $4000. We therefore speculate that the use of 3D printing in this context provides a more cost-effective method for prosthesis production while realizing superior end results.

5.22.4 Conclusions

This study has presented a novel technique by which the use of high-precision 3D modeling and printing technologies can be integrated into the process chain for auricular prosthesis production. Using our approach, we could successfully reproduce the patient's uncompromised anatomy directly from medical CT scans and mirror this anatomy to form a template for casting of a direct silicone replacement using high-resolution 3D printing. It was found that the low shore hardness, 100% TangoPlus 3D printed template was optimal for conversion of the anatomic features into an alginate gel. This was then successfully used to form the typical wax model used in the prosthesis casting process. Ultimately it was found that the use of 3D modeling and printing for anatomic reproduction realized superior results when compared with traditional techniques both in terms of shape, size, and overall symmetry while dramatically reducing the time and cost for production by as much as 30%—35%. Formation of an accurate wax reproduction of the patient's anatomy is arguably the most challenging part of the production phase. Therefore, the use of 3D modeling and printing to address this limitation provides the greatest current benefit from this technology while issues relating to the 3D printed material biocompatibility, surface finish, or color realism develop to deliver acceptable results. In the present study, the patient's orientation during the scan necessitated additional model modifications to regain a natural position for the final prosthesis. It is appreciated that our intention to perform the 3D modeling was unknown to the radiologist performing the scan, so should this technique be developed for routine use, patient orientation must be considered to streamline prosthesis production. Finally, we concluded that the presented technique holds considerable potential for routine implementation within a clinical setting, streamlining prosthesis production and providing great levels of patient care.

References

[1] Jack RE, Schyns PG. The human face as a dynamic tool for social communication. Current Biology 2015;25(14):R621—34.

[2] Koster META, Bergsma J. Problems and coping behaviour of facial cancer patients. Social Science and Medicine 1990;30(5):569—78.

[3] Mardani MA, et al. Prosthetic rehabilitation of a patient with partial ear amputation using a self-suspension technique. Prosthetics and Orthotics International 2011;35(4):473—7.

[4] Mantri SS, Thombre RU, Pallavi D. Prosthodontic rehabilitation of a patient with bilateral auricular deformity. The Journal of Advanced Prosthodontics 2011;3(2):101—5.

[5] Penkner K, et al. Fabricating auricular prostheses using three-dimensional soft tissue models. The Journal of Prosthetic Dentistry 1999;82(4):482—4.

[6] Shifman A, et al. Prosthetic restoration of orbital defects. The Journal of Prosthetic Dentistry 1979;42(5):543—6.

[7] Jain S, et al. Nasal prosthesis rehabilitation: a case report. Journal of Indian Prosthodontic Society 2011;11(4):265—9.

[8] Kretlow JD, McKnight AJ, Izaddoost SA. Facial soft tissue trauma. Seminars in Plastic Surgery 2010;24(4):348—56.

[9] Ranganath K, Hemanth Kumar HR. The correction of posttraumatic Pan facial residual deformity. Journal of Maxillofacial and Oral Surgery 2011;10(1):20—4.

[10] Mannoor MS, et al. 3D printed bionic ears. Nano Letters 2013;13(6):2634—9.

[11] Jung-Seob L, et al. 3D printing of composite tissue with complex shape applied to ear regeneration. Biofabrication 2014;6(2):024103.

[12] Visscher DO, et al. Advances in bioprinting technologies for craniofacial reconstruction. Trends in Biotechnology 2016;34(9):700—10.

[13] Ozturk AN, Usumez A, Tosun Z. Implant-retained auricular prosthesis: a case report. European Journal of Dentistry 2010;4(1):71—4.

[14] Mohammed MI, Tatineni J, Cadd B, Peart P, Gibson I. Applications of 3D topography scanning and multi-material additive manufacturing for facial prosthesis development and production. Proceedings of the 27th Annual International Solid Freeform Fabrication Symposium 2016:1695—707.

[15] Mohammed MI, et al. Advanced auricular prosthesis development by 3D modelling and multi-material printing. KnE Engineering 2017;2(2):37—43.

[16] Bibb R, et al. An investigation of three-dimensional scanning of human body surfaces and its use in the design and manufacture of prostheses. Proceedings of the Institution of Mechanical Engineers - Part H: Journal of Engineering in Medicine 2000;214(6):589—94.

[17] Lim SK, Tong J, Cheng AC. Maxillofacial prosthetic management of an auricular defect for a young patient with hemifacial microsomia: a clinical report. Singapore Dental Journal 2011;32(1):33—8.

[18] Eggbeer D, et al. Evaluation of direct and indirect additive manufacture of maxillofacial prostheses. Proceedings of the Institution of Mechanical Engineers - Part H: Journal of Engineering in Medicine 2012;226(9):718—28.

[19] Jani RM, Schaaf NG. An evaluation of facial prostheses. The Journal of Prosthetic Dentistry 1978;39(5):546—50.

[20] Parthasarathy J. 3D modeling, custom implants and its future perspectives in craniofacial surgery. Annals of Maxillofacial Surgery 2014;4(1):9—18.

[21] Cohen A, et al. Mandibular reconstruction using stereolithographic 3-dimensional printing modeling technology. Oral Surgery, Oral Medicine, Oral Pathology, Oral Radiology and Endodontics 2009;108(5):661—6.

[22] Mohammed MI, Fitzpatrick AP, Malyala SK, Gibson I. Customised design and development of patient specific 3D printed whole mandible implant. Proceedings of the 27th Annual International Solid Freeform Fabrication Symposium 2016:1708—17.

[23] Paterson AM, et al. Computer-aided design to support fabrication of wrist splints using 3D printing: a feasibility study. Hand Therapy 2014;19(4):102—13.

[24] Fitzpatrick AP, et al. Design of a patient specific, 3D printed arm cast. KnE Engineering 2017;2(2):135—42.

[25] Mohammed MI, Fitzpatrick AP, Gibson I. Customised design of a patient specific 3D printed whole mandible implant. KnE Engineering 2017;2(2):104—11.

[26] Palousek D, Rosicky J, Koutny D. Use of digital technologies for nasal prosthesis manufacturing. Prosthetics and Orthotics International; 2013.

[27] Subburaj K, et al. Rapid development of auricular prosthesis using CAD and rapid prototyping technologies. International Journal of Oral and Maxillofacial Surgery 2007;36(10):938—43.

[28] Liacouras P, et al. Designing and manufacturing an auricular prosthesis using computed tomography, 3-dimensional photographic imaging, and additive manufacturing: a clinical report. The Journal of Prosthetic Dentistry 2011;105(2):78—82.

[29] Bos EJ, et al. Developing a parametric ear model for auricular reconstruction: a new step towards patient-specific implants. Journal of Cranio-Maxillofacial Surgery 2015;43(3):390—5.

[30] Ciocca L, et al. CAD/CAM ear model and virtual construction of the mold. The Journal of Prosthetic Dentistry 2007;98(5):339—43.

[31] He Y, Xue GH, Fu JZ. Fabrication of low cost soft tissue prostheses with the desktop 3D printer. Scientific Reports 2014;4:6973.
[32] Stratasys, polyjet materials data sheet. 2014.

Further reading

[1] Mobbs RJ, et al. The utility of 3D printing for surgical planning and patient-specific implant design for complex spinal pathologies: case report. Journal of Neurosurgery: Spine 2017;26(4):513—8.
[2] Moradiellos J, et al. Functional chest wall reconstruction with a biomechanical three-dimensionally printed implant. The Annals of Thoracic Surgery 2017;103(4):e389—91.

CHAPTER 5.23

Prosthetic rehabilitation applications case study 12—Customized design and development of patient-specific 3D printed whole mandible implant*

5.23.1 Introduction

The minimization of time for surgical procedures and patient recovery are highly pivotal in reducing financial burdens on healthcare providers and patients while also improving patient outcomes. This goal has led to many recent advancements and innovations in medicine, particularly in the area of patient-specific treatment options. Disruptive technologies, such as additive manufacturing and digital 3D modeling have positively impacted areas of preoperative planning and treatment [1,2], and have led to the creation of patient-specific assistive and implantable devices, such as surgical resection guides [3–5], planning models [3,6,7], anatomic teaching aids [8,9], and prosthetics [10–15]. More recently, the US Food and Drug Administration (FDA) has increased their approval of 3D printed implants under the 510k (premarket notification) approval system, which permits the

* The work described in this chapter was first reported in the references below and is reproduced here with permission of the organizing committee of the Solid Freeform Fabrication Symposium.
Mohammed MI, Fitzpatrick AP, Malyala SK, Gibson I. Customised design and development of patient specific 3D printed whole mandible implant. Proceedings of 27th Solid Freeform Fabrication Symposium, Austin TX, USA; 2016. 1708–17.

Medical Modeling
ISBN 978-0-323-95733-5
https://doi.org/10.1016/B978-0-323-95733-5.00028-4

use of additive manufactured parts in routine and complex surgical procedures [16]. It is suspected that recent approvals and the wider acceptance of additive manufacturing is likely the result of increasing maturity of this technology, which has allowed the direct processing of an increasing number of biocompatible materials with ever increasing printing precision.

A mandibulectomy is the removal of large sections of the lower jaw and is performed following either physical trauma or severe disease infiltration (cancer, etc.) [17,18]. Current surgical rehabilitation is performed using a guide plate, screw fixtures, and grafted donor bone tissue (hip/femur) inserted into the void space to support the jaw. However, such methods are prone to short-term failure (approx. ≤1 year), guide plates are predominantly manipulated during surgery to approximately match the patient's contours, and treatment arguably does not provide a solid base to allow osseointegration of a patient's bone tissue. It has been more recently found that fixation methodologies traditionally used are leading to complications such as mandibular fractures [6,19,20] resulting from fixation in structurally compromised bone tissue. Therefore, there is increasing evidence to support the use of whole mandible replacement implants.

In this study, we assessed the potential of creating a custom replacement mandible implant following a mandibulectomy. The implant was created using patient-specific medical imaging data (CT DICOM) alongside state-of-the-art 3D design/manipulation software, before realizing the implant using low-cost fused deposition modeling (FDM) to assess and finish the model (Fig. 5.163). It is noted that the FDM printed models are for preliminary verification purposes and would not be used as the final implantable device. This case study was performed on a patient who presented with cancer infiltration through the entire right side of the mandible. CT scan images were used to generate a three-dimensional representative model of the patient's corrupted mandible, before an initial full jaw replacement implant was constructed. Mandible replacement devices have been attempted previously; however, this work differs from previously reported designs as we attempt to retain the major contours of the patient, so the implant retains the original facial structural contours of the patient. This contrasts favorably to other such examples of this form of treatment where the replacement mandible is generally much smaller than the original bone structure or requires bone grafts for reconstruction. We examined and compared the efficacy of reconstruction using anatomy mirroring and by direct mandible digital reconstruction. Ultimately, this work provides a framework for the design criteria to create an optimized patient-specific,

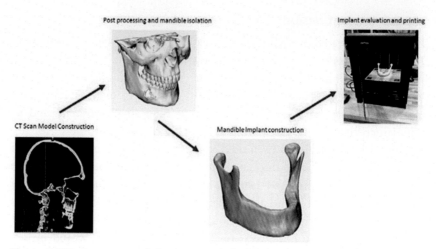

Figure 5.163 A summary of the process chain for the construction of the complete mandible implant.

full-mandible implant. The techniques used in this work could equally be applied to alternative patient anatomies, providing a generic methodology for implant design. Such work will provide a powerful tool for clinicians to overcome the shortcomings of typical fixation plate-based treatment options, thereby improving overall patient outcomes.

5.23.2 Methodology

5.23.2.1 Anatomic data modeling

Representative models of the patient's anatomic data were constructed based on Digital Imaging and Communications in Medicine (DICOM) data from CT scans using a Somatom scanner (Siemens Healthcare, Germany). In the DICOM format, the data are presented as a series of slices through the patient's anatomy, each approximately 0.6 mm thick. The software package Mimics (Materialise, Belgium) was used to compile the DIOCM data into axial, sagittal, and coronal planes. Using Mimics, the various grayscales can be selectively marked to isolate a particular anatomic tissue type (i.e., bone, muscle, etc.) and from this to construct a 3D model of this tissue type. The Mimics software has a series on inbuilt thresholds that allow for the rapid isolation of anatomic tissues, and the inbuilt thresholds for bone were used to construct the final model, in additional to manual region growing.

5.23.2.2 Implant design

The model created in Mimics was exported to 3-Matic (Materialise, Belgium) for further processing and construction of the mandible model. Prior to any operations, the CT scan—extracted model underwent an initial phase of error correction. This is required to remove excess data points, duplicate triangles, and inverted normals, to filter small subshells, and to unify larger shells into a single model, allowing the model to be fit for 3D printing. The software was then used to segment the skull model to isolate the jaw. In this study, two distinct approaches were examined whereby the mandible was reconstructed using a mirroring of the patient's healthy portion of the mandible and also by direct reconstruction of the cancerous section. Both approaches could potentially offer a high-quality matching of the patient's anatomy, but it was unclear which technique would be superior, so both approaches were assessed.

5.23.2.3 Implant prototyping and manufacturing

In this study, the various design iterations of the jaw model were realized using FDM printing in ABS plastic. FDM printing is a low-cost technique for part production, and evaluation prints were realized using a Zortrax M200 (Olsztyn, Poland) printer. This methodology allowed for the rapid printing of the proposed implant design iterations to evaluate model build integrity and patient fitment/sizing against the original jaw anatomy.

5.23.3 Results

5.23.3.1 Anatomic modeling

Using the inbuilt threshold function within Mimics (Bone CT, 226—3071), the bone tissue was predominantly isolated. However, several features of the soft tissue were also captured at this grayscale value, in additional to several uncaptured data points of the bone. Therefore, additional manual processing of the model was required to remove unwanted data points and to add missing bone data point in the CT scan slices across all three-dimensional planes. Once this had been achieved to a satisfactory point, the highlighted dataset was then converted into a representative model of the patient's wider skull, which contained the corrupt mandible.

This constructed model comprised a segment of the overall skull, the maxilla, and the mandible. Primarily, the mandible was the area of interest in the jaw reconstruction; however inclusion of the skull and maxilla and realization into a 3D printed model allowed for analysis of the jaw size and

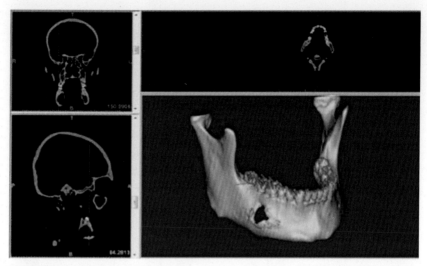

Figure 5.164 An image of the various planes of the patients CT scan data and the constructed model of the isolated bone tissue segment, containing the corrupt mandible.

fit compared with the original anatomy of the patient. Fig. 5.164 shows the patient's skeletal data and the resulting model that was formed for the mandible. Despite the automated threshold function and manual datapoint addition/removal, some elements can be missing, in additional to stray noise being present in the final digital model. In such instances a wrapping function can be performed to improve the final quality of the model and without adjusting the size threshold of the model. Following wrapping, the model was visually inspected to check that the major features were adequately rendered before the model was exported as an STL file. The model was then loaded into 3-Matic to perform error corrections, segmentation, reconstruction of the corrupt region of the mandible, and additional postprocessing.

5.23.3.2 Implant design

When the initial model was imported into 3-Matic, error corrections were initially performed to ensure the integrity of the model prior to implant production. Next the mandible was isolated to assess the degree of cancer infiltration in the patient. Fig. 5.165a shows a transparent digital model of the mandible where the cancerous region can be seen as a large cavity. At the extremities, the cancerous region was estimated to be 25 mm in the vertical and 50 mm in the lateral direction.

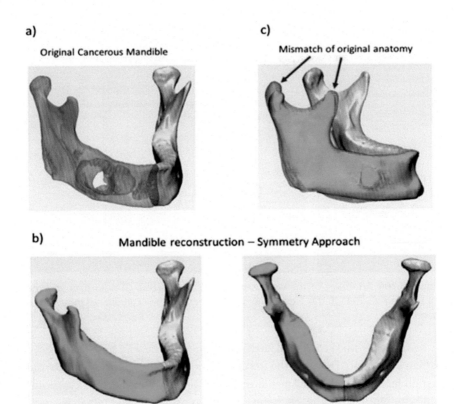

a)

Original Cancerous Mandible

c)

Mismatch of original anatomy

b)

Mandible reconstruction – Symmetry Approach

Figure 5.165 (a) The original corrupt mandible, (b) outcomes of the symmetry-based approach for mandible reconstruction, and (c) illustration of the anatomy mismatch of the final model against the original mandible.

Two distinct approaches were examined to ascertain the best methodology for the implant production and which comprised reconstruction through mirroring of the uncompromised area and direct reconstruction of the cancerous jaw segment. To assess the fit of the resulting mandible design, a model of the patient's maxilla and temporomandibular joint was created from the patient's DICOM data and printed using an FDM printer. Additionally, the corrupt mandible model was printed to compare against the corrected mandible model.

5.23.3.3 Anatomy mirroring

The main area of cancerous infiltration was restricted to the left portion of the patient's mandible. Therefore, the mandible was divided into two sections, and the right half was mirrored to construct the final implant.

When attempting to section the mandible, the teeth were retained to help guide the location for sectioning. Upon closer examination of the mandible, there were several natural contours in the front of the jaw that could act as a guide point for the sectioning process. Several locations were examined as the reference point to produce the complete implant, with promising results, as shown in Fig. 5.165b. Cross-referencing of the original compromised section of the jaw revealed that due to asymmetry present in the original jaw, the location of the ramus, condylar process, and coronoid process were all out of alignment, Fig. 5.165c. Such misalignment would likely result in placement issues during implantation as well as compromising postsurgical rehabilitation and movement. Several attempts to readjust the spatial orientation of the mirrored mandible segments were unsuccessful in achieving a matching orientation to the original mandible anatomy. Based on these results, it was determined that anatomy mirroring in these particular circumstances of execution was not a robust technique for digital reconstruction of the corrupted mandible.

5.23.3.4 Segment reconstruction

In the segment reconstruction approach, the corrupt portion of the mandible was isolated and reconstructed manually to create a solid portion that matched the original contours of the patient's jaw. The advantage of this approach is that the majority of the shape and geometry of the original mandible are preserved. Initially, the mandible model was made transparent to assess and visualize the extent of the cancer infiltration, as shown in Fig. 5.165a. This methodology allowed for the complete visualization of the cancerous growth, which aided the digital reconstruction process. This compromised region of the mandible was segmented for additional processing.

To aid the design process, the mandible was segmented into two halves about an approximate midpoint, in a similar manner to anatomy mirroring approach. The uncompromised half of the mandible was then mirrored and used as a reference template for the reconstruction of the segment, to achieve the desired thicknesses and surface contours. The segmented model in its native form comprised a hollow part with openings at both sides of the lateral sections of the mandible. It was also noted that the cancerous growth had caused an expansion of the surrounding jaw line, so in addition to removal of the hollow sections, the segment required reworking to this lateral displacement. Several trimming and smoothing operations were performed to bring the side of the segment back to a similar dimension to that of the mirrored reference. It was additionally noticed that surface

a)

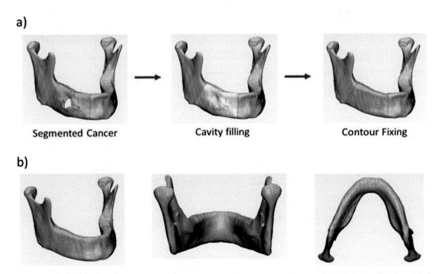

Segmented Cancer   Cavity filling   Contour Fixing

b)

Figure 5.166 (a) Stages of the segment reconstruction strategy and (b) the final model of the mandible.

contours of the uncompromised region contained a variety of holes, beyond the natural cavities containing the mental nerve, which were required to be filled to ensure the maximum integrity of the final model. These holes were present on the patient's original CT scan data, so the holes/cavities were the result of digital artifacts during the modeling process. This potentially implied that the mandible was structurally compromised, so should a traditional fixation plate approach be employed, this could result in mandibular fracturing. These findings further confirmed the necessity for a total mandibular replacement. The final segment and its placement in the sectioned mandible are shown in Fig. 5.166, where it can be seen that, overall, a very good reproduction of the original symmetry could be achieved. This final model also matched the original locations of the ramus, condylar process, and coronoid process due to only working on the cancer-infiltrated region of the mandible, and therefore this reconstruction technique was considered the superior approach for this specific case study.

5.23.3.5 Mandible 3D printing

When constructing models based on data ascertained from medical imaging data, there could be facets of the model, such as low thickness tolerances in regions of the part, that could lead to compromised structural integrity. Given the complexity of such models, it may not always be possible to perform adequate thickness analysis and finite element analysis to validate mechanical properties. Additionally, as such parts are intended as an implantable device, a rigorous and robust evaluation procedure is desirable. Therefore, a preliminary low-cost polymer printing approach can help provide qualitative validation of a part's structural integrity, before the final, more costly printing in metallic materials.

The final model was then 3D printed initially on an FDM printer to assess the structural integrity of the final models and also to ensure correct dimensional accuracy of the jaw against both the corrupt mandible and partial skull/maxilla models. It was found that the model printed very well with no perceivable issues with its structural integrity. Fig. 5.166 shows the printed corrupt and reconstructed mandible, alongside the reconstructed mandible implant placed into a representative model of the skull/maxilla. It can be seen that the reconstruction approach of the jaw was highly effective at reproducing the patient's original jaw line, with minimal impact to the overall structural contours. The resulting mandible models proved an equally excellent fit into the representative model of the patient skull, as can be seen in Fig. 5.167. These qualitative tests confirmed the efficacy of the

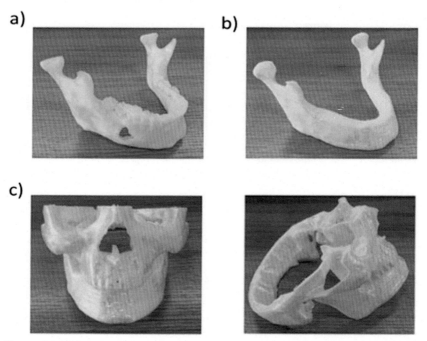

Figure 5.167 Representative FDM models of (a) the original cancerous mandible, (b) the reconstructed mandible implant, and (c) the developed implant placed onto a model of the patient's skull and maxilla.

design and the structural integrity when realizing the mandible model by 3D printing processes. We are therefore confident that the design could be equally rendered by SLM processed in titanium, a certified material for medical implantation, and this will be the subject of future studies.

5.23.4 Conclusions

In this study, we have examined the use of medical CT scan data to reconstruct a digital model of a patient's skeletal anatomy and, from this, a full replacement mandible to treat a cancerous infiltration. The final model and the patient's surrounding skeletal anatomy were then realized using an FDM 3D printing process, to evaluate the size and geometric precision of the final implant. Investigating various design techniques, it was found that a mandible reconstruction approach offered superior fit compared to the anatomy mirroring approach. Ultimately, this study demonstrates an optimized approach for patient-specific mandible reconstruction that could have significant potential for use as a strategy in patient-specific, custom implant design.

References

[1] Rengier F, et al. 3D printing based on imaging data: review of medical applications. International Journal of Computer Assisted Radiology and Surgery 2010;5(4):335—41.
[2] Chae MP, et al. Emerging applications of bedside 3D printing in plastic surgery. Frontiers in Surgery 2015;2:25.
[3] Parthasarathy J. 3D modeling, custom implants and its future perspectives in craniofacial surgery. Annals of Maxillofacial Surgery 2014;4:9—18.
[4] Ma L, et al. 3D-printed guiding templates for improved osteosarcoma resection. Scientific Reports 2016;6:23335.
[5] Fu J, et al. (Use of four kinds of three-dimensional printing guide plate in bone tumor resection and reconstruction operation). Zhongguo Xiu Fu Chong Jian Wai Ke Za Zhi 2014;28(3):304—8.
[6] Fabris V, Bacchi A. Fixation of a severely resorbed mandible for complete arch screw-retained rehabilitation: a clinical report. The Journal of Prosthetic Dentistry 2016;115(5):537—40.
[7] Cohen A, et al. Mandibular reconstruction using stereolithographic 3-dimensional printing modeling technology. Oral Surgery, Oral Medicine, Oral Pathology, Oral Radiology, and Endodontology 2009;108(5):661—6.
[8] McMenamin PG, et al. The production of anatomical teaching resources using three-dimensional (3D) printing technology. Anatomical Sciences Education 2014;7(6):479—86.
[9] Adams JW, et al. 3D printed reproductions of orbital dissections: a novel mode of visualising anatomy for trainees in ophthalmology or optometry. British Journal of Ophthalmology 2015;99(9):1162—7.

[10] Jain S, et al. Nasal prosthesis rehabilitation: a case report. Journal of Indian Prosthodontic Society 2011;11(4):265—9.

[11] Palousek D, Rosicky J, Koutny D. Use of digital technologies for nasal prosthesis manufacturing. Prosthetics and Orthotics International; 2013.

[12] Bos EJ, et al. Developing a parametric ear model for auricular reconstruction: a new step towards patient-specific implants. Journal of Cranio-Maxillofacial Surgery 2015;43(3):390—5.

[13] Kuru I, et al. A 3D-printed functioning anatomical human middle ear model. Hearing Research 2016;340.

[14] Subburaj K, et al. Rapid development of auricular prosthesis using CAD and rapid prototyping technologies. International Journal of Oral and Maxillofacial Surgery 2007;36(10):938—43.

[15] Eggbeer D, Bibb R, Evans P. Assessment of digital technologies in the design of a magnetic retained auricular prosthesis. 2006.

[16] FDA. 3D printing of medical devices. 2016. Available from: http://www.fda.gov/MedicalDevices/ProductsandMedicalProcedures/3DPrintin gofMedicalDevices/default.htm.

[17] Guerra MFM, et al. Rim versus sagittal mandibulectomy for the treatment of squamous cell carcinoma: two types of mandibular preservation. Head and Neck 2003;25(12):982—9.

[18] Barttelbort SW, Bahn SL, Ariyan S. Papers of the society of head and neck surgeons rim mandibulectomy for cancer of the oral cavity. The American Journal of Surgery 1987;154(4):423—8.

[19] Raghoebar GM, et al. Etiology and management of mandibular fractures associated with endosteal implants in the atrophic mandible. Oral Surgery, Oral Medicine, Oral Pathology, Oral Radiology, and Endodontology 2000;89(5):553559.

[20] Chrcanovic BR, Custódio ALN. Mandibular fractures associated with endosteal implants. Oral and Maxillofacial Surgery 2009;13(4):231—8.

Orthotic applications

CHAPTER 5.24

Orthotic applications case study 1—A review of existing anatomic data capture methods to support the mass customization of wrist splints*

5.24.1 Introduction

Mass customization (MC) is an increasingly common and popular manufacturing approach in many disciplines, particularly when combined with additive manufacturing (AM). The MC—AM relationship is expanding and progressing, from the consumer market through to the medical industry for development of surgical planning, teaching, preparation aids, and assistive technologies [1]. MC is intended to collect the needs and requirements of the end user to make their life easier and more comfortable through customized fit, function, performance, and aesthetics; the European initiative project CUSTOM-FIT illustrates the need for customization, with particular interest in assistive technologies for disabled

* The work described in this chapter was first reported in the reference below and is reproduced here, in part or in full, with the permission of both CRDM and Taylor and Francis Publishers.

Paterson, AMJ, Bibb, RJ, Campbell, RI, "A review of existing anatomic data capture methods to support the mass customisation of wrist splints," Virtual and Physical Prototyping, 2010, 5(4):201—207, https://doi.org/10.1080/17452759.2010.528183

Paterson, AMJ, Bibb, RJ, Campbell, RI, "A review of existing anatomic data capture methods to support the mass customisation of wrist splints," in: 11th National Conference on Rapid Design, Prototyping & Manufacture, eds. Jacobson, D, Bocking, CE, Rennie, AEW, 2010, CRDM, Ltd., High Wycombe, pp 97—108, ISBN 978-0-9566643-0-3.

Medical Modeling
ISBN 978-0-323-95733-5
https://doi.org/10.1016/B978-0-323-95733-5.00029-6

users [2]. Therefore, anatomic data capture is essential for the manufacture of customized assistive devices.

According to E. Donnison (personal communication, February 4, 2010), the prescription of wrist splints inevitably requires a certain amount of customization to suit the patient's individual fit and requirements, regardless of whether the splints are fully custom-fitted or prefabricated "off the shelf" splints. This is particularly relevant for patients with degenerative diseases such as rheumatoid arthritis, as patients often need splints that will be comfortable, robust, and long lasting. A wide range of literature suggests the suitability of anatomic imaging equipment such as magnetic resonance imaging (MRI) and computed tomography (CT) for the design of pros-thetics and orthotics, but very little research has been focused into the data acquisition of hand and wrist geometry for the MC of wrist splints [3–5].

A US patent application by Fried [6] stated ownership of a process for creating wrist splints, from data acquisition of anatomic features of wrists and hands through to splint fabrication using AM. However, data acqui-sition of the wrists and hands is not a straightforward process. As of yet, there is no standardized method for collecting topographical skin surface data for the wrists and hands, and numerous problems emerge with various data collection methods. Therefore, this paper evaluates the strengths and weaknesses of four different data acquisition methods: CT, MRI, 3D laser scanning, and anthropometrics. The most suitable method will be identified to support the digitization process of customized wrist splint design and manufacture for small-scale research. It should be noted that this report is not a comprehensive study into all data acquisition methods, nor does it imply that the chosen method is suitable for any or all clinical applications, but it reviews the most common data acquisition types discussed in other case studies within the field of medicine to support future doctorate research.

5.24.2 Data acquisition methods

5.24.2.1 Non-contact data acquisition—Computed tomography

CT is widely used within medicine, typically for diagnosis and surgical planning. CT has the ability to generate both 3D and 4D images, along with quality, volume-rendered imaging. The patient is placed between an X-ray tube and a detector array; X-rays pass through the patient, and the level of attenuation is detected by the detector array and then logged by a computer [7]. Measurements are taken from various angles to produce a series of axial slices and are illustrated in grayscale to demonstrate different densities within the body, ranging from black to represent air and white to represent the densest bone [4]. There are now two main types of CT: sequential and spiral [7]. Spiral CT can capture data in real time to produce high-quality 4D imagery, and it can be used for functional analysis procedures such as CT—coronary angiography and diagnosis of joint instability [8,9]. This is particularly beneficial for patients who are unable to maintain a still position [4,8]. Other advantages of CT are high image resolution between soft tissue, bone, and air and the ability to improve contrast and decrease structure noise. Users can also focus on a specified field of view to produce more accurate scans [4,7]. For these reasons, CT within anaplastology for the design of prosthetics is now standard practice and is often linked with AM for prosthetic fabrication [4,10].

However, there are several drawbacks to CT imaging. Radiation is the biggest concern; CT imaging was dismissed by Eggbeer [11] for prosthetic design due to radiation exposure. Radiation exposure is directly proportional to the duration of scanning, so higher image resolution and larger area coverage will require longer scanning times, thus exposing patients to greater radiation dosage [5,9]. Bibb and Winder [5] state that radiologists should rationalize the acceptable level of accuracy and resolution from CT imaging, so scanning times may be balanced with radiation exposure. Also, resolution between different soft tissues is often poor because collected images are divided by pixel shade; two different densities that "share" a pixel create an intermediate density, known as the "partial pixel effect," which can create a blurred boundary [4]. Minns et al. [3] identified issues in lack of definition when combining CT imagery with AM; CT slice distances can range from 0.5 mm upward, whereas certain AM equipment can be capable of creating 0.15-mm-thick layers [4]. Therefore, intermediate sections between CT data layers are required. Also, a large field of view can demonstrate poor resolution when scanning small, intricate detailing [12]. However, Tay et al. [9] justify the use of CT for wrist function analysis, since they claim that the wrist is not sensitive to radiation.

5.24.2.2 Non-contact data acquisition—Magnetic resonance imaging

Approximately 500 MRI machines are in use within the United Kingdom and over 20,000 worldwide [13]. MRI uses three different magnetic fields, a static magnetic field, a switched gradient field, and a pulsed radio frequency field. The combination of these fields causes hydrogen atoms within the body to align, so the equipment can differentiate between re-actions and physical attributes [5,13,14]. MRI image "slices" can be used solely for diagnoses or combined through suitable medical software into a 3D virtual form for design and fabrication of surgical guides and prosthetics [4,15]. The captured data illustrate different densities; areas with a high concentration of hydrogen atoms, such as soft tissue, are represented in light grayscales, whereas areas with low hydrogen atom concentration appear darker [4]. Therefore, MRI is excellent at differentiating between different soft tissues and air in close proximity and can be used for capturing skin surfaces [4].

Various strengths are associated with MRI such as high resolution of soft tissue and bone morphology, and unlike CT, patients are not exposed to radiation [16]. However, there are several disadvantages. Over one million examinations are performed using MRI equipment in the United Kingdom each year. Therefore, the equipment is in high demand among a wide variety of medical specialties, and waiting times can be lengthy [16]. This is particularly relevant when considering patients who require more urgent access to MRI equipment. The mechanical design of the equipment can also deter patients suffering from claustrophobia, and the noise created may be unnerving [4,17,18]. MRI scanning can also be time consuming, and movement can cause distortion and shadowing. Various studies have investigated the administration of general anesthesia and patient sedation during MRI screening to gather "undisturbed" data imagery, but these methods may come with concerns, particularly for children and individuals with compromised health conditions [17,19,20]. Patients with pacemakers, cardioverter-defibrillators, and metal fabricated implants also face diffi-culties, as exposure to strong magnetic fields could be detrimental to patient health, although Roguin [21] discusses these concerns and rationalizes MRI usage after careful clinical evaluation.

To date, no literature has been found with regard to collecting skin surface topography of the wrist and hand using MRI. This may be because volume-rendered imaging is rarely performed on MRI scanning due to the time involved in scanning and processing data, and most diagnoses can be performed with single accurate slice images.

5.24.2.3 Non-contact data acquisition—Optical-based systems

Optical-based systems are becoming more prominent across a range of fields in medicine. Close-range triangulation laser scanning, for example, involves a laser source and one or more sensors (Fig. 5.168). A laser is

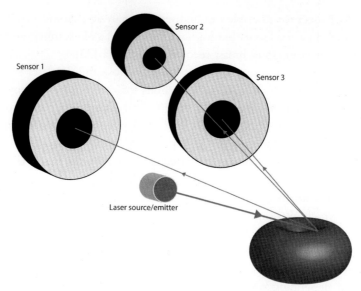

Figure 5.168 3D laser scanning principles.

emitted from the source and pointed toward the artifact; the reflected light is then captured by the sensor(s). Since the angle of the source relative to the sensors is known, the distance of the reflected source can be calculated using triangulation to form a 3D virtual representation of the artifact.

Structured light systems also have a light-emitting source, but in this case, patterns such as moiré are emitted across a wider area; these patterns alternate in shape, size, and position over a set period of time. The reflected data captured of the emitted patterns are compared in complimenting software to create a 3D representation. Photogrammetry tends to refer to a single-mounted structured light system and will capture data only in its field of vision. Stereophotogrammetry, however, involves the use of multiple light-emitting sources and sensors, strategically placed in different positions relative to one another to ideally surround the artifact to be scanned. Each unit captures a different angle of the object. These scans can then be combined together in supporting software, and depending on the positions of the light source and sensor, they can give a complete 360-degree 3D representation of the artifact.

Uptake of the technology has been limited in the past due to the cost of the equipment, particularly high-resolution stereophotogrammetry equipment. However, recent developments in technology show promise for opening up a new market for low-cost optical-based systems for the future;

the Xbox Kinect for example can be used for analyzing biomechanics [22], but it can also be used to gather topographic data as a 3D scanner, and it has been used for comparing breast reconstruction data [23].

In terms of applications, Chua et al. [11] chose the 3D laser scanning method over other "conventional" methods for prosthesis modeling, such as plaster of paris impressions, MRI, and CT. Laser scanning has also been used to assess anatomic changes in facial morphology over time to demonstrate its suitability [24].

A significant benefit of optical-based data capture systems is that only the patient's skin surface is captured.

This substantially reduces data file sizes compared with MRI and CT imagery, which also capture imaging of internal anatomy [11]. Processing time is also significantly reduced; point cloud data can be converted to a polygonal mesh within 3D CAD based software. Another benefit is the speed of data collection. Kau et al. [24] reported facial scanning times of 2.5 s, and depending on the optical equipment used, accuracy can be as high as 0.05 mm between data points [11,25]. The financial benefits of optical methods are also an incentive when compared with CT and MRI, offering affordable hardware and software and minimal training requirements, which correspond to ease of use, availability, accessibility, efficiency, validity, and repeatability [24].

However, one significant problem with any optical-based system is the inability to capture wanted internal structures and intricate surfaces due to line-of-sight limitations [25,26]. This is particularly relevant to the application of prosthetics and orthotics. Certain topographic sections of the human anatomy have intricate creases and folds, particularly between fingers and the thenar webbing when the hand is in a neutral position; scanning these areas can result in void data and unwanted point convergence, as demonstrated in Fig. 5.169 [11,12,27]. Various sources suggest using reverse engineering software capable of postprocessing to produce a "watertight" model by repairing and resculpturing void data [11,25,26]. However, this process can be time consuming and may not be a true representation of the scanned object once corrected [25]. Bibb et al. [25] suggested another approach to capture shadowed data by collecting and combining several overlapping scan data files to produce a complete model. However, data point density would increase, and alignment of point clouds may be difficult [25,26]; time and expertise in reverse engineering software are also significant factors to combine scans effectively; software such as Geomagic Design X (https://www.oqton.com/geomagic-designx/) is available where the process is

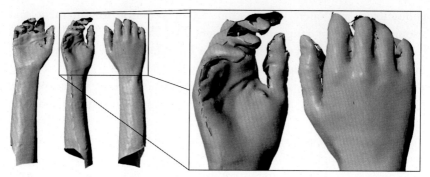

Figure 5.169 Direct 3D laser scanning, data voids, and convergence of points between fingers.

simplified for the user through more complex software algorithms, but at a cost. Another limitation to 3D laser scanning is inaccurate data acquisition due to involuntary movement, causing noise and distortion [4], and this concern is particularly relevant to this investigation. Two studies have suggested using custom-made position jigs to prevent movement and to achieve better scanning results with limited success [25,27].

In contrast, Direct Dimensions [28] identified an accurate technique of capturing anatomic attributes to support the fabrication of a lifelike prosthetic hand using AM. The process involved scanning a plaster cast model of a patient's hand in a neutral position [28]. Two laser scanners were mounted to a Faro Arm and a motorized, precision, coordinate-measuring machine. The setup was capable of capturing intricate detail such as pores and creases [28]. The case study demonstrates that scanning can be effective when scanning inanimate objects, because concerns over involuntary movement are diminished. This method is supported by Tikuisis et al. [29], who reported highly accurate resolution of scanned hand data. Based on the conclusion of Surendran et al. [26], plaster of paris would be a very effective material for scanning, due to its matte appearance, even color tone and opacity, and its ability to capture fine detail, but casting can be messy, time consuming, and uncomfortable for patients.

5.24.3 Contact data acquisition

5.24.3.1 Anthropometrics

A report by Greiner (1990) has been stated as a reliable resource for accurate anthropometric measurement techniques for hands [30,31]. Calipers and goniometers are often used within orthopedics for measurement and assessment of joint range of movement [32]. Williams [33] designed a technique for creating customized gloves by collecting key anthropometric measurements using calipers and goniometers. The values were entered into a parameter table, and the parametric 3D CAD hand model was adapted to meet these dimensions. The parametric model was converted into a suitable format for AM. Li et al. [34] investigated the concept of extracting anthropometric data from scanned resin hand casts, stating that this would be a successful approach in future anthropometrics.

However, collecting enough anthropometric data to construct a wrist splint would be costly in time and labor, and it would potentially demonstrate high error if measurements were entered incorrectly into a parametric model within a 3D CAD program. Particular difficulties would come from measuring prominent bony landmarks and contours within the wrist such as the radial styloid process, ulnar styloid process, and pisiform [27,35]. Since the hand and wrist have many degrees of freedom, any movement within the joints could affect the measurements, which in turn could invalidate other dependent measurements.

5.24.4 Conclusion and future work

This review has illustrated that there is no standardized method to capture the entire surrounding skin surface topography of the wrist and hand to support customized splint design and manufacture. Table 5.13 lists the ideal characteristics for data acquisition methods targeted at splint design.

A common problem among all data acquisition methods is voluntary and involuntary movement, and this may be particularly problematic for patients who suffer from medical conditions such as Parkinson's disease. Therefore, methods of immobilization through jigging or casting may be the most effective method to support data collection. However, given the range of movement capable from the joints within the wrists and hands, jigging may not be the most appropriate method, as the jig itself would need to cover joints and may interfere with data collection, particularly with optical-based methods like 3D laser scanning. Jigging would be most appropriate for capturing anatomic contouring of the wrist and forearm, but if the patient requires additional support around the digits, i.e., to support fingers from ulnar drift, then a polygonal mesh between the digits may be necessary to create this additional support; this may be difficult and time consuming to capture.

Table 5.13 Comparative advantages and disadvantages of data acquisition methods.

Data capture method / Ideal characteristics	CT/MRI	Anthropometry	Direct 3D laser scanning	Indirect 3D laser scanning (support jig)	Indirect 3D laser scanning (plaster cast)
No trained specialists required	X	✓	✓	✓	✓
Affordable for low-budget projects	X	✓	✓	✓	✓
Quick, easy access	X	✓	✓	✓	✓
Low/no waiting times	X	✓	✓	✓	✓
Minimum preparation time	✓	✓	✓	X	✓
Quick data collection	MRI = X CT = ✓	X	✓	✓	✓
Low/no discomfort for patients	✓	✓	✓	✓	✓
No health/safety risks to patients	X	✓	✓	✓	✓

Table 5.13 Comparative advantages and disadvantages of data acquisition methods.—cont'd

Data capture method Ideal characteristics	CT/MRI	Anthropometry	Direct 3D laser scanning	Indirect 3D laser scanning (support jig)	Indirect 3D laser scanning (plaster cast)
Low risk of data capture error	✓	X	X	Dependent on the suitability of the jig	✓
Suitable accuracy	✓	✓	✓	✓	✓
Suitable resolution	✓	✓	✓	✓	✓
Data capture unaffected by movement	MRI = X CT = ✓	X	X	Dependent on the suitability of the jig	✓
No line-of-sight limitations	✓	✓	X	X	X
Little/no concern over positioning	✓	X	X	X	✓
Reliable repeatability	✓	X	X	Dependent on the suitability of the jig	✓
Direct export into 3D CAD software	✓	X	✓	✓	✓

Findings conclude that in terms of accuracy, resolution, patient safety, cost, speed, and efficiency, optical-based acquisition methods appear to be the most suitable to meet all needs.

In the context of gathering test data for continued research and development of a digitized approach to splinting, one may combine optical methods with plaster casting to reduce collection of ambiguous data while providing a repeatable and reliable data source for validation purposes. From an orthopedic aspect (E. Donnison, personal communication February 4, 2010), it is important to note that hands should be cast in the neutral or intrinsic plus position if possible, or a rested comfortable position if the splint is for a patient with a degenerative disease such as rheumatoid arthritis. It is an accepted fact that collecting scanned data between fingers is difficult, particularly when the hand is in a neutral position. Therefore, plaster casts could be placed and scanned in several different orientations if using fixed position optical-based scanners, until acceptable data is collected between the digits. Hand-held scanners will also allow free movement around the plaster cast to capture greater detail of intricate surfaces from different angles. This will form the focus for future research, since a cast can provide and maintain a constant reference position suitable for repeatability testing.

Also, by using the technique suggested by Li et al. [34], anthropometric measurements could also be taken from a plaster cast if required, without further consultation with the patient. Although the casting process can be time consuming and messy, physical and digital versions of the patient's hand could be kept on record, which may be useful for orthopedics when assessing disease progression. The process may be similar to the case study by Kau et al. [24] with regard to changes of facial structure captured using laser scanning. How accurate the scan data would need to be to fabricate a comfortable form around the patient's wrist and hand is yet to be estab- lished. However, it is assumed that point density should vary throughout the model, with a denser point count around complex geometry, and a sparse point count to represent simple geometry. Reverse engineering software capable of point cloud manipulation could be used to tailor scan data to a suitable level of accuracy to reduce file sizes and processing time. It may also be unnecessary to capture anatomic data between the fingers, but consultation with different disciplines (e.g., occupational therapy, physio- therapy) may demonstrate different requirements for each individual patient depending on the severity of the condition and their individual needs with regard to levels of wrist and hand immobilization.

Acknowledgments

Special thanks to Ella Donnison and Lucia Ramsay for their help, advice, and contribution.

References

[1] Webb PA. A review of rapid prototyping (RP) techniques in the medical and biomedical sector. Journal of Medical Engineering & Technology 2000;24(4):149—53. https://doi.org/10.1080/03091900050163427.

[2] Gerrits A, Jones CL, Valero R. Custom-fit: a knowledge-based manufacturing system enabling the creation of custom-fit products to improve the quality of life. In: Rapid product development conference, 2004 October 12—13: Portugal; 2004.

[3] Minns RJ, Bibb R, Banks R, Sutton RA. The use of a reconstructed three-dimensional solid model from CT to aid the surgical management of a total knee arthroplasty: a case study. Medical Engineering & Physics 2003;25(6):523—6. https://doi.org/10.1016/S1350-4533(03)00050-X.

[4] Bibb R. Medical modelling: the application of advanced design and development techniques in medicine. Cambridge: Woodhead Publishing; 2006, ISBN 978-1845691387.

[5] Bibb R, Winder J. A review of the issues surrounding three-dimensional computed tomography for medical modelling using rapid prototyping techniques. Radiography 2010;16(1):78—83. https://doi.org/10.1016/j.radi.2009.10.005.

[6] Fried S. Splint and or method of making same. 2007. US Patent application 200700 16323 A1.

[7] Goodenough DJ. Tomographic imaging. In: Beutel J, Kundel HL, Van Metter RL, editors. Handbook of medical imaging, volume 1. Physics and psychophysics. Bellingham, Washington: SPIE - The International Society for Optical Engineering; 2000, ISBN 978-0819477729. p. 511—54.

[8] Feyter PJD, Meijboom WB, Weustink A, Van Mieghem C, Mollet NRA, Vourvouri E, et al. Spiral multislice computed tomography coronary angiography: a current status report. Clinical Cardiology 2007;30(9):437—42. https://doi.org/10.10 02/clc.16.

[9] Tay SC, Primak AN, Fletcher JG, Schmidt B, Amrami KK, Berger RA, et al. Four-dimensional computed tomographic imaging in the wrist: proof of feasibility in a cadaveric model. Skeletal Radiology 2007;36(12):1163—9. https://doi.org/10.1007/s00256-007-0374-7.

[10] Eggbeer D. The computer aided design and fabrication of facial prostheses. PhD Thesis. University of Wales; 2008.

[11] Chua CK, Chou SM, Lin SC, Lee ST, Saw CA. Facial prosthetic model fabrication using rapid prototyping tools. Integrated Manufacturing Systems 2000;11(1):42—53. https://doi.org/10.1108/09576060010303668.

[12] Bibb R, Eggbeer D, Evans P. Rapid prototyping technologies in soft tissue facial prosthetics: current state of the art. Rapid Prototyping Journal 2010;16(2):130—7. https://doi.org/10.1108/13552541011025852.

[13] Institute of Physics. MRI and the physical agents (EMF) directive. London: Institute of Physics; 2008.

[14] Pickens D. Magnetic resonance imaging. In: Beutel J, Kundel HL, Van Metter RL, editors. Handbook of medical imaging, volume 1. Physics and psychophysics. Bellingham, Washington: The International Society for Optical Engineering; 2000, ISBN 978-0819436214. p. 373—461.

[15] Fitzpatrick JM, Hill DL, Shyr Y, West J, Studholme C, Maurer CRJ. Visual assessment of the accuracy of retrospective registration of MR and CT images of the brain. IEEE Transactions on Medical Imaging 1998;17(4):571—85. https://doi.org/10.1109/42.730402.

[16] NHS. MRI-scan - advantages and disadvantages. 2009. Available at: http://www.nhs.uk/Conditions/MRI-scan/Pages/Advantages.aspx.

[17] Laurence AS. Sedation, safety and MRI. British Journal of Radiology 2000;73(870):575—7. https://doi.org/10.1259/bjr.73.870.10911777.

[18] Wiklund ME, Wilcox SB. Human factors roundtable. In: Designing usability into medical products. Boca Raton, FL: CRC Press; 2005, ISBN 978-0849328435. p. 31—54.

[19] Low E, O'Driscoll M, MacEneaney P, O'Mahony O. Sedation with oral chloral hydrate in children undergoing MRI scanning. Irish Medical Journal 2008;101(3):80—2.

[20] Lawson GR. Sedation of children for magnetic resonance imaging. Archives of Disease in Childhood 2000;82(2):150—3. https://doi.org/10.1136/adc.82.2.150.

[21] Roguin A. Magnetic resonance imaging in patients with implantable cardioverter-defibrillators and pacemakers. Journal of the American College of Cardiology 2009;54(6):556—7. https://doi.org/10.1016/j.jacc.2009.04.047.

[22] Bonnechère B, Jansen B, Salvia P, Bouzahouene H, Omelina L, Moiseev F, et al. Validity and reliability of the Kinect within functional assessment activities: comparison with standard stereophotogrammetry. Gait & Posture 2014;39(1):593—8. https://doi.org/10.1016/j.gaitpost.2013.09.018.

[23] Henseler H, Kuznetsova A, Vogt P, Rosenhahn B. Validation of the Kinect device as a new portable imaging system for three-dimensional breast assessment. Journal of Plastic, Reconstructive & Aesthetic Surgery 2014;67(4):483—8. https://doi.org/10.1016/j.bjps.2013.12.025.

[24] Kau CH, Zhurov A, Bibb R, Hunter L, Richmond S. The investigation of the changing facial appearance of identical twins employing a three-dimensional laser imaging system. Orthodontics and Craniofacial Research 2005;8(2):85—90. https://doi.org/10.1111/j.1601-6343.2005.00320.x.

[25] Bibb R, Freeman P, Brown R, Sugar A, Evans P, Bocca A. An investigation of three-dimensional scanning of human body surfaces and its use in the design and manufacture of prostheses. Proceedings - Institution of Mechanical Engineers H 2000;214(6):589—94. https://doi.org/10.1243/0954411001535615.

[26] Surendran NK, Xu XW, Stead O, Silyn-Roberts H. Contemporary technologies for 3D digitization of Maori and Pacific Island artifacts. International Journal of Imaging Systems and Technology 2009;19(3):244—59. https://doi.org/10.1002/ima.20202.

[27] Li Z, Chang C, Dempsey PG, Cai X. Refraction effect analysis of using a hand-held laser scanner with glass support for 3D anthropometric measurement of the hand: a theoretical study. Measurement 2008;41(8):842—50. https://doi.org/10.1016/j.measurement.2008.01.007.

[28] Direct Dimensions I. Case study: using 3D imaging to create high-res prosthetic hand. 2010. Available at: http://directdimensions.blogspot.com/2010/01/case-study-using-3d-imaging-to-create.html.

[29] Tikuisis P, Meunier P, Jubenville C. Human body surface area: measurement and prediction using three-dimensional body scans. European Journal of Applied Physiology 2001;85(3):264—71. https://doi.org/10.1007/s004210100484.

[30] Greiner T. Hand anthropometry of U.S. Military personnel. Washington D.C.: United States Department of Defense; 1990.

[31] Wilcox SB. Finding and using data regarding the shape and size of the user's body. In: Wiklund ME, Wilcox SB, editors. Designing usability into medical products. Boca Raton, FL: CRC Press; 2005, ISBN 978-0849328435. p. 77—83.

[32] Gajdosik RL, Bohannon RW. Clinical measurement of range of motion: review of goniometry emphasizing reliability and validity. Physical Therapy 1987;67(12): 1867—72.

[33] Williams GL. Improving fit through the integration of anthropometric data into a computer aided design and manufacture based design process. PhD Thesis. Loughborough University; 2007.

[34] Li Z, Chang CC, Dempsey PG, Ouyang L, Duan J. Validation of a three-dimensional hand scanning and dimension extraction method with dimension data. Ergonomics 2008;51(11):1672—92. https://doi.org/10.1080/00140130802287280.

[35] Srinivas Reddy R, Compson J. Examination of the wrist—surface anatomy of the carpal bones. Current Orthopaedics 2005;19(3):171—9. https://doi.org/10.1016/j.cuor.2005.02.008.

Orthotic applications case study 2—comparison of additive manufacturing systems for the design and fabrication of customized wrist splints*

5.25.1 Introduction

Wrist splints provide multifaceted treatment outcomes to patients, including pain relief through immobilization of affected joints [1,2]. Wrist immobilization splints, for example, are designed to immobilize the wrist while allowing mobility of all digits to promote endurance to everyday tasks [3].

There are two main categories of splints: prefabricated "off-the-shelf" splints and custom-made splints. Prefabricated splints can be bought from a variety of stores, such as pharmacies, but may also be prescribed by splinting practitioners, such as occupational therapists or physiotherapists. Prefabricated splints may come in a range of sizes (e.g., small, medium, and large), which assumes a "one-size-fits-all" strategy, which is not necessarily tailored to suit an individual unless adjusted by the user or a splinting practitioner. Alternatively, custom-made splints are produced and distributed exclusively by splinting practitioners to suit each individual patient's

* The work described in this chapter was first reported in the reference below and is reproduced here, in part or in full, with the permission of the *Rapid Prototyping Journal*. Paterson AM, Bibb RJ, Campbell RI and Bingham GA. Comparison of Additive Manufacturing Systems for the Design and Fabrication of Customized Wrist Splints. Rapid Prototyping Journal, 2014; Vol. 21, Issue 3, pp 230—243, DOI:10.1108/RPJ-10-2013-0099.

Medical Modeling
ISBN 978-0-323-95733-5
https://doi.org/10.1016/B978-0-323-95733-5.00030-2

lifestyle, as well as anatomic demands relative to their condition. Custom-made splints offer superior fit and comfort and, in many circumstances, can be less bulky than off-the-shelf items. They can also be made to accommodate extremes of size and deformity that is not always possible with off-the-shelf items, which inevitably have limits on their adjustability. Custom-made splints maintain their shape at all times, while off-the-shelf items need to be adjusted each time they are put on, and it is not always possible to replicate the adjustment precisely on each occasion. Consequently, off-the-shelf splints cannot accommodate every patient, and there will always be a need for custom-made splints.

This chapter focuses on the creation of custom-made splints, since their end use is considered synonymous to fundamental benefits of additive manufacturing in terms of mass customization with regard to anatomic fit, function, and appearance. While splints can be designed for many purposes, this study focused on wrist immobilization splints intended to alleviate the symptoms of rheumatoid arthritis as an exemplar application. An image of an immobilization splint can be found in Fig. 5.170. If prescribed by a splinting practitioner, patients are typically provided with wear instructions, including how to don/doff the splint and the expected wear schedule. Each wear schedule is tailored to suit the patient, their lifestyle, and the suggested optimal treatment approach to their condition, as determined by their therapist [4].

A process model for designing and fabricating a custom-made wrist immobilization splint can be found in Fig. 5.171, deduced from several

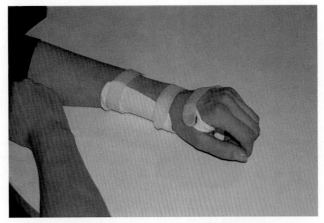

Figure 5.170 Custom-made wrist immobilization splint.

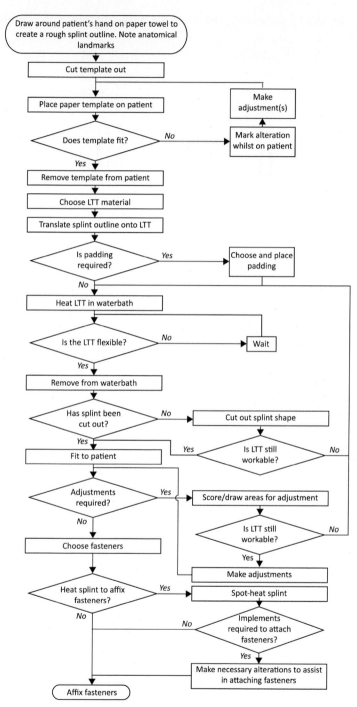

Figure 5.171 Traditional splint fabrication process [5]. *(Image courtesy of SFF symposium.)*

sources [6—8]. In summary, custom-made wrist immobilization splints are typically handmade; they are formed from sheets of thermoplastic, which are cut, heated, molded to the patient, adjusted, and then finished with fasteners (such as Velcro) to ensure a secure fit to the patient. Fundamentally, the splinting process is inherently a combination of designing and making in a single process. As a result, limitations of materials and fabrication processes impede the design for fit and function. Consequently, several factors affect patient compliance, such as discomfort and poor aesthetics, often resulting in a reduced willingness to wear splints to match the prescribed wear regime. The aesthetics of splints can have implications on the suggested duration and location of wear [9], since splints typically look clinical and unattractive despite the best efforts of clinicians to finish splints to a high standard and to suit patient preference. For example, patients are encouraged to choose different Velcro colors in a bid to improve compliance [10]. However, choice is limited to the material stock available to the clinic and the associated properties of the thermoplastic.

As the manufacturing process is entirely manual and skill dependent, the splint may also be poorly fitted, resulting in shear stress, directional misalignment, and pressure over bony prominences, which in turn can induce pressure sores [11]. Furthermore, the presence of a thermoplastic splint with uniform thickness and limited perforation can induce excessive perspiration, which can collect within the porous elements of padding (if present). In turn, this harbors bacteria, resulting in an odorous, unhygienic, yet often compulsory form of treatment for patients [12]. Furthermore, splints are difficult to keep clean, particularly if a padded lining is present.

In response to these issues, the opportunity for using additive manufacturing (AM) for upper extremity splinting was considered a viable option for future splint fabrication. Campbell [13] states that AM can account for functional, environmental, ergonomic, aesthetic, emotional, and user fit requirements, and as such, it is a proven, viable method for the design and fabrication for customized body-fitting items. The scope for AM applications continues to widen in a broad range of disciplines; the use in the medical and dental industry, for example, continues to be the world's third largest serving industry (15.1%) in the AM sector for the past 11 years [14]. AM has already been explored in a range of exoskeletal assistive devices, ranging from clubfoot treatment methods [15—17], spinal braces [18], and ankle—foot orthoses [19—22], the latter being the focus of a

Framework 7 European project called *A-Footprint*. The majority of these works have focused on the use of laser sintering (LS), the benefits of which include relatively low part cost compared with other AM processes and the ability to retrieve unsintered powder for future use. Furthermore, the fact that unsintered powder subsequently behaves as support for down-facing structures is also a significant advantage, since this enables the creation of complex geometries without incurring significant clean-up time and subsequent costs associated with manual postprocessing, while reducing the build time. Additive manufactured textiles (AMTs) described by Bingham et al. [23] demonstrate benefits of AM in terms of part consolidation and assembly builds while proposing functional articulating wearable structures. Various AMT linkage designs and arrays have been explored for stab-resistant body armor, demonstrating its capabilities in generating functional constructs to enable movement of the intended wearer [24]. Furthermore, AMTs look visually appealing, and although their integration into artifacts to date remains a niche topic, the scope for integrated aesthetical yet discrete functional AM textiles into custom-fitting wearable devices is entirely feasible.

Extending the use of AM in the context of assistive devices, Gibson et al. [19] investigated the suitability of a low-cost extrusion-type 3D printer, their justification being a low cost, in-clinic approach to fabrication through the use of a RapMan 3D print extrusion system (3D Systems, Rock Hill, SC, USA). Furthermore, Mavroidis et al. [21] used stereolithography (SLA) (3D Systems, Rock Hill, SC, USA) for ankle–foot orthoses, since a range of materials could be used to offer different properties; for example, Arptech's DSM Somos 9120 Epoxy (Arptech Pty Ltd., Victoria, Australia) offered a flexible solution while being biocompatible. SLA also demonstrates a high level of surface quality and resolution when compared with original STL data. In the context of using material jetting systems for fabrication of assistive devices, Smith [25] describes the developments of the Miraclefeet organization and North Design Labs with Objet Connex systems into custom-made pediatric clubfoot orthoses. The most significant benefit of the approach was through exploiting the multimaterial build capabilities on offer, enabling heterogeneous builds including a range of Shore hardnesses alongside more rigid materials to incorporate functional parts, such as flexible hinges and soft edge features for improved functionality and comfort.

Despite the previous studies into AM for lower limb prosthetics and orthotics, the suitability of AM technologies in the context of upper extremity (i.e., hand, wrist, and forearm) splinting is yet to be compared and evaluated, with suitable courses of action established for future development. To date, there has only been limited research and development into AM upper extremity splinting. The application has been implied by several additional sources [26–28], and although no commercial approach has been proposed yet, a number of research institutes and individuals have explored the feasibility of AM splinting. For example, Fraunhofer IPA used an EOS P100 LS system with PA2200 powder (EOS GmbH Electro Optical Systems, Krailling, Germany) to offer a single example of an attractive solution to traditional splinting [29]. The benefit of LS in this instance was the ability to introduce intricate locking mechanisms that would have required less postprocessing to remove unsintered powder acting as support structures. While Fraunhofer IPA have explored the integration of Voronoi patterns to their prototype [30], Evill [31] has proposed the use of LS for splinting wrist fractures, incorporating a honeycomb structure with interlocking fasteners. Palousek et al. [32] also explored AM for splinting, but the intent was limited to reverse engineering a single existing splint that had been designed for traditional manufacturing techniques, and this attempt failed to recognize or explore the fundamental opportunities of design for AM principles such as improved aesthetics and or lightweight lattice-type structures.

In terms of material jetting technologies, *Carpal Skin* by Oxman [33] explored the multiple material build capabilities available with Objet Connex systems (Stratasys, Eden Prairie, MN, USA). A range of Shore hardnesses were incorporated alongside stiffer materials in one build, resulting in a heterogeneous splint. The term *"synthetic anisotropy"* was established to communicate the effect of directional influencers in the form of a specified pattern distribution to allow or restrict movement, the dispersion of which was dictated by an algorithm sourced from a pain map defined by an individual [34,35]. While this approach explores the opportunities of design for AM, no clinical validity has been published to date. Taking an alternative stance to performance, functionality, and design strategies of integrated features, Paterson et al. [5] explored the potential integration of multiple materials into wrist splints under the direction of qualified and experienced clinicians who specialize in the design and fabrication of custom-fitted wrist splints within the National Health Service (NHS) in the United Kingdom. The research focused on developing a specialized workflow designed for splinting practitioners to allow them to design splints in a virtual environment to support AM; the manuscript focused specifically on the intent for placing multiple materials to behave as hinges or cushioned features as opposed to traditional fabrication processes where a similar approach would be impossible to replicate.

Having identified the relative strengths of four different AM systems along with existing work into other custom-fitted devices, the authors chose to explore opportunities in upper extremity splinting for improved fit, functionality, and aesthetics. Features that were considered as potentially beneficial included a *"best fit first time"* approach to provide a customized fit for a patient relative to their anatomic and rehabilitation needs and AMT elements for hinges using LS for easier donning and doffing. Furthermore, the authors chose to explore the opportunities for integrating varying Shore hardnesses into wrist splints using the Objet Connex system, to include functional features such as integrated elastomer hinges or cushioned features over bony prominences. These new opportunities were impossible to deliver in traditional splinting but are now entirely feasible as a result of AM. By exploring new, novel integrated applications into upper extremity splinting, future avenues for product development could be pursued. It was anticipated that with these potential benefits, patient compliance would be improved.

5.25.2 Aim and objectives

The research reported in this paper represents one stage of a long-term research project exploring the whole process of digital splint design and manufacture, from data acquisition through to data manipulation in three-dimensional (3D) computer-aided design (CAD) to support AM. This paper describes the exploration of AM prototypes through comparison and subsequent suitability of different AM processes based on the digital design workflow developed and described by Paterson et al. [36]. By investigating previous research activity and potential AM process benefits, several design characteristics were planned for upper extremity splint integration for this investigation. Particular focus was placed on improving aesthetics, fit, and function of splints and by exploring existing design features in the context for upper extremity splinting.

The aim of this chapter is to demonstrate and evaluate the suitability of a range of AM processes to deliver custom-fitting wrist splints, using the following objectives:

Objective 1. To evaluate the design and fabrication of homogeneous AM splints;

Objective 2. To evaluate the design and fabrication of heterogeneous (multiple material) AM splints with functional features, e.g., an elastomer hinge and soft edges;

Objective 3. To demonstrate varied part consolidation, resulting from AM processes through integrated fastener features;

Objective 4. To evaluate the integration of a textile hinge to demonstrate assembly build capabilities;

Objective 5. Establish relative strengths and limitations of each AM process relative to recognized best practice in splint design;

Objective 6. Establish areas for future research and development relating to design of wrist splints for AM.

5.25.3 Method

One of the fundamental requirements of the AM splinting approach was to deliver a customized fit to the patient, since this is a standard requirement in traditional custom-made splinting. To deliver this, the patient's skin surface topography would be required to capture the patient's anatomic data. These data could then be used to extract and subsequently generate a 3D virtual form of a splint to match their topography.

Before AM splints could be fabricated for evaluation, 3D CAD splint models had to be generated. The organic topography of a forearm and hand was required to generate an accurate profile for CAD manipulation. Scan data were acquired from a healthy volunteer by 3D laser scanning a plaster cast, as described by Paterson et al. [37]. A plaster cast was used to eradicate concerns with noise that would otherwise be collected through involuntary movement and tremor during scanning. The plaster cast provided a repeatable, static data source in case a repeat scan was required later in the research process [36]. This form of data acquisition was to enable this research only and is not suggested for future clinical practice. The plaster cast was scanned with a ZCorp ZScanner 800 hand-held 3D laser scanner (3D Systems, Rock Hill, SC, USA). The same scan data were used for all subsequent splint designs, and therefore each demonstrated almost identical anatomic topographies (excluding integrated features such as lattices and fasteners), which would therefore fit the participant used to generate the plaster cast. Using this scan data, six different splints were designed in a range of 3D CAD software packages and plugins:

- Autodesk Maya 2011 (Autodesk, San Rafael, CA, USA)
- Geomagic Studio 2012 (3D Systems, Rock Hill, SC, USA)

- McNeel Rhinoceros Version 4.0 (Robert McNeel and Associates, Seattle, WA, USA)
  - o Grasshopper plugin (Robert McNeel and Associates, Seattle, WA, USA)
- PTC Pro/Engineer Wildfire 5.0 (PTC, Needham, MA, USA)
- Geomagic FreeForm Modeling Plus (3D Systems, Rock Hill, SC, USA)

The justification for using a wide range of CAD programs was related to the underlying research; a separate and previous stage in the overall research project was to develop a specialized CAD workflow that effectively replicated the splint design process used by splinting practitioners/clinicians. Therefore, there was a need to establish suitable tools that could ideally replicate and/or improve upon traditional fabrication methods and techniques in a virtual environment within 3D CAD programs. Several different CAD packages were used during this phase of the research since each program had different tools, offering different strengths and limitations that were considered appropriate for splinting applications. Tools and CAD strategies were developed in that phase of the research to create a customized software workflow, and these methods were taken forward to the design and fabrication of the customized AM splints described in this paper. The design workflow and exploration of CAD approaches are described fully by Paterson et al. [36] and are the subject of publications pending. In short, this workflow translates into a three-dimensional CAD environment with the same design intent as the traditional design approach. Therefore the "design rules" for the design of splints is adhered to in the digital workflow, and this has been evaluated and approved by a number of qualified splinting professionals. The design steps include, for example, defining the boundary of the splint following anatomic landmarks, defining the material thickness, rounding edges, and alleviating pressure on sensitive areas. Over and above the traditional approach, the choice of lattice pattern is introduced to reduce weight, improve ventilation, and enhance aesthetics. This workflow has been developed to facilitate recognized clinical practice within a digital environment and has been critically evaluated by experienced and qualified clinical professionals.

Four different 3D CAD models were created in total, designed for homogeneous AM fabrication using the following:
- Laser sintering (LS): EOS P100 Formiga, made with EOS PA2200 50:50 powder (EOS GmbH Electro Optical Systems, Krailling, Germany). This process was used to explore opportunities to integrate textile elements into a splint for added functionality.

- Fused deposition modeling (FDM): Stratasys Dimension SST1200es, made with acrylonitrile butadiene styrene (ABS) (Stratasys, Eden Prairie, MN, USA).
- Stereolithography (SLA): 3D Systems 250, made with Accura Xtreme resin (3D Systems, Rock Hill, SC, USA).
- PolyJet matrix material jetting: Objet Connex 500, made with FullCure 515 and FullCure535 to generate RGD5160-DM (Stratasys, Eden Prairie, MN, USA), displaying ABS-like properties.

Finally, a heterogeneous splint was designed to exploit the multiple material capabilities of Objet Connex technologies, by using Tango-BlackPlus and VeroWhitePlus to generate DM9840FLX and DM9850FLX material ranges (Stratasys, Eden Prairie, MN, USA). Although outputs were similar to Oxman [38], the combination with integrated aesthetic lattice structures was also targeted as an output in this context. With the exception of the Objet Connex heterogeneous build, material choice was not considered important within this investigation since the research focused on highlighting potential differences in AM systems and their specific capabilities. However, the authors acknowledge that material choice is crucial in many aspects in assessing part quality and delivering for the intended need. Material choice to the extent of being able to vary properties in the heterogeneous build was, however, important for the multimaterial splint since this was considered the most important characteristic to display in this context.

In the next section, each of these designs will be described in more detail, corresponding to the AM process considered most appropriate for the design. Different designs were fabricated on different AM systems to demonstrate the versatility of AM systems relative to their particular strengths. Furthermore, LS was only used for the fabrication of a textile splint, since previous studies described in Section 5.25.1 have already demonstrated homogeneous splint prototypes using LS; the development of a textile element in this case was therefore considered a potential novel contribution to knowledge.

5.25.3.1 Homogeneous AM textile splint

The AM textile splint featured an AMT element along the ulnar aspect. This was incorporated to consolidate splint parts into one assembly. The AMT element was designed to behave as a hinge to enable the user to open the splint for easier donning and doffing. Furthermore, the AMT element was designed to follow contours of the upper extremity geometry, demonstrating the drapability and free movement described by Bingham et al. [23]. The repeating units in the AMT element were formed to follow the topography of the scan data and were aligned using a custom mesh array algorithm devised by Bingham and Hague [39] in MATLAB (MathWorks, Natick, MA, USA). The textile element was then incorporated into the remaining splint geometry using uniform spaced links generated in Grasshopper, a generative modeling plugin for Rhinoceros to enable quick adjustments within set parameters. The automated linkages were united with the main structures of the splint using a Boolean union function.

The AMT element was specifically designed to exploit the freedom of form available through LS (Fig. 5.172), since support structures commonly required by other AM systems would not be required in this instance. Since unsintered powder provides support for overhangs, LS was considered the most effective approach since overhangs in AMT linkages were abundant. Clean-up time was also considered, since the unsintered powder would only require removal via high-pressure air jets and vibration to remove excess powder between linkages. Other processes could be used such as 3D jetting/printing techniques for example. However, depending on the exact process and material combination used, support material removal could

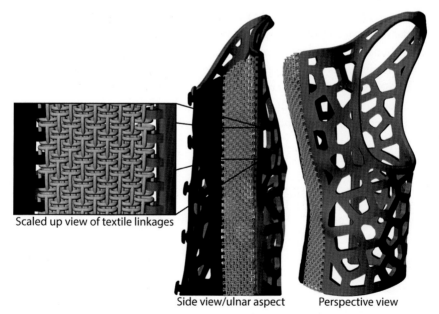

Scaled up view of textile linkages

Side view/ulnar aspect   Perspective view

Figure 5.172 Splint prototype with textile hinge, modeled using MATLAB, Rhino, and Grasshopper for linkage integration.

result in additional time and in some cases could damage the AMT linkages. SLA would also prove ineffective for this approach because supports would be required for links within the textile element, which would prove difficult and time consuming to remove.

In addition, mushroom-like fasteners were integrated into the design to demonstrate part consolidation of fasteners within the build, as opposed to detached fasteners used on traditional splints such as Velcro.

5.25.3.2 Homogeneous circumferential build designs

Two circumferential splint designs were modeled in 3D CAD (Fig. 5.173). The splints were designed in two corresponding parts to enable donning and doffing. The design was intended to behave as a "pinch-splint," where the user could pinch the palmar region laterally to separate the two halves when taking the splint off. This pinch design demonstrated the ability to integrate subtle, discrete fasteners into the splint while still being functional, subsequently highlighting part consolidation compared with traditional fastening methods (e.g., Velcro, D-rings).

5.25.3.3 Heterogeneous splint using Objet Connex technologies

A splint was designed for heterogeneous AM system fabrication, as described by Paterson et al. [5]. The underlying intent of this approach was to enable the practitioner to specify and localize areas where softer materials might benefit the patient to relieve pressure (e.g., elastomer elements over bony prominences or areas prone to pressure sores). In traditional splinting, practitioners may have to create cavities over bony prominences or integrate separate gel discs [12], but this approach affects the topography and subsequent aesthetics of the splint. However, the use of the Objet Connex

Figure 5.173 Circumferential, homogeneous two-part splints for SLA (Voronoi) and Objet ("Swirl") builds respectively.

would enable subtle integration of elastomer features that would not drastically affect the topography of the splint, therefore creating a less cumbersome appearance that may be more conducive to complying with wearing regimes for the patient.

Fig. 5.174 shows the 3D CAD model developed specifically for the Objet Connex 500 system to exploit its multimaterial capabilities. Various closed shells were required within the 3D CAD model to allocate different digital materials prior to fabrication. The shells were created by trimming the initial scan mesh (generated from cloud data) in Geomagic Studio before manipulating and thickening the geometry in other CAD software such as McNeel Rhinoceros. Elements labeled "1" were soft elastomer edges to provide a comfortable interface between the skin and the rigid splint structures (labeled "2"). A soft elastomer cushion [3] was located over a bony prominence (pisiform), and a flexible hinge (element 4) was integrated along the ulnar aspect of the splint to enable donning and doffing.

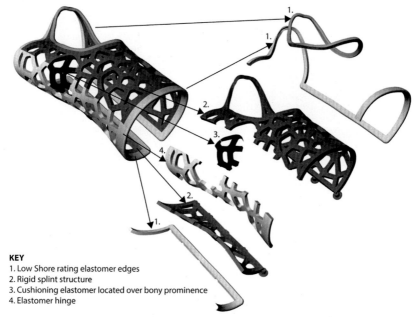

KEY
1. Low Shore rating elastomer edges
2. Rigid splint structure
3. Cushioning elastomer located over bony prominence
4. Elastomer hinge

Figure 5.174 Closed shell distribution and intent of heterogeneous wrist splint.

5.25.4 Results

All of the splint prototypes were built according to the suppliers' recommended parameters using commercially available materials and machines. Each AM process and subsequent build has been reviewed in the following sections.

5.25.4.1 Homogeneous AM textile splint

The LS splint shown in Fig. 5.175 proved successful in capturing its intended outcome of an integrated AMT hinge to enable easier donning

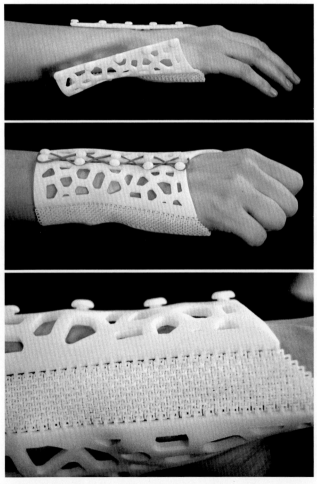

Figure 5.175 Laser sintered splint with AMT linkage hinge. *(Courtesy of Dr. C. Majewski, University of Sheffield, UK.)*

and doffing; the links offered sufficient freedom to enable this. The links proved strong enough during a preliminary wearer trial to maintain their structure without failing when the splint was worn. Furthermore, the AMT element added a unique aesthetic quality to the splint. The union of AM textiles and upper extremity splinting with the aim as a medical intervention was considered a world first.

However, a small number of links remained fused together due to residual unsintered powder trapped between linkages. It is anticipated that the porous nature of the surfaces would also inherently affect the hygiene of the splint by absorbing dirt, sweat, sebum, dead skin cells, etc., as described by Bibb et al. [40]. Furthermore, the AMT element exacerbates these concerns. The small links and tight textile design used in this example could also potentially catch on vellus and/or terminal arm hair, causing discomfort if extracted from hair follicles (i.e., trapping and pulling out body hairs). However, larger links would reduce this risk. Cleaning such a splint would also prove problematic unless immersed in a detergent or washed with an automated process/system such as a dishwasher, as proposed by Fried [28]. LS parts can withstand dishwashing, and this has been discussed with reputable LS suppliers, although to date, this has not been rigorously tested or reported. As a preliminary test, the researchers placed the splint in a dishwasher at various temperatures (45°C, 50°C, and 65°C) along with branded dishwasher detergent, with no visible aftereffects.

5.25.4.2 Homogeneous circumferential splints

The FDM splint shown in Fig. 5.176 demonstrated comparatively poor surface quality, with obvious layering and stepping; these factors affected the aesthetics, which also could affect the comfort of the splint at the edges. Pitted areas between layers and tracks could collect waste products as described earlier and therefore could be an unhygienic solution for a splinting application. However, the ABS material is relatively robust and a widely used material in domestic and wearable products such as frames of eyeglasses, and subsequently, it is reasonable to anticipate that the ABS splint can withstand mechanical cleaning with mild detergents.

The SLA splint shown in Fig. 5.177 was oriented to reduce the requirements for supports as can be seen in Fig. 5.178. Similar to the homogeneous ABS-like splint, the design of the splint was effective in allowing donning and doffing by pinching the palmar element. Overall,

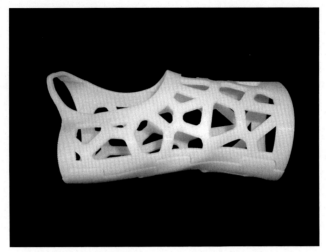

Figure 5.176 FDM splint.

Figure 5.177 Accura Xtreme splint, built on a 3D systems 250 [5]. *(Prototype courtesy of Dr. D. Eggbeer, Product Design Development Research (PDR), Cardiff. Image courtesy of SFF symposium.)*

the surface quality of the SLA splint was considered the highest of all the AM processes used in this investigation. The smooth surfaces facilitate cleaning and minimize hygiene risks. However, despite side and up-facing surfaces being smooth, down-facing surfaces demonstrated abrasive imperfections where supports had been removed (Fig. 5.178). Such imperfections could cause discomfort for a patient if left untreated. Manual postprocessing with abrasives would be required, as described by Bibb

Figure 5.178 SLA build, showing support lattice structures. *(Courtesy of S. Peel, PDR.)*

et al. [41], adding cost in labor and resources if the approach were implemented for clinical application. However, this could be minimized by using patterns that formed self-supporting structures, which would eliminate the majority of the supports and the issues encountered in their removal. A simple example is shown in Fig. 5.179. The disadvantage of this approach would be a reduction in aesthetic possibilities and limited patient choice of pattern.

Finally, a fourth homogeneous splint built on the Objet Connex is shown in Fig. 5.180. An initial visual and tactile interpretation of surface

Figure 5.179 Self-supporting pattern design for SLA.

quality was considered acceptable compared with other AM processes described in this paper. In addition, the fastener design and overall splint structure were able to perform and withstand their intended functions in flexing to allow the user to don and doff by pinching the palmar aspect of the splint, and they subsequently demonstrated part consolidation in the context of AM splint fabrication. However, a significant amount of support material (FullCure 705) was required, but it was removed easily and relatively quickly with a high-pressure water jet.

5.25.4.3 Heterogeneous splint using Objet Connex technologies

Prior to the building of this example, material choices were specified in Objet Studio (Stratasys, Eden Prairie, MN, USA). The software interface was used to define variables for Objet Connex build systems. The splint made with the DM98 material range provided a stronger color contrast, as shown in Fig. 5.181.

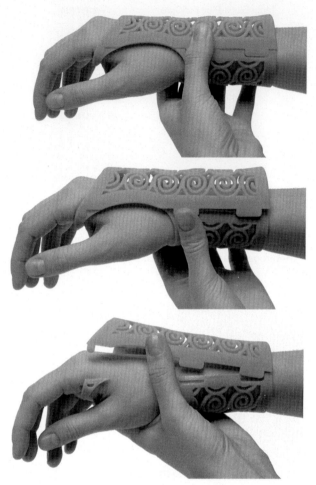

Figure 5.180 FullCure515/FullCure535 with ABS-like properties, built on the Objet Connex 500 model.

The 9850 Shore 50 elastomer placed over the bony prominence (pisiform) expanded when pressure was applied within the splint, which demonstrated that a small amount of expansion could accommodate swelling if required. The hinge was also functional, although several failures occurred over a period of 12 months (Fig. 5.182). This was due to repeated flexion when opening and closing the splint. The splints

TangoBlack Plus FullCure   VeroWhite

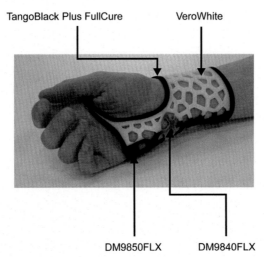

DM9850FLX   DM9840FLX

Figure 5.181 Heterogeneous splint; assorted materials within one AM build, using the Objet Connex.

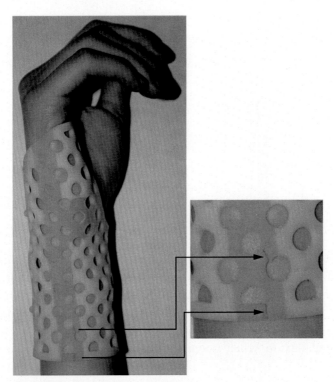

Figure 5.182 Failures in the multimaterial elastomer hinge [5]. *(Image adapted, courtesy of SFF symposium.)*

were used as proof-of-concept prototypes, and therefore, they have been handled by a large number of people to demonstrate the intent of the research. This particular splint prototype has been handled by a large number of individuals in presentations and demonstrations and consequently has undergone more than 50 openings and closings over a 12-month period. Many people handled the prototype with little knowledge of its physical limitations, and it is possible that its physical limits were exceeded by careless handling. In addition, the prototype has been displayed in exhibitions and has been subject to extended periods of exposure to strong light sources that may have affected its physical properties over time. It was also noted that the position and shape of the flexible hinge was not ideal. The position was along the ulnar aspect of the forearm, and the intended wearer had to use significant force while adjusting the posture of their hand and wrist to don the splint correctly. Simultaneously, a significant amount of force was applied to the hinge while in an open position while donning, and therefore the elastomer elements were more susceptible to compression and tension forces, resulting in split lines. An additional hinge could have been placed along the radial aspect so both sides of the splint could have been opened to help donning. Similarly, the shape of the splint hinge was inappropriate as it was formed in 3D CAD by two parallel planes intersecting the splint geometry. If one considers a living hinge in a polypropylene DVD case, for example, rotation around a single axis results in a uniform distribution of compression and tension exerted throughout the length of the hinge structure. However, because of the organic topography of the splint, compressive and tensile stresses varied throughout the structure, subsequently having a higher tensile concentration toward the borders of the splint, while a higher compression concentration was demonstrated on the inner region of the hinge. Therefore, the shape of the hinge element may have benefitted from being a varied shape to suit the topography. Although not formally documented, a very slight level of creep was also observed over time, resulting in a slight twist in the splint. This is most likely due to the fact that the splint has been stood upright on display for extended periods, and it would be reasonable to assume that were the splint worn for extended periods, it would be more likely to retain the intended shape. Therefore, future research would be required to assess the extent of creep, as well as establishing design rules to reduce local strain and therefore avoid failures at the hinge. However, surface quality and resolution were considered adequate in comparison to FDM and LS.

Unfortunately, the multimaterial build required a large volume of support material (FullCure 705), as shown in Fig. 5.183. Much like the ABS-like splint, the support material was removed with a high-pressure water jet. Not only does this increase cost in terms of material consumption but also costs relating to labor time. The prototypes' properties and costs are summarized in Table 5.14.

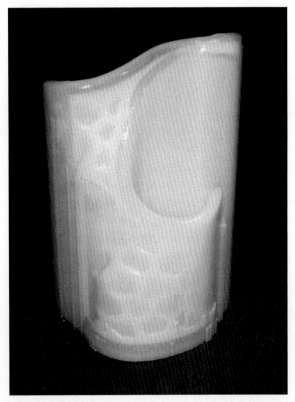

Figure 5.183 Support material required for the Objet multimaterial build in upright position.

Table 5.14 Prototypes' properties and costs.

AM process Evaluation criteria	SLA	LS (textile splint)	FDM	PolyJet (homogeneous)	PolyJet (heterogeneous)
System	3D systems 250	EOS P100 Formiga	Stratasys dimension SST1200es	Objet Connex 500	Objet Connex 500
Material(s)	Accura Xtreme	PA2200	ABS	FullCure515/FullCure535 (ABS-like)	TangoBlackPlus VeroWhitePlus DM9850FLX DM9840FLX
Unique benefits	—	Textile element	—		Heterogeneous
Build speed	Poor (draining and postcuring required)	Poor (heat-up and cool-down periods required)	Poor	Adequate	Adequate
Surface quality	Good	Poor	Poor	Adequate	Adequate
Accuracy and resolution	Good	Good	Poor	Good	Good
Z plane build weaknesses?	Mild	Mild	Severe	Mild	Mild
Rigid build supports required?	Yes	No	Yes	Yes	Yes
Postcuring required?	Yes	No	No	No	No
Rigid support removal required?	Yes, removal by hand, followed by surface sanding	No, although removal of powder between linkages is required	Yes, although WaterWorks supports are soluble	Yes, using high-pressure water jet	Yes, using high-pressure water jet
Comparative material cost (£/1 kg) (excl. VAT and shipping)	High £179.00	Lowest ~£55	High £290	Highest £246 and £267 respectively	High TangoBlackPlus = £245 VeroWhite = £200

5.25.5 Conclusions and future work

Each process displayed benefits and limitations in the context of upper extremity splinting. When compared, the most inappropriate AM process was considered to be FDM. Despite the advantages of robust materials, improvements in surface quality would be needed. SLA proved to have good surface quality and reasonably robust materials, the effects of cleaning notwithstanding. LS and PolyJet material jetting each displayed unique advantageous characteristics, made feasible by the AM process. Previous studies and prototypes that used LS such as Fraunhofer IPA [29] and Evill [31] demonstrated the ability to integrate aesthetically pleasing structures, but incorporating a textile element has not previously been reported in the context of splinting. This paper further describes additional features that could be beneficial in the future. In addition, the multimaterial splints do demonstrate similarities to Oxman [33]. However, the underlying ethos of placing different materials was a different approach, allowing clinicians to specify materials with varying Shore hardness where they would consider clinically appropriate, whereas Oxman [33] developed an automated approach to integrate materials to direct or restrict the patient's movement. Both strategies demonstrate strengths, and ideally, both would be made available features in specialized 3D CAD for clinicians to explore in the future.

In terms of integrated functionality, the heterogeneous splint was the most versatile and could be exploited in a range of situations, as highlighted by Paterson et al. [5]. If the digitized splinting approach were to be introduced as a realistic option for clinics, the choice of AM process would be dependent on the needs of the patient, as prescribed by their therapist, but a thorough understanding of different AM processes and relevant AM materials would be needed by the therapists.

Despite the aesthetic and functional advantages displayed in the results, several developments would be required before such processes may be feasible for adoption in clinical situations. Firstly, the development of suitable materials would need to be explored further, taking into consideration the long-term exposure to the skin. Although a number of AM materials such as Objet's MED610 transparent, rigid material and Stratasys' ABS-M30i claim ISO 10993 (Biological evaluation of medical devices) and/or USP 23 class VI approval in a range of conditions such as irritation and hypersensitivity, regulations involve standardized tests that may not necessarily take into account specific design and manufacturing processes.

While this certification demonstrates that the materials are inherently low in toxicity, the requirements of the European Medical Device Directive and the various international equivalents require clinical trials that prove the safety of the entire design, materials, and manufacturing process.

Cost analysis must also be performed to determine which process, if any, can be cost effective in terms of clinical demands. Although current practice involves the use of cheap materials, the labor costs could be cut dramatically. This is especially the case when fabricating duplicate splints, for example. In current practice, the entire crafting process must be repeated for each and every splint—there are no economies of scale. When using AM, repeat splints would incur materials costs only. It is also important to consider the hidden costs of time and travel involved in clinic appointments for the patient as well as the clinician. The AM process can eliminate repeat prescription time; if a patient requires a replacement/duplicate splint as a result of previous failure or the desire for a different aesthetic design, the patient will not need a repeat clinic session with their practitioner to fabricate a new splint. Instead, a request could be logged and a duplicate splint ordered instantaneously. Such an order could be added to a queue for manufacture and then dispatched when the build is complete, a similar approach to the latter stages of the proposed automated process by Fried [28]. This reduction in clinic time could reduce demands on the clinic, potentially reducing waiting times and patient waiting lists. In turn, this could improve patient satisfaction in the healthcare system. The economic advantages of an AM approach are predicated on replacing a high-labor-cost manual crafting process with a much more efficient design then manufacture process. Consequently, the economic factors are very much context dependent. It can be envisioned that in high-labor-cost regions, the AM approach may greatly reduce labor costs to the extent that the higher material costs of AM are more than compensated for. This work has been conducted within a UK NHS context, but future costing will enable a more direct calculation of economic benefit in different contexts around the world. Similar arguments for the potential benefits of a digital design and manufacture process have also been explored in maxillofacial prosthetics by Eggbeer (2012) [42]. Another implication in cost effectiveness is improved compliance, leading to improved patient outcomes in the longer term. The impact of patient involvement (i.e., choosing patterns and colors) and enhanced aesthetics on compliance is the subject of current research by the authors.

One of the aesthetic considerations available in clinics, which could not be demonstrated fully, was the customization of color choice. Color ranges are currently limited in AM processes; the 3D Systems ZPrinter 450 (3D Systems, Rock Hill, SC, USA) can provide multicolor builds, while a range of FDM machines can offer various single-color builds. However, within the scope of this investigation, color choice among LS, SLA, FDM, and PolyJet material jetting is limited. The Objet Connex gave the widest variety, allowing for an integrated range from black to white. Other colors are achievable with the Objet Connex such as VeroBlue (RGD840), green (ABS-like RGD810), and transparent (VeroClear RGD810) (Stratasys, Eden Prairie, MN, USA), although Holmes [43] speculates multicolor build capabilities in the future in response to work by Oxman [38]. Some service providers are able to supply color-dyed LS parts, but typically, color choice is limited, and color fastness is an issue.

It was hypothesized that the mechanical failures of the multimaterial splints were due to a number of factors, including the following.

- limited tensile strength, elongation, and tear resistance of flexible Objet materials, e.g., 0.5—1.5 MPa, 150%—170%, and 4-6 kg/cm respectively for FLX9840-DM material (figures according to Stratasys [44]); similar issues have been highlighted elsewhere [42,45] (Fig. 5.184);
- degradation of material over a period of time (including creep);

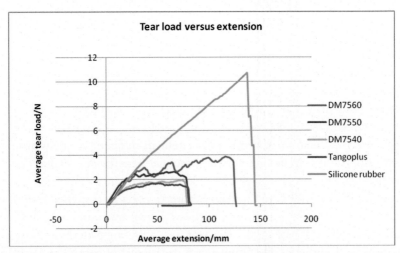

Figure 5.184 Tear load versus extension of Objet Connex materials and silicone rubber for maxillofacial prosthetics. *(Adapted from Eggbeer et al. [42]. See also color section.)*

- suboptimal location/shape of the flexible hinge relative to the organic geometry of the splint and the force exerted when opening and closing the splint when donning and doffing.

However, further research is required to explore these areas further, with opportunities to establish interventions to overcome these limitations. Since the underlying research focused on the 3D CAD processes to design splints for AM, performing mechanical tests on the prototypes was outside the remit of this investigation. The authors have performed research using finite element analysis on a number of the proposed designs to enable comparison of the mechanical properties with traditional practice, and this work will be published in due course. Structural analysis of homogeneous and heterogeneous splint builds will be performed by the authors to compare results with splints made with traditional fabrication methods and techniques. Concerns relating to UV exposure of photopolymer resin processes such as material jetting and SLA builds also need further exploration in an attempt to resolve breakdown of elastomer elements, as suggested by Eggbeer et al. [42]. Similar efforts should be performed on prevention of creep and discoloration of such materials and the AM processes used, particularly for PolyJet builds. Work is also being performed on the effects of material exposure to mechanical cleaning and everyday household chemicals, such as washing powders, liquids, and other detergents, as this could potentially lead to suggestions into cleaning techniques for splints. As described in Section 5.25.4.1, the authors have begun preliminary testing of cleaning splints in a dishwasher, and as such, this will form further research to determine suitable variables for extending the life of a splint.

In Section 5.25.1 (Introduction), the authors identified that previous researchers had limitations relating to clinical validity for the use of upper extremity splinting. Similarly, the authors also recognize and acknowledge that significant future research is required with respect to their own documented work before conducting clinical trials to consider the efficacy of the approach for end-use applications. This includes the exploration of analytical studies relating to all design work to consider material properties and structural integrity. The authors also acknowledge that further work must be explored into suitable data capture methods to support the digitized splinting approach for suggested improvements to fit and function. In addition, comments relating to surface finish and surface roughness were not quantified but preliminary in this case. Therefore, further research is required to quantify the surface roughness of AM splints and the potential effect with the skin if worn by a patient, such as abrasion. Lastly, the current research is limited in terms of the usability of splints, since developments to date have not yet allowed for clinical trials. Further work will be required to assess the usability of different splint designs relative to a range of manual tasks (e.g., driving), which patients can compare and contrast against their previously prescribed splints made with traditional methods.

Acknowledgments

Many thanks to Dr. Candice Majewski of the University of Sheffield and Dr. Dominic Eggbeer and Sean Peel of the National Center for Product Design and Development Research, Cardiff, for building the SLS and SL prototype splints, respectively. Thanks to Nigel Bunt and Sarah Drage of HK Rapid Prototyping Ltd. for building the black and white Objet Connex splint, to Mark Tyrtania at LaserLines for the FDM splint, and to Phil Dixon, Loughborough University, for assistance with the building of numerous Objet Connex prototypes. The research was carried out as part of a Ph.D. research project at Loughborough University.

References

[1] Callinan NJ, Mathiowetz V. Soft versus hard resting hand splints in rheumatoid arthritis: pain relief, preference and compliance. American Journal of Occupational Therapy 1996;50(5):347—53. https://doi.org/10.5014/ajot.50.5.347.

[2] Jacobs M. Splint classification. In: Jacobs M, Austin N, editors. Splinting the hand and upper extremity: principles and process. Baltimore: Lippincott Williams and Wilkins; 2003. p. 2—18. ISBN-13: 978-0683306309.

[3] Pagnotta A, Korner-Bitensky N, Mazer B, Baron M, Wood-Dauphinee S. Static wrist splint use in the performance of daily activities by individuals with rheumatoid arthritis. Journal of Rheumatology 2005;32(11):2136—43.

[4] Lohman H, Poole SE, Sullivan JL. Clinical reasoning for splint fabrication. In: Coppard BM, Lohman H, editors. Introduction to splinting: a clinical and problem-solving approach. 2nd ed. St Louis, MO: Mosby, Inc.; 2001. p. 103—38. ISBN-13: 978-0323009348.

[5] Paterson AM, Bibb RJ, Campbell RI. Evaluation of a digitised splinting approach with multiple-material functionality using Additive Manufacturing technologies. In: Bourell D, Crawford RH, Seepersad CC, Beaman JJ, Marcus H, editors. Proceedings of the 23rd annual international solid freeform fabrication symposium—an additive manufacturing conference, Austin, TX: 6—8; August 2012. p. 656—72.

[6] Lohman H. Wrist immobilisation splints. In: Coppard BM, Lohman H, editors. Introduction to splinting: a clinical reasoning and problem-solving approach. 2nd ed. St Louis, MO: Mosby, Inc.; 2001. p. 139—84. ISBN-13: 978-0323009348.

[7] Jacobs M, Austin N. Splint fabrication. In: Jacobs M, Austin N, editors. Splinting the hand and upper extremity: principles and process. Baltimore: Lippincott Williams and Wilkins; 2003. p. 98—157. ISBN-13: 978-0683306309.

[8] Austin NM. Process of splinting. In: Jacobs M, Austin NM, editors. Splinting the hand and upper extremity: principles and process. Baltimore: Lippincott Williams and Wilkins; 2003. p. 88—99. ISBN-13: 978-0683306309.

[9] Veehof MM, Taal E, Willems MJ, van de Laar MAFJ. Determinants of the use of wrist working splints in rheumatoid arthritis. Arthritis Care and Research 2008;59(4):531−6. https://doi.org/10.1002/art.23531.

[10] Austin NM. Equipment and materials. In: Jacobs M, Austin NM, editors. Splinting the hand and upper extremity: principles and process. Baltimore: Lippincott Williams and Wilkins; 2003. p. 73−87. ISBN-13: 978-0683306309.

[11] Coppard B. Anatomical and biomechanical principles of splinting. In: Coppard BM, Lohman H, editors. Introduction to splinting: a clinical-reasoning and problem-solving approach. 2nd ed. St Louis, MO: Mosby, Inc.; 2001. p. 34−72. ISBN-13: 978-0323009348.

[12] Coppard BM, Lynn P. Introduction to splinting. In: Coppard BM, Lohman H, editors. Introduction to splinting: a clinical reasoning and problem-solving approach. 2nd ed. St Louis, MO: Mosby, Inc.; 2001. p. 1−33. ISBN-13: 978-0323009348.

[13] Campbell RI. Customer input and satisfaction. In: Hopkinson N, Hague JM, Dickens PM, editors. Rapid manufacturing: an industrial revolution for the digital age. Chichester: John Wiley and Sons, Ltd.; 2006. p. 19−38. ISBN-13: 978-0470016138.

[14] Wohlers TT. Wohlers report 2012: additive manufacturing and 3D printing state of the industry. Annual worldwide progress report. Colorado: Wohlers Associates, Inc.; 2012.

[15] V. Gervasi, D. Cook, R. Rizza and S. Kamara and Liu, X. "Fabrication of custom dynamic pedorthoses for clubfoot correction via additive-based technologies", In: Bourell, D., Proceeding of the 20th annual international solid freeform fabrication symposium, Austin, TX: pp. 652-661.

[16] Smith S. Miraclefeet and Objet bring innovation to braces. Available at: http://www.deskeng.com/articles/aabamt.htm. [Accessed 25 May 2011].

[17] Cook D, Gervasi V, Rizza R, Kamara S, Xue-Cheng L. 'Additive fabrication of custom pedorthoses for clubfoot correction. Rapid Prototyping Journal 2010;16(3):189−93. https://doi.org/10.1108/13552541011034852.

[18] Summit S, Trauner KB. Custom braces, casts and devices having limited flexibility and methods for designing and fabricating. United States Patent US008613716B2 2013.

[19] Gibson KS, Woodburn J, Porter D, Telfer S. Functionally optimised orthoses for early rheumatoid arthritis foot disease: a study of mechanisms and patient experience. Arthritis Care and Research 2014;66(10):1456−64. https://doi.org/10.1002/acr.22060.

[20] Faustini MC, Neptune RR, Crawford RH, Stanhope SJ. Manufacture of passive dynamic ankle-foot orthoses using selective laser-sintering. IEEE Transactions on Biomedical Engineering 2008;55(2):784−90.

[21] Mavroidis C, Ranky R, Sivak M, Patritti B, DiPisa J, Caddle A, et al. Patient specific ankle-foot orthoses using rapid prototyping. Journal of NeuroEngineering and Rehabilitation 2011;8(1):1. https://doi.org/10.1186/1743-0003-8-1.

[22] J. H. P. Pallari, J. Dalgarno, J. Munguia, et al. "Design and additive fabrication of foot and ankle-foot orthoses", In: D. Bourell (ed), Proceedings of the twenty first annual international solid freeform fabrication proceedings—an additive manufacturing conference, August 9−11: pp. 834-845

[23] Bingham GA, Hague RJM, Tuck CJ, Long AC, Crookston JJ, Sherburn MN. Rapid manufactured textiles. International Journal of Computer Integrated Manufacturing 2007;20(1):96−105. https://doi.org/10.1080/09511920600690434.

[24] Johnson A, Bingham GA, Wimpenny DI. 'Additive manufactured textiles for high-performance stab resistant applications. Rapid Prototyping Journal 2013;19(3):199−207. https://doi.org/10.1108/13552541311312193.

[25] Smith S. Micraclefeet and Objet bring innovation to braces. Available at:. 2011. 2012, http://www.deskeng.com/articles/aabamt.htm. [Accessed 7 August 2012].

[26] Summit S, Trauner KB. Custom braces, casts and devices and methods for designing and fabricating. World Intellectual Property Organization WO2010/054341A1 2010.

[27] Fried S, Michas L, Howard J. Method of providing centralized splint production. United States Patent Application US20050015172A1 2005.

[28] Fried S. Splint and or method of making same. United States Patent Application US20070016323A1 2007.

[29] Grzesiak A. Fraunhofer additive manufacturing alliance—highlights, current RTD activities and strategic topics. In: Fraunhofer, editor. Proceedings of the additive manufacturing international conference, Loughborough, UK; July 08, 2010. p. 14.

[30] Breuninger J. Voronoi wrist splint [conversation] (personal communication, 26 November 2010). 2010.

[31] Evill J. Cortex. 2013. Available at: http://jakevilldesign.dunked.com/cortex. [Accessed 1 August 2013].

[32] Palousek D, Rosicky J, Koutny D, Stoklasek P, Navrat T. Pilot study of the wrist orthosis design process. Rapid Prototyping Journal 2013;20(1):27—32. https://doi.org/10.1108/RPJ-03-2012-0027.

[33] Oxman N. Material-based design computation. Ph.D. Thesis. Massachusetts Institute of Technology; 2010.

[34] PopTech. Neri Oxman: on designing form. Available at: http://www.youtube.com/watch?v=txl4QR0GDnU. [Accessed 17 April 2010].

[35] Oxman N. Variable property rapid prototyping. Virtual and Physical Prototyping 2011;6(1):3—31. https://doi.org/10.1080/17452759.2011.558588.

[36] Paterson AM. Digitisation of the splinting process: exploration and evaluation of a computer aided design approach to support additive manufacture. PhD Thesis. Loughborough University; 2013.

[37] Paterson AM, Bibb RJ, Campbell RI. A review of existing anatomical data capture methods to support the mass customisation of wrist splints. Virtual and Physical Prototyping Journal 2010;5(4):201—7. https://doi.org/10.1080/17452759.2010.528183.

[38] Oxman N. Imaginary beings @ Centre pompidou. Available at: http://materialecology.blogspot.co.uk/2012/05/imaginary-beings-centre-pompidou.html. [Accessed 18 June 2012].

[39] Bingham GA, Hague RJM. Efficient three-dimensional modelling of additive manufactured textiles. Rapid Prototyping Journal 2013;19(4):269—81. https://doi.org/10.1108/13552541311323272.

[40] Bibb R. Medical modelling: the application of advanced design and development techniques in medicine. Cambridge: Woodhead Publishing; 2006. ISBN-13: 978-1845691387.

[41] Bibb R, Eggbeer D, Evans P, Bocca A, Sugar A. Rapid manufacture of custom-fitting surgical guides. Rapid Prototyping Journal 2009;15(5):346—54. https://doi.org/10.1108/13552540910993879.

[42] Eggbeer D, Bibb R, Evans P, Ji L. Evaluation of direct and indirect additive manufacture of maxillofacial prostheses. Proceedings of the Institution of Mechanical Engineers Part H: Journal of Engineering in Medicine 2012;226(9):718—28. https://doi.org/10.1177/0954411912451826.

[43] Holmes S. Objet printers get colour treatment. 2012. Available at: http://develop3d.com/blog/2012/05/objet-printers-get-colour-treatment. [Accessed 18 June 2012].

[44] Stratasys. Objet digital materials^{TM} data sheets [online PDF]. 2013. Available at: http://www.stratasys.com/materials/polyjet/~/media/879CBF2F6582406C9C1B629F4F9E05D5.ashx. [Accessed 4 April 2013].

[45] Moore JP, Williams CB. Fatigue characterisation of 3D printed elastomer material. In: Bourell D, Crawford RH, Seepersad CC, Beaman JJ, Marcus H, editors. Proceedings of the twenty third annual international solid freeform fabrication symposium—an additive manufacturing conference; August 6—8, 2012. p. 641—55.

CHAPTER 5.26

Orthotic applications case study 3—Evaluation of a digitized splinting approach with multiple-material functionality using additive manufacturing technologies*

5.26.1 Introduction

Rheumatoid arthritis (RA) is a chronic, systemic autoimmune disease, which typically affects joints within the hands, wrists, ankles, and feet [1,2]. Symptoms can include inflamed synovia and tendon sheaths and destruction of cartilage and bone, resulting in pain and discomfort [2]. Borenstein et al. [3] state that the approach to RA treatment is a multilayered pyramid, consisting of "education, physical and occupational therapy, rest and nonsteroidal antiinflammatory drugs" (pp. 545). Occupational therapy in particular addresses limitations that patients may encounter during everyday activities in an attempt to circumvent the limitations and to improve

* The work described in this chapter was first reported in the references below and is reproduced here, in part or in full, with the permission of The University of Texas at Austin.
Paterson AM, Bibb RJ, Campbell RI, "Evaluation of a Digitized Splinting Approach with Multiple Material Functionality using Additive Manufacturing Technologies." In: Bourell D., Crawford RH, Seepersad CC, Beaman JJ and Marcus H., Proceedings of the Twenty-Third Annual International Solid Freeform Fabrication Symposium—An Additive Manufacturing Conference. Austin, TX: University of Texas at Austin, 2012; pp. 656—672.

Medical Modeling
ISBN 978-0-323-95733-5
https://doi.org/10.1016/B978-0-323-95733-5.00031-4

wellbeing and quality of life. One method of intervention is splint prescription, the perceived benefits of which are multifaceted [2,4–6].

i. relieves pain through immobilization and protection of affected joints

ii. protects painful contractures from impacts, scarring, and excessive movement

iii. promotes movement of stiff joints through immobilization of more mobile joints

iv. encourages healing of fractures and contractures

v. prevents or corrects deformities and contractures

vi. rests affected joints

vii. provides support

Practitioners such as occupational therapists and physiotherapists may prescribe either custom-made or "off-the-shelf" prefabricated splints. However, this paper will describe and discuss the design and fabrication methods of custom-made static wrist immobilization (SWI) splints in particular (shown in Fig. 5.185), since they are one of the most commonly prescribed splints among a range of conditions [7].

Therapists must take into account the likelihood of their patients adhering to their splint-wearing regimes. Unfortunately, Sandford et al. [8] found that two-thirds of patients reported nonadherence and advise therapists to be aware of and acknowledge low adherence levels. There are several reasons for poor patient adherence in terms of wear duration and frequency [2,4,8–10].

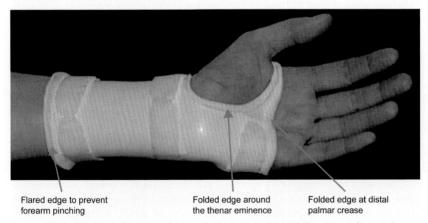

Flared edge to prevent   Folded edge around   Folded edge at distal
forearm pinching   the thenar eminence   palmar crease

Figure 5.185 Traditionally manufactured static wrist immobilization splint and common characteristics.

- The splint does not address the patient's condition.
- The patient has received insufficient information about their condition.
- The patient has not had sufficient information justifying the need for the splint.
- The splint is unattractive.
- The splint is difficult to don/doff.
- The splint is uncomfortable to wear.
- The splint is impractical in certain environments or for certain tasks.
- The patient may not be interested or informed on the potential beneficial outcome of wearing their splint.

Hygiene issues may also contribute to low adherence; open cell padding within splints can absorb moisture, such as perspiration, which can lead to odors and collection of bacteria [11]. More generally and in terms of assistive devices, Louise-Bender Pape et al. [12] reviewed literature linking perceived social stigma to the association of assistive devices and can contribute to nonuse of such devices. In addition, 78% of 27 participants reported immobilization splints as unwieldy [10].

5.26.1.1 The splinting process

Please refer back to Fig. 5.185 in the previous case study that depicts the splint design and fabrication workflow, deduced from the literature [13–15]. The therapist must choose the low temperature thermoplastic (LTT) based on a variety of properties prior to splint forming, including contour conformability, thickness, and color. Color selection by the patient may be encouraged in a bid to improve adherence [11]. However, choice is often dictated by stock available in clinics or properties that are deemed more important by the therapist, such as thickness. In addition, therapists are advised to consider and apply padding *before* forming the LTT, to avoid inducing pressure to prone areas [4,11,16]. Padding types can vary from sheets and stockinettes to silicone gel discs or pads, which can be placed over bony prominences for cushioning [11]. Other common SWI splint characteristics are proximal edge flaring to avoid forearm pinching and folding of both the distal and thenar edges to provide a more comfortable edge against the skin while adding rigidity to the splint (Fig. 5.185).

Although many of the characteristics described are integrated to address functional needs, results may appear unwieldy, voluminous, and unsightly. For example, if cavities are integrated over bony prominences to relieve pressure, the appearance of the altered topology can be compromised. In addition, folded or rolled edges add volume to the palmar region, potentially affecting palmar grasp capacity. In terms of the fabrication process, there is little room for error; the farther the therapist progresses down the fabrication workflow, the more challenging adjustments may be to make. If additional or replacement splints are required, the whole process must be repeated.

5.26.1.2 Additive manufacturing for upper extremity splint fabrication

Several research studies have explored the use of additive manufacturing (AM) for upper extremity splint fabrication. Fraunhofer IPA fabricated a prototype SWI splint using an EOS P100 Laser Sintering (LS) system (EOS GmbH, Krailling, Germany) with polyamide (PA2200) powder 17,18; the benefits were improved aesthetics, bespoke fit, and integral fasteners to exploit the AM capabilities of part consolidation. A hand immobilization splint prototype developed by Materialise (fabricated using an EOS P730 LS machine with PA2200 powder) gave a reduction in weight [19]. Oxman [20] fabricated *"Carpal Skin"* using Objet Connex500 technologies (Objet Geometries, Rehovot, Israel), by exploiting the system's capabilities to enable multiple-material builds [21]. The structures within the splint-like gloves were dictated by a predefined pain map to allow or restrict movement via a reaction-diffusion pattern [20]. Not only does the use of AM in the context of upper extremity splinting allow for exploitation of bespoke fit and function, but both prototypes by Oxman [20] and Fraunhofer IPA also demonstrated improved aesthetics as a result of geometric complexity and freedom, which are synonymous characteristics only viable as a combination through AM.

Sketches shown in Fig. 5.186 by Bibb [22] show the intent of AM for splint fabrication with lattice feature integration. The lattice structures were intended to look aesthetically pleasing while promoting airflow to the skin in a bid to reduce perspiration. An additional intent was to enable patients to personalize their splints by choosing their own perforation patterns in an attempt to improve patient adherence.

The investigators were also interested in exploring additional and novel approaches for multiple-material integration into splints, particularly the use of elastomeric regions for a variety of new design features as a result of Objet Connex capabilities. For example, radial and/or ulnar-based elastomer hinges could be integrated into the splint to allow for easier donning and doffing. In addition, specific elastomer regions could be placed over bony prominences (similar to gel discs and pads in traditional splinting). Both concepts could be achieved without affecting the overall topology of

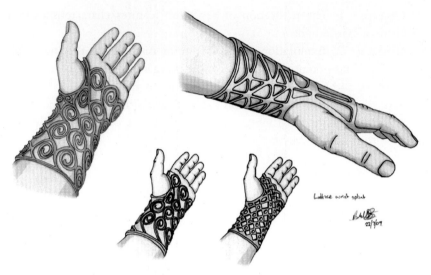

Figure 5.186 AM splint concepts.

the splint. It was anticipated that regional patches could protect and cushion bony prominences while providing dynamic pressure over areas prone to fluctuating edema with an aim to prevent edema pooling.

However, if such a design and manufacturing approach is to be realized for upper extremity splinting in a clinical domain, supporting software technologies need to be developed and tailored to suit the target user. Pallari et al., Rogers et al., and Knoppers and Hague [23–25] support the need for product-specific computer-aided design (CAD) software for practitioners to enable the intended users to adopt quickly and easily without dedicating time to learn. In addition, Smith [26] justifies the need for customized software, as it gives the intended users the tools to achieve what they want and need while removing unnecessary functions.

5.26.2 Research aim and objectives

A number of weaknesses in traditional splinting have been identified, as well as the strengths and feasibility of AM in splint fabrication. The need for customized three-dimensional (3D) CAD software to support the use of AM has also been noted. Therefore, the aim of this paper is to evaluate the feasibility of new design features for upper extremity splinting only made viable through AM, as well as the digitization of traditional splint characteristics. Objectives to achieve the aim are as follows.

Objective 1. The replication of key splint features/characteristics in a virtual environment.

Objective 2. Exploration of new features only viable through AM, which could potentially address concerns regarding aesthetics, hygiene, comfort, and form. Features include the following.

2.1. Aesthetically appealing and personalized lattice integration.

2.2. Multiple-material integration (using Objet Connex technologies) to do the following:

2.2.1. Imitate gel discs or pads.

2.2.2. Potentially replace the need to create bulbous features in splints.

2.2.3. Provide dynamic pressure over areas susceptible to fluctuating edema.

2.2.4. Integrate flexible hinges for easier donning/doffing.

Objective 3. Evaluation into the feasibility of the digitized splinting approach via a specialized CAD software prototype.

3.1. Refine replicated characteristics into a virtual workflow.

3.2. Devise a representation of the digitized approach via a software prototype.

3.3. Evaluate the digitized approach, specifically new features as a result of AM.

It should be noted that the scope of this investigation is purely related to the data manipulation of acquired patient scan data within 3D CAD to support AM splint fabrication. The investigation does not address data acquisition methods suitable for clinical environments, although Paterson et al. [27] review data acquisition methods to generate test data in support of this investigation. Similarly, the process does not investigate the use of finite element analysis (FEA), although it can be assumed that this would be a crucial feature if the digitized approach were to be realized. It should also be noted that the digitized splinting approach is not intended to make the splinting profession obsolete, nor does it imply the redundancy of splinting practitioners. The intended proposition is to provide therapists with an additional toolset to capture their design intent quickly and easily while addressing the needs and concerns of patients through improved aesthetics, fit, and functionality.

A systematic approach was adopted to develop a sequence of functions to meet the objectives. Each objective will be addressed individually.

5.26.3 Methods

Objective 1—Replication of key splint features/characteristics in a virtual environment

The critical splint characteristics required for the specialized splinting workflow were tested iteratively in a range of 3D CAD software packages, including Geomagic Studio [28] and McNeel Rhinoceros [29], using an action research strategy. The approach was intended to test and refine different CAD tools and CAD strategies to capture and replicate the design intent of typical splint features, such as pressure relief cavities. An example of the iterative testing is described by Paterson et al. [30].

Objective 2—Explore new features only viable through AM

Objective 2.1—Aesthetically appealing and personalized lattice integration

Similar to addressing objective 1, an action research strategy was adopted to explore lattice applications to a previously defined splint surface. After iterative testing, it became apparent that an additional feature within the workflow was required to form a splint border, which would encase the lattice structure. This would address inherent issues of sharp protrusions, which could otherwise cause lacerations if borders were not implemented into the proposed splints.

Objective 2.2—Multiple-material integration/heterogeneous structures

As mentioned previously, the purpose of using Objet Connex technologies was to explore the viability of new features in the context of splinting, making the transition from homogeneous to heterogeneous splints. Objet Connex technologies can deposit materials in predefined combinations, which dictate regional material proportions to form so called *digital materials* (DMs). For example, to create variations in shore hardness among the Objet flexible (FLX) FLX97-DM range, the primary material (TangoPlus) and secondary material (VeroWhitePlus) are interspersed by varying dual-jet distribution using Objet's PolyJet Matrix Technology. The material depositions are then cured using ultraviolet light [31,32].

There were two main approaches to consider for multiple-material integration: continuous functional grading (CFG) versus stepped functional grading (SFG) (Fig. 5.187). CFG proposes a gradual transition of one

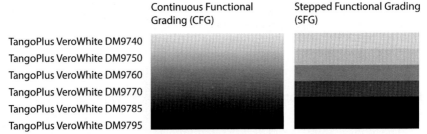

Figure 5.187 Continuous functional grading (CFG) versus specific boundary representation for stepped functional grading (SFG).

or more properties across a defined volume and has been of significant interest across a range of disciplines. However, the implementation of such a feature in terms of CAD modeling strategies, file export, and AM systems is still under development [33]. Knoppers et al. [25] and Siu and Tan [34] highlight the difficulties in representing functionally graded geometries in boundary representation modeling, due to the geometries being defined by a series of surfaces to determine the topology of a closed volume. If functional grading were to be a viable feature in splints, an alternative or additional modeling strategy within the CAD methodology (e.g., voxel-based modeling) would need to be adopted. A consequential limitation, however, is the increased memory power, processing consumption, and modeling complexity during the design phase [20,33]. This would add unnecessary complexity to the geometry and CAD strategies. In addition, current Objet Connex technologies only offer a finite number of pre-defined deposition variations, which limit the CFG approach. As a result, CFG was considered excessively complex and unnecessary for this particular application, and it was decided that SFG would be sufficient to deliver the finite variation that Objet Connex technologies offer.

The disadvantage of SFG is that it creates boundary lines between adjacent materials' properties, which, depending on the differences in the adjacent properties, could lead to stress concentrations at the boundary or abrupt changes in flexural modulus that could cause unwanted folds or creases at the boundary. The worst-case scenario can be imagined to be a choice of only two materials, for example very rigid and very flexible. In that circumstance, the abrupt change at the boundary is likely to present problems. This would be mitigated by the ability to specify a greater number of regions and a sufficiently high number of materials' property choices. It can be assumed that, for a given application, a sufficiently high number of materials' choices could satisfactorily approximate continuous variation.

Objective 2.2.1—Imitate gel discs or pads

This objective was to explore the possibility of using multimaterial capability to reproduce the function of gel pads. In existing practice, the rigid shell of the splint is raised in local areas to provide space to insert soft pads. These are required to reduce pressure on bony prominences and improve comfort. The ability to define material properties within a localized region enables the body of the splint to be made soft in the required area. This eliminates the pads and reduces bulk of the splint.

Objective 2.2.2—Potentially replace the need to create bulbous features in splint

This is essentially the same function as 2.2.1 except that in existing practice, some areas of the splint may be raised to provide space for sensitive areas or bony prominences but without placing a gel disc or pad into the space.

Objective 2.2.3—Provide dynamic pressure over areas susceptible to fluctuating edema

This function is currently very difficult to accommodate with a single material rigid splint. The ability to alter the flexibility of the splint, possibly in specific regions, would enable a conformal fit while allowing some degree of expansion that would accommodate temporary or fluctuating swelling. This would be a unique advantage of the AM approach over existing practice.

Objective 2.2.4—Integration of flexible hinges for easier donning/doffing

With existing practice, the rigid splint has to incorporate large openings to allow donning and doffing. This leaves large regions unsupported. In addition, straps and fasteners have to be added into or onto the splint, which adds construction time, increases bulk, and adds features that will be more susceptible to damage and wear. The ability to build in flexible joints or hinges enables a splint design with greater coverage and therefore increased support. The ability to build in hinges and fasteners reduces the bulk and the time required to add those features after forming the splint. This approach has many other potential advantages in the elimination of purchasing and stocking fasteners, straps, etc. In addition, it is frequently straps and fasteners such as Velcro that wear, fail, or become unacceptably soiled in use earlier than the splint itself.

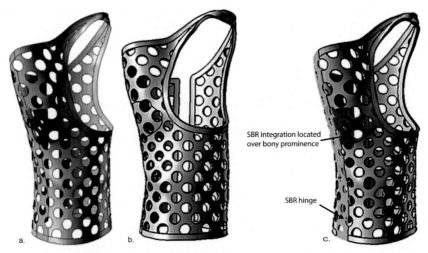

Figure 5.188 Example of elastomer integration over a bony prominence and medial/ulnar hinge: (a) three separate surfaces present, (b) thickened splint, (c) and regions of splints, depicting different materials.

In terms of 3D CAD strategy testing to support the digitized splinting approach, the refined strategy for enabling multiple-material builds was to create a curve on the splint surface to act as a trim boundary prior to pattern application. After applying the pattern to the surface, the splint surface could then be trimmed (Fig. 5.188a). Separate patches could then be thickened (Fig. 5.188b and c) and exported as separate shells in a single STL file. Each shell can subsequently be assigned a digital material in the Objet preparation software. For example, in Fig. 5.188c, there are three shells. The predominant gray shell could be a rigid material, the green shell could be an intermediate flexible material, and the red shell could be very soft.

Objective 3—Evaluation of approach via a specialized splinting CAD software prototype

The requirements established after CAD testing were refined into a specialized CAD software workflow (objective 3.1) (Fig. 5.189). The order was determined by consideration of the traditional splinting workflow depicted in Figure 5.171 (see previous case study 5.25), as well as constraining/best practice CAD approaches (e.g., detailed application to surfacing prior to thickening). An important feature to note in the digitized approach was the intent for traversal operation: the ability to move back and forth between different features independent of one another. This, fundamentally, was anticipated as an important step for splint fabrication

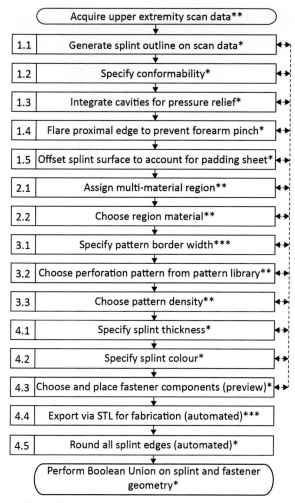

Figure 5.189 Refined digitization workflow: * = existing feature/activity, ** = new feature/activity, *** = necessary feature as a result of a new feature/activity.

when compared with traditional splint fabrication, where the further one progresses through design and fabrication, the more difficult and time consuming it may be to make adjustments.

Having established a refined workflow, a high-fidelity concept software prototype was developed within Microsoft Access 2010 [35] using Visual Basic for Applications 2010 [36], which could be used to depict the workflow for final evaluation (objective 3.2). The prototype featured a viewport and a series of slider controls and drop-down boxes to control

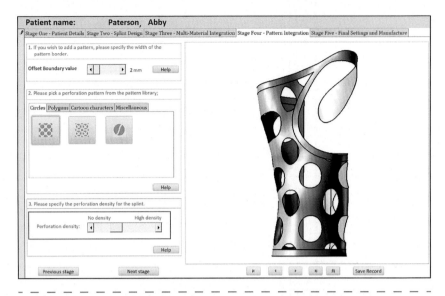

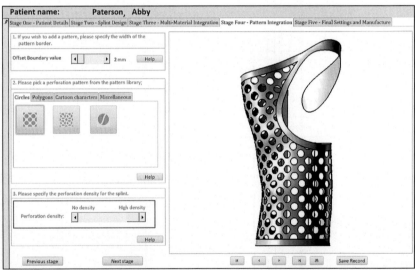

Figure 5.190 Prototype interface example and resulting changes from altering pattern perforation density.

features such as pattern density (Fig. 5.190). When controls were adjusted, the image in the viewport would change to suit the recent adaptation to provide direct visual feedback to the user. The prototype also featured a pattern library and a separate fastener library, which allowed users to browse different perforation patterns and fasteners. The user would then be able to select the desired options from the libraries for automatic placement to the splint geometry in the viewport.

Ten splinting practitioners within the United Kingdom were invited to participate in evaluation sessions (physiotherapist: n = 2, occupational therapist: n = 8). Eight participants had one-to-one sessions, while two participants joined in one session for their convenience. Each evaluation session comprised four activities: a briefing into the intent of AM for splint fabrication, a demonstration of the prototype, user trials of the software prototype, and a semistructured interview. Qualitative data were captured regarding the digitized approach, together with specialist feedback regarding multiple-material integration into splints. Participants completed a demographic questionnaire to gather data including the number of years of splinting experience. In addition, participants signed informed consent forms agreeing to the capture of audio and computer screen—capture recordings while testing the prototype. The recordings were used to identify trends in opinions and were transcribed into NVivo [37] for coding and trend identification. In addition, proof-of-concept splints were manufactured using various AM processes to give physical, tactile representations of the intended output (Fig. 5.191). The purpose was to show participants what could be achieved using AM, to ensure that the intent was conveyed effectively to participants in the context of upper extremity splinting.

5.26.4 Results and discussion

A wide range of positive comments and suggestions were gathered during the evaluation sessions. The approach not only highlighted many exciting new avenues for exploration within splint design and fabrication, but it also expanded the window of opportunity for Objet Connex technologies. Participants mentioned that the integration of multiple materials within a single splint was a completely new and exciting toolbox for therapists, which could not be achieved using existing fabrication methods. Participants were interested in new applications as a result of multiple-material capabilities, such as protection and cushioning of bony prominences without compromising the splint topography. Of the three participants who were asked, all three participants favored "mono-splints" with elastomer hinges (as shown in Fig. 5.188), to make donning and doffing easier for patients with restricted dexterity; two participants even suggested having two hinges, placed along the radial and ulnar aspects of the wrists, to assist further. Three participants were interested in using elastomer materials to

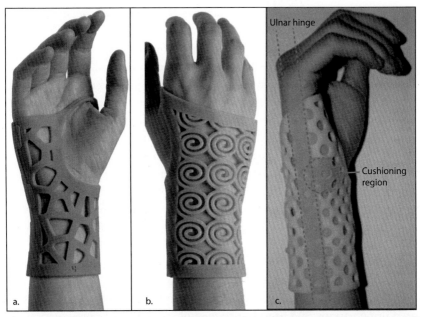

Figure 5.191 Proof-of-concept AM splints: (a) fabricated using 3D systems SLA 250 (3D-Systems Xtreme material), (b) fabricated using Objet Connex500 (RGD5160-DM), and (c) fabricated using Objet Connex500, with multimaterial regions (ulnar hinge: FLX9760-DM; cushioning region: FLX9740-DM; remainder of splint: VeroWhitePlus FullCure 835).

apply dynamic pressure over areas prone to fluctuating edema. In extension to these applications, one participant suggested the use of elastomer materials in splints for treatment/management of burns and scars and protection of postsurgical metalwork (i.e., alignment screws and brackets). Another participant suggested integration of elastomer borders on splint edges, which could aid in pressure distribution and could provide a softer interface between the more rigid structure of the splint and the skin. Therapists also expressed their interests in allowing partial dynamic features to allow or restrict movement, similar to *Carpal Skin* by Oxman [20].

Many participants suggested additional creative features to be integrated within the digitized workflow, such as multicolor integration; one participant asked whether photographs could be applied to splints. Multicolor and multimaterial single builds are now possible with the newest release of the Objet Connex3 [38], but the material properties require further development before they may be suitable for this particular application. Therapists were also keen on part consolidation with regard to fastener integration. Participants felt that the workflow depicted in the software prototype demonstrated the workflow used in traditional fabrication, with the exception of new features and placement of color and thickness controls. In response to traversal movement within the workflow, eight participants felt this was a useful feature, particularly since variables could be altered independently from one another, which is a desirable novelty when compared with traditional splinting.

The intent of edge filleting in the CAD workflow (Fig. 5.189, flow chart item 4.4) was to relieve the need for rolling of splint edges, consequently reducing cumbersome structures presented in traditional splinting. Therapists were surprised at the strength and rigidity of the prototype splints without having folded or rolled edges within the palm. One participant was interested to know whether this would improve palmar grip capacity, since material volume was reduced. However, some participants felt they would still want the opportunity to produce the effect of rolled and folded edges within the splinting software.

Despite the abundance of positive responses, participants also raised concerns regarding data acquisition methods, structural integrity, fabrication time, and cost of the approach. Two participants were concerned that edematous, inflamed regions might protrude through the lattice structure, potentially causing discomfort for the patient. Another participant felt that patient adherence would not only be improved as a result of improved aesthetics, but the level of improved wear duration could potentially be detrimental; they were concerned that patients would be wearing their splints for longer durations than is necessary or recommended, potentially resulting in splint dependency. Cooper states that static splints should not be worn more than necessary as they contribute to stiffness, atrophy, and joint disuse [16]. Therefore, a more stringent wearing regime might be required to combat this concern if AM splints were to be eventually prescribed.

5.26.5 Future work

Although the investigation highlighted many new positive and exciting applications within AM splint fabrication, there is much to be done before the approach can be used as a clinically viable treatment method. To date, the software prototype described in this paper is purely a visual representation for evaluative purposes. Future development is required to create fully functional specialized splinting CAD software. New features must also be investigated for feasibility into workflow integration, such as edge folding. The ability to reuse previously defined splint files to make replacement splints must also be an integral feature in the CAD approach, as this would allow therapists to fabricate duplicate splints without incurring additional design time. The opportunity to adjust previously defined splint files should also be explored, so therapists may alter one or many design variables where required. Alternative file format exports are also required to support depiction of complex organic geometry while allowing for multiple-material and mono/multicolor capabilities; the development of STL 2.0 or Additive Manufacturing File Format (AMF) is promising, particularly for the composition of complex geometries and multiple-material builds in support of the Objet Connex capabilities [39]. In terms of CAD process efficiency, Objet has developed CADMatrix, a plugin for material assignment within selected CAD applications prior to automated STL creation [40]. If the proposed software prototype were to be developed, this approach could be applied within the custom software.

Cost was a concern expressed by the participants that demands further work. It is too early in the development of the CAD approach to predict costs realistically. However, the advantages of the CAD/AM approach for future cost reduction are based on two premises. Firstly, there is very little scope for cost reduction in existing practice. The materials costs incurred in current practice are minimal, and by far the greater proportion of cost is attributed to time and salary costs for the professionals involved. There is little scope for cost reduction of the existing materials, and significant cost reduction would only be possible by speeding up work or reducing salaries, neither of which is likely to be acceptable, and any savings would be small and incremental. A shift to a CAD/AM approach would achieve cost savings by eliminating physical work (especially for remakes or replacements) and enabling a much faster design stage to be completed by the professionals. The separation of the manufacture from the design enables designs to be done at any time, potentially without the patient being physically present, while manufacture does not incur salary costs. Although AM machine and materials costs are currently high, it is reasonable to assume that these costs will reduce significantly over time, as has already been demonstrated by other technologies. The potential advantages of a CAD/AM approach in the provision of custom-fitting medical devices have been recognized in other researches, such as the provision of maxillofacial prosthetics [41].

Further development of medical-grade materials conforming to ISO 10993 category standards will also be necessary for this application, in preparation for clinical trials. In response to burns/scar treatment, further research into suitable materials would be required. The multiple-material splint shown in Fig. 5.187c also suffered a significant amount of warping and deformation at room temperature; these issues will also need addressing.

Although not within the focus of this investigation, FEA should also be investigated to address concerns regarding structural integrity: both the way in which it may be applied to the splint geometry in 3D CAD but also the way in which the user would interact with the feature. The approach would need to be easy to use with the potential for automated alterations through analysis and regeneration of problem areas. Therapists expressed interest into multiple-material splints to either permit or restrict movement, termed "synthetic anisotropy" by Oxman [20]. Therefore, not only does research need to be performed into how to achieve such results in terms of 3D CAD modeling and specification of Objet Connex materials but also the most efficient way to integrate such features within the specialized splinting CAD software.

In response to therapists' concerns into edematous regions protruding through lattice structures, displacement mapping could be used on the outer splint surface to overcome such problems. This could be used instead of or in addition to material-to-perforation ratio alterations. Not only would this aim to resolve issues related to edema, but it would also be a step forward in resolving concerns regarding structural integrity.

Acknowledgments

Our sincerest thanks to Lucia Ramsey at the University of Ulster, Belfast, Ella Donnison at Patterson Medical Ltd., Sutton-in-Ashfield, Jason Watson at Queens Medical Center, Nottingham, and Dr. Dominic Eggbeer at The National Center for Product Design and Development Research, Cardiff, for their help, support, and advice throughout the project. Extended thanks to all participants involved in the study, for their time, expertise, and valued feedback. This research was funded by Loughborough University.

References

[1] Biese J. Arthritis. In: Cooper C, editor. Fundamentals of hand therapy: clinical reasoning and treatment guidelines for common diagnoses of the upper extremity. St Louis, MO: Mosby Inc., Elsevier Inc.; 2007. p. 349—75.

[2] Melvin JL. Rheumatic disease: occupational therapy and rehabilitation. 2nd ed. Philadelphia: F. A. Davis Company; 1982.

[3] Borenstein DG, Silver G, Jenkins E. Approach to initial medical treatment of rheumatoid arthritis. Archives of Family Medicine 1993;2(5):545—51. https://doi.org/10.1001/archfami.1993.01850060113018.

[4] Taylor E, Hanna J, Belcher HJCR. 'Splinting of the hand and wrist. Current Ortho-paedics 2003;17(6):465—74. https://doi.org/10.1016/j.cuor.2003.09.001.

[5] Jacobs M. Splint classification. In: Jacobs M, Austin N, editors. Splinting the hand and upper extremity: principles and process. Baltimore: Lippincott Williams & Wilkins; 2003, ISBN 978-0683306309. p. 2—18.

[6] Colditz JC. Principles of splinting and splint prescription. In: Peimer CA, editor. Surgery of the hand and upper extremity. New York: McGraw-Hill; 1996. p. 2389—409.

[7] Stern EB. Wrist extensor orthoses: dexterity and grip strength across four styles. American Journal of Occupational Therapy 1991;45(1):42—9. https://doi.org/10.5014/ajot.45.1.42.

[8] Sandford F, Barlow N, Lewis J. A study to examine patient adherence to wearing 24-hour forearm thermoplastic splints after tendon repairs. Journal of Hand Therapy 2008;21(1):44—52. https://doi.org/10.1197/j.jht.2007.07.004.

[9] Callinan NJ, Mathiowetz V. Soft versus hard resting hand splints in rheumatoid arthritis: pain relief, preference and compliance. American Journal of Occupational Therapy 1996;50(5):347—53. https://doi.org/10.5014/ajot.50.5.347.

[10] Spoorenberg A, Boers M, Linden S. Wrist splints in rheumatoid arthritis: a question of belief? Clinical Rheumatology 1994;13(4):559—63.

[11] Coppard BM, Lynn P. Introduction to splinting. In: Coppard BM, Lohman H, editors. Introduction to splinting: a clinical reasoning & problem-solving approach. 2nd ed. St Louis, MO: Mosby, Inc.; 2001, ISBN 978-0323009348. p. 1—33.

[12] Louise-Bender Pape T, Kim J, Weiner B. The shaping of individual meanings assigned to assistive technology: a review of personal factors. Disability & Rehabilitation 2002;24(1—3):5—20.

[13] Lohman H. Wrist immobilisation splints. In: Coppard BM, Lohman H, editors. Introduction to splinting: a clinical reasoning & problem-solving approach. 2nd ed. St Louis, MO: Mosby, Inc.; 2001, ISBN 978-0323009348. p. 139—84.

[14] Jacobs M, Austin N. Splint fabrication. In: Jacobs M, Austin N, editors. Splinting the hand and upper extremity: principles and process. Baltimore: Lippincott Williams & Wilkins; 2003, ISBN 978-0683306309. p. 98—157.

[15] Kiel JH. Basic hand splinting techniques: a pattern-designing approach. 1st ed. Lippincott Williams & Wilkins; 1983, ISBN 978-0316491778.

[16] Cooper C. Fundamentals of clinical reasoning: hand therapy concepts and treatment techniques. In: Cooper C, editor. Fundamentals of hand therapy: clinical reasoning and treatment guidelines for common diagnoses of the upper extremity. St Louis, MO: Mosby Inc., Elsevier Inc.; 2007. p. 3—21.

[17] Breuninger J. Wrist splint [email]. Message to A.M. Paterson (a.m.paterson@lboro.ac.UK). 2012. Sent 11 July 2012, 14:31.

[18] Fraunhofer IPA. Fraunhofer institute for manufacturing engineering and automation [online]. 2012. Available at: http://www.ipa.fraunhofer.de/index.php?L=2. [Accessed 10 December 2012].

[19] Pallari J. A-footprint project [Email]. Sent to A.M. Paterson (a.m.paterson@lboro.ac.UK). Sent 9 August 2011, 16:13. Former technical product manager at materialise, Leuven, Belgium. All correspondence now to J. Pallari, Peacocks medical Group, Newcastle upon Tyne, UK. 2011. 2011.

[20] Oxman N. Variable property rapid prototyping. Virtual and Physical Prototyping 2011;6(1):3—31. https://doi.org/10.1080/17452759.2011.558588.

[21] Objet Ltd. Objet Connex Family. Available at: http://www.objet.com/3D-Printer/Objet_Connex_Family/; 2012.

[22] Bibb RJ. Lattice within splints. Drawing; July 22, 2009.

[23] Pallari JHP, Dalgarno KW, Woodburn J. Mass customisation of foot orthoses for rheumatoid arthritis using selective laser sintering. IEEE Transactions on Biomedical Engineering 2010;57(7):1750—6.

[24] Rogers B, Stephens S, Gitter A, Bosker G, Crawford R. 'Double-Wall, transtibial prosthetic socket fabricated using selective laser sintering: a case study. Journal of Prosthetics and Orthotics 2000;12(3):97—100.

[25] Knoppers R, Hague RJM. CAD for rapid manufacturing. In: Hopkinson N, Hague RJM, Dickens PM, editors. Rapid manufacturing: an industrial revolution for the digital age. West Sussex: John Wiley & Sons, Ltd.; 2006. p. 39—54.

[26] Smith MF. Software prototyping: adoption, practice and management. Maidenhead: McGraw-Hill Book Company (UK) Ltd.; 1991.

[27] Paterson AM, Bibb RJ, Campbell RI. A review of existing anatomical data capture methods to support the mass customisation of wrist splints. Virtual and Physical Prototyping Journal 2010;5(4):201—7. https://doi.org/10.1080/17452759.2010.528183.

[28] Geomagic. Geomagic studio version 12 [software]. Durham, North Carolina: Geomagic; 2010.

[29] Robert McNeel & Associates. McNeel Rhinoceros® [software]. Version 4.0 service release 9. Seattle, Washington: Robert McNeel & Associates; 2011.

[30] Paterson AM, Bibb RJ, Campbell RI. Digitisation of the splinting process: development of a CAD strategy to support splint design and fabrication. In: Rennie AEW, Bocking CE, editors. 12th conference on rapid design, prototyping and manufacturing. Lancaster University; June 17, 2011. p. 97—104.

[31] Objet L. New Objet materials: the power behind your 3D printer. [Datasheet PDF]. 2012. Available at: http://www.objet.com/Portals/0/docs2/New%20materials%20 data%20sheets_low%20res.pdf.

[32] Objet Ltd. PolyJet MatrixTM 3D printing technology. 2012. Available at: http:// www.objet.com/products/polyjet_matrix_technology/.

[33] Erasenthiran P, Beal V. Functionally graded materials. In: Hopkinson N, Hague RJM, Dickens PM, editors. Rapid manufacturing: an industrial revolution for the digital age. Chichester: John Wiley & Sons, Ltd.; 2006. p. 103—24.

[34] Siu YK, Tan ST. Representation and CAD modeling of heterogeneous objects. Rapid Prototyping Journal 2002;8(2):70—5.

[35] Microsoft Corporation. Microsoft® Access® 2010 [software]. Version 14.0.6112.5000 (64bit) ed. Redmond, WA: Microsoft Corporation; 2010.

[36] Microsoft Corporation. Microsoft® Visual Basic® for applications 2010 [software]. Version 7.0 1625. 2010.

[37] NVivo. Software. Version 9 ed. Doncaster, Victoria, Australia: QSR International Pty Ltd.; 2011.

[38] Stratasys. Objet500 Connex3 - Vivid color and multi-material 3D printing. 2014. Available at: http://www.stratasys.com/3d-printers/design-series/precision/objet500-connex3.

[39] Hiller JD, Lipson H. STL 2.0: a proposal for a Universal multi-material additive manufacturing file format. In: Bourell D, editor. Proceedings of the twentieth annual international solid freeform fabrication symposium, August 3-5, University of Texas at Austin: University of Texas at Austin; 2009. p. 266—78.

[40] Objet Ltd. CADMatrix™ - Objet add-in for CAD Solidworks®. Available at: http:// objet.com/sites/default/files/CADMatrix_solidworks_A4_IL_low.pdf; 2009.

[41] Bibb R, Eggbeer D, Evans P. Rapid Prototyping technologies in soft tissue facial prosthetics: current state of the art. Rapid Prototyping Journal 2010;16(2):130—7. https://doi.org/10.1108/13552541011025852.

CHAPTER 5.27

Orthotic applications case study 4—Digitization of the splinting process: development of a CAD strategy for splint design and fabrication*

5.27.1 Introduction

Designing and fabricating custom-made splints is a highly skilled, creative process. Orthotists must understand anatomic and biomechanical principles of the upper extremities, the best strategy for addressing different ailments, and the biomechanical capabilities of each patient [1]. Splint designs and materials can vary depending on a patient's condition(s), daily lifestyle, and activity levels [2]. However, there are direct and indirect factors resulting in poor patient compliance in terms of wear duration and frequency.

The splinting process can be time consuming and awkward for both the patient and the orthotist, particularly as the complexity of the splint increases; mistakes or weakened areas of splints often require additional support material to maintain the desired structural integrity but inevitably reduce the aesthetic appeal of the splint [3]. Splints should be low profile, light, and finished professionally, as poor-looking splints will deter patients from wearing them, particularly in social environments due to the desire to "fit in" [4,5].

---

* The work described in this chapter was first reported in the references below and is reproduced here, in part or in full, with the permission of CRDM, Ltd.
Paterson AM, Bibb RJ, Campbell RI, "Digitisation of the splinting process: development of a CAD strategy to support splint design and fabrication." In: Bocking C, Rennie AEW, Twelfth conference on Rapid Design, Prototyping and Manufacturing, CRDM Ltd.: High Wycombe, 2011, pp. 97—104.

Medical Modeling
ISBN 978-0-323-95733-5
https://doi.org/10.1016/B978-0-323-95733-5.00032-6

Poorly fitted splints also result in poor compliance, as they can cause discomfort and pain during everyday activities [2]. The splinting design and fabrication process can also be painful for some patients, as the orthotist may have to apply force onto thermoplastic templates while draping it onto the patient's extremities to fit it correctly. The design and fabrication of other custom-fitting assistive devices such as hearing aids have been digitized, using three-dimensional (3D) patient scan data acquisition methods through to fabrication using additive manufacturing (AM). However, the success of these advances is partially due to the development of suitable 3D computer-aided design (CAD) software strategies, specifically developed for specialized applications. To date, no CAD strategy for splint design has been published. It is anticipated that CAD use would reduce splinting time and provide a more comfortable clinic experience. An appropriate CAD strategy must be able to cater to the orthotists' creative skills and clinical decision-making. However, it should be simple, intuitive, and not require additional training to use it effectively [6]. Therefore, this paper outlines part of a CAD strategy (Fig. 5.192) that will directly manipulate 3D patient scan data to design and fabricate a more aesthetically pleasing, comfortable, and bespoke wrist splint using AM. In particular, the paper focuses on the most suitable CAD tools to assign a splint shape for a static wrist immobilization splint (Fig. 5.193). The most suitable CAD methods had to replicate or potentially improve conventional splinting methods that are currently used to form a splint outline, as highlighted in Fig. 5.193c. Various CAD methods have been analyzed, with strengths and weaknesses identified.

Figure 5.192 Proposed digitized splinting process. This paper addresses the substages "assign a splint outline" and "make alterations if necessary."

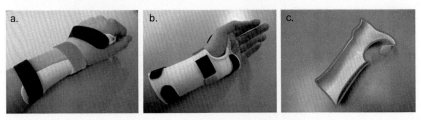

Figure 5.193 A custom-made static wrist immobilization splint in use (a and b) and the outlined border of a static wrist immobilization splint (c).

5.27.2 Current splinting techniques

To understand the needs of orthotists, a thorough understanding of the splinting processes is required. Conventional splint fabrication involves a trial-and-error approach involving various steps, dependent on the splint required for condition. A static wrist immobilization splint can be fabricated in 19 steps [7]; in summary, the patient's hand is placed palm down on a piece of A4 paper, and the orthotist draws around the patient's hand and forearm. The orthotist marks the anatomic landmarks on the paper (Fig. 5.194), such as the radial styloid (marked A), ulnar styloid (B), second metacarpophalangeal (MCP) joint head (C), and the fifth MCP joint head (D). The remaining points in Fig. 5.194 are offset from other anatomic landmarks. By intersecting the points, a rough splint template is drawn. The template is cut out and drawn around on a sheet of thermoplastic. Next, the orthotist places the thermoplastic material in a heated water bath until it reaches its glass transition temperature. At this point, the thermoplastic is removed, cut to shape, and then draped on the patient's extremity.

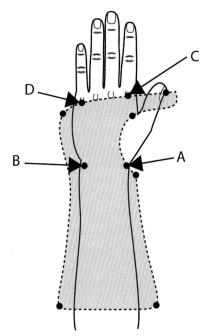

Figure 5.194 Volar-based static wrist immobilization splint pattern. *(Adapted from Lohman [7].)*

Depending on the thermoplastic characteristics, the orthotist may have to apply pressure to form it correctly. While in its flexible state, excess material is removed from the thermoplastic using scissors or other suitable implements. Adjustments must be made until the orthotist is satisfied with the fit, which may require repeated heating in the water bath. The orthotist can then add fastenings such as Velcro and/or pop rivets.

As described, assigning a splint outline requires structural consideration, but the main pitfall of conventional splinting is that the orthotist must approximate the overall form of the splint using only a few guidelines on a 2D plane, not within 3D space. As a result, the orthotist must make many adjustments during the fitting process. This is often costly in terms of time, labor, materials, and funds available. A common problem when designing static wrist immobilization splints on a 2D plane is that the drawing of the forearm trough can often be sketched too narrow [7]. Consequently, when the 2D thermoplastic sheet is cut out and fitted, the trough will be too narrow for the patient and will not provide adequate support. If this is the case, it must be discarded and started again. However, by digitizing the splinting process, the limitations of conventional splinting may be reduced. Orthotists can manipulate the scan data and view alterations directly to produce a suitable form in fewer stages, without approximation and without material waste. The splinting experience may also be significantly less painful for patients, since the orthotist will not need to apply pressure to the thermoplastic during fitting, nor will the patient be expected to place their hand palm down on paper (dependent on the scan data acquisition method), since this is a painful experience if severe deformities are present. In fact, at this design stage, the patient need not be present at all.

5.27.3 Experimental procedures

Various modeling strategies were explored including mesh modeling, surface modeling, solid modeling, and haptic modeling. The following software packages were used during testing.

- Geomagic Studio Version 12 (Geomagic, 3200 East Hwy 54, Cape Fear Building, Suite 300, Research Triangle Park, NC 27709, USA)
- Pro Engineer Wildfire 4.0 (PTC Corporate Headquarters, 140 Kendrick Street, Needham, MA 02494, USA)
- Rhinoceros 4.0 (McNeel North America, 3670 Woodland Park Ave N, Seattle, WA 98103, USA)
- Maya 2011 (Autodesk, Inc., 111 McInnis Parkway, San Rafael, CA 94903, USA)

- FreeForm Modeling Plus Version 10 (SensAble Technologies, 181 Ballardvale Street Wilmington, MA 01887, USA)

Different tools within the software packages were trialed and compared for suitability and similarity to conventional splinting techniques. The methods offered potentially desirable splint fabrication characteristics, such as easy assignment of a splint outline, while offering additional benefits like being able to adjust the outline position and size interactively.

The trial CAD methods were divided into three categories.

a. Curve on surface category

Method i. Sketch on Polygon Mesh and *Trim* tools (Fig. 5.195a). This method is performed in Rhinoceros. The user draws free-hand sketches on the scan mesh surface by using the left mouse button and dragging the mouse across the mesh. This creates a nonuniform rational B-spline (NURBS) curve. Initially, the curve automatically attaches to the mesh. However, the position of the curve is view orientation dependent, so the user may have to make more than one curve when the view orientation is changed to optimize curve position. The user must then manually close any open loops by using the *Match* command to attach anchor points or nearby NURBS curves. Alterations can be made by moving the control points/knots, but the curve detaches from the surface as a result; if this is the case, the user must reattach the edited curve back onto the scan mesh. Once a closed loop is created, the user can trim the scan mesh with the curve, using the *Trim* tool.

Method ii. Trim with Curve tool (Fig. 5.195b). This is a tool within Geomagic Studio. The user creates a NURBS curve on the scan mesh by placing numerous control points on the mesh. The control

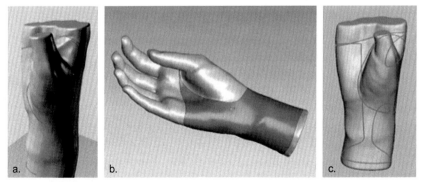

Figure 5.195 Rhinoceros "Sketch on Polygon Mesh" (a); Geomagic Studio "Trim with Curve" (b); and FreeForm Modeling Plus "Trim Mesh with Curve" (c).

points are constrained to the surface, so the user can alter the points easily if required. Once the user is satisfied with the shape and position of the NURBS curve, the NURBS curve is used to trim the mesh (providing the curve is a closed loop).

Method iii. Trim Mesh with Curve (Fig. 5.195c). This method combines two tools within FreeForm Modeling Plus; the user must create a closed loop on the scan mesh and then use a mesh trim tool. The method is a similar approach to *Trim with Curve* within Geomagic Studio, except the user must use a PHANTOM haptic stylus to place the points on the scan data.

b. Import category

Method i. Boolean Intersection (Fig. 5.196a and b). This method was tested using the Pro Engineer Interactive Surface Design Extension (ISDX). Four imported surfaces are placed accordingly to intersect the scan data. The user must then use the *Intersect* tool to create an intersecting curve for each intersecting region. Each curve is then used to trim the scan data, and the unwanted surface(s) can be deleted.

Method ii. Drape (Fig. 5.196c and d). This method uses *nCloth nDynamics* in Maya. A template "cloth" is imported and draped under gravitational and magnetic influence onto the scan data, to replicate the thermoplastic draping techniques in current splinting methods. The virtual "cloth" is then developed, and the scan data is redundant.

Method iii. Shrinkwrap (Fig. 5.196e and f). This method uses the Multi-Tool 1.0.1 Maya plug-in (M. Richter, http://www.creativecrash.com/maya/downloads/scripts   -plugins/utility-external/export/c/multi-tool). A surrounding mesh (i.e., cylinder) is imported, reorientated in a suitable position around the scan mesh, and then "shrink-wrapped" onto the scan data. Once complete, the scan data is redundant, and the shrink-wrapped mesh is developed further.

c. Surface patching category

Method i. RhinoReverse (Fig. 5.197a). The user places numerous curve control points on the scan mesh to construct a series of four-sided NURBS curve boundaries, which are then surfaced automatically. The curves are constrained to the scan mesh, and once they are surfaced with NURBS patches, the scan mesh is redundant, and the NURBS surfaces are developed instead.

Method ii. Patching (Fig. 5.197b). This method is similar to RhinoReverse patching, except the method uses a PHANTOM haptic stylus and FreeForm Modeling Plus.

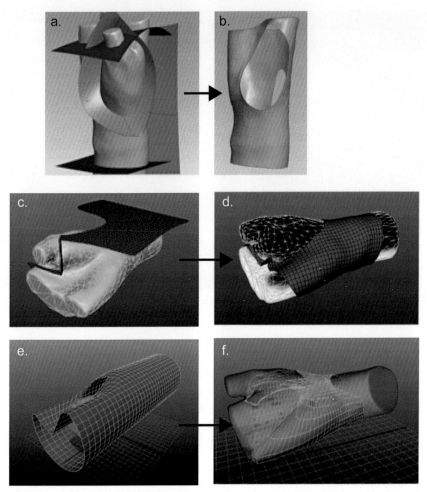

Figure 5.196 Pro engineer ISDX Boolean intersection (a and b), Maya nCloth drape (c and d), and Maya shrinkwrap plug-in (e and f); see also color section.

5.27.3.1 Results

Each category was assessed and compared; strengths and weaknesses were identified and evaluated.

The *curve on surface* category displayed many strengths; the user can locate anatomic landmarks that are used in current splinting methods and tailor the splint shape directly around them. The user also has freedom over the placement of control points for a more controlled fit The resulting advantages of this are that the user can see the shape of the splint as they progress and can adjust accordingly and can see directly where the mesh will be

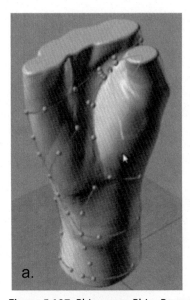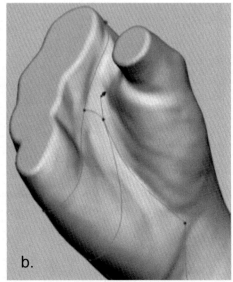

Figure 5.197 Rhinoceros RhinoReverse plug-in (a) and FreeForm Modeling Plus curve tool for surface boundary creation (b).

trimmed before performing a *Trim* command. This means the user can capture their design intent of the splint outline easily and quickly. The scan data can also be used directly as the splint surface, so the contours of the patient's upper extremities can be complemented closely by the proposed splint. There are also no limits on the hand position from the scan data acquisition method, as the user can place points on any part of the scan mesh and from any angle (view orientation dependent). As a result, the scan data does not need reorienting with respect to a reference. No weaknesses were encountered when compared with conventional splinting techniques.

The *import* category strengths included a potential for automation; assuming a library of different splint templates is available, the user could simply browse through the library and then apply a suitable template to the scan mesh. The process would also save the specialist from redesigning the same splint from scratch.

However, *import* weaknesses were significant. The methods result in a limited view of the scan data, restricting the ability to identify anatomic landmarks and the ideal splint placement. In addition, the shape of the splint outline is difficult to interpret until the import geometry has been used. Therefore, it is highly likely that the applied import feature will need adjusting to suit each individual. In some cases, the templates may not suit the need of the patient and their condition, so the user would need to adapt an existing template. Also, to account for anthropometric variations of different patients, the import features may require rescaling using anthropometric measurements to account for extremes of size. Success of the import features is also dependent on the quality of the scan data in terms of digit position; the thumb orientation in particular can obstruct the correct operation of the import features, which can lead to overlapping/intersecting geometry. Reorientation of either the scan data or the import features will be required to ensure the correct regions of the scan data are used and manipulated. This will ultimately result in another stage in the digitization process.

Surface patching category strengths are similar to that of the *curve on surface category,* where curves are easy to apply and adjust on the scan mesh. *However,* the weaknesses were also significant. Each curve on surface must have the same number of control points, and each surface boundary must consist of four sides only. As a result, there is no obvious reason to novice users as to why each surface must have a four-sided boundary. The process can also be repetitive and time consuming, particularly if the orthotist requires a high-quality replication of the scan mesh; a higher number of four-sided NURBS surfaces will offer higher accuracy but requires more time in curve construction. The user may also have difficulty constructing a splint with four-sided boundaries, either through forgetting the fundamental limitations or being unable to link up boundaries with four sides to create a smooth splint outline. Another limitation is that the scan data is not used directly; more time is required to gain better scan accuracy, as the user must create more patches. However, the fundamental disadvantage of this category is that the methods are an unnecessary extension to the *curve on surface* category.

5.27.4 Conclusion

After identifying the strengths and weaknesses of each category, the *curve on surface* category would be the most suitable for orthotists to allow the level of creativity that they are familiar with, since it has no weaknesses.

The main limitation of the *import* category is that the organic geometry of every patient is different; with regard to precut splint patterns, "generic patterns rarely fit persons correctly without adjustments" [3]. The same principle applies in this digitized context; having generic patterns/templates with a "one-size-fits-all" approach to manipulate the scan data would not necessarily be a faster alternative to the orthotist. Anthropometric measurements could be taken to adjust the template to a more suitable size, but this would add another stage prior to the splinting process in taking the anthropometric measurements. The *import* geometry may not satisfy the needs of the patient's condition either. The traditional splinting fabrication method allows the orthotist to adapt splints to suit the patient's individual needs and to cater for more than one ailment in some cases. To account for this, a large database would be required with many different splint types with minor differences. In this case, the orthotist would have to search through many import options and may end up selecting an unsuitable splint by mistake. Alternatively, the orthotist may have to adapt existing templates using more advanced CAD tools, which would add time and complexity to the process.

Out of all three categories, *surface patching* would be the most inappropriate, due to the surfacing requirements. If the fundamentals of NURBS surfacing are met, i.e., four-sided surfaces from curves with equal control point count, the reasons for this are likely to be unclear to orthotists. In addition, the greatest weakness to the *patching* category is that it is an extension to the *curve on surface* category. Not only does it take longer to construct the patch boundaries, but another stage is required for surface boundaries. Conformability of the topological detail using patching is also significantly reduced.

Out of the three methods within the *curve on surface* category, *Sketch on Polygon Mesh* would be unsuitable since the curves must be drawn free-hand. Trimming a free-hand curve would create a rough splint edge and could be uncomfortable for the patient. Therefore, the *Trim with Curve* method in Geomagic Studio and *Trim Mesh with Curve* method in Free-Form Modeling Plus remain. As mentioned previously, the main difference between these methods is the haptic hardware required for FreeForm Modeling Plus, allowing the user to "feel" anatomic landmarks. However, the haptic hardware is costly, and the tactile sensation of the PHANTOM stylus may not make splint design easier; it can be assumed that many orthotists are familiar with how to use a computer mouse but have had little to no interaction with haptic hardware. Therefore, if a haptic system were to be implemented, then orthotists may have to dedicate time to learn how to use the haptic equipment effectively, which is an undesirable consequence. After consulting with a senior hand therapist,[1] the favored

---

[1] Ella Donnison, clinical specialist and senior hand therapist, Patterson Medical Ltd (Homecraft Roylan), Nottingham. Interview date March 17, 2011, 5:00 p.m.

method of assigning the splint outline is the *Trim with Curve* tool within Geomagic Studio. The method requires only two actions: to place a curve on the surface using control points and to trim the scan data using that curve, reducing the splinting process from 19 steps to 2 steps in its entirety.

Acknowledgments

Many thanks to Ella Donnison for her time and contribution to the research project and to Loughborough University for the funding and support provided.

References

[1] Coppard B. Anatomical and biomechanical principles of splinting. In: Coppard BM, Lohman H, editors. Introduction to splinting: a clinical-reasoning & problem-solving approach. 2nd ed. St Louis, MO: Mosby, Inc.; 2001. p. 34—72.
[2] Lohman H, Poole SE, Sullivan JL. Clinical reasoning for splint fabrication. In: Coppard BM, Lohman H, editors. Introduction to splinting: a clinical & problem-solving approach. 2nd ed. St Louis, MO: Mosby, Inc.; 2001. p. 103—38.
[3] Coppard BM, Lynn P. Introduction to splinting. In: Coppard BM, Lohman H, editors. Introduction to splinting: a clinical reasoning & problem-solving approach. 2nd ed. St Louis, MO: Mosby, Inc.; 2001. p. 1—33.
[4] Doman C, Rowe P, Tipping L, Turner A, White E. Tools for living. In: Turner E, Foster M, Johnson SE, editors. Occupational therapy and physical dysfunction: principles, skills and practice. 5th ed. Churchill Livingstone, Elsevier; 2005. pp. 165—210-178.
[5] Louise-Bender Pape T, Kim J, Weiner B. The shaping of individual meanings assigned to assistive technology: a review of personal factors. Disability and Rehabilitation 2002;24(1—3):5—20.
[6] Pallari JHP, Dalgarno KW, Woodburn J. Mass customisation of foot orthoses for rheumatoid arthritis using selective laser sintering. IEEE Transactions on Biomedical Engineering 2010;57(7):1750—6.
[7] Lohman H. Wrist immobilisation splints. In: Coppard BM, Lohman H, editors. Introduction to splinting: a clinical reasoning & problem-solving approach. 2nd ed. St Louis, MO: Mosby, Inc.; 2001. p. 139—84.

CHAPTER 5.28

Orthotic applications case study 5—Evaluation of a refined three-dimensional computer-aided design workflow for upper extremity splint design to support additive manufacture*

5.28.1 Introduction

Patients with chronic conditions such as rheumatoid arthritis are often prescribed custom-made static wrist immobilization (SWI) splints (commonly known as cockup splints) (refer back to Figure 5.185 or 5.193 for examples). The intended outcomes of such splints include reduced pain and inflammation, prevention of deformities and contractures, and promoting movement of stiffer joints along with lost motor function [1–4]. SWI splints are designed to enable full dexterity of the digits to allow perseverance in various everyday tasks [2]. There are various fundamental features built into each splint, some of which are bespoke to the patient's needs. For example, cavities are often integrated over pressure-prone areas to avoid forming pressure sores [5]. In addition, the proximal edges of splints are flared for a more comfortable edge against the skin [4].

---

* The work described in this chapter was first reported in the references below and is reproduced here, in part or in full, with the permission of CRDM, Ltd.
Paterson, AM, Bibb RJ and Campbell RI, "Evaluation of a refined three-dimensional Computer Aided Design workflow for upper extremity splint design to support Additive Manufacture." In: Bocking, C, Rennie, AEW, 13th Conference on Rapid Design, Prototyping and Manufacturing, High Wycombe: CRDM Ltd., 2012, pp 61–70.

Medical Modeling
ISBN 978-0-323-95733-5
https://doi.org/10.1016/B978-0-323-95733-5.00033-8

However, evidence suggests that patient adherence is affected by the following.

- **Fit and Function**. If the splint is not fabricated correctly (such as insufficient attention given to pressure-prone areas), the splint can be uncomfortable to wear. Pressure can result in soft tissue damage such as pressure sores, ulceration, and persistent edema [6,7]. As a result, patients may discard their splints [6]. Similarly, if donning and doffing of the splint is difficult or painful for the patient, patients may, again, feel reluctant to wear them. In addition, dissatisfaction occurs due to the fasteners used and their tendencies to snag on clothing [8]. Velcro is commonly used as a fastener for SWI splints. If the splint is worn for the specified amount of time (dictated by their therapist) or worn for rigorous, dirty activities, the Velcro deteriorates in terms of its ability to latch correctly.

- **Aesthetics**. Despite the efforts of practitioners to fabricate splints to a high quality, the functional appearance of splints can affect patients' willingness to wear them in social situations, such as visiting friends and relatives [8]. Pape et al. describe patients' reluctance to use assistive devices within the public domain, since they feel assistive devices "threaten their sense of fitting in" [9]. In addition to Velcro's tendency to snag fabrics, Velcro also attracts and collects other loose items such as hair and fabric fibers within the hooks, reducing the aesthetics further. Co-design of the splint is also limited between the patient and therapist due to the materials available. Patients may have the option to pick the color of Velcro straps and low-temperature thermoplastic (LTT) wherever possible in an attempt to encourage adherence [5,10].

In addition, the fabrication process is laborious and hands-on. There is heavy emphasis within splinting textbooks that prior planning is critical to avoid wasting time and resources; for example, therapists are encouraged to consider padding prior to forming the splint, as delaying padding integration can result in pressure points [5]. This can lead to further discomfort if issued to the patient. Alternatively, the therapist may have to reform the splint to account for padding, but this is costly in terms of labor and clinic time.

Austin [11] uses the acronym PROCESS to describe splint design as a systematic workflow.

1. "Pattern creation"—up to and including evaluation of the patient, their condition(s), and treatment choice. The splint pattern is created by

drawing around a patient's forearm and hand on a piece of paper/towel and noting relevant landmarks (e.g., bony prominences).

2. "Refine pattern"—the fit of the paper pattern is tested on the patient, and adjustments are made where necessary.
3. "Options for materials"—LTT, padding solutions, and suitable fasteners are chosen. At this point, the paper/towel template is transferred onto the chosen LTT.
4. "Cut and heat"—the LTT is placed into a heated water bath until pliable. The LTT is then cut to shape using scissors. Padded features are also applied during this stage.
5. "Molding and Evaluating fit"—The LTT is fitted to the patient, and adjustments are made.
6. "Strapping and components"—fasteners are fixed to the formed splint.
7. "Splint finishing touches"—Austin [11] claims that edge flaring/rolling should be within this phase. However, Lohman [2] states it should be prior to adding fasteners.

In a bid to resolve issues with adherence, an alternative fabrication approach has been investigated to exploit the geometric freedom offered through additive manufacturing (AM). Concept sketches by Bibb [12] (refer back to Figure 5.186) suggest integration of aesthetically pleasing lattice structures that could also potentially improve ventilation to reduce perspiration and its effects (such as maceration). In addition, the approach had scope to encourage co-design between the practitioner and the patient; the patient would be able to personalize their splint by specifying a bespoke pattern to promote user-centered design.

However, to exploit the geometric freedom of AM, geometric manipulation within three-dimensional (3D) computer-aided design (CAD) software is necessary. Introducing mainstream CAD software to splinting practitioners would be an inappropriate solution, since practitioners would need to dedicate a significant amount of time to learn how to use the software effectively. In response to this issue, Rogers et al. [13] and Pallari et al. [14] suggest the need for specialized software for physicians, to allow them to capture their design intent quickly and with minimal training for prosthetic and orthotic fabrication using AM.

Therefore, the aim of this investigation was to devise a way to improve patient adherence by introducing AM for splint fabrication, but more importantly, providing the CAD tools required to achieve this. The objectives required to meet the aim were to do the following.

i. Complete a task analysis of conventional splinting to identify key heuristics displayed in traditional custom-made splint design and fabrication.

ii. Capture the key fabrication characteristics in a virtual environment using suitable CAD packages and tools, by manipulating virtual upper extremity scan data.

iii. Identify the most suitable CAD tools/strategies for new features (i.e., lattices).

iv. Reform the strategies identified in *ii* and *iii* into a logical workflow for practitioners.

v. Translate the workflow devised in *iv* into a specialized CAD software prototype.

vi. Evaluate the software prototype generated in *v* through user trials and interviews.

It is important to note that the proposed approach is not intended to replace the expertise of splinting practitioners, nor is it intended to make their profession obsolete. Instead, the investigation aims to develop a new toolset to empower therapists in future splint design and fabrication. In addition, the scope of this paper and associated PhD research is to address the development of a specialized splinting CAD workflow, with the intent to represent it as specialized 3D CAD software. The authors acknowledge that a suitable data acquisition method is yet to be identified to support the digitized splinting approach within a clinical environment. However, Paterson et al. [15] have explored a range of data acquisition methods to capture test data to support this investigation. Furthermore, this paper focuses on a small yet vital element within a larger explorative study into the macro-digitization of the splinting process, as described by Paterson et al. [16].

5.28.2 Method

The objectives set out on the previous page were addressed in the following manner.

Objective i

Initially, the needs and requirements of practitioners were identified through different media, attendance of a splinting class, clinical observation, and through a literature survey. The intention was to identify key fabrication methods shared among therapists for design and fabrication of SWI splints. In addition, the decision-making process of clinicians was vital to ensure the proposed workflow corresponded with their expectations in the design process and to make the transition from conventional to digital design and fabrication easier. As a result, the devised task analysis workflow of traditional splinting is shown in Figure 5.171 (please refer back to case study 5.25) as a baseline for future development. It became apparent that the main disadvantage to the workflow was the need to repetitively heat the LTT to make adjustments, and that making alterations in the latter stages of the workflow was more costly in time and resources.

Objective ii

Having identified the key features and behaviors of SWI splint fabrication, a variety of CAD tools and approaches were tested to capture the design intent of required features in a virtual environment. Features such as creating the splint outline [16], flaring edges, and adding cavities over bony prominences were investigated. An action research strategy was adopted to plan, test, evaluate, and refine each approach. Testing was performed on STL scan data acquired through indirect 3D laser scanning of an upper extremity plaster cast [15], using a ZCorporation ZScanner 800 3D laser Scanner (3D Systems, 333 Three D Systems Circle, Rock Hill, SC 29730, USA) [17]. A range of 3D CAD software packages were used during testing, as described by Paterson et al. [16].

Objective iii

To facilitate the integration of lattice structures in the workflow, CAD tools and approaches were investigated for lattice pattern repetition. A similar action research strategy to *objective ii* was adopted for exploring lattice integration.

Objective iv

Where applicable, different CAD tools and approaches devised in *objectives ii* and *iii* were compared and evaluated against set criteria, namely being able to capture the initial design intent. Successful CAD strategies were taken forward for workflow refinement. The workflow is shown in Figure 5.189 (please refer back to case study 5.26).

Three key points about the refined workflow were as follows.

- The integration of lattice feature capabilities.
- Discretionary workflow operation. Traditional splinting involves several mandatory stages to make both major and minor adjustments. As a solution, the workflow gives users the ability to alternate to and from

different stages, so they are not restricted should they feel the need to make adjustments.

- The change in "color" and "thickness" option location. Choosing the color and thickness of LTT in traditional splinting is often one of the first decisions to be made. However, there is no opportunity to change the LTT color or thickness further down the process chain. In addition, the color of the LTT was viewed as a minor concern compared with "key features," such as flaring edges. Therefore, it was placed at the end of the refined workflow as a final adjustment.

Objective v

Once the workflow had been refined, qualitative feedback was required from experienced splinting practitioners. The most appropriate data collection method was to introduce therapists to the workflow via an interactive software prototype (Fig. 5.198). The prototype was a robust method to introduce splinting practitioners to features typically found in CAD software, such as a viewport. By making the prototype interactive, direct feedback could be obtained on the effect of the participants' decision-making by viewing the changes in the viewport. The prototype also replicated the likely interface features if the workflow were to be developed into fully functional software. A high-fidelity (hi-fi) throwaway prototype was developed within Microsoft Access, with customized features

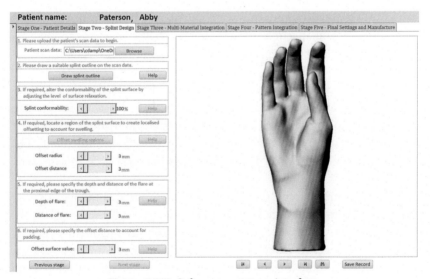

Figure 5.198 Software prototype interface.

created via Visual Basic for Applications. The majority of parameters were adjustable using slider controls. In addition, the pattern integration stage featured a library of different lattices; the concept intended to show that both the therapist and patient could browse for a pattern that would suit the patient's preference.

Objective vi

Ten participants were involved in the evaluation phase. All participants were certified splinting practitioners within the United Kingdom, either within the field of occupational therapy ($n = 8$) or physiotherapy ($n = 2$), with splinting experience ranging from 1 year to over 10 years. None of the participants had prior CAD experience. Participants were given a briefing into the aims and objectives of the investigation, supporting information about the benefits of AM, and the need for specialized CAD software to create AM splints. To explain the capabilities of AM, proof-of-concept splint prototypes were created (Fig. 5.199). Participants were then invited to test the software prototype, followed by a semistructured interview. Probing questions were asked concerning the workflow through direct comparison with traditional splinting methods, as well as the extent of capturing design intent. Voice recordings were used to capture verbal feedback, which were then used to identify trends in opinions. Participants were asked to sign an informed consent form prior to the evaluation session.

Figure 5.199 Proof-of-concept AM splints (left and right, splints kindly manufactured by PDR).

5.28.3 Results and discussion

Despite the fact that none of the participants had previous experience with using 3D CAD software, participants commented on how easy the prototype was to use. All participants were fully engaged with the software prototype, with many suggesting additional creative tools to be integrated, such as multicolor options and lattice perforation motifs in the shape of branded logos. Therapists were in favor of the digitized approach and their ability to alternate to and from different stages in the prototype to make adjustments. This compares favorably to conventional splint fabrication, where the further down the process chain one goes, the more complex and potentially more expensive and time consuming it is to make adjustments. The majority of participants were not concerned by the overall workflow order demonstrated in the prototype; when discussing the change in location of the "color" and "thickness" options, eight participants were not concerned with the change. They felt that "color" was better suited toward the end of the workflow because it was seen as a "final touch" and less important than key features such as edge flaring or cavity integration, which appeared earlier in the workflow. In response to perceived benefits offered through AM, all participants were excited by the prospect of integrating lattice patterns. One participant felt that cartoon character perforations, for example, would be ideal for pediatrics. However, one participant was concerned that issues with adherence might be reversed, to the point where patients would be wearing their splints too much. This could lead to muscle atrophy, so a more stringent wearing regime would be required to avoid this issue. Two participants had concerns regarding window edema as a result of the lattice structures, with a chance that edema could protrude through perforations. On a related note, all participants had concerns over the structural integrity of the splints resulting from material removal. Most participants expressed concerns over initial data acquisition of the upper extremity and the way in which they could capture desired positions without therapist intervention. AM fabrication time and cost of the approach was also a concern.

5.28.4 Conclusions and further work

The introduction of a new splinting toolset was well received by all participants, providing a method to promote co-design between the patient and therapist. In addition, the prospect of decision-making during splint fabrication made the digitized approach a welcoming prospect to the participants, with many suggestions for future feature integration.

In response to the participants' concerns regarding window edema and compromised structural integrity, further research is required into material-to-perforation ratios and the potential for surface displacement mapping instead. Finite element analysis should be investigated for integration within the workflow, either to highlight problem areas or to automatically adjust the geometry to make splints structurally sound. This could be achieved through varying thicknesses, similar to work demonstrated by Rogers et al. [18], or alterations in pattern material-to-perforation ratio. However, it is likely to be a challenging development to map due to the effects of different pathologies on extremity mobility, particularly on an individual basis with unique deformities and range of movement. In addition, a major concern among most participants was the ability to capture the desired posture of the patient's upper extremity, as the most appropriate data collection method in this context is ill defined for clinical applications. Therefore, future research of data acquisition in a clinical context is vital to support the use of AM for upper extremity splint fabrication.

Acknowledgments

Many thanks to Dr. Dominic Eggbeer, Sean Peel, and staff at The National Center for Product Design and Development Research (PDR), Cardiff for manufacturing proof-of-concept stereolithography splints. In addition, our sincerest thanks to Lucia Ramsay (University of Ulster), Ella Donnison (Pulvertaft Hand Clinic, Derby), and all participants for their contribution throughout the investigation.

References

[1] Cooper C. Fundamentals. In: Cooper C, editor. Fundamentals of hand therapy: clinical reasoning and treatment guidelines for common diagnoses of the upper extremity. St. Louis, MO: Mosby Inc., Elsevier Inc.; 2007. p. 3–21.
[2] Lohman H. Wrist immobilisation splints. In: Coppard BM, Lohman H, editors. Introduction to splinting: a clinical reasoning & problem-solving approach. 2nd ed. St Louis, MO: Mosby, Inc.; 2001. p. 139–84.
[3] Schultz-Johnson K. Splinting the wrist: mobilization and protection. Journal of Hand Therapy 1996;9(2):165–77.
[4] Jacobs M, Austin N. Splint fabrication. In: Jacobs M, Austin N, editors. Splinting the hand and upper extremity: principles and process. Baltimore: Lippincott Williams & Wilkins; 2003. p. 98–157.
[5] Coppard BM, Lynn P. Introduction to splinting. In: Coppard BM, Lohman H, editors. Introduction to Splinting: a clinical reasoning & problem-solving approach. 2nd ed. St Louis, MO: Mosby, Inc.; 2001. p. 1–33.

[6] Lohman H, Poole SE, Sullivan JL. Clinical reasoning for splint fabrication. In: Coppard BM, Lohman H, editors. Introduction to splinting: a clinical & problem-solving approach. 2nd ed. St Louis, MO: Mosby, Inc.; 2001. p. 103–38.

[7] Coppard B. Anatomical and biomechanical principles of splinting. In: Coppard BM, Lohman H, editors. Introduction to splinting: a clinical-reasoning & problem-solving approach. 2nd ed. St Louis, MO: Mosby, Inc.; 2001. p. 34–72.

[8] Veehof MM, Taal E, Willems MJ, van de Laar MAFJ. Determinants of the use of wrist working splints in rheumatoid arthritis. Arthritis Care & Research 2008;59(4):531–6.

[9] Louise-Bender Pape T, Kim J, Weiner B'. The shaping of individual meanings assigned to assistive technology: a review of personal factors. Disability & Rehabilitation 2002;24(1–3):5–20.

[10] Austin NM. Equipment and materials. In: Jacobs M, Austin NM, editors. Splinting the hand and upper extremity: principles and process. Baltimore: Lippincott Williams & Wilkins; 2003. p. 73–87.

[11] Austin NM. Process of splinting. In: Jacobs M, Austin NM, editors. Splinting the hand and upper extremity: principles and process. Baltimore: Lippincott Williams & Wilkins; 2003. p. 88–99.

[12] Bibb RJ. Lattice within splints. Drawing; July 22. 2009.

[13] Rogers B, Stephens S, Gitter A, Bosker G, Crawford R, 'Double-Wall. Transtibial prosthetic socket fabricated using selective laser sintering: a case study'. Journal of Prosthetics and Orthotics 2000;12(3):97–100.

[14] Pallari JHP, Dalgarno J, Munguia J, Muraru L, Peeraer L, Telfer S, et al. Design and additive fabrication of foot and ankle-foot orthoses. In: Proceedings of the twenty first annual international solid FreeForm fabrication proceedings - an additive manufacturing conference; August 9-11; 2010. p. 834–45.

[15] Paterson AM, Bibb RJ, Campbell RI. A review of existing anatomical data capture methods to support the mass customisation of wrist splints. Virtual and Physical Prototyping Journal 2010;5(4):201–7.

[16] Paterson AM, Bibb RJ, Campbell RI. Digitisation of the splinting process: development of a CAD strategy to support splint design and fabrication. In: 12th conference on rapid design, prototyping and manufacturing; 17th June 2011. CRDM Ltd; 2011.

[17] 3D Systems. ZScanner® 800. 2012. Available at: http://www.zcorp.com/documents/182_ZScanner800-tearsheet-v05wb.pdf.

[18] Rogers B, Bosker GW, Crawford RH, Faustini MC, Neptune RR, Walden G, et al. Advanced trans-tibial socket fabrication using selective laser sintering. Prosthetics and Orthotics International 2007;31(1):88–100.

CHAPTER 5.29

Orthotic applications case study 6—Design optimization of a thermoplastic splint*

5.29.1 Introduction

The use of orthotics devices postsurgery is an important part of the overall patient recovery process, particularly with respect to cases involving the upper limbs, such as hand/arm amputations [1] and shoulder joint replacements [2]. The orthosis allows for vital load bearing support of the patient's residual appendages while their remaining ligaments/musculature recover [3]. Traditionally, such orthotic devices are custom made, using thermoforming plastics, conforming to the natural anatomy of the patient, providing a configuration to stabilize the residual appendages in a desired orientation for directed recovery [4].

Modern 3D technologies are now allowing for unprecedented accuracy and flexibility for product development, with benefits to be seen in terms of 3D data acquisition [5], how design iterations can be explored and developed, and the final manufacturing of the prototype and functional devices [6,7]. Given the overwhelming potential of this technology as a transformative product development paradigm, we have seen adoption across multiple fields ranging from sports technology, prototyping of commercial products, and with a considerably large uptake for medical applications

---

*The work described in this chapter was first reported in the references below and is reproduced here with permission of the organizing committee of the Solid Freeform Fabrication Symposium.
Fitzpatrick AP, Mohammed MI, Collins P, Gibson I. Design optimisation of a thermoplastic splint. Proceedings of 28th solid freeform fabrication symposium, Austin TX; 2017. p. 2409–2418.

Medical Modeling
ISBN 978-0-323-95733-5
https://doi.org/10.1016/B978-0-323-95733-5.00034-X

[8,9]. More recently, the use of such technology for orthotic development and production is a highly emerging area [10–15], but also it is a largely underexploited area of interest. We predict with the rise in the net global, and in particular aging, populations, such developments will become of growing interest and necessity in the coming decades.

In this study, we aim to develop a streamlined methodology for next-generation orthotics production that leverages the advantages of emerging 3D technologies. As a case study, we examine an existing orthotic device made for a patient who has undergone a partial hand amputation, assess the merits and limitations both in terms of efficacy, ergonomics, and aesthetics, and apply a combination of optical scanning, computer-aided design (CAD), and additive manufacturing to redevelop the part to address these limitations. We discovered optical scanning allows for rapid virtual reproduction of the orthosis (1–2 min). The use of CAD can allow for the incorporation of intricate open structures alongside part thickness tolerance considerations, something not readily achievable using traditional fabrication techniques. Finally, additive manufacturing using Acrylonitrile Butadiene Styrene (ABS) material extrusion (ME) printing readily re-produces the intricacies of the CAD design, producing a robust and superior final orthosis. We believe our findings would provide guidance to orthotists to streamline partial amputation orthotics and potentially provide a range of alternative orthotics for both the upper and lower limbs.

5.29.2 Original orthotic assessment

In the present case study, we examined an orthotic device made for a patient who had undergone a partial amputation of their hand (thumb and index finger). The devised splint was required postsurgery to help support and fix the mobility of the ligaments, nerves, and tendons to aid in the recovery process. Such devices are normally made using a thermoformable plastic, made of cured polyester resin, which has been the industry standard for several decades (Fig. 5.200). The plastic is available in a range of thicknesses (1.6–5.0 mm) and can have small holes regularly spaced within the bulk of the material, with the intention of providing porosity for moisture release.

It is typically found, however, that such orthotic devices made from this material are generally quite uncomfortable for patients, due to both weight and excessive moisture build-up (even when using the perforated version), which causes issues such as chaffing. Prolonged excessive moisture build-up

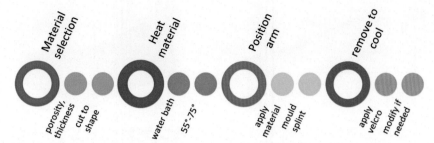

Figure 5.200 A diagram illustrating the typical phases of traditional upper limb orthotic production.

and chaffing can result in ulceration [16]. Additionally, the aesthetics of the device generally makes patients feel highly self-conscious of their condition, resulting in issues with respect to wider societal inclusion. We therefore identified that modifications could be made to the design to reduce the weight (in this case approximately 110 g) to reduce strain on already compromised musculature while introducing design elements, such as open structures, to both minimize moisture retention and improve the overall aesthetics.

5.29.3 Methodology

5.29.3.1 Optical scanning and digitization

The current thermoplastic splint was digitized (3D scanned) using a Kreon Solano 3D laser scanner (Kreon, Belgium), attached to a seven-axis rotatable arm for ease of scanner translation. The scanner uses a red laser line, with a wavelength of 610–630 nm [17], and a triangulation method of measurement. To perform a scan, the laser line is translated over the object at a fixed distance to build up the point cloud data profile for a respective surface. Once an exposed surface has been recorded, the orthosis was reorientated to scan additional surfaces until the entire profile was obtained. The initial scan is shown in Fig. 5.201a i.

The scanner creates a point cloud data file within its proprietary software, which requires postprocessing to remove any outlying data points and any unwanted additional scan data from the surrounding environment. Once postprocessing is complete, the point cloud forms the basis of the final 3D model. After cleaning, the dataset was solidified into an STL (stereolithography) format; see Fig. 5.201a ii. The STL model was measured virtually and compared with measurements taken from the physical model.

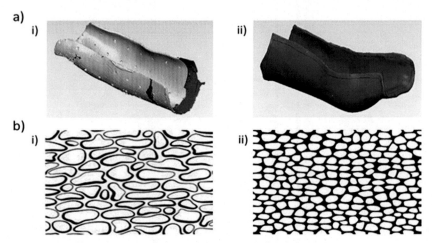

Figure 5.201 (a) (i) The initial scan of the splint and (ii) the resultant cleaned model; (b) (i and ii) design variants chosen by the end user.

The exported STL file was then taken into 3Matic STL (Materialise, Belgium) for further error correction to ensure model integrity for 3D printing. Following this procedure, the virtual model was again checked for dimensional accuracy.

5.29.4 Orthotic design

The exported model, once taken into 3-Matic STL, is in the same orientation as the scan was taken. Designs were developed using the original orthosis topography as the basis. At this stage, the user is able to choose a parametric tile pattern (Fig. 5.201b) that can be applied into the bulk of the digital structure. The pattern is scaled according to strength and aesthetic requirements, and it must be ensured that there is an effective boundary. To create this surface map, a surface contour must be defined. This surface is used to create a patterned map [18]. Using the pattern mapping process, the chosen design can be applied to the surface in 3-Matic STL [11]. Using the applied map, the design can be subtracted from the blank model using the texture function, removing any unwanted sections. As the applied map is an image file, the borders of the pattern edges are jagged; this is due to the subtraction process as well as the tessellation of the model prior to this operation. This is fixed using the surface smooth function in the software.

Next, the model is again checked for accuracy to ensure the design process has not moved any of the external faces that may affect comfort. Following this, finite element analysis is performed on the design to ensure mechanical integrity for the intended application. The variables that could change within the design are the splint thickness and the spar thickness. This could be an iterative process that would involve a dialogue between the designer and the client. This is then verified for 3D printing and exported to the slicing software.

5.29.5 Manufacturing

We opted for polymer-based ME as the preferred additive manufacturing method, which allowed for accurate reproduction of the developed design while also being suitably low cost to serve as a method for mainstream orthosis production. The final design was printed using ABS plastic, due to its low cost and suitable mechanical and biocompatibility properties, and with the orthotic device orientated upright in the Z build direction. Initially, preliminary design iterations were printed using a Zortrax M200 ME printer (Zortrax, Poland). However, due to build volume constraints, the optimum orientation could not be reached, in addition to printed models being relatively weak. It did, however, allow a test fit model to be created at a much lower cost than the final splint. To overcome the mechanical and size constraints, a Fortus 450M (Stratasys, USA) FDM printer was utilized to produce the final orthosis.

5.29.6 Results

5.29.6.1 Virtualization

The time taken to 3D scan the splint accurately was around 45 min. This was because the scanning arm was unwieldy to use. A turntable would have made it easier, but it would not have worked with the current scanner as it uses the arm to register its position rather than the surface topology. The aim was to get the splint scanned in one pass, rather than stitching multiple scans together. Although more time-consuming, this made the result more accurate. The accuracy of the virtual model produced was ±0.1 mm using a scanner resolution of around 16 µm, measured within 3-Matic. The process of digitization was made easier as it was scanning an inanimate object. It would be made inherently more difficult if the scan were of a human subject as any movement would be shown in the scan.

5.29.6.2 Design process

The integration of end user input is an important aspect in customized designs, such as what we hope to achieve with the developed splint design. This is a trend that has been increasing within other fields of engineering [19]. However, this is a concept that has had a hesitant uptake in the medical devices industry [20], but there are indications that this is beginning to change [10]. Having control over the design of the treatment device has been shown to increase the quality of life [21].

To examine facets of user-based design, we selected two arbitrary, yet aesthetically pleasing tile patterns (Fig. 5.201b) from the perspective of the researchers in this study and assessed for viability for the splint. From preliminary design evaluations, the pattern in Fig. 5.201b ii was chosen for the final design as it provided greater strength throughout the splint. Additionally, this design was advantageous as manufacturing required less support material and could be achieved comparatively more quickly.

The mapping process first requires the definition of the surface, leaving a border around the external contours of the virtual splint that remain unmodified. This process prevents open patterns at the splint extremities to ensure its strength, stability, and comfort when worn by a user. In the present design, the border was applied manually to an approximate thickness of 3—5 mm.

The pattern was then applied to the UV map. This was then resized, positioned, and rotated until it was in the correct orientation. Shown in Fig. 5.202a, the 2D projection map of the pattern is shown alongside the wrapped 3D model. This allowed the accurate positioning of the pattern and to ensure the borders had adequate support. The thinking behind the border was to minimize the sectioned holes, as they would be filled in the next step to increase the size of the splint border and to integrate the design seamlessly.

Following the texture application, the model was modified using the 2D to 3D texture function in 3-Matic STL. The function gives the user the options of offsetting the surfaces positively or negatively in the direction normal to the surface. Using the external surface of the splint for the texture mapping, the white sections were offset through the splint, while the blue surfaces remained in a fixed position, as shown in Fig. 5.202b. At this stage, the model was error checked again to ensure no digital anomalies (inverted normal, overlapping triangles, etc.). The resulting splint model is shown in Fig. 5.202b ii.

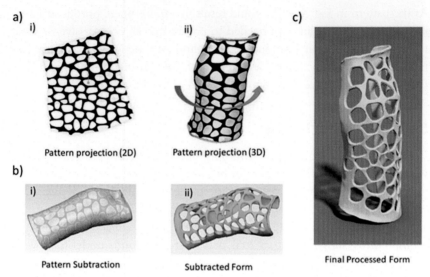

a)

i)

ii)

Pattern projection (2D)   Pattern projection (3D)

b)

i)

ii)

Pattern Subtraction   Subtracted Form

c)

Final Processed Form

Figure 5.202 (a) (i) Section of splint to be patterned converted to a 2D surface and (ii) the projected patterned area converted to 3D. (b) (i) Splint overlaid with a textured model and (ii) the resulting subtracted form. (c) The final postprocessed design.

The accuracy of the model following the design application stays the same as the scanned model. This is because borders and datum points were used to ensure any design variations would not affect the applicability of the model. Any changes that were made to the splint were completed on a duplication of the original scan. This was to assess the accuracy of the model through the design period. This is completed in both 3-Matic STL as well as Magics (Materialise, Belgium), allowing for a smaller file size and the slicing software to quickly process the model. Additive manufacturing allows more complex designs to be manufactured. Using ME to print the model, the design needs to be as self-supporting as possible to minimize the material used in manufacture. Following exporting the STL files, renders were done in Showcase (Autodesk, USA) for color and design approval, shown in Fig. 5.202c. Additive manufacturing allows the production of multiple pieces in a variety of colors, giving the end user the ability to pair the splint to their outfit for example.

The current splint design was constrained by the aesthetic and ergonomic properties of the thermoplastic splint, which performed relatively satisfactorily from a biomechanical standpoint. For this reason, we aimed to retain these aspects such that the overall longitudinal rigidity, rotational stiffness, and latitudinal compressive strength were equivalent in the developed splint.

As the pattern in Fig. 5.201b would result in a more open, porous structure in the bulk of the splint, we needed to ensure that all compressive or shear forces would be even distributed throughout the design. As our primary intension in this study was to demonstrate the design and product realization facets of the splint, we hope to address this in future studies.

5.29.7 Manufacturing

The splint took 23 h to print, with 58 g of ABS. The print was completed with the splint lying flat of the print bed. This was due to height restrictions in the Fortus. Because of the orientation, the splint required support. The weight of support used was 50 g, contributing to a slower build time.

The initial model, printed on the Zortrax, and the final part, completed on the Fortus (shown in Fig. 5.203), acted very differently under load, with the Zortax model fracturing relatively early. This could have been due to the orientation at which it was printed. More work will be completed to characterize each of the designs and the mechanical properties that are affected by design changes.

5.29.8 Discussion and future work

Comparing the thermoplastic and 3D printed splints, some of the notable differences between these approaches are as follows.
- the reduction in weight, with a possible 60% reduction;
- the breathability of the AM splint should be considerably better in comparison to the thermoplastic, due to the open pore design;

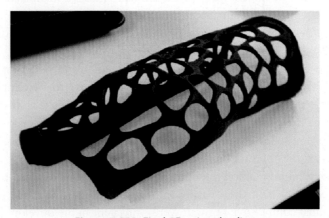

Figure 5.203 Final 3D printed splint.

- the comfort of the splint, due to the reduction in weight and sweat build-up, should be considerably better with the AM splint.

Furthermore, indications are that the splint should chaff less as the design does not appear to move in relation to the body as much. The aesthetic appeal of the splint can be adjusted to suit the user, with even the possibility of having a number of variants to suit aesthetic requirements. Future work planned will take a more quantitative perspective, looking at the engineering properties of the design variants, material properties, stress, fracture load, etc. Being able to characterize the splint properties by varying the design constraints should provide a better quality of life for the end user, as well as to make it easier to manufacture the splints. A variety of designs can be assessed, allowing the end user to have a wider range of patterns that can be easily designed and manufactured while ensuring the correct care is provided.

The major problem to be solved is reducing the overall time to create the orthosis from start to finish. A more effective scanning method is being developed that will scan the upper limbs in real time. We are also working on optimizing the design process to streamline the scan data manipulation while incorporating the aesthetic design components and maintaining mechanical integrity. Finally, we have chosen a Delta configuration ME system [22] that will allow the construction of larger orthoses to suit the upper limb with minimal supports. All of these approaches will considerably speed up the design and build process.

5.29.9 Conclusion

Concluding this study, it was found that the current thermoplastic splint, although effective, does not seem to be a perfect solution. The use of 3D printed splints can enable tailoring to suit the end user's preferences, potentially increasing the quality of life during the treatment process. This can aid in the end user's recovery as well as allowing them to return to normality without the stresses of wearing something with which they are not content.

References

[1] Carl R, Chudnofsky. SEB 2010, 'splinting techniques'. 5th ed. Roberts: clinical procedures in emergency medicine, vol. 5. Saunders/Elsevier. p. 28.

[2] Browner BD, Jupiter JB, Krettek C, Anderson PA. Skeletal trauma E-book. Elsevier Health Sciences; 2014.

[3] *Manufacturing guidelines: upper limb orthoses — physical rehabilitation programme — international Committee of the Red Cross* (ICotR). 2014.

[4] MacDonald K, Chinchalkar SJ, Pipicelli J. Forearm positioning and its functional implications. Journal of Hand Therapy 2016;29(3):376.

[5] Treleaven P, Wells J. 3D body scanning and healthcare applications. Computer 2007;40(7):28—34.

[6] Bagaria V. Technical note: 3D printing and developing patient optimized rehabilitation tools (port) - a technological leap. International Journal of Neurorehabilitation 2015;02(03):1085—9.

[7] Mohammed MI, Fitzpatrick AP, Malyala SK, Gibson I. Customised design and development of patient specific 3D printed whole mandible implant. In: Proceedings of 27th Solid Freeform fabrication symposium. Austin, TX; 2016. p. 1708—17.

[8] Mohammed MI, Fitzpatrick AP, Gibson I. Customised design of a patient specific 3D printed whole mandible implant. KnE Engineering 2017;2(2):104—11.

[9] Mohammed MI, Tatineni J, Cadd B, Peart G, Gibson I. Advanced auricular prosthesis development by 3D modelling and multi-material printing. KnE Engineering 2017;2(2):37.

[10] de Jesus Faria AST. Additive manufacturing of custom-fit orthoses for the upper limb. 2017.

[11] Fitzpatrick AP, Mohammed MI, Collins PK, Gibson I. Design of a patient specific, 3D printed arm cast. KnE Engineering 2017;2(2):135—42.

[12] Kelly SP, A. Bibb R. A review of wrist splint designs for Additive Manufacture. In: Rapid design, prototyping and manufacture conference, Loughbrough, Great Britain; 2015. p. 12.

[13] Lasane A. 3D-Printed Exoskeletal Cast Provides All the Support with None of the Funk. Complex. 2013. https://www.complex.com/style/a/andrew-lasane/jake-evil-3d-printed-exoskeletal-cast.

[14] Paterson A, Bibb RJ, Campbell RI, Bingham GA. Comparing additive manufacturing technologies for customised wrist splints. Rapid Prototyping Journal 2015;21(3):13.

[15] Scott Summit. Wrist fracture. 2007. Available from: 27 June 2017, http://www.summitid.com/#/anais/.

[16] Gregory P, Guyton MD. An analysis of iatrogenic complications from the total contact cast. Foot & Ankle International 2005;26(11):5.

[17] George AF, Al-waisawy S, Wright JT, Jadwisienczak WM, Rahman F. Laserdriven phosphor-converted white light source for solid-state illumination. Applied Optics 2016;55(8):1899—905.

[18] Policarpo b, Oliveira MM, Jo, Comba oLD. Real-time relief mapping on arbitrary polygonal surfaces. In: Paper presented to Proceedings of the 2005 symposium on Interactive 3D graphics and games, Washington, District of Columbia; 2005.

[19] Collins LG. Industry case study: rapid prototype of mountain bike frame section. Virtual and Physical Prototyping; 2016. p. 295—303.

[20] Money AG, Barnett J, Kuljis J, Craven MP, Martin JL, Young T. The role of the user within the medical device design and development process: medical device manufacturers' perspectives. BMC Medical Informatics and Decision Making 2011;11(1):15.

[21] Fayers PM, Machin D. Quality of life: the assessment, analysis and interpretation of patient-reported outcomes. John Wiley & Sons; 2013.

[22] Anzalone GC, Eujin Pei D, Wijnen B, Pearce JM. Multi-material additive and subtractive prosumer digital fabrication with a free and open-source convertible delta RepRap 3-D printer. Rapid Prototyping Journal 2015;21(5):506—19.

CHAPTER 5.30

Orthotic applications case study 7—Design and additive manufacturing of a patient-specific polymer thumb splint concept*

5.30.1 Introduction

The use of splint devices is the traditional benchmark treatment option for patients suffering from a large range of musculoskeletal conditions of the upper limb, both acute and chronic [1,2]. Health professionals, such as occupational therapists, physiotherapists, and orthotists with specialized training use several processes to manufacture custom-made splints designed to meet the clinical requirement of a specific condition. Failure to produce a splint that meets these requirements can impact recovery and can result in long-standing deficits to the patient. Ultimately, reduced function of the upper limb may impact the patient's ability to complete activities of daily living (ADLs), maintain employment, purse leisure activities, and even their participation in the community.

---

* The work described in this chapter was first reported in the references below and is reproduced here with permission of the organizing committee of the Solid Freeform Fabrication Symposium. This research was conducted by the Australian Research Council Industrial Transformation Training Center in Additive Biomanufacturing (IC160100026). Ethics approval for this project was granted by the Deakin University Ethics Committee (HEAG-H 55_2018) following conformity to the requirements of the National Statement on Ethical Conduct in Human Research (2007).
Mohammed MI, Fay P. Design and additive manufacturing of a patient specific polymer thumb splint concept. Proceedings of 29th solid freeform fabrication symposium, Austin, TX; 2018. p. 873–886.

Medical Modeling
ISBN 978-0-323-95733-5
https://doi.org/10.1016/B978-0-323-95733-5.00035-1

The current clinical practice is to immobilize the target anatomy using a splint device. Immobilization is typically achieved using plaster of paris, which can be highly restrictive and uncomfortable for the patients. Splints can comprise either an "off-the-shelf" premade variant or a custom device made from a thermoplastic material. The choice of which is implemented is typically based on the judgment of the healthcare professional. Fabrication of a custom-made upper limb splint can be a complex process by the necessity to tailor requirements to the individuals needs with respects to anatomic variances, aesthetic appearance, and cost. From the practitioner's perspective, additional factors relating to cost, ease of manufacture, and the suitability of available materials to achieve the desired end result play a part in the decision-making process. Unfortunately, even with the provision of custom-made splints, using the current techniques and materials, the rate of nonadherence is relatively high, with comfort, splint appearance, and influence on ADLs being cited as areas of concern [3]. As the devices are also handmade, there is the issue of variability in the device and its effectiveness between different clinics, which further confound this issue. Therefore, the current accepted process for manufacturing upper limb custom-made splints has the opportunity to be challenged, with new technology, design, and processes implements to better meet the needs of the people who require these types of splints.

Additive manufacturing (AM) is an emerging fabrication methodology that allows for the construction of products using a systematic "additive" layer-by-layer build process and in complex configurations that would not be otherwise attainable using traditional manufacturing processes [4,5]. Indeed, AM allows for the relatively straightforward realization of highly complex digitally generated models, with minimal user inputs to the manufacturing process or intimate knowledge of the materials from which they are fabricated. Such versatility in the build process is resulting in a disruption of traditional manufacturing paradigms, with impact in several major industries ranging across the automotive [7−9] to the medical sectors [7−9]. With respect to the medical industry, AM's potency to disrupt existing practices stems for the ability to reproduce even the most complex of human forms in what has been described as patient-specific devices, with practical examples being demonstrated for patient-centric bone replacement implants [8], prosthetics [10], and rehabilitation/splint-based devices [11−14]. The potential of this technology is enhanced further by the alignment of technologies to digitally reproduce the complexity of the human form rapidly, with demonstrations using both optical surface scanning [15,16] and medical imaging [7,10] data sets.

More recently, AM technologies have been utilized toward the development and production of custom splint devices [12,15,17—19]. Researchers are aiming to leverage the advantages found in noncontact data digitization, advanced computer-aided design (CAD), and AM to realize more patient-centric designs that overcome limitations of current devices and provide a less intrusive experience for the patient during the orthosis creation. With respect to current limitations of current splint devices, insights have revealed a range of factors relating to the nonadherence of wearing an orthosis [3,8]. Factors include difficulties keeping splints clean/dry, poor aesthetical qualities, discomfort due to poorly fitting devices, compromised ability to perform routine tasks, and odor-related issues. It is believed that the versatility of CAD and AM can be applied to address these issues, realizing more ergonomic and aesthetically appealing designs, tailored to the individual's unique needs and priorities.

In this study, we examine the use of high-resolution optical surface scanning, CAD, and AM to develop a new patient-centric thumb splint device. We initially use optical scanning to rapidly capture a person's hand anatomy while they maintain the position for the final form of the thumb splint. We then convert the captured data into a three-dimensional digital model, which is used as the template to form the splint device. We examine the ideal tolerances of the device alongside design features to ensure ease of fit while still maintaining potential therapeutic effectiveness desired by a standard thumb orthosis. An open, porous design is also examined to create an aesthetically appealing device, which more readily allows for the release of trapped moisture on the skin compared to a tradition splint [12,19—21]. Once a final design was derived, the model was manufactured using fused filament fabrication (FFF) in acrylonitrile butadiene styrene (ABS) polymer to assess the final device and its potential for use in place of a traditional thumb splint. Overall, we believe we have developed a robust methodology for thumb orthosis design and manufacture in a form that is more ergonomic and provides better fit and aesthetic qualities.

5.30.2 Existing thumb splint fabrication

The short thumb spica splint is a commonly used splint that can be used across a number of musculoskeletal conditions including acute ligament repair, bone fracture, and osteoarthritis of the thumb. This splint was specifically chosen to challenge the technology being studied as the desired immobilization position of the hand is one of the more challenging orientations to achieve in a respective device. The thumb position for the splint is vital to ensure the correct anatomic position is achieved, while also allowing functional use of the hand by maintaining freedom of the meta-carpophalangeal joints of the fingers and interphalangeal joint of the thumb, while not restricting the natural action of the wrist. A low-temperature thermoplastic (Aquaplast, OPC Health, Australia) is used to manufacture this splint, a material that is both strong and flexible enough to meet the requirements of this type of design, and it is the typical material used by the orthotist assisting in this study. An overview of the fabrication workflow can be seen in Fig. 5.204.

In this study, we asked a local orthotist to design a thumb splint for the volunteer in this study, which would act as a comparison for the device we will develop in this study. The technique described is a relatively typical example of how a thermoplastic thumb splint is created. The initial stage of the manufacturing process requires the health professional to create a template of the hand, and this is commonly done by tracing an outline of the hand on paper/card and marking significant anatomic landmarks, such as the fingers, thumb, etc. A preliminary splint design is then traced over this template, and if the health professional is happy with the design, this is

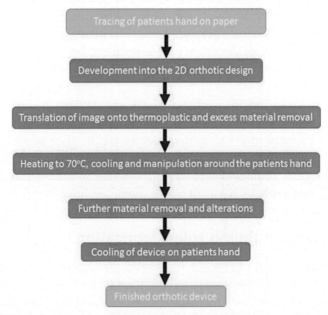

Figure 5.204 A diagram of the typical workflow performed by an occupational therapist to design and fabricate a thumb splint.

transferred onto the thermoplastic. The thermoplastic is cut to size and then heated in a hydroculator (heat bath) to the specified temperature of the thermoplastic, which in this study was approximately 70°C. The health professional then places the patient's upper limb into the required position ensuring that the thumb, fingers, and wrist are in the correct orientation. This position is often held by the patient for several minutes while the health professional removes the thermoplastic from the water, lets it cool to a safe temperature, and then places this directly onto the patient to mold the splint to the patient's hand. This process relies on the skill and knowledge of the health professional to be able to mold the splint as required, in addition to the ability of the patient to maintain the desired position. Once the thermoplastic has cooled, it begins to harden and can be removed from the patient for any additional alterations. Alterations can include cutting away of excess material, addition of straps to secure the splint, or any other modifications as required. If further alterations are required, the thermoplastic can be reheated, but there is a risk it may lose its shape.

5.30.3 Methodology

The primary elements of this study comprise the acquisition of a person's data using optical scanning, construction of a 3D anatomic model of the hand, the development of the orthosis from this scan data, the manufacturing of the device, and the final evaluation. A summary of a typical workflow can be seen in Fig. 5.204.

5.30.3.1 Optical scanning

Surface topography scans were obtained using a light reflectance scanning system (Spider, Artec, Luxembourg), which is minimally invasive and utilizes visible light, meaning scans can be performed with no health risks to the person being scanned. The scanner has a scan resolution of approximately 50–100 µm, which allows for both the major anatomic form and minor details such as creases in the skins surface to be resolved. In this study, the scanner was set to a resolution of 500 µm, which was considered suitable for the level of detailed required to be captured. The scanner is designed to work alongside a propriety software package (Artec Studio 10, Artec, Luxembourg), which streamlines the data acquisition phase and allows for real-time visualization of the capture process.

To capture a surface of an object, the scanner is activated and held as a constant distance of approximately 50 cm from the object, before being

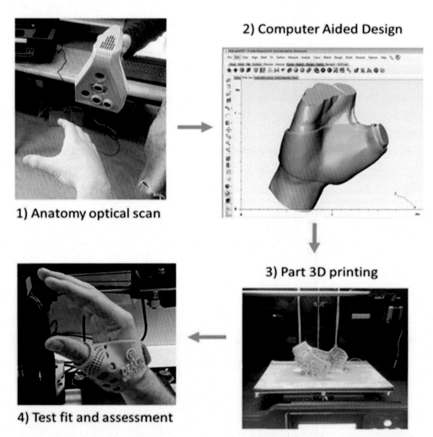

1) Anatomy optical scan

2) Computer Aided Design

3) Part 3D printing

4) Test fit and assessment

Figure 5.205 A diagram illustrating the workflow from data capture to final production of the thumb splint device.

translated around the object to build up a surface profile (Fig. 5.205-1). Care must be taken to not translate the scanner too quickly (>10 cm/s) or to move too far away from the optimal imaging distance or the scanner loses tracking of the object. A typical scan procedure can be completed within 3–5 min. The scanner software also allows for the data processing of the scanned surfaces to remove spurious data points, noise, and unwanted scan data. The software is also capable of rudimentary smoothing and fixing processes to create an enclosed 3D model, but generally it does so with some degree of data loss of finer details. Therefore, such operations were performed independently using CAD software to retain the model details.

5.30.3.2 Computer-aided design

The scanner data was postprocessed using the CAD software 3-Matic STL 10.0 (Materialise, Belgium). Initially, the model was checked for digital errors before some minor smoothing operations to improve the consistency of the data (Fig. 5.205-2). Following this, the model was remeshed to improve the consistency of the overall mesh before the data were suitable for construction of the thumb splint. The splint is realized using an approach of applying multiple surface projections to create hand models that were 0.5 and 3.5 mm thicker than the original hand model. The hollow form of the model was created using Boolean operations, followed by selective trimming to create the basic form of the model. In this study, we examined an approach comprising a single-piece splint design, which can be strapped to the person using a single piece of Velcro. This matches the typical configuration of traditional thumb splint device.

Beyond formation of the general shape of the device, we examined a technique of projecting a two-dimensional pattern on the surface of the model, before extruding a three-dimensional shape into the segments of the pattern as we have described previously [12,19]. This variant of design was realized using 3-Matic STL (v10, Materialise, Belgium).

5.30.3.3 Additive manufacturing

Once a device design had been finalized, the resulting models were 3D printed to create the final thumb splint model (Fig. 5.205-3). Designs were manufactured using FFF using a Flash Forge Creator Pro (FlashForge Corporation, Zhejiang, China) in standard ABS material, which provides adequate material stiffness to manufacture functional models. When printing, a hot end temperature of 232°C, bed temperature of 105°C, and layer resolution of 0.2 mm were employed. When printing, parts were oriented in the z-direction (Fig. 5.204) to ensure the parts were printed with minimal support material and achieved the best cosmetic finish with minimal surface roughness (Fig. 5.205-4).

5.30.4 Results

5.30.4.1 Patient-specific data acquisition

The primary intention of this study was to develop a methodology for streamlined data capture and model development, so scanning procedures were performed on a healthy individual who suffered no complications with respect to the thumb and its mobility. For therapeutic effectiveness, the correct orientation for immobilization of the thumb is in a position that was described by the orthotist as the shape the hand would make "holding a soft drink can." A picture of the hand position can be seen in Fig. 5.205-1, where the fingers are curved inward in an arc shape while the thumb holds a position parallel to the fingers, equally with an arc shape. The volunteer was asked to hold this position with their arm partially extended to allow for easy access of the scanner. Once in position the scanner was translated through various orientations, at a fixed distance, to construct the various surface profiles of the volunteer's hand. It was found that the majority of the hand could be easily resolved with a single translation of the scanner. However, additional adjustment of the scanner capture acquisition time and image saturation had to be suitably adjusted to ensure reflection-based aberrations were reduced to produce a well-defined image for capture. It was also found that the inner contours of the hand proved to be more challenging to scan due to scan acquisition issues and the awkward positions that the scanner had to be rotated into to capture the lower portion of the hand. In this instance, we believe scanning issues stemmed from interference of the reflected light due to natural shadows of the hand or reflectance of the light away from the scanner detectors. To compensate for these issues, several adjustments were made to the scanner to perform scan using the "fast fusion" function, which disregarded skin pigmentation data, as only topologic data were required.

Following optimal scanner setup, the data acquisition phase was not without its own challenges. Firstly, minor movements of the volunteer's hand resulted in alignment discrepancies during the postscan image processing. For the most part, this was a minor issue as the scanner software was found to adequately compensate for this. Secondly, and perhaps more importantly, there was a certain degree of discomfort described by the volunteer in having to keep their hand and arm still for the duration of the scanning while we were optimizing the best technique for data acquisition. Once our scanning techniques had been optimized, a typical scan could be performed within the space of 2–3 min. We therefore acknowledge that should this technique be implemented in real circumstances, there will no doubt be a period of user training to refine optimal scanning technique by a practitioner. Comparing this to traditional methods of capturing a person's anatomic data using thermal polymer molding, the examined technique is both noninvasive, quicker, and potentially provides greater accuracy in reproduction. We hope to further qualify these differences in further studies. Ultimately, despite both techniques suffering some limitations, we concluded that the scanning approach was the superior of the two techniques.

Fig. 5.206a illustrates segments of the scan data captured of the volunteer's hand, where it was found that the final model could be adequately constructed from three separate scan data segments. It was also found that due to complications with data acquisition, the final form could not be adequately resolved to the point where an enclosed 3D model could be constructed, so final finishing of the surface data was performed manually using CAD software.

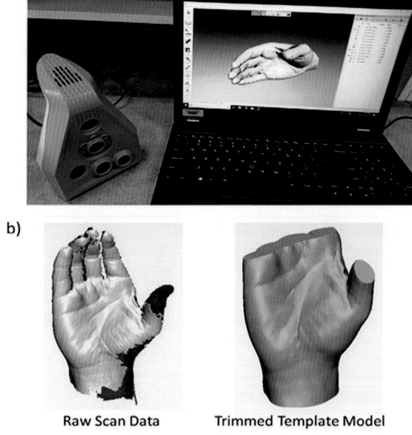

a)

b)

Raw Scan Data   Trimmed Template Model

Figure 5.206 (a) A picture of the scanner and image data processing equipment illustrating segments of the hand surface being stitched together and (b) stages of the raw data from the scanner software to the final closed 3D model template data used for splint construction.

5.30.4.2 Orthosis design

Initially the surface data of the volunteer was manipulated to completed regions that were unable to be captured during the scanning phase. This was achieved in the 3-Matic software using a manual triangulation bridging, where regions of the model mesh were manually reconstructed to complete the form of the model. Fig. 5.206b shows the compiled raw surface data that were output from the Artec software alongside the final, trimmed 3D model of the hand.

As a primary objective in this study was to create a more patient-oriented splint, we use the model as the template to form the device. When manufacturing the splint, it is desirable to obtain a design that has adequate stiffness to resist the typical forces the patient may impose on the device to avoid breakages. Feasibly, the part could be deliberately made to larger thicknesses to increase the overall stiffness, thereby avoiding breakages. However, it is desirable to achieve a compromise, so the part is made as light and as thin as possible to increase the user's comfort when wearing the device. From preliminary in-house testing, a thickness of 3 mm was considered the ideal minimum thickness so was used as the thickness tolerance for the final splint design. We hope to provide a more comprehensive evaluations of the stiffness characteristics resulting from varying part thicknesses in future studies.

The initial stages of forming the hand splint comprised the trimming down of the hand model to form a template of how the splint will be located on the hand. To guide the trimming process, the distal palmar and interphalangeal creases of the hand were used, where a clearance of 10 mm from the distal palmar and 5 mm from the interphalangeal creases were used to ensure that the potential device would not restrict movement of the upper portion of the thumb, fingers, and bending about the distal palmar. The lower portion of the splint was trimmed approximately 10 mm above the crease of the wrist to ensure that when the splint was worn mobility was not compromised. From these template data, the surface of the models was extruded to 0.5 mm and a second model to 3.5 mm, before a Boolean subtraction was performed to create a hollowed form of the splint. Then a region (65−70 mm long and 25−35 mm wide) relative to the back of the hand was trimmed, and two recess holes were created for the placement of a Velcro strap. An image of the final template concept can be seen in Fig. 5.207a. The digital template was now ready for application of the porous pattern and additional design stylizations.

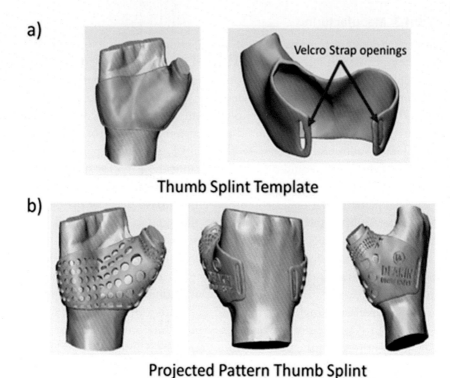

Figure 5.207 Various designs phases illustrating (a) the thumb splint concept template and (b) the final design of the projected pattern style splints.

To construct porous design, an approach using a 2D projected pattern was applied, similar to what has been described previously [12]. Firstly, the model is segmented to leave a border of 5 mm around the edges of the model, alongside a patch around the back of the thumb. This border ensured that the porous pattern would not be applied around the edges of the model, which could result in stress concentration points, in addition to points that could potentially press into the wearer's skin. The region behind the thumb was to act as a platform onto which we would incorporate an example of a custom design feature to demonstrate the mass customization potential of such splints. In this instance, we incorporated the Deakin University logo onto the thumb model, as can be seen in Fig. 5.207b. For the remaining portion of the splint, a 2D pattern was projected onto the surface of the template and adjusted to give a complex, graded size to the pattern, further exemplifying the versatility of this approach. Once the projection was achieved, a generic circle pattern was then extruded throughout this pattern, resulting in the final design found in Fig. 5.207b.

5.30.4.3 Manufacturing

The splint design was printed in ABS plastic using 100% infill to ensure the most robust part possible and printed using a raft material to optimize print bed adhesion. As the part is to be worn, keeping the manufacturing surface roughness as low as possible was desirable to maximize comfort for the wearer. It was recognized that due to the complex form of the orthosis that support material would be required to ensure the integrity of the model during the build process. Despite this, initial tests were performed to examine if the design could be printed without the need for support material to both reduce the need for additional postprocess and to reduce additional material that could compromise the surface finish of the part, as the printer utilized the print material as the support material. All such printing attempts failed, so it was deemed necessary to use supports.

It was unknown what configuration of support material would be ideal, so two commonly used configurations were tested comprising "rectangular" and "tree-like" branching types. Our settings for printing supports comprised overhang angles less than 40° and with a structure thickness of 3 mm. Both configurations provided adequate support to reproduce the part in its entirety. Of the two structures, the tree-like supports minimized the contact area with the part and were found to the easiest to remove so were considered the preferred choice of supports. It was found that in some instances, when the supports were removed, there was residual material left behind on the model. To remedy this, a craft knife was used to scrape the bulk of the material off, followed by a light sanding using 120-grit sandpaper, until a smooth finish was achieved. Following this methodology, a part was ready for use within 3—5 min following printing. Fig. 5.208 shows the 3D printed part as it appeared on the printer, following postprocessing and in its final form containing the Velcro strap and placed upon the wearer's hand. As a comparison, we fabricated a thermoplastic splint using traditional methodologies, which can be seen next to the developed 3D printed splint in Fig. 5.208c. Superficially, the shapes of both matched one another, resulting in the hand of the wearer being oriented in approximately the same position. From a qualitative external inspection, it can be seen that the 3D printed splint only uses the material necessary for production. By contrast the traditional splint contains sizable amounts of excess material folded over the outer surfaces around the opening for the thumb and the area in contact with the palm of the hand. Additional design

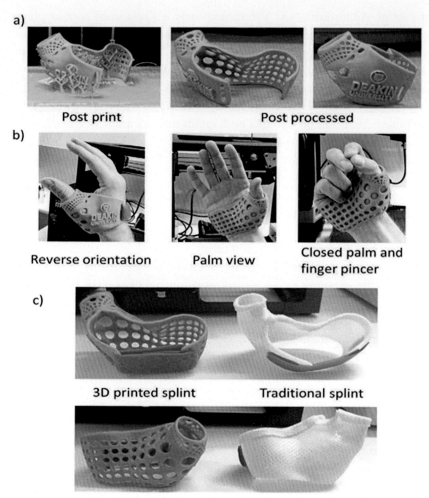

Figure 5.208 Images of the manufactured splint designs (a) postprint and post-processed, (b) being worn and used by the volunteer, and (c) compared with a traditionally designed splint.

benefits comparing the devices in this study are that the 3D printed design incorporates openings for the Velcro strap, which contrast with the traditional splint that required adhesive Velcro pads to incorporate the strap. Indeed, the Velcro fastening mechanism highlights additional limitations to the existing design. Firstly, the adhesive pads for the Velcro are often prone to loosing adhesion when exposed repeatedly to moisture, causing failure of the splint. Additionally, the Velcro strap only has two small contact points to maintain the tension of the device. By contrast the 3D printed splint, with the Velcro strapping round the person's hand does not require any adhesive pads and can accommodate a much larger region of Velcro hooks to keep the strap in place. Ultimately, this leads to a much firmer fit that is less prone to failure. Further comparisons of the two designs can be found in Table 5.15. It can be seen that both splints have identical weights, but the new splint offers greater stiffness and a 3−12 factor increase in porosity, meaning a greater capacity for moisture release. In principle, the higher the porosity is, the greater is the capacity for a splint to release moisture, neglecting a material's moisture-wicking capability. We hope in future work to develop concepts to investigate elements of moisture release in greater detail. It is worth noting that a negative to the 3D printing process is the fact that the splint took approximately eight to nine times longer to manufacture, which is still acceptable but less than desirable with respect to it potentially meaning an additional visitation by a potential patient. We believe that as printing technology develops, the turnaround time for a part will begin to decrease, so this limitation may be overcome in the near future for FFF printing technology. Alternatively, new forms of printing could be employed, such as continuous liquid interface production, which possesses greater print resolution, with vastly improved speed of printing, potentially reducing production time by a factor of up to 50.

5.30.4.4 Qualitative assessment

In a recent review conducted by Kelly et al. 2015, various user-based metrics were described relating to the adherence of wearing an upper body splint. Factors included difficulties in maintaining cleanliness and hygiene, aesthetics, comfort, and ergonomics. However, such factors are arguably qualitative in nature and relate to the individual user's experience and perception of the device. Therefore, performance metrics relating to such devices can only be qualitative in nature. In an attempt to assign metrics to assess such factors with respect to splints, research groups have taken an approach of user-centered experience questionnaires to either validate designs or provide suggestions for improvements [22]. In this study, we apply a similar methodology to assess the effectiveness and ergonomics of the thumb splint. We asked the volunteer of the study to wear each

Table 5.15 Comparison of the various thumb splint manufacturing times and properties taking into account data capture, design, manufacturing, and postprocessing times.

Design	Manufacturing time (h)	Weight (g)	Approximate porosity (%)
Traditional splint	1	30	2—8
Hole-patterned splint	8—9	30	24

device for a period of 2—3 min and perform tasks to bend the palm of the hand, attempt to move their thumb in a circular motion, bend their thumb, rotate the wrist, and finally to perform a pincer action with the thumb and fingers, as can be seen in Fig. 5.208b. Once motions had been completed, we asked the volunteer to rank each device on a scale of 1—10 (1 being low and 10 being high) for a series of user-centered questions. These included the following.

(Q1) How do you rate the aesthetic qualities of the splint?

(Q2) Would you feel happy wearing this device in public?

(Q3) How comfortable is the device to wear?

(Q4) How stiff would you say the device feels when worn?

(Q5) How much mobility of your hand would you say you have remaining to perform everyday tasks?

(Q6) How would you rate the device for ease of cleaning your hand and for release of moisture while being worn?

Following the questions, we obtained a score for each device with a maximum of 60. Based on these findings, the highest scoring device would be considered optimal. Results for each question are found in Table 5.16, where it can be seen the 3D printed splint scored highest and is therefore considered optimal in this study. It is noted that the results are highly subjective and only based on the single user analysis, so findings will no doubt vary between larger cohorts of users. We hope to develop these initial questions, suggested hand motion evaluations of the participant, wear time and user metrics, and to apply them to a larger cohort size in future studies. However, these very preliminary findings hint at the potential of AM devices to offer user-evaluated improvements over traditional devices.

Table 5.16 User-based qualitative assessment of the various splints.

Design	Q1	Q2	Q3	Q4	Q5	Q5	Total score
Traditional splint	3	5	6	4	6	2	26
Projected patterned splint	7	9	8	8	8	6	46

5.30.5 Conclusion

Findings from this study have proven the efficacy of using noninvasive optical scanning to readily derive patient-specific anatomy. Following a brief period to refine our scanning techniques, the digitized data can easily be acquired using the Artec system, converted into a 3D hand model, and be used as template from which to form a patient-specific thumb splint concept. We have found there to be considerable scope for design freedom, and we demonstrate the efficacy of creating a design incorporating a generic, porous pattern. The design was evaluated by a trial user and stated to provide an excellent fit, while being both an aesthetically pleasing and ergonomic configuration to allow for ease of hand mobility. We note that the user evaluation is largely opinion based, so to affirm the efficacy of the splint, a significantly larger cohort study (>25 people) more accurately must be performed and done so over extended periods of user wear time. It is also noted that the turnaround time for fabrication of the new device is several hours longer than for the traditional device. We believe that this challenge will be addressed over time as printing technology develops. However, more importantly, the improvements AM offers may lead to greater user adherences of such devices, which ultimately would outweigh the negatives of longer fabrication time with regard to patient outcomes. We hope to better understand this facet in future studies and encourage other groups to explore the potential of AM for upper limb splints toward improved patient adherence.

References

[1] Ahern M, Skyllas J, Wajon A, Hush J. The effectiveness of physical therapies for patients with base of thumb osteoarthritis: systematic review and meta-analysis. Musculoskeletal Science and Practice 2018;35:46—54.

[2] Benjamin HJ, Hang BT. Common acute upper extremity injuries in sports. Clinical Pediatric Emergency Medicine 2007;8:15—30.

[3] O'Brien L. Adherence to therapeutic splint wear in adults with acute upper limb injuries: a systematic review. Hand Therapy 2010;15:3—12.

[4] Mohammed MI, Gibson I. Design of three-dimensional, triply periodic unit cell scaffold structures for additive manufacturing. Journal of Mechanical Design 2018;140:071701.

[5] Gibson I, Rosen D, Stucker B. Additive manufacturing technologies: 3D printing, rapid prototyping, and direct digital manufacturing. 2 ed. New York: Springer-Verlag; 2015.

[6] Richardson M, Haylock B. Designer/maker: the rise of additive manufacturing, domestic-scale production and the possible implications for the automotive industry. Computer-Aided Design & Applications PACE 2012;2:33—48.

[7] Mohammed MI, Ridgway MG, Gibson I. Development of virtual surgical planning models and a patient specific surgical resection guide for treatment of a distal radius osteosarcoma using medical 3D modelling and additive manufacturing processes. In: Presented at the proceedings of the 28th solid freeform fabrication symposium, Austin, TX; 2017.

[8] Wang X, Xu S, Zhou S, Xu W, Leary M, Choong P, et al. Topological design and additive manufacturing of porous metals for bone scaffolds and orthopaedic implants: a review. Biomaterials 2016;83:127—41.

[9] Mitsouras D, Liacouras P, Imanzadeh A, Giannopoulos AA, Cai T, Kumamaru KK, et al. Medical 3D printing for the radiologist. RadioGraphics 2015;35:1965—88.

[10] Mohammed MI, Cadd B, Peart G, Gibson I. Augmented patient-specific facial prosthesis production using medical imaging modelling and 3D printing technologies for improved patient outcomes. Virtual and Physical Prototyping 2018;13:164—76.

[11] David P, Jiri R, Daniel K, Pavel S, Tomas N. Pilot study of the wrist orthosis design process. Rapid Prototyping Journal 2014;20:27—32.

[12] Fitzpatrick A, Mohammed M, Collins P, Gibson I. Design optimisation of a thermoplastic splint. In: Proceedings of the 28th solid freeform fabrication symposium, Austin, TX; 2017. p. 2409—18.

[13] Paterson A, Bibb R, Campbell R. "Orthotic rehabilitation applications," Medical modelling: the application of advanced design and rapid prototyping techniques in medicine. New York: Woodhead Publishing; 2014. p. 283.

[14] Portnova AA, Mukherjee G, Peters KM, Yamane A, Steele KM. Design of a 3D-printed, open-source wrist-driven orthosis for individuals with spinal cord injury. PLoS One 2018;13:E0193106.

[15] Volonghi P, Baronio G, Signoroni A. 3D scanning and geometry processing techniques for customised hand orthotics: an experimental assessment. Virtual and Physical Prototyping 2018;13:105—16.

[16] Pilley MJ, Hitchens C, Rose G, Alexander S, Wimpenny DI. The use of non-contact structured light scanning in burns pressure splint construction. Burns 2011;37:1168—73.

[17] Paterson AMJ, Bibb RJ, Campbell RI. A review of existing anatomical data capture methods to support the mass customisation of wrist splints. Virtual and Physical Prototyping 2010;5:201—7.

[18] Kelly S, Paterson A, Bibb RJ. A review of wrist splint designs for additive manufacture. In: Presented at the proceedings of 2015 14th rapid design, prototyping and manufacture conference (RDPM 14). Great Britain: Loughborough; 2015.

[19] Fitzpatrick AP, Mohanned MI, Collins PK, Gibson I. Design of a patient specific, 3D printed arm cast. KnE Engineering 2017;2:135—42.

[20] Paterson AM, Donnison E, Bibb RJ, Ian Campbell R. Computer-aided design to support fabrication of wrist splints using 3D printing: a feasibility study. Hand Therapy 2014;19:102—13.

[21] Paterson AM, Bibb R, Campbell I, Bingham G. Comparing additive manufacturing technologies for customised wrist splints. Rapid Prototyping Journal 2015;21:230—43.

[22] van der Wilk D, Hijmans JM, Postema K, Verkerke GJ. A user-centered qualitative study on experiences with ankle-foot orthoses and suggestions for improved design. Prosthetics and Orthotics International 2018;42:121—8.

CHAPTER 5.31

Orthotic applications case study 8—Digital design and fabrication of a controlled porosity, personalized lower limb ankle foot orthosis[a]

5.31.1 Introduction

Ankle foot orthosis (AFO) is a classification of externally worn medical devices that support the lower limbs of a person to treat an underlying physical impairment, from conditions such as compromised musculature, drop foot, stroke, and cerebral palsy [1]. Often users experience poor fit of an AFO, complaining of pain and discomfort, the aesthetics of a device, and chaffing due to inadequate moisture release. This results in the user not wearing the AFO for the prescribed daily durations, thereby reducing the therapeutic usefulness. To improve user conformity, devices ideally need to

---

[a] The work described in this chapter was first reported in the references below and is an open access article distributed under the terms of the Creative Commons Attribution License (http://creativecommons.org/licenses/by/4.0), which permits unrestricted use, distribution, and reproduction in any medium, provided the original work is properly cited. The authors would like to thank Daniel Wilson for his assistance with flexural tests and Ben McMurtrie (Geelong Orthotics) for his insights on commercially used ankle foot orthoses. The research related to human use complies with all the relevant national regulations and institutional policies and has been approved by the authors' institutional review board or equivalent committee.

Mohammed M I and Elmo F "Digital design and fabrication of controlled porosity, personalized lower limb AFO splints" Transactions on Additive Manufacturing Meets Medicine, Vol. 2, Issue. 1, 2020, Article ID 013. DOI: 10.18416/AMMM.2020.2009013.

Medical Modeling
ISBN 978-0-323-95733-5
https://doi.org/10.1016/B978-0-323-95733-5.00036-3

be more ergonomic, incorporate great levels of airflow, and have the flexibility to incorporate custom design aesthetics.

Traditionally, AFO devices are handcrafted using either plaster of paris or thermoplastic materials using time-consuming methods, which require several patient visitations. Digital scanning, computer-aided design (CAD), and additive manufacturing (AM) workflows are emerging as an advantageous alternative to traditional fabrication methods. This approach allows considerable potential for AFO production, as a patient's anatomy can be recorded precisely digitally using surface scanning and stored for future use [2]. Equally, design iterations are assessed virtually, reducing the potential number of patient visitations. Finally, the use of AM provides a considerable number of design freedoms to realize complex patient-specific device geometries, localized mechanical properties, and unique aesthetic qualities matching patient preferences [3]. Indeed, such technologies have been demonstrated effectively for upper limb splints [2,3], and more recently, attention has been moving to focus on applications for the lower limb [1].

We examine the use of digital design and fabrication toward the construction of functionally personalized AFO for drop foot. Ideally, devices need a balance of stiffness, flexibility, and porosity while minimizing thickness to create a discrete AFO that can be worn easily within a person's existing footwear. Most studies to date focus on design elements, we now investigate how the flexural mechanical properties evolve based upon the introduction of open, porous design features. We conducted and compared standard three-point flexural bend tests on various 3D printed samples with varying porosity against commercial splint thermoplastic material. We then assessed the feasibility of designing these porous features into a drop foot AFO concept based upon using a participant's anatomic data.

5.31.2 Material and methods

5.31.2.1 Flexural tests

Flexural tests were conducted using three-point bend tests as stipulated in ASTM standard D790 for flexural testing. This comprised five test repeats using rectangular test coupons with dimensions $3.2 \times 12.7 \times 127$ mm. All flexural tests were performed using an Instron 5960 Dual Column 50 kN universal test system (MA, USA), set a with strain rate of 0.01 (mm/mm)/minute.

5.31.2.2 AFO design and manufacturing

Person-specific data were obtained scanning a volunteer's foot placed at approximately 5 degrees of dorsiflexion, which mimics preloading. Scan data were obtained using an Artec Spider scanner and postprocessed using Artec Studio 10 software (Artec, Luxembourg), producing a template STL file for further design manipulation, as previously described [2]. AFO designs were creating using 3-Matic STL software (Materialise, Belgium) using the individual's scan data as a template to create the final concept. The AFO was 3D printed using a Creator Pro FFF system (FlashForge, China) in ABS polymer, using a 0.2 mm layer thickness, print temperature of 235°C, print bed temperature of 100°C, and print speed of 60 mm/s. AFO bindings were realized using Velcro straps that were slotted into two recesses, designed into the AFO.

5.31.3 Results and discussion

Several porous flexural samples were examined to determine the flexural stress strain relationship or configurations that could be incorporated into the AFO design. Fig. 5.209a) illustrates the sample porosities examined for regularly spaced circular pores with a diameter of 7.5 mm. Fig. 5.209b illustrates the measured flexural stress for increasing porosity of samples, alongside a measurement using standard 3.2-mm AFO thermoplastic materials, for reference. It was found that the ABS material, even at the highest porosity examined (38%), was found to be approximately twice the flexural stiffness than standard thermoplastic material. This is a favorable result as increased porosity reduces mass and increases the moisture release capacity of a respective design. Additionally, initial results suggest we could use ABS material at porosities >38% and maintain an appropriate stiffness. We hope to investigate this further in future studies.

Fig. 5.210 illustrates a basic overview of the digital manufacturing process elements. The basic form of the AFO was created through various Boolean subtraction, digital trimming, and smoothing operations, before

a)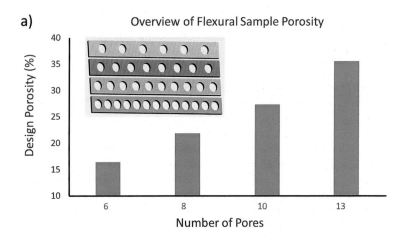

b)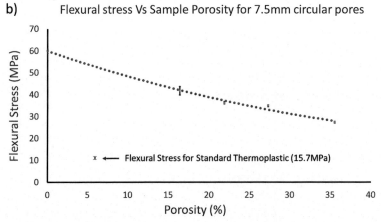

Figure 5.209 3D printed sample flexural data. (a) Design porosity for increased number of pores on a standard ASTM rectangular flexural test coupon and (b) Measured flexural stress for various porous 3D printed flexural test samples compared to the flexural stress of standard medical splint porous thermoplastic.

circular patterns were projected through the resulting model. Based upon the flexural data, it was observed that with the maximum pore density examined (38% porosity), the stiffness of ABS still surpasses that of the thermoplastic material. We therefore utilized this density to create the final AFO CAD concept, which can be seen in Fig. 5.210. The resulting CAD model was then 3D printed and test fit on the study volunteer. Feedback from the volunteer was that the AFO fitted very well and felt comfortable

Scan Data   CAD AFO Design

3D Printed AFO Concept   AFO worn within footwear

Figure 5.210 Various stages of AFO development.

to wear, even when worn within their current footwear. While much more work needs to be conducted, these preliminary results show the potential of our approach to create a functionally personalized AFO.

5.31.4 Conclusions

We demonstrate a robust method of determining the ABS flexural stiffness for a given porosity. These results can be applied to inform the design of a lower limb AFO, which has been realized using digital manufacturing. Using this approach, we believe we could realize a generation of AFOs that have greater design and mechanical conformities for potential clinical use.

References

[1] Wojciechowski E, et al. Feasibility of designing, manufacturing and delivering 3D printed ankle-foot orthoses: a systematic review. Journal of Foot and Ankle Research 2019;12(1):11.

[2] Mohammed MI, Fay P. Design and additive manufacturing of a patient specific polymer thumb splint concept. In: Proceedings of Solid Freeform Fabrication Symposium; 2018. p. 873—86.

[3] Paterson AM, et al. Computer-aided design to support fabrication of wrist splints using 3D printing: a feasibility study. Hand Therapy 2014;19(4):102—13.

CHAPTER 5.32

Orthotic applications case study 9—A review of 3D printed patient-specific immobilization devices in radiotherapy[a]

5.32.1 Introduction

Radiotherapy is one of the most common treatments to stop the proliferation of cancerous cells, with almost half of cancer patients receiving radiotherapy during some period of their treatment [1–3]. Despite good therapeutic results, radiotherapy can have negative side effects, with the mechanism used to kill cancerous cells also capable of damaging healthy tissue. As a result, immobilization devices are frequently used to minimize patient movements during radiation treatments, thereby ensuring the radiation dose is localized predominantly on the tumor site. This approach also limits exposure of healthy cells to radiation, while also allowing

[a] The work described in this chapter was first reported in the references below and is reproduced here and published by Elsevier B.V. On behalf of European Society of Radiotherapy and Oncology. This is an open access article under the CC BY license (http://creativecommons.org/licenses/BY/4.0/). This review article was supported by the Australian Research Council Industrial Transformation Training Center in Additive Biomanufacturing, Australia (Grant ID: IC160100026) http://www. additivebiomanufacturing.org. Additionally, the support of the Gross foundation is acknowledged. Funding sources were not involved in the study design or other aspects of this research. Supplementary data in Tables S1 and S2 for this article can be found online at
https://doi.org/10.1016/j.phro.2020.03.003.
Asfia A, Novak J I, Mohammed M I, Rolfe B, Kron T "A review of 3D printed patient specific immobilisation devices in radiotherapy" Physics and Imaging in Radiation Oncology 13 (2020) 30–35. https://doi.org/10.1016/j.phro.2020.03.003.

Medical Modeling
ISBN 978-0-323-95733-5
https://doi.org/10.1016/B978-0-323-95733-5.00037-5

reproducibility of setup on a day-to-day basis [4]. Immobilization devices can broadly be classified into two categories: invasive and noninvasive. Invasive fixation requires surgical fitting to the patient and has been found to accurately hold a patient in position; for example, cranial devices have been reported to hold patients within 1 mm of the desired position [5,6]. However, invasive fixation suffers from issues of patient discomfort, potential for infection, and the need for treatment to be conducted in one session.

Noninvasive fixation has become the preferred method of treatment, mitigating the issues of invasive fixation [7]. Considering the head and neck region, this method typically involves creating a custom-fitting mask made from a thermoplastic sheet molded directly over the patient prior to treatment planning, with a thickness in the range of 1.6—3.1 mm [8]. The custom-fitting device can immobilize a patient through a close fit with their anatomy and a secure fixture to a treatment table or other device. However, there remain several disadvantages to these masks including the accuracy of fitting being dependent on operator experience when thermoforming the mask, allowing at least some patients to maintain some motion during treatment [9]. Changes to facial geometry during the course of treatments due to weight loss can also be an issue with the fit of masks [8], and the mask fabrication process itself can be time-consuming, uncomfortable, claustrophobic, and distressing for patients [10—12]. While some studies report a 1-mm accuracy using thermoplastic masks [5,13—15], others argue that high accuracy comparable to invasive fixation is not possible [16].

To address these challenges, technologies such as 3D printing have seen increased research focus due to its ability to manufacture complex, customizable forms. Also known as additive manufacturing (AM) [17,18], 3D printing is a layer-by-layer manufacturing process that directly produces an object from 3D digital data. AM has found clinical applications as varied as patient-specific prosthetics manufacturing, anatomic models, implants, pharmaceutical research, and organ and tissue creation [19]. In particular, AM has the capacity to reproduce the complexities of the human form [20—22], resulting in a new generation of patient-tailored medical solutions. More recently, AM has been applied to upper body splint-based applications [23—25], and there is growing interest for uses in radiotherapy treatments [10,11]. As a result, it has gained the attention of researchers as a new method for producing noninvasive immobilization devices utilizing 3D patient models, such as those captured during computed tomography (CT) or magnetic resonance imaging (MRI) [22,26–28], combined with advanced computer-aided design (CAD), 3D scanning [23,25], and virtual planning software.

With the rapid developments of 3D printing technology, and the growing interest in its application to improve immobilization devices, this review paper systematically analyzed the academic literature on 3D printed immobilizers for radiation therapy published between January 2000 to January 2019. The primary goal was to provide researchers with an understanding of major developments, trends, and opportunities in this field, providing insight into the advantages and disadvantages of the technology for the specific application of immobilization. The secondary goal was to highlight gaps in the research for future inquiry.

5.32.2 Method

Initial scoping of literature began in February 2018. Thirty-eight databases and journals were included in the search to capture literature on the topic. In February 2019, the final systematic literature review was conducted using the following keywords: additively manufactured immobilizers, 3D printed thermoplastic masks, 3D printed masks, thermoplastic masks, head and neck thermoplastic masks, and 3D printed immobilizers. A limitation was placed on the publication date of literature from January 2000 to January 2019 due to initial scoping search results and the growth of AM technology, which was predominantly used only for prototyping prior to 2000.

Manual screening of the titles and abstracts was performed to include only papers consistent with the application of 3D printing techniques for the creation of immobilization devices used in radiotherapy treatments. Given the breadth of the search, this meant many results were excluded due to various factors, including the use of 3D printing for purposes other than immobilization, the use of 3D printing to prototype a concept before it was manufactured using another technology, or medical devices such as masks that were not additively manufactured. Duplicates were also removed.

Final screening was performed through reading of full-text papers and removing any that did not specifically discuss 3D printed immobilization devices for radiotherapy treatments. Results were analyzed by all authors to confirm suitable literature selection. Key information from each article was recorded in a spreadsheet to allow comparison, with categories including the type and number of test subjects, the specific part of the body examined and type of immobilization, the objectives of the study, and the testing techniques. Technical fabrication details were also recorded including the type of printer, the utilized materials, and the implemented software during the design procedure.

5.32.3 Results

A total of 4152 papers were identified through the search. Supplementary Table S1 identifies the 38 databases and journals, along with the number of initial results collected. Following the exclusion strategy, 4080 results were removed, and a further 54 duplicates were also removed. The result was 18 papers that were appropriate for this study and analyzed in detail.

Only one journal article was dated before 2014, occurring in 2002, before a 12-year gap where no academic literature was published on the topic of additively manufactured immobilizers for radiotherapy. One journal article was then published in 2014, and literature has been consistently published in the range of three to five annual articles between 2015 and 2018. Overall, 55% (n = 10) of publications were journal articles, 28% (n = 5) posters, 11% (n = 2) conference papers, and one thesis.

Summary information for each article is provided in Table 5.17, with an expanded table of results included in Supplementary Table S2. Sixty-one percent (n = 11) of studies involved human tests, while 22% (n = 4) involved animals, 11% (n = 2) involved phantoms, and one was experimental without subject testing. The head and neck regions of the body were the most studied with 78% (n = 14) of articles, one of which focused on the oral region, while 11% (n = 2) immobilized the whole body (of animals), one study focused on the breast, and one was experimental utilizing test pieces only.

Data captured from the studied articles showed that a range of technologies were used to capture geometry for designing the immobilizers, with 45% (n = 8) utilizing CT scans, 22% (n = 4) MRI, 11% (n = 2) optical 3D scanning, and 22% (n = 4) undisclosed or not utilizing any form of scanning. For the head and neck specifically, CT scans were used in 36% (n = 5) of studies, MRI was used in 29% (n = 4) of studies, optical 3D scanning was used in one study, and 29% (n = 4) were undisclosed or did not use scans to capture geometry.

Data captured from the literature also reveal that 78% (n = 14) of studies used 3D printing to fabricate the end-use immobilizer or headrest, while 11% (n = 2) used 3D printing to produce a replica of a head over which a traditional thermoformed mask immobilizer was created [29,30], and one

Table 5.17 Summarized results of the literature review categorized by body part and test subject.

Body part	Test subject	References	Results
Head	Human, n = 1	Sanghera et al. [11]	The 3D printed face mask exhibited similar performance under X-rays to traditional polystyrene based masks.
	Human, n = 1	Laycock et al. [10]	Confirmation that 3D printed masks are feasible for treatment with similar qualities to Orfit masks.
	Human, n = 8	Unterhinninghofen et al. [31]	The final product reportedly had high positioning accuracy equal to, or better than, traditional devices.
	Human, n = 10	Chen et al. [32]	The proposed segmentation method can produce masks with high accuracy, with a small segmentation mean error of 0.4 mm.
	Human, n = 0	Márquez-Graña et al. [9]	Authors claim that the proposed immobilizer is less invasive and more comfortable than a thermoformed or surgical immobiliser.
	Human, n = 17	Robertson et al. [33]	The external reproducibility of 3D printed beam directional shells and thermoplastic equivalents had no considerable differences.
	Human, n = 11	Pham et al. [29]	The 3D printed head was accurate enough to be used for molding the thermoplastic masks onto.
	Human, n = 30	Luo et al. [36]	3D printed headrests had higher transmittance (98.89%) compared with standard SRS headrests (98.51%).
	Human, n = 8	Haefner et al. [34]	3D printed masks provided a high setup accuracy.
	Phantom, n = 1	Fisher et al. [35]	A lack of adequate fitting of the 3D printed mask was found.
	Phantom, n = 1	Sato et al. [38]	3D printed masks have almost the same positional accuracy to that of conventionally made devices.
	Animal, n = 10	Zarghami et al. [39]	The mean irradiation targeting error was 0.14 ± 0.09 mm.
	Animal, n = 7	Slater et al. [30]	Surgical head posts were found to reduce movement more than the masks.

Continued

Table 5.17 Summarized results of the literature review categorized by body part and test subject.—cont'd

Body part	Test subject	References	Results
Whole body	Animal, n = 6	McCarroll et al. [37]	The 3D printed model decreased the setup variation considerably.
	Animal, n = 3	Steinmetz et al. [41]	Animals were not distressed while they were immobilised.
Oral	Human, n = 1	Wilke et al. [43]	3D printing eliminated several time-consuming steps in the fabrication of oral stents, minimising treatment delays.
Breast	Human, n = 10	Chen et al. [40]	The 3D printed breast immobiliser considerably reduced the radiation exposure to the lungs and heart.
Experimental	Experimental	Meyer et al. [42]	The effect on dosimetry requires investigation before clinical implementation of a 3D printed immobiliser. This study provides a framework for completing this testing.

study used 3D printing to produce several immobilizers as well as a head for thermoforming a traditional mask over [10]. Six of the studies examined the accuracy of the 3D printed immobilizers, with four studies finding a good level of accuracy in face masks on human volunteers [31−34], and one study finding a high degree of accuracy compared with the initial CAD model [11]. However, one study performed on a phantom found discrepancies in face mask geometry between 4 and 14 mm [35] and would not be suitable for immobilizing a patient. This failure was attributed to improper thresholding values when converting the CT scan of the phantom into a 3D model.

The results in Table 5.17 also reveal that studies analyzing the setup accuracy of 3D printed immobilizers found a high degree of accuracy locating patients for treatment and reducing movement. In particular, several studies found that 3D printed immobilizers provided better accuracy than traditional methods [31,36,37], while others found a good level of accuracy that may be comparable to existing thermoformed methods [34,38−40]. Only one study comparing surgical head posts to vacuum-formed masks produced over 3D printed heads (molds) for monkeys found the surgical head posts provided better accuracy by reducing movement during treatments [30]. However, this study also found that the addition of a mouth opening in the vacuum-formed mask to allow food rewards encouraged voluntary engagement by the monkeys, reducing stress and anxiety. Studies also suggested that 3D printed immobilizers improve targeted treatment of tumors and reduced the radiation exposure to surrounding tissue [40,41], although further research is required to confirm this hypothesis.

In relation to the beam attenuation of 3D printed materials, several studies found similarity with traditional thermoplastic immobilizers when made of a similar thickness [10,11,37]. Within a study on headrests for stereotactic radio surgery (SRS), 3D printed headrests had higher transmittance (98.89%) than standard SRS headrests (98.51%) [36]. Two studies demonstrated a methodology to measure beam attenuation with variable thickness and infill patterns [10,42].

Summarizing the literature on 3D printed immobilization devices, the identified advantages of the technology included improved patient comfort [9,31,34,39], reduced patient visits to a clinic [29,35,43], elimination of the stressful thermoforming process for face masks [31,33,34,41], high accuracy and tolerance to the patient [11,31−34], repeatable positional accuracy [31,34,36−38,40], less damage to surrounding healthy tissue [40,41], similar

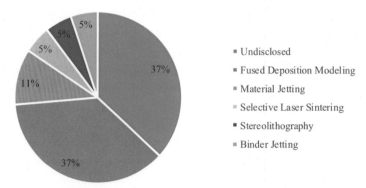

Figure 5.211 AM technologies utilized in literature.

beam attenuation properties to traditional polymer masks [10,11,36,37], and the opportunity to consider additional features to improve fit or patient engagement [30,39]. Disadvantages that were identified in the literature included the potential for the 3D printing process to negatively influence the material behavior of a device [32], inaccuracies due to conversion of scan data (e.g., CT) to 3D model [35], the slow process of 3D printing, which is unsuitable for rapid deployment or modification [11,29], lower accuracy compared with surgical immobilization [30], and costs that may be the same or higher than traditional immobilizers [30].

According to ISO/ASTM 52900 standards for classifying AM technologies, there are seven broad categories, five of which are represented in the 18 papers collected in this study. As shown in Fig. 5.211, fused filament fabrication (FFF) is the most utilized of the AM technologies. It is important to note that 37% (n = 7) of studies did not specify the print technology, making future replication of studies difficult.

5.32.4 Discussion

During radiotherapy, high doses of radiation are delivered to specific localized areas of the patient; as a result, targeting accuracy is vital to minimize negative effects to surrounding healthy tissue [7,34]. Immobilization devices are critical to this treatment process, minimizing patient movement and ensuring repeatability of the treatment over as many as 40 sessions [32].

While there is a general consensus that 3D printed immobilizers are a promising replacement for traditional immobilizers, particularly thermoformed face masks, the low volume of studies, and the low number of human, animal, or phantom subjects included in the studies, means that further research is required to validate claims of providing improved performance or comfort. In total, 97 human patients and volunteers have been featured in 11 studies, which is on average around nine subjects per study. This remains a small sample size, and further research on human subjects is required to properly evaluate 3D printed immobilizers.

Given that research has only gained momentum in the last 5 years, it appears that academics and medical practitioners are still in the early experimental phase of developing 3D printed immobilization technology, and this recent growth may be due to several factors: firstly, costs of 3D printers and their materials have rapidly declined as the technology has become more mainstream [44,45], making them accessible to researchers within medical and other disciplines [46]. Secondly, 3D printing technology has shifted from a predominantly rapid prototyping technology to an end-use manufacturing technology as materials have matured, making the production of functional parts, such as immobilization devices, possible. A similar growth in medical research using AM technology was noted in a systematic review of surgical orthopedic guides [46], with an increase in research observed shortly after 2009 when key FFF patents expired. FFF is an extrusion process also known as fused deposition modeling, and it has become synonymous with affordable desktop machines, although there are also high-end commercial varieties of this technology [47,48]. A broader systematic review of 3D printing across medical fields confirmed this trend [49], and while little evidence exists to correlate the patent expiry with the increased use of FFF within research, the results from this study align with recent systematic reviews that indicate a rapid growth in medical research utilizing 3D printing within the last 10 years [46,49].

However, the 12-year period without any new research into 3D printed immobilization devices between 2002 and 2014 is interesting, particularly because the results of the 2002 study were positive [11]. Several factors may account for this gap, including the relatively niche application of 3D printing for immobilization compared with other aspects of medical 3D printing, such as surgical guides and implants, which dominate practice and literature. The technologies required for an immobilizer were also expensive and slow; for example, the 3D scanning system in the 2002 study reportedly cost $\sim$£100,000 [11], and the 3D printing machine was similarly expensive with a maximum print speed of 20 mm/s requiring 5 days to produce a single mask. It is only recently that costs have rapidly declined in line with Moore's law [45,50], with today's cheap ∼£800 desktop FFF machines capable of speeds of ∼100 mm/s for the same material and layer thickness. Lastly, the large size of immobilization devices makes them more challenging to 3D print than smaller medical devices like implants and surgical guides, which can be produced in a matter of hours and on a broader range of machines.

These factors remain challenging, with the first stage of creating a patient-specific immobilization device requiring patient geometry to be digitized. Studies have used CT, MRI, optical 3D scanning, and more manual CAD methods, with no consensus about the optimal method at this time. However, utilization of CT scanners for capturing the digital data of the patient's anatomy may not be the best technique as it increases the absorbed radiation dose by the patient if several CT scans are required. Designers also need to deal with the poor resolution of the CT scanners [51] compared with using high-quality optical scanners that are safer for patients and provide higher resolution of the external patient anatomy. While CT scanning was used in nearly half of the studies, given the relatively small number of studies in this review, and the focus on the end product rather than the technical aspects of digitizing patient anatomy, further research is required to specifically examine and quantify the optimal method for digitizing patient data for the development of an immobilization device. It will become increasingly important for practitioners to learn how to manipulate DICOM files within 3D CAD software to progress this field of research and may require collaborations between clinicians and surgeons with expertize in human anatomy and designers and engineers with expertize in advanced CAD software systems.

The second phase of the workflow requires a digital design of an immobilization device to be developed for each specific patient. It is clear that there is even less consensus about this stage, with a variety of software being implemented that ranges from high-end commercial design programs used by engineers and designers to freely available software. Many studies do not disclose the software used to develop immobilization devices, which may limit progress in this field as researchers and engineers are left to trial various methods rather than building upon those established across the literature. It is the recommendation of this research that future studies more rigorously report the details of the design process to establish protocols that others may build from, with scope for studies specifically examining the suitability of different CAD systems for the purposes of integrating patient scans with the design of custom immobilizers.

The final stage of the workflow is the production of an immobilization device or mold using 3D printing. However, the printing quality of the final product needs to be evaluated through inspection for defects and radiologic suitability. One of the challenges in 3D printing is the inherent uncertainty that exists in the process of 3D printing. The final printed products could have different densities even if they are printed with the same printer and the same material from the same manufacturer, which could eventually lead to a variation in radiologic properties and dose uncertainty of the printed components [52].

Fig. 5.210 identified that FFF is the most utilized 3D printing method in the literature, which is most likely due to the affordability of machines. However, there is a breadth of FFF technologies used, from industrial machines [31,34] that costs tens of thousands of dollars to desktop machines costing a few thousand dollars [37,41]. These differences, combined with the different materials being used, and the four other 3D printing technologies found in the literature, make it challenging to objectively compare results since each technology, and each individual 3D printer, plays a significant role in the quality of the 3D printed outcome. Future research must aim to assess the most appropriate 3D printing technologies and materials for patient immobilization and converge toward standardized methods of producing immobilizers that are quantifiable and reproducible. The large number of studies that do not disclose the 3D print technology or material at this time inhibits this agenda, and researchers should prioritize these data alongside data related to the results of patient trials.

The results of this study also highlight gaps in knowledge surrounding the costs of 3D printed immobilization devices compared with traditional methods. One study claimed a cost of $\sim$\$350CAD to produce a whole body immobilizer for a mouse [39], whereas another study with monkeys reported an initial setup cost of $\sim$\$3223USD, including the 3D printing of a head (mold) and vacuum forming of an immobilizer [30]. Each additional mask would then only cost $\sim$\$212USD since the mold could be reused, although the study also found that over the course of a year, three of the monkeys required one to three replacement molds and immobilizers due to changes in their body weight and head geometry. These costs are comparable to surgical methods [30]. Costs were not recorded in other literature, and it is unclear how economical 3D printed immobilizers are compared with other methods, especially across the range of print technologies and materials employed in literature to date.

Production time also remains unclear, with the 2002 study claiming 5 days to 3D print a mask [11], while a more recent 2018 study recorded 36 hours print time for a human head to thermoform a mask over [29]. These times are significantly longer than more immediate traditional methods of thermoforming or surgically attaching immobilizers; however, a lack of information reported in literature does not provide enough evidence to provide a clear understanding of the time to 3D print immobilization devices.

In conclusion, this literature review found that AM technology has gained attention for the production of immobilizers for use in radiotherapy, emerging over the last 5 years as 3D printing technology has become more affordable, accessible, and capable of producing functional parts. Research has found that the main advantages of the technology include the ability to manufacture immobilizers from digital patient data, removing the uncomfortable process of thermoforming directly over a patient, or surgically attaching fixation devices. Good levels of accuracy have also been reported, both in terms of matching the patient's unique body geometry as well as allowing for repeatable setup for treatment. Through the systematic review process, this study also highlighted several areas for future research, in particular the need for larger samples sizes in human testing to validate claims supporting the use of 3D printing to produce immobilization devices. Furthermore, a lack of published information was noted in relation to the technical aspects of additively manufacturing immobilizers including their design, cost, production, and associated settings, as well as materiality. These issues must be addressed as researchers continue to pursue this application of AM.

References

[1] Barton MB, Jacob S, Shafiq J, Wong K, Thompson SR, Hanna TP, et al. Estimating the demand for radiotherapy from the evidence: a review of changes from 2003 to 2012. Radiotherapy and Oncology 2014;112:140−4. https://doi.org/10.1016/j.radonc.2014.03.024.

[2] Delaney G, Jacob S, Featherstone C, Barton M. The role of radiotherapy in cancer treatment: estimating optimal utilization from a review of evidence-based clinical guidelines. Cancer: Interdisciplinary International Journal of American Cancer Society 2005;104:1129−37. https://doi.org/10.1002/cncr.21324.

[3] Delaney G, Barton M. Evidence-based estimates of the demand for radiotherapy. Clinical Oncology 2015;27:70−6. https://doi.org/10.1016/j.clon.2014.10.005.

[4] Lee HT, Kim SI, Park JM, Kim HJ, Song DS, Kim HI, et al. Shape memory alloy (SMA)-based head and neck immobilizer for radiotherapy. Journal of Computer Design Engineering 2015;2:176−82. https://doi.org/10.1016/j.jcde.2015.03.004.

[5] Ramakrishna N, Rosca F, Friesen S, Tezcanli E, Zygmanszki P, Hacker F. A clinical comparison of patient setup and intra-fraction motion using frame-based radiosurgery versus a frameless image-guided radiosurgery system for intracranial lesions. Radiotherapy and Oncology 2010;95:109–15. https://doi.org/10.1016/j.radonc.2009.12.030.

[6] Leksell L. The stereotactic method and radiosurgery of the brain. Acta Chirurgica Scandinavica 1951;102:316–9.

[7] Bichay TJ, Mayville A. The continuous assessment of cranial motion in thermoplastic masks during cyberknife radiosurgery for trigeminal neuralgia. Cureus 2016;8. https://doi.org/10.7759/cureus.607.

[8] Dieterich S, Ford E, Pavord D, Zeng J. Chapter 6 - immobilization techniques in radiotherapy. In: Dieterich S, Ford E, Pavord D, Zeng J, editors. Practical radiat oncol phys. Philadelphia. Elsevier; 2016. p. 87–94.

[9] Márquez-Graña C, McGowan P, Cotterell M, Ares E. Qualitative feasibility research to develop a novel approach for a design of a head support for radiosurgery or stereotactic radiosurgery (SRS). Procedia Manufacturing 2017;13:1344–51. https://doi.org/10.1016/j.promfg.2017.09.112.

[10] Laycock S, Hulse M, Scrase C, Tam M, Isherwood S, Mortimore D, et al. Towards the production of radiotherapy treatment shells on 3D printers using data derived from DICOM CT and MRI: preclinical feasibility studies. Journal of Radiotherapy in Practice 2015;14:92–8. https://doi.org/10.1017/S1460396914000326.

[11] Sanghera B, Amis A, McGurk M. Preliminary study of potential for rapid prototype and surface scanned radiotherapy facemask production technique. Journal of Medical Engineering and Technology 2002;26:16–21. https://doi.org/10.1080/03091900 110102445.

[12] McKernan B, Bydder S, Deans T, Nixon M, Joseph D. Surface laser scanning to routinely produce casts for patient immobilization during radiotherapy. Australasian Radiology 2007;51:150–3. https://doi.org/10.1111/j.1440-1673.2007.01686.x.

[13] Sahgal A, Ma L, Chang E, Shiu A, Larson DA, Laperriere N, et al. Advances in technology for intracranial stereotactic radiosurgery. Technology in Cancer Research and Treatment 2009;8:271–80. https://doi.org/10.1177/153303460900800404.

[14] Tryggestad E, Christian M, Ford E, Kut C, Le Y, Sanguineti G, et al. Inter-and intrafraction patient positioning uncertainties for intracranial radiotherapy: a study of four frameless, thermoplastic mask-based immobilization strategies using daily cone-beam CT. International Journal of Radiation Oncology, Biology, Physics 2011;80:281–90. https://doi.org/10.1016/j.ijrobp.2010.06.022.

[15] Pasquier D, Dubus F, Castelain B, Delplanque M, Bernier V, Buchheit I, et al. Repositioning accuracy of cerebral fractionated stereotactic radiotherapy using CT scanning. Cancer Radiotheraphy: Journal of Social Francaise Radiotheraphy Oncology 2009;13:446–50. https://doi.org/10.1016/j.canrad.2009.05.003.

[16] Bichay T, Dieterich S, Orton CG. Submillimeter accuracy in radiosurgery is not possible. Medical Physics 2013;40. https://doi.org/10.1118/1.4790690.

[17] Gibson I, Rosen DW, Stucker B. Additive manufacturing technologies. Springer; 2014.

[18] Redwood B, Schffer F, Garret B. The 3D printing handbook: technologies, design and applications. 3D Hubs 2017. ISBN: 978-90-827485-0-5.

[19] Ventola CL. Medical applications for 3D printing: current and projected uses. Pharmacology and Therapeutics 2014;39:704.

[20] Mohammed MI, Tatineni J, Cadd B, Peart G, Gibson I. Advanced auricular prosthesis development by 3D modelling and multi-material printing. KnE Engineering 2017;2:37–43. https://doi.org/10.18502/keg.v2i2.593.

[21] Mohammed M, Tatineni J, Cadd B, Peart P, Gibson I. Applications of 3D topography scanning and multi-material additive manufacturing for facial prosthesis development and production. In: Proceedings of the 27th Annual International Solid Freeform Fabrication Symposium; 2016. p. 1695−707.

[22] Mohammed M, Fitzpatrick A, Malyala S, Gibson I. Customised design and development of patient specific 3D printed whole mandible implant. In: Proceedings of the 27th Annual International Solid Freeform Fabrication Symposium; 2016. p. 1708−17.

[23] Mohammed MI, Fay P. Design and additive manufacturing of a patient specific polymer thumb splint concept. In: Proceedings of the 29th Annual International Solid Freeform Fabrication Symposium; 2018. p. 873−86.

[24] Paterson AM, Bibb R, Campbell RI, Bingham G. Comparing additive manufacturing technologies for customised wrist splints. Rapid Prototyping Journal 2015;21:230−43.

[25] Fitzpatrick A, Mohammed M, Collins P, Gibson I. Design optimisation of a thermoplastic splint. In: Proceedings of the 28th Annual International Solid Freeform Fabrication Symposium; 2017. p. 2409−18.

[26] Matsumoto JS, Morris JM, Rose PS. 3-dimensional printed anatomic models as planning aids in complex oncology surgery. JAMA Oncology 2016;2:1121−2. https://doi.org/10.1001/jamaoncol.2016.2469.

[27] Salmi M. Possibilities of preoperative medical models made by 3D printing or additive manufacturing. Journal of Medical Engineering 2016;2016. https://doi.org/10.1155/2016/6191526.

[28] Mohammed MI, Cadd B, Peart G, Gibson I. Augmented patient-specific facial prosthesis production using medical imaging modelling and 3D printing technologies for improved patient outcomes. Virtual and Physical Prototyping 2018;13:164−76. https://doi.org/10.1080/17452759.2018.1446122.

[29] Pham QVV, Lavallée AP, Foias A, Roberge D, Mitrou E, Wong P. Radiotherapy immobilization mask molding through the use of 3D-printed head models. Technology in Cancer Research and Treatment 2018;17. https://doi.org/10.1177/1533033818809051.

[30] Slater H, Milne AE, Wilson B, Muers RS, Balezeau F, Hunter D, et al. Individually customisable non-invasive head immobilisation system for non-human primates with an option for voluntary engagement. Journal of Neuroscience Methods 2016;269:46−60. https://doi.org/10.1016/j.jneumeth.2016.05.009.

[31] Unterhinninghofen R, Giesel F, Wade M, Kuypers J, Preuss A, Debus J, et al. OC0412: 3D printing of individual immobilization devices based on imaging ñ analysis of positioning accuracy. Radiotherapy and Oncology 2015;115:S199−200.

[32] Chen S, Lu Y, Hopfgartner C, Sühling M, Steidl S, Hornegger J, et al. 3-D printing based production of head and neck masks for radiation therapy using CT volume data: A fully automatic framework. In: IEEE 13th International Symposium; 2016. p. 403−6. https://doi.org/10.1109/ISBI.2016.7493293.

[33] Robertson FM. A Comparison of thermoplastic and 3D printed beam directional shells on viability for external beam radiotherapy and user experience. Queen Margaret University; 2017.

[34] Haefner MF, Giesel FL, Mattke M, Rath D, Wade M, Kuypers J, et al. 3D-Printed masks as a new approach for immobilization in radiotherapy—a study of positioning accuracy. Oncotarget 2018;9:6490. https://doi.org/10.18632/oncotarget.24032.

[35] Fisher M, Applegate C, Ryalat M, Laycock S, Hulse M, Emmens D, et al. Evaluation of 3D printed immobilisation shells for head and neck IMRT. Open Journal of Radiology 2014;4:322−8. https://doi.org/10.4236/ojrad.2014.44042.

[36] Luo L, Deng G, Zhang P, Dai P, Wang X, Wang D, et al. Comparision of 3D printed headrest for pediatric patients and SRS standard headrest in radiation therapy. International Journal of Radiation Oncology, Biology, Physics 2018;102:e477. https://doi.org/10.1016/j.ijrobp.2018.07. 1364.

[37] McCarroll RE, Rubinstein AE, Kingsley CV, Yang J, Yang P. 3D-printed small-animal immobilizer for use in preclinical radiotherapy. Journal of American Association Lab Animal Science 2015;54:545—8.

[38] Sato K, Takeda K, Dobashi S, Kishi K, Kadoya N, Ito K, et al. OC-0271: positional accuracy valuation of a three dimensional printed device for head and neck immobilisation. Radiotherapy and Oncology 2016;119:S126—7.

[39] Zarghami N, Jensen MD, Talluri S, Foster PJ, Chambers AF, Dick FA, et al. Immunohistochemical evaluation of mouse brain irradiation targeting accuracy with 3D-printed immobilization device. Medical Physics 2015;42:6507—13. https://doi.org/10.1118/1.4933200.

[40] Chen T, Chung M, Tien D, Wang R, Chiou J, Chen K, et al. Personalized Breast Holder (PERSBRA): a new cardiac sparing technique for left-sided whole breast irradiation. International Journal of Radiation Oncology, Biology, Physics 2017;99:E646. https://doi.org/10.1016/j. ijrobp.2017.06.2161.

[41] Steinmetz A, Platt M, Janssen E, Takiar V, Huang K, Zhang Y, et al. Design of a 3D printed immobilization device for radiation therapy of experimental tumors in mice. International Journal of Radiation Oncology, Biology, Physics 2017;99:E618. https://doi.org/10.1016/j.ijrobp. 2017.06.2090.

[42] Meyer T, Quirk S, D'Souza M, Spencer D, Roumeliotis M. A framework for clinical commissioning of 3D-printed patient support or immobilization devices in photon radiotherapy. Journal of Applied Clinical Medical Physics 2018;19:499—505. https://doi.org/10.1002/acm2.12408.

[43] Wilke CT, Zaid M, Chung C, Fuller CD, Mohamed AS, Skinner H, et al. Design and fabrication of a 3D—printed oral stent for head and neck radiotherapy from routine diagnostic imaging. 3D. Print Medicine 2017;3:12.

[44] Novak JI, Bardini P. The popular culture of 3D printing: when the digital gets physical. In: Ozgen O, editor. Handbook research consumption, media, popular culture global age. Hershey, PA, USA: IGI Global; 2019. p. 188—211. https://doi.org/10.4018/9781-5225-8491-9.ch012.

[45] Novak JI. Self-directed learning in the age of open source, open hardware and 3D printing. In: Ossiannilsson E, editor. Ubiquitous inclusive learn digital era. Hershey, PA, USA: IGI Global; 2019. p. 154—78. https://doi.org/10.4018/978-1-52256292-4.ch007.

[46] Popescu D, Laptoiu D. Rapid prototyping for patient-specific surgical orthopaedics guides: a systematic literature review. Proceedings - Institution of Mechanical Engineers, Part H: Journal of Engineering Medicine 2016;230:495—515. https://doi.org/10.1177/0954411916636919.

[47] Gibson I, Rosen D, Stucker B. Additive manufacturing technologies: 3D printing. In: Rapid prototyping, and direct digital manufacturing. 2 ed. New York: Springer; 2015.

[48] Novak JI, O'Neill J. A design for additive manufacturing Case study: fingerprint stool on a BigRep ONE. Rapid Prototyping Journal 2019;25:1069—79.

[49] Tack P, Victor J, Gemmel P, Annemans L. 3D-printing techniques in a medical setting: a systematic literature review. BioMedical Engineering Online 2016;15:115.

[50] Benson CL, Triulzi G, Magee CL. Is there a moore's Law for 3D printing? 3D Printing and Additive Manufacturing 2018;5:53—62. https://doi.org/10.1089/3dp.2017.0041.

[51] Dipasquale G, Poirier A, Sprunger Y, Uiterwijk JWE, Miralbell R. Improving 3Dprinting of megavoltage X-rays radiotherapy bolus with surface-scanner. Radiation Oncology 2018;13:203.

[52] Craft DF, Kry SF, Balter P, Salehpour M, Woodward W, Howell RM. Material matters: analysis of density uncertainty in 3D printing and its consequences for radiation oncology. Medical Physics 2018;45:1614—21. https://doi.org/10.1002/mp.12839.

CHAPTER 5.33

Orthotic applications case study 10—case reports of three-dimensional printed assistive technology[a]

5.33.1 Introduction

A person who has differences in their body structures and body functions may experience limitations to activities, such as mobility and activities of daily living as well as limitations to participation, such as education and employment [1]. Assistive technologies (ATs), items that are purchased and/or customized, have been used to improve independence in the functional ability of people with disabilities [2]. AT rehabilitation team members match the AT to the needs of the human completing a specific activity within the context of their environment, such as with the human, activity, assistive technology (HAAT) model, shown in Fig. 5.212 [3]. The human—technology interface (also called user interface) can either promote or inhibit the person's ability to use a device.

---

[a] The works described in the case studies described below were first reported in the following:
Thelander Hill, M., Salatin, B., and Gronseth, B. (2021, April—June) Poster 922: 3D Printing to Restore Cycling for Veterans with Spinal Cord Injury. AOTA Inspire 2021. American Occupational Therapy Association. Online Conference. United States. https://www.youtube.com/watch?v=ks32qKArh24
Ripley, B. (2017). VA center using 3D printing to create devices to help veterans feel whole. VAntagepoint. Retrieved on 5/29/2022 from https://blogs.va.gov/VAntage/36409/vas-center-innovation-using-3d-printing-create-devices-help-veterans-feel-whole/.

Medical Modeling
ISBN 978-0-323-95733-5
https://doi.org/10.1016/B978-0-323-95733-5.00038-7

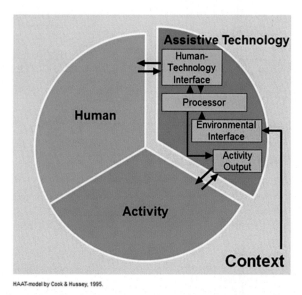

HAAT-model by Cook & Hussey, 1995.

Figure 5.212 The human, activity, assistive technology model [3].

5.33.2 Three-dimensional printed assistive technology

Many peer-reviewed reports of three-dimensional printed (3DP) ATs document the support of a person's body or limbs, such as with orthoses (splints) that may improve motion, reduce pain, and/or reduce deformity [4–7]. AT may also include external supports that adapt the task demands by modifying the built environment or equipment so that the end user is able to complete their daily activities [8]. 3DP offers novel solutions to improve independence in activities and participation through AT applications such as with writing, typing, and eating [9,10], taking prescribed medicine [11], and compensating for movement deficits related to amyotrophic lateral sclerosis [12]. The utility of AT to modify mobility equipment for people with spinal cord injury (SCI) will be discussed in the following case reports.

5.33.3 Case study: Power mobility wheelchair user interface

"Elijah" (named changed to protect anonymity) experienced functional mobility constraints consistent with an ASIA A, C6 SCI [13]. Elijah's preliminary rehab and initial wheelchair evaluation was at a prominent neurorehabilitation hospital in the western United States where he received a power wheelchair. Elijah was able to use the power wheelchair independently except for the power and mode portions of the joystick control user interface. The user interface consisted of two low-profile selector switches that Elijah could not consistently control with his available hand function (Fig. 5.213, Permobil Joystick User Interface www.permobil.com/us/wp-content/uploads/2017/03/UM_US_M3-Corpus.pdf). To improve his independence, an occupational therapist at the neurorehabilitation hospital created small, custom thermoplastic splinting material shells to cover the power and profile/speed selector switches. The switch cover concept worked well and allowed Elijah to change his speed and wheelchair function through the user interface, but the thermoplastic shells would fall off so frequently that they were connected to Elijah's wheelchair with string. Elijah did not have the required hand dexterity to return the thermoplastic shells to the selector switches. This decreased his independence in moving his wheelchair independently and caused Elijah to be dependent on his children or a care attendant to replace the selector switch shell covers. Just as importantly, it meant that he could not reliably access vital features such as tilt-in-space, which allowed for changes in his body position to prevent painful and potentially deadly pressure injuries.

Figure 5.213 Permobil joystick user interface. Item A is the power and profile selector; item B is the speed selector. *(Source: https://www.permobil.com/.)*

To improve Elijah's ability to consistently use the selector switches, a clinical rehabilitation engineer at a New Mexico specialty spinal cord rehabilitation unit designed a more robust selector switch cover. The clinical rehabilitation engineer at the facility measured the dimensions of the selector switch on the wheelchair control interface and found that it was tapered in multiple directions. The first iteration of the selector switch cover was fabricated on a material extrusion 3D printer using a rigid plastic, acrylonitrile butadiene styrene (ABS). The design fitted but not snugly enough to resist getting pried off if bumped so was only a partial success (Fig. 5.214, ABS selector switch cover). The next iteration of the design was also fabricated with a material extrusion 3D printer but used a flexible plastic, thermoplastic polyurethane (TPU) instead, which provided a higher-friction "grip" on the tapered selector switch. Additionally, the soft TPU plastic provided a soft touch surface for the end user who would use the dorsal (back) side of his hand to manipulate the selector switch due to his hand function limitations. Resulting from the 3DP TPU selector switch cover, Elijah was able to successfully use all wheelchair features for mobility and changing his body position (Fig. 5.215, TPU selector switch cover).

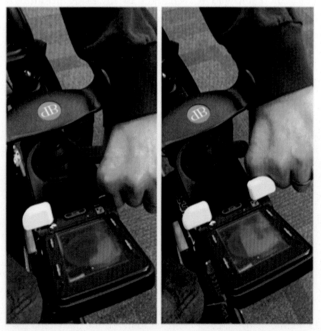

Figure 5.214 Selector switch without cover and with ABS 3D printed cover.

Figure 5.215 Final selector switch cover 3D printed from flexible TPU plastic.

5.33.4 3D printing in mobility for recreation and leisure

This section will describe the initial design of a 3DP item of AT to enhance mobility equipment that was created for one person and the design process used to adapt it for the needs of another person with similar needs. Mobility equipment may include manual or power wheelchairs (as with Elijah's case study), may include ambulation equipment such as canes or walkers, or may include recreational equipment, such as recumbent bicycles [14]. The next case series will describe a collaborative effort between an occupational therapist (OT), a physical therapist (PT), and a clinical rehab engineer to create an item of 3DP AT to promote recumbent cycling for people who survived incomplete SCIs.

"Grant" (named changed to protect anonymity) demonstrated functional ability consistent with a C3 Asia D SCI [13] related to cervical spinal stenosis (narrowing of the spinal canal). Grant's lower body motion was spared by the injury. Though he did have some balance loss, he was able to walk, sit, and stand. Grant did experience paralysis of some upper body muscles, especially those related to spinal cord function for the C3 nerve root. This resulted in very weak trapezius muscle and chronic neck and shoulder pain. Wishing to encourage physical activity that can enhance mental and physical health [14] and decrease risk of cardiovascular complications common with SCI [15], Grant's PT had been working with Grant to use a recumbent tricycle for exercise. Recumbent tricycles allow propulsion with lower extremities and steering with the upper body while the user is low to the ground in a seated position (Fig. 5.216, example of recumbent cycling position). A person, such as Grant, who experiences decreased dynamic sitting and standing balance can use a recumbent tricycle safely.

The recumbent tricycle user interface, the handlebars controlled by his hands, had a small foam wrist support (Figs. 5.217 and 5.218). Even with the wrist support, Grant experienced extreme pain in his shoulders and neck on his left side due to the static positioning required to hold the right handlebar and brake mechanism. The pain limited him from being able to use the bicycle for more than 10 minutes. Grant's OT, PT, and clinical rehab engineer collaborated to design an arm trough that could offer more support to Grant to allow cycling for longer periods.

First, the OT created a mock-up of an arm trough out of moldable thermoplastic splinting material for proof of concept (Fig. 5.219).

Figure 5.216 Example of recumbent cycling position [14].

Figure 5.217 Recumbent tricycle user interface included a small, padded wrist support. *(Photograph by Ben Salatin.)*

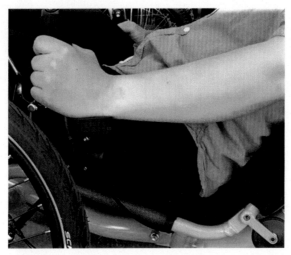

Figure 5.218 Available recumbent tricycles offer a small hand rest with no vertical adjustment options. *(Photograph owned by Ben Salatin.)*

Then referencing the prototype, a 3D computer model was created by the clinical rehab engineer and fabricated using a material extrusion 3D printer and polylactide thermoplastic (PLA) filament material (Fig. 5.220).

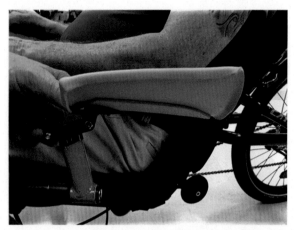

Figure 5.219 The OT fabricated a thermoplastic mock-up. The PT constructed a mounting bracket from available bike parts. The team tested the mock-up for compatibility with body function and structure. *(Photograph owned by Ben Salatin.)*

Figure 5.220 The rehab engineer created 3D CAD model from thermoplastic splinting material mock-up and 3D printed trough with plastic. The design was iterated in plastic first before the final design was printed in metal. *(Photograph owned by Ben Salatin.)*

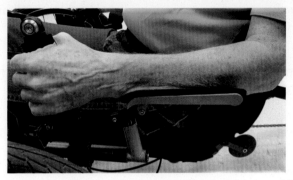

Figure 5.221 Final first edition trough prototype 3D printed in metal for maximum robustness. *(Photograph owned by Ben Salatin.)*

Once testing of the plastic arm trough was completed to ensure anthropomorphic compatibility, minor changes were made to the design to make it more lightweight. The PT designed and fabricated a bracket that would attach to the handlebars and secure an arm trough. For maximum robustness of the final design, it was fabricated in metal (titanium) leveraging a relationship with another hospital that owned a metal-capable powder bed fusion 3D printer. A gel cover was designed by the OT and created by Gelovations to make the arm trough comfortable (Fig. 5.221). Grant has been able to cycle on trails for 5—10 miles and at least 30 minutes due to the intervention of this team. Grant uses the arm trough and gel cover independently. The complete design iteration for the first edition trough is pictured in Fig. 5.221.

5.33.5 Adapting the design for another user

At the same SCI rehabilitation center, another user presented with similar body function and structure concerns that could be remediated by the same 3DP concept. "Bill's" functional ability was consistent with an Asia A C3 central cord spinal injury. Central cord cervical CSIs often result in more motor impairment in upper extremities than lower; they are the most common form of SCI [16]. Because Bill's lower extremity motor function was largely spared with his SCI, and he showed some recovery of upper extremity motor function from his exemplary efforts with his OT, his interdisciplinary team worked to restore his recreational passion: cycling. Bill worked with the PT to determine which model of recumbent cycle would suit his needs and to complete on-road training. Bill loved cycling but continued to experience pain in his right shoulder that limited him to very short rides or avoiding riding his recumbent tricycle.

5.33.6 Small design changes for a second-edition trough

Though Grant and Bill had similar physical ability and pain in their shoulders, Bill's right upper extremity was primarily affected. The clinical rehab engineer created a mirrored version of the original design file to match the opposite arm. Other design lessons from Grant's use case were incorporated into the digital design model for Bill such as rounding the side of the trough, adding a cutout for better comfort when making sharp turns, and lightening the weight of the trough's solid surface by reducing it to a lattice (Fig. 5.222).

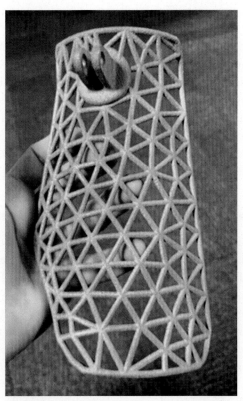

Figure 5.222 The design was mirrored and updated with a lattice for another user. *(Photograph owned by Ben Salatin.)*

As with the prior use, Bill's PT designed and fabricated a bracket that would attach to the handlebars and secure an arm trough to Bill's model of recumbent tricycle using commercially available bicycle parts. After application of the arm trough, Bill was able to use the arm trough independently and was able to ride his bike for at least 60 minutes (10 to 15 miles) and reported 0/10 pain when cycling.

5.33.7 Conclusion

In these case reports, the design process from thermoplastic splinting material mock-up to 3D printed prototypes to final 3D printed AT items was described. For these case reports, the items that were modified using 3D printing were already AT items procured for people who experience decreased movement related to specific circumstances of their SCIs. Considering the user's needs with the HAAT [3], the team altered the user interfaces of a power wheelchair and two recumbent cycles using custom designed and 3D printed alterations.

AT offers opportunities for improved safety and independence for people with wide-ranging mobility differences. 3D printing improved the outcomes for use for the individuals described in these case reports, so one person could independently change his body position and utilize power wheelchair controls, and two people could use adaptive cycling equipment to improve their wellness.

References

[1] World Health Organization. International classification of functioning, disability and health (ICF). Geneva: World Health Organization; 2001.
[2] Bondoc S, Goodrich B, Githow L, Smith R. Assistive technology and occupational performance. American Journal of Occupational Therapy 2016;70(s1−s9).
[3] Cook A, Polgar J. Cook and Hussey's assistive technologies: principles and practice. 3rd ed. London: Elsevier Health Sciences; 2013.
[4] Georgiou Tomás A, Asnaghi D, Liang A, Agogino AM. The sparthan three-dimensional printed exo-glove: a preliminary evaluation of performance via case study. Journal of Medical Devices 2019;13(3). https://doi.org/10.1115/1.4043976.
[5] Kim SJ, Kim SJ, Cha YH, Lee KH, Kwon J-Y. Effect of personalized wrist orthosis for wrist pain with three-dimensional scanning and printing technique: a preliminary, randomized, controlled, open-label study. Prosthetics and Orthotics International 2018;42(6):636−43. https://doi.org/10.1177/0309364618785725.
[6] Khas KS, Pandey PM, Ray AR. Development of an orthosis for simultaneous three-dimensional correction of clubfoot deformity. Clinical Biomechanics 2018;51:67−75. https://doi.org/10.1016/j.clinbiomech.2017.12.002.

[7] Patterson RM, Salatin B, Janson R, Salinas SP, Mullins MJS. A current snapshot of the state of 3D printing in hand rehabilitation. Journal of Hand Therapy 2016;33(2):156−63. https://doi.org/10.1016/j.jht.2019.12.018.

[8] Lunsford C, Grindle G, Salatin B, Dicianno BE. Innovations with 3-dimensional printing in physical medicine and rehabilitation: a review of the literature. PM&R 2016;8(12):1201−12. https://doi.org/10.1016/j.pmrj.2016.07.003.

[9] Lee KH, Kim DK, Cha YH, Kwon JY, Kim DH, Kim SJ. Personalized assistive device manufactured by 3D modeling and printing techniques. Disability and Rehabilitation: Assistive Technology 2019;14(5):526−31. https://doi.org/10.1080/17483107.2018.1494217.

[10] Degerli YI, Dogu F, Oksuz C. Manufacturing an assistive device with 3D printing technology - a case report. Assistive Technology January 2, 2022;34(1):121−5. https://doi.org/10.1080/10400435.2020.1791278.

[11] Schwartz J, Ballard DH. Feasibility of customized pillboxes to enhance medication adherence: a randomized controlled trial. Archives of Physical Medicine and Rehabilitation April 14, 2022. https://doi.org/10.1016/j.apmr.2022.03.018. S0003-9993(22)00341-0.

[12] Rasmussen KM, Stewart BC, Janes WE. Feasibility of customized 3D-printed assistive technology within an existing multidisciplinary amyotrophic lateral sclerosis clinic. Disability and Rehabilitation: Assistive Technology February 11, 2022:1−7. https://doi.org/10.1080/17483107.2022.2034996.

[13] Rupp R, Biering-Sørensen F, Burns SP, Graves DE, Guest J, Jones L, et al. International standards for neurological classification of spinal cord injury. Topics in Spinal Cord Injury Rehabilitation 2021;27(2):1−22. https://doi.org/10.46292/sci2702-19.

[14] McDaniel J, Lombardo LM, Foglyano KM, Marasco PD, Triolo RJ. Setting the pace: insights and advancements gained while preparing for an FES bike race. Journal of NeuroEngineering and Rehabilitation 2017;14:118. https://doi.org/10.1186/s12984-017-0326-y.

[15] Arul K, Ge L, Ikpeze T, Baldwin A, Mesfin A. Traumatic spinal cord injuries in geriatric population: Etiology, management, and complications. Journal of Spine Surgery 2019;5(1):38−45. https://doi.org/10.21037/jss.2019.02.02.

[16] Kretzer RM. A clinical perspective and definition of spinal cord injury. Spine 2016;41:S27. https://doi.org/10.1097/BRS.0000000000001432.

Dental applications

CHAPTER 5.34

Dental applications case study 1—the computer-aided design and additive manufacture of removable partial denture frameworks[a]

5.34.1 Introduction

Computer-aided design, manufacture, and rapid prototyping (CAD/CAM/RP) techniques have been extensively employed in the product development sector for many years and have also been extensively used in maxillofacial technology and surgery [1,2]. In addition, CAD/CAM technologies have been introduced into dentistry, particularly to the manufacture of crowns and bridges [3], but there has been little research into the use of such methods in the field of removable partial denture (RPD) framework fabrication. This may in part be attributed to the lack of

[a] The work described in this chapter was first reported in the references below and is reproduced here in part or in full with the permission of the Council of the Institute of Mechanical Engineers and Quintessence Publishing Ltd.

Eggbeer D, Bibb R, Williams R, "The Computer Aided Design and Rapid prototyping of Removable Partial Denture Frameworks", Proceedings of the Institute of Mechanical Engineers Part H: Journal of Engineering in Medicine, 2005, Volume 219, Issue Number H3, pages 195—202.

Eggbeer D, Williams RJ, Bibb R, "A Digital Method of Design and Manufacture of Sacrificial Patterns for Removable Partial Denture Metal Frameworks", Quintessence Journal of Dental Technology, 2004, Volume 2, Issue Number 6, pages 490—499.

The authors would like thank Frank Cooper at the Jewellery Industry Innovation Center (JIIC) in Birmingham, UK, who kindly supplied the Perfactory and Solidscape RP patterns and Kevin Liles at 3D Systems Inc. who supplied the Amethyst RP pattern.

Medical Modeling
ISBN 978-0-323-95733-5
https://doi.org/10.1016/B978-0-323-95733-5.00039-9

suitable, dedicated software. Recent pilot studies have showed that CAD/RP methods of designing and producing a sacrificial pattern for the production of metal components of RPD metal frameworks could have promising applications [4,5]. These studies explored the application of computer-aided technologies to the surveying of digital casts and pattern design and the subsequent production of sacrificial patterns using RP technologies.

The potential advantages offered by the introduction of advanced CAD/CAM/RP into the field of RPD framework fabrication include automatic determination of a suggested path of insertion, almost instant elimination of unwanted undercuts (reentry points), and the equally rapid identification of useful undercuts. At another stage, components of an RPD could be stored in a library and "dragged and dropped" in place on a scanned and digitally surveyed cast from icons appearing on the screen, allowing virtual patternmaking to be carried out in a much faster time than is achieved by current techniques. The quality assurance of component design can also be built into the software. Since RP machines build the object directly, scaling factors may also be precisely imposed to compensate for shrinkage in casting. In addition to the potential time savings, the CAD/RP process also delivers inherent repeatability, which may help to eliminate operator variation and improve quality control in the dental laboratory.

The current chapter reports on the application of CAD/RP methods to achieve the stages of surveying and design using a software package that provides a virtual sculpting environment, Geomagic FreeForm Plus (Geomagic Solutions, 430 Davis Drive, Suite 300, Morrisville, NC 27560, USA). It also discusses the application of RP technologies to produce sacrificial patterns for casting the definitive chromium—cobalt framework component. The advantages, limitations, and future possibilities of these techniques are concluded.

5.34.2 Materials and methods

5.34.2.1 3D scanning

A three-dimensional (3D) scan of a partially dentate patient's dental cast was obtained using a structured white light digitizer (Comet 250; Steinbichler Optotechnik GmbH, AM Bauhof 4, D-83115 Neubeuern). This particular type of scanner is used in high-precision engineering applications and has been used in maxillofacial technology [6]. Multiple overlapping scans were used to collect point cloud data that were aligned using Polyworks software

(InnovMetric Software Inc., 2014 Jean-Talon Blvd. North, Suite 310, Sainte-Foy, Quebec G1N 4N6, Canada). Spider software (Alias-Wavefront Inc., 210 King Street East, Toronto, Ontario M5A 1J7, Canada) was used to produce a polygon surface, STL (Manners, C. R., 1993, "STL File Format" available on request from, 3D Systems Inc., 333 Three D Systems Circle, Rock Hill, SC 29730, USA) model file that could be imported into FreeForm.

5.34.2.2 FreeForm modeling

FreeForm is a CAD package with tools analogous to those used in physical sculpting. A haptic interface (Phantom Desktop haptic interface) incorporates positioning in 3D space and allows rotation and translation in all axes while translating hand movements into the virtual environment (Fig. 5.223). It also allows the operator to feel the object being worked on in the software. The combination of tools and force feedback sensations mimic working on a physical object and allows shapes to be designed and modified in an arbitrary manner.

Figure 5.223 The phantom stylus.

Objects being worked on are referred to as virtual "clay," which can be rotated and viewed from any angle on the screen. A "buck" setting prevents a model being unintentionally modified, but it allows "clay" to be added or copied. The STL cast was imported into FreeForm as a "buck" model.

5.34.2.3 Surveying

Surveying is undertaken in dental technology laboratories to identify useful dental features in order for the RPD design to be retained in the oral cavity effectively. The "parting line" (also known as "split line") function within FreeForm was used to delineate up- and down-facing surfaces, thus identifying areas of undercut in a different color to the "buck" model. The effect is identical to the physical technique of using dental survey lines to identify and mark the most bulbous areas of teeth with a pencil line (highlighted in Fig. 5.224a). The undercuts were assessed to establish the best path of insertion and possible points for active clasp termination, and the model rotated accordingly.

A visual comparison (Figs. 5.224a and 5.224b) was made between the physically surveyed cast and the same model cast surveyed using the software. Once a suitable angle was chosen, the model was reexported as an STL file.

5.34.2.4 Removing unwanted undercuts

When creating an RPD, most undercuts are removed so that the resulting framework can be inserted and removed in a comfortable manner. The STL file of the rotated cast was imported into FreeForm but this time using the "extrude to plane" option. When the cast was viewed from above, this

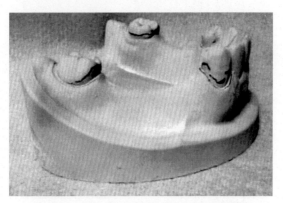

Figure 5.224a The physically surveyed cast.

Figure 5.224b The digitally surveyed "buck" cast. Undercuts are shown as dark areas.

option took the maximum extents of the profile and extruded them down by a user-defined distance. This effectively removed undercuts and replaced them with vertical surfaces (Figs. 5.225a and 5.225b).

5.34.2.5 Identifying useful undercuts

FreeForm's "ruler" tool was used to measure the distance between the original cast model and the version with undercuts removed. The useful undercuts were marked with a line for use in the design stages. RPDs provide a firm location on the existing dentition by using flexible clasps. The clasp components of the RPD open on initial contact during insertion and removal and return to their original position within the undercut on final seating, thus providing secure retention.

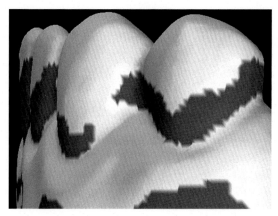

Figure 5.225a Undercuts are shown as dark areas.

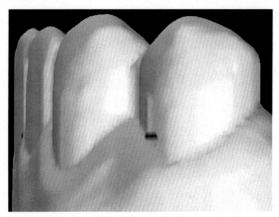

Figure 5.225b Undercuts have been removed and replaced by vertical surfaces.

5.34.2.6 Creation of relief

The areas without teeth require a spacer, known as relief, to prevent the framework resting on the surface of the gums. Relief was created by selecting and copying an area from the cast with undercuts removed and then pasting this as a new piece of clay. This was then offset to the outside by 1 mm. The results of this process are highlighted in Fig. 5.226.

The entire modified model was saved as an STL file and then reimported using the "buck" setting to avoid unintentional modification during the next stages of RPD design.

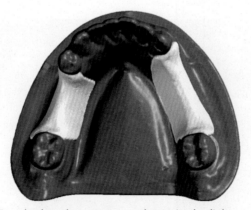

Figure 5.226 Relieved edentulous areas are shown in the lighter color on the dark cast.

5.34.2.7 Framework design

The RPD design employed in this study was based on recognized dental technology methods emphasizing simplicity, aesthetics, and patient comfort [7]. Some of the key design features outlined in the design stages are labeled in Fig. 5.227.

The entire framework was designed on the relieved "buck" cast with undercuts removed, with the exception of the clasp components. The clasps use the undercuts to function, and they were therefore designed on the original "buck" cast. The following techniques were used in the framework design.

Occlusal rests (label a in Fig. 5.227): A combination of two-dimensional drawing and three-dimensional creation and manipulation tools were used to create pieces of clay that were copied and located where required on the teeth.

Polymeric retention framework (b in Fig. 5.227), lingual bar (c in Fig. 5.227), acrylic line (d in Fig. 5.227), and nonactive clasps (e in Fig. 5.227): The "draw" tool was used locate curves directly onto the cast surface. These formed the center of the framework's profile (Fig. 5.228). The "groove" tool was used to define and create the exact oval and square sectional dimensions as clay.

Guide plates (f in Fig. 5.227): Guide plates were created using the same method as relief creation. The "attract" and "smudge" tools were also used to build up plate areas and blend them onto the framework sections.

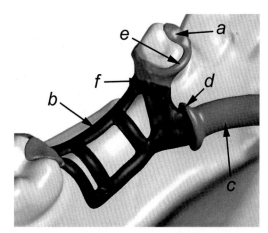

Figure 5.227 a = occlusal rest, b = polymeric retention frame, c = lingual bar, d = acrylic line, e = nonactive clasp, f = guide plate.

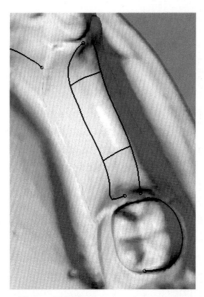

Figure 5.228 Construction curves.

Finishing: "Smooth," "attract," and "smudge" tools were used to blend the components together. The "buck" cast was removed, acting as a Boolean cutting tool to leave just the clay framework.

Active clasps: The clasps were designed in the same manner as the nonflexible parts of the framework but using the "buck" cast with undercuts. The construction lines were joined to the termination point previously marked in the undercut measurement stage.

The "buck" cast was removed leaving the clasps. These were joined to the main framework and blended in. Fig. 5.229 shows the final, virtual design. The entire framework was exported as an STL file.

5.34.2.8 Pattern manufacture

Four RP methods were compared: stereolithography (SL) (3D Systems Inc, 26081 Avenue Hall, Valencia, CA 91355, USA), ThermoJet (3D Systems Inc.), Solidscape T66 (Solidscape Inc. 316 Daniel Webster Highway, Merrimack, NH 03054–4115, USA), and Perfactory (EnvisionTEC GmbH, Elbestraße 10, D-45768 Marl, Germany). Two SL resins were compared: DSM Somos 10110 (Waterclear) (2 Penn's Way, Suite 401, New Castle, DE 19720, USA) and Accura Amethyst (3D Systems Inc.). Both of the SL patterns were an epoxy-based polymer, the ThermoJet was

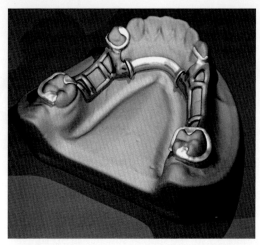

Figure 5.229 The complete FreeForm design.

TJ88 grade wax polymer, the Solidscape was a soft thermoplastic, and Perfactory an acrylate-based polymer. The Waterclear and ThermoJet patterns were manufactured at International Centre for Design Research (PDR), and the others were prepared and built by external suppliers. The Amethyst, Solidscape, and Perfactory materials are used by the jewelry industry to produce sacrificial patterns.

SLA-250 in Waterclear example:

The STL framework design was prepared using 3D Lightyear (3D Systems Inc.) with a "fine point" support structure (Fig. 5.230). The framework was oriented with the fitting surfaces facing upward to avoid the rough finish created by the support structures affecting fit.

Two build styles were compared: standard 0.1000-mm-thick layers and high-resolution 0.0625-mm layers. Once completed, the patterns were carefully removed from the machine platform and cleaned in isopropanol. They were then postcured in UV light to ensure full polymerization. The other patterns were produced according to the supplier specifications.

5.34.2.9 Pattern comparison

Of the four RP processes compared in this study, the SL processes provided the most suitable patterns. The SL patterns were accurate and robust and had an acceptable surface finish, but they did require relatively lengthy cleaning and finishing when removing support structures. The ThermoJet build preparation was simpler and faster than SL, and both the ThermoJet

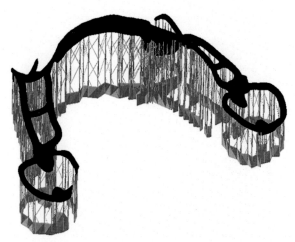

Figure 5.230 The support structure in 3D lightyear.

and Solidscape processes produced accurate patterns with a good surface finish that required minimal finishing. These wax patterns were, however, extremely fragile and could not be cast. The Perfactory produced pattern showed a very smooth surface finish but was also extremely flexible and was easily distorted when handled.

5.34.2.10 Casting

The SL and Perfactory patterns were cast in chrome cobalt without using a refractory cast. A slow mold heating cycle was used to avoid cracking. Fig. 5.231 shows the unfinished cast from the SL, Amethyst pattern. This shows that air inclusions from the casting process did not adhere to the pattern surface.

Although casts were obtained from the SL and Perfactory patterns, it proved difficult to add sprues due to the thin framework sections. To improve casting, the design was thickened in FreeForm, and revised SL patterns were produced and cast. This improved the pattern's strength and the casting reliability.

5.34.2.11 Finishing

The casts produced from the original, thin Amethyst and thicker Water-clear patterns were polished and test fitted to the original, physical cast. These were all judged satisfactory. Fig. 5.232 shows the finished RPD framework that was cast from the high-resolution Waterclear, SLA-250 pattern.

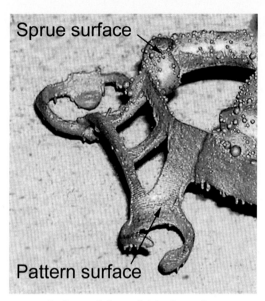

Figure 5.231 Surfaces of the unfinished amethyst pattern cast.

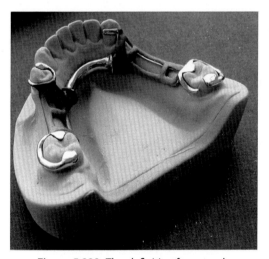

Figure 5.232 The definitive framework.

5.34.3 Conclusions

The design stages of this technique rely on having an accurate 3D scan of a patient cast and an understanding of both RPD framework design and CAD techniques. This meant that the time taken to produce castable patterns using the technology described is considerable. However, this would be significantly reduced with familiarity and practice.

The most suitable choice of RP process was determined primarily by accuracy and part strength. The ThermoJet and Solidscape patterns, although accurate, were too fragile and were therefore not suitable for the tasks associated with spruing and casting. Although the Perfactory pattern cast well, the accuracy was poor due to distortion inflicted on the flexible pattern during handling. The stiffer patterns produced by SL were easy to handle and accurate and produced satisfactory results. The layer effect exhibited by all RP processes was not evident after finishing, and the difference between the high-resolution and standard SLA-250 patterns was negligible.

The techniques undertaken and described above outline a stage in the development of machine-produced RPD frameworks and point to many possible advances that can be achieved in the future. The application of CAD would allow access to new RP technologies that build parts directly in metal alloys, including chromium—cobalt and stainless steel. Sacrificial pattern manufacture and casting may be eliminated altogether. This will be explored in future studies.

The introduction of digital design and RP production into current practices would present a significant change in the field of dentistry, and this is unlikely to happen quickly. Studies so far have shown how CAD and RP may be applied, and some principles have been developed and established. Possible future benefits and the potential shortfalls have also been discussed.

References

[1] Hughes CW, Page K, Bibb R, Taylor J, Revington P. The custom-made titanium orbital floor prosthesis in reconstruction for orbital floor fractures. British Journal of Oral and Maxillofacial Surgery 2003;41:50—3.
[2] Bibb R, Brown R. The application of computer aided product development techniques in medical modeling. Biomedical Sciences Instrumentation 2000;36:319—24.
[3] Van der Zel J, Vlaar S, de Ruiter W, Davidson C. The CICERO system for CAD/CAM fabrication of full ceramic crowns. The Journal of Prosthetic Dentistry 2001;85:261—7.
[4] Williams R, Bibb R, Rafik T. A technique for fabricating patterns for removable partial denture frameworks using digitized casts and electronic surveying. The Journal of Prosthetic Dentistry 2004;91(1):85—8.
[5] Williams R, Eggbeer D, Bibb R. CAD/CAM in the fabrication of removable partial denture frameworks: a virtual method of surveying 3-dimensionally scanned dental casts. Quintessence Journal of Dental Technology 2004;2(3):268—76.

[6] Bibb R, Freeman P, Brown R, Sugar A, Evans P, Bocca A. An investigation of three-dimensional scanning of human body surfaces and its use in the design and manufacture of prostheses. Proceedings of the Institution of Mechanical Engineers - Part H: Journal of Engineering in Medicine 2000;214(6):589—94.

[7] Budtz-Jorgensen E, Bocet G. Alternate framework designs for removable partial dentures. The Journal of Prosthetic Dentistry 1998;80:58—66.

CHAPTER 5.35

Dental applications case study 2—trial fitting of a removable partial denture framework made using computer-aided design and additive manufacturing[a]

5.35.1 Introduction

Computer-aided design and manufacture (CAD/CAM) techniques have been adopted as a method of fabrication for fixed partial denture restorations [1,2], and CAD/CAM and rapid prototyping (RP) have been extensively used in maxillofacial technology and surgery [3–5]. The application of the principles of CAD and RP to the fabrication of removable partial denture frameworks (RPDs) is in the early stage of development, but already the potential advantages are clear and have been discussed [6–8]. Developments achieved so far have included electronic surveying of a three-dimensionally scanned dental cast [7] and the production of successful castings from plastic patterns produced by RP technologies [6,8]. The potential future benefits include a rapid and semiautomated method of digital surveying, a "drag and drop" system of

[a] The work described in this chapter was first reported in the references below and is reproduced here in part or in full with the permission of Sage Publishing.
Bibb R, Eggbeer D, Williams RJ, Woodward A, "Trial fitting of a removable partial denture framework made using computer-aided design and rapid prototyping techniques," Proceedings of the Institute of Mechanical Engineers Part H: Journal of Engineering in Medicine 2006; 220(7): 793-797, ISSN: 0954–4119, http://doi.org/10.1243/09544119JEIM62.

Medical Modeling
ISBN 978-0-323-95733-5
https://doi.org/10.1016/B978-0-323-95733-5.00040-5

617

virtual patterning using onscreen icons of RPD components, which could be dragged onto the computer model of a scanned cast of a patient.

However, although the castings fabricated so far by CAD/CAM-produced patterns have been judged to be acceptable for clinical presentation, to date none has been trial fitted to a patient. This case report follows on from previously published work by providing details of the first fitting to a patient of an RPD framework produced by CAD and RP technologies.

5.35.2 Methods

5.35.2.1 The case

A female patient presented to the University Hospital of Wales Dental School with all lower anterior teeth present along with the lower left premolar (referred to as a Kennedy class 1 case). There was some retroclination, especially of the second incisors, the canines, and the premolar. The most viable form of treatment was considered to be the provision of a cobalt–chromium metal alloy framework RPD. A design based on the "Rest, Proximal plate, I clasp" (RPI) [9] principle was formulated.

5.35.2.2 Data capture and digital RPD design

The master cast of the patient's mandibular dental structures was three-dimensionally scanned using a structured white light digitizer. The device used is an optical system that utilizes a projected fringe pattern of light and digital camera technology to capture approximately 140,000 points in three dimensions on the surface of the object (Comet 250, Steinbichler Optotechnik GmbH, Am Bauhof 4, D-83115 Neubeuern, Germany). The data points collected are referred to as a "point cloud." The fringe pattern can be seen in Fig. 5.233.

Software (PolyWorks, InnovMetric Software Inc., 2014 Jean-Talon Blvd. North, Suite 310, Sainte-Foy, Quebec G1N 4N6, Canada) was used to automatically combine multiple scans (which overcomes the problems caused by line of sight) by aligning overlapping areas of scan data. A further software package (Spider, Alias-Wavefront Inc., 210 King Street East, Toronto, Ontario M5A 1J7, Canada) was used to produce a triangular faceted surface model from the point cloud data, shown in Fig. 5.234.

The surface of the computer model created from the point cloud was produced using triangular polygons in the form of a stereolithography (STL) file, which is a suitable format for importing into the virtual sculpting

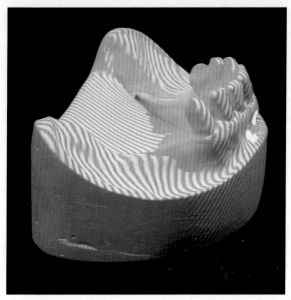

Figure 5.233 Lateral view of master cast undergoing scanning.

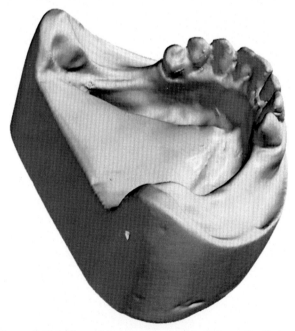

Figure 5.234 The solid computer model of the patient's cast.

environment, FreeForm. This facility was used to electronically survey the scanned model according to the principles outlined by Williams et al. [7]. Although no undercuts were present in the areas of clasp engagement on the abutment teeth, FreeForm is able to measure undercuts, as discussed by Williams et al. [7].

Once surveying was completed, the model was saved in a protected manner, so that it could not be altered inadvertently during the next stage of virtual patterning. A pattern was designed "on screen" to the design discussed above on the digitally scanned model. The design followed the principles described more fully in previous work [8,10]. Again, the package FreeForm provided excellent facilities for this process. Accurately defined, semicircular profiles such as the lingual bar were built up using construction curves, and then a "groove" tool was used to create a raised section. "Smudge" and "smooth" tools were used to merge the components together. The process is illustrated in Figs. 5.235a–5.235c.

5.35.2.3 Rapid prototyping and investment casting

Previous work had indicated that stereolithography was capable of producing suitable sacrificial patterns of RPDs [8,10]. Compared to other RP techniques, stereolithography patterns were found to possess a good balance of properties, being rigid enough to hold their shape, thus maintaining accuracy while being tough enough to allow handling and investment casting without inadvertent damage. Therefore, in this case, once the digital

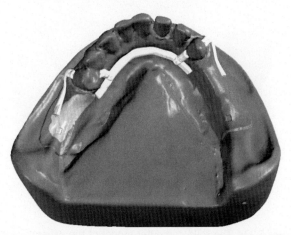

Figure 5.235a Illustration of the major and minor connectors, gingivally approaching clasps, and area of retention for acrylic defined.

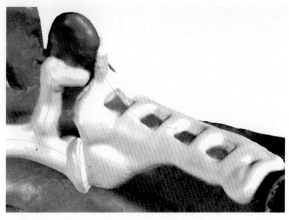

Figure 5.235b Right side of framework at a later stage of development with components joined.

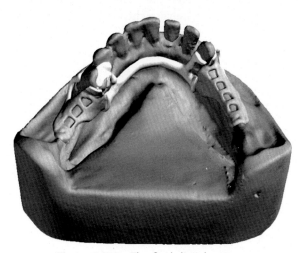

Figure 5.235c The final digital pattern.

design had been finalized, a stereolithography RP machine (SLA 250/40; 3D Systems Inc, 26081 Avenue Hall, Valencia, California 91355, USA) was used to build a physical pattern in epoxy-based resin (WaterClear 10110; DSM Somos, New Castle, Delaware, USA).

The "fine-point" supporting structures of the sacrificial pattern were thin and easily removed with a scalpel. The pattern then had wax sprues attached and was investment cast according to the procedures typically used in dental technology with the exception of the use of a refractory model.

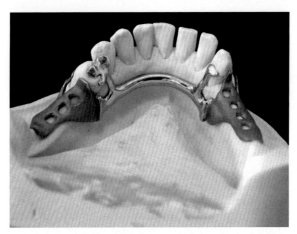

Figure 5.236 Metal alloy framework ready for clinical presentation.

The pattern was cast directly in cobalt–chromium alloy and finished by grit blasting and polishing in the normal manner. The framework was then test fitted to the master cast, shown in Fig. 5.236. Test fitting indicated a good fit, and the framework was forwarded to the dental clinic.

This project was undertaken as an elective and was submitted for approval to the internal and external supervisors at the Welsh National School of Medicine. In this case, informed consent was required and subsequently obtained from the patient to undertake a single trial fitting of the experimental framework. The patient was not inconvenienced further and was provided with a partial denture produced using standard techniques.

5.35.3 Results

The framework was prepared for test fitting to the patient in the usual manner. The framework was test fitted on the patient by a dentist (clinical supervisor). When fitted to the patient, on initial insertion, there were some discrepancies, but with some adjustment with reference to the patient in the dental clinic, the framework fitted satisfactorily, as shown in Fig. 5.237. It is not unusual for a cast RPD framework to require minor adjustment in order to fit the patient perfectly, as there may be slight differences between the patient anatomy and the cast. The clinical supervisor also confirmed that the alloy framework was satisfactory and that it could proceed to the next stage of construction, that of adding acrylic bases and artificial teeth.

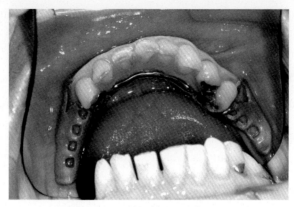

Figure 5.237 The framework in situ.

5.35.4 Discussion

The concept of introducing CAD/CAM into the fabrication of RPD frameworks is taken a step further than previous work [8,10]. The successful trial fitting of the cast cobalt—chromium framework to an actual patient in the clinic has indicated that the accuracy, tolerances, and overall "fitness for purpose" suggested in previous research [10] can be achieved in an actual clinical case. Although considerable time was involved in producing a framework by this method, it is foreseeable that with more research, the time could be greatly reduced, making the computer-aided method competitive with existing conventional techniques.

The fact that the framework was judged to be acceptable by an independent dentist (clinical supervisor) indicates that the application of CAD/RP techniques to this field stands comparison with existing casting techniques and suggests that there is potential for further investigation. Although the resulting framework required some minor adjustment to achieve a perfect fit, the nature and extent of this adjustment was comparable to that normally undertaken with cast frameworks produced by existing techniques. This suggests that the stereolithography casting pattern was of comparable accuracy to the wax patterns produced by existing techniques.

As this trial case has proved promising, further cases will be undertaken to verify that the technique is repeatable and consistent, and test fitting will also be assessed by physical measurement.

Future research will explore multiple clinical cases of finished prostheses, as well as the application of direct RP technologies such as selective laser melting that could be used to build frameworks in appropriate metal alloys, thus eliminating the casting stage altogether.

5.35.5 Conclusions

This paper demonstrates that a clinically satisfactory RPD framework can be designed and produced by CAD and an RP-built sacrificial pattern. A framework fabricated by the methods described was clinically verified and found to be acceptable and suitable to proceed to the next stage of RPD construction and clinical use.

Acknowledgments

The authors are grateful to Mrs. Rowena Bevan for her invaluable support.

References

[1] van der Zel J, Vlaar S, de Ruiter W, Davidson C. The CICERO system for CAD/CAM fabrication of full ceramic crowns. The Journal of Prosthetic Dentistry 2001;85:261–7.
[2] Mormann WH, Bindl A. The CEREC 3–a quantum leap for computer-aided restorations: initial clinical results. Quintessence International 2000;31:699–712.
[3] Hughes CW, Page K, Bibb R, Taylor J, Revington P. The custom-made titanium orbital floor prosthesis in reconstruction for orbital floor fractures. British Journal of Oral and Maxillofacial Surgery 2003;41:50–3.
[4] Bibb R, Brown R. The application of computer aided product development techniques in medical modeling. Biomedical Sciences Instrumentation 2000;36:319–24.
[5] Bibb R, Bocca A, Evans P. An appropriate approach to computer aided design and manufacture of cranioplasty plates. The Journal of Maxillofacial Prosthetics and Technology 2002;5:28–31.
[6] Williams R, Bibb R, Rafik T. A technique for fabricating patterns for removable partial denture frameworks using digitized casts and electronic surveying. The Journal of Prosthetic Dentistry 2004;91:85–8.
[7] Williams R, Eggbeer D, Bibb R. CAD/CAM in the fabrication of removable partial denture frameworks: a virtual method of surveying 3-dimensionally scanned dental casts. Quintessence Journal of Dental Technology 2004;2:242–67.
[8] Eggbeer D, Williams R, Bibb R. A digital method of design and manufacture of sacrificial patterns for removable partial denture metal frameworks. Quintessence Journal of Dental Technology 2004;2:490–9.
[9] Kratchovil FJ. Influence of occlusal rest position and clasp design on movement of abutment teeth. The Journal of Prosthetic Dentistry 1963;13:114–24.
[10] Eggbeer D, Bibb R, Williams R. The computer-aided design and rapid prototyping fabrication of removable partial denture frameworks. Proceedings of the Institution of Mechanical Engineers, Part H 2005;219:195–202.

CHAPTER 5.36

Dental applications case study 3—direct additive manufacture of removable partial denture frameworks[a]

5.36.1 Introduction

Over the last decade, computer-aided design, computer-aided manufacture, and rapid prototyping (CAD/CAM/RP) techniques have been employed in dentistry but predominantly for the manufacture of crowns and bridges [1—3]. However, there has been little research into the use of such methods in the field of removable partial denture (RPD) framework fabrication. Although rapid prototyping and additive manufacture (AM) techniques have proved successful in other dental applications, the lack of suitable design software has restricted their application in producing RPD frameworks. Recent studies have established a valid approach to the computer-aided surveying of digital casts, framework design, and the subsequent production of sacrificial patterns using RP technologies [4—7].

The potential advantages offered by the introduction of CAD in the field of RPD framework design include automatic determination of a suggested path of insertion, instant elimination of unwanted undercuts, and the equally rapid identification of useful undercuts, which are all crucial in

---

[a] The work described in this chapter was first reported in the reference below and is reproduced here in part or in full with the permission of the copyright holders.
Bibb R, Eggbeer D, Williams R, "Rapid manufacture of removable partial denture frameworks", Rapid Prototyping Journal 2006; 12(2): 95—99, ISSN: 1355—2546, http://doi.org/10.1108/13552540610652438.

Medical Modeling
ISBN 978-0-323-95733-5
https://doi.org/10.1016/B978-0-323-95733-5.00041-7

dental technology. The potential advantages of an AM approach are reduced manufacture time, inherent repeatability, and elimination of interoperator variation.

5.36.2 Methodology

5.36.2.1 Step 1: 3D scanning

A three-dimensional (3D) scan of a partially dentate patient's dental cast was obtained using a structured white light digitizer (Comet 250; Steinbichler Optotechnik GmbH, AM Bauhof 4, D-83115 Neubeuern, Germany, www.steinbichler.de). Multiple overlapping scans were used to collect point cloud data that were aligned using Polyworks software (InnovMetric Software Inc., 2014 Jean-Talon Blvd. North, Suite 310, Sainte-Foy, Quebec G1N 4N6, Canada, www.innovmetric.com). Spider software (Alias-Wavefront Inc., 210 King Street East, Toronto, Ontario M5A 1J7, Canada, www.alias.com) was used to produce a polygon surface in the STL file format [8].

5.36.2.2 Step 2: Design of the RPD framework

The CAD package used in this study, called FreeForm, was selected for its capability in the design of complex, arbitrary, but well-defined shapes that are required when designing custom appliances and devices that must fit human anatomy. The software has tools analogous to those used in physical sculpting and enables a manner of working that mimics that of the dental technician working in the laboratory (3D Systems company - 3D Systems Inc., Rock Hill, South Carolina, United States, www.3dsystems.com). The software utilizes a haptic interface (Touch or Touch X haptic interface) that incorporates positioning in 3D space and allows rotation and translation in all axes, transferring hand movements into the virtual environment. It also allows the operator to feel the object being worked on in the software. The combination of tools and force feedback sensations mimic working on a physical object and allows shapes to be designed and modified in a natural manner. The software also allows the import of scan data to create reference objects or "bucks" onto which fitting objects may be designed. The RPD metal frameworks used in this study were designed according to established principles in dental technology using this CAD software and based on a 3D scan of a patient's cast [9]. The computer-aided design of RPD frameworks using this software has been described previously [5—7]. The finished design used in this case is shown in the screen capture shown in Fig. 5.238.

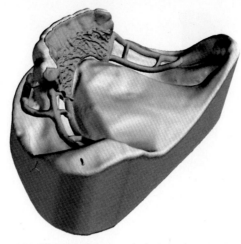

Figure 5.238 The RPD framework designed in FreeForm CAD.

5.36.2.3 Step 3: Additive manufacture

In a previous study, the application of RP methods was investigated for the production of sacrificial patterns that were used to investment cast RPD frameworks in cobalt—chrome alloy [7]. Four RP methods were compared: stereolithography (SL) (3D Systems Inc.), ThermoJet (3D Systems Inc.), Solidscape T66 (Solidscape Inc., 316 Daniel Webster Highway, Merrimack, NH 03054—4115, USA, www.solid-scape.com), and Perfactory (EnvisionTEC GmbH, Elbestraβe 10, D-45768 Marl, Germany, www.envisiontec.de). These various RP processes are described more fully in Chapter 4 — Physical reproduction.

In this study, direct manufacture was attempted with the aim of eliminating the time- and material-consuming investment-casting process. The development of selective laser melting (SLM) technology showed potential application for dental technologies due to the ability to produce complex shaped objects in hard-wearing and corrosion-resistant metals and alloys directly from CAD data. SLM is described in Chapter 4, Section 4.5.1.

To build the RPD framework successfully using the SLM Realizer machine (supplied at the time by MCP Tooling Technologies Ltd.), adequate supports had to be created using Magics software (Version 9.5, Materialise N.V., Technologielaan 15, 3001 Leuven, Belgium, www.materialise.com). The purpose of the supports is to provide a firm base for the part to be built onto while separating the part from the substrate plate. In addition, the supports conduct heat away from the material as it

Figure 5.239 RPD oriented and supported to avoid the fitting surfaces.

melts and solidifies during the build process. Inadequate support results in incomplete parts or heat-induced curl, which leads to build failure. As the supports need to be removed with tools, the part was oriented such that the supports avoided the fitting surface of the RPD, as shown in Fig. 5.239. This meant that the most important surfaces of the resultant part would not be affected or damaged by the supports or their removal.

5.36.2.4 First experiment

316L stainless steel was selected for the first experiment for its excellent corrosion resistance, making it suitable for dental applications. In addition, the SLM machine manufacturers have shown that the material is well suited to processing by SLM. The part and support files were "sliced and hatched" using the SLM Realizer software with a layer thickness of 0.050 mm. The material used was 316L stainless-steel spherical powder with a maximum particle size of 0.045 mm (particle size range 0.005–0.045 mm) and a mean particle size of approximately 0.025 mm (Sandvik Osprey Ltd., Red Jacket Works, Milland Road, Neath SA11 1NJ, UK, www.smt.sandvik.com/osprey). The laser had a maximum scan speed of 300 mm/s and a beam diameter 0.150–0.200 mm. The first two parts attempted were partially successful due to insufficient support and erroneous slice data. These errors resulted in incomplete RPDs. The third attempt was prepared with more

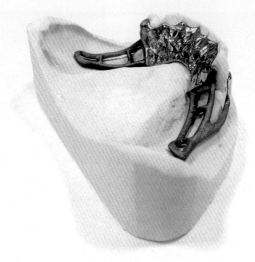

Figure 5.240 316L stainless-steel RPD framework fitted to patient cast.

support, and the data was sliced using different software (VisCAM RP, Marcam Engineering GmbH, Fahrenheitstrasse 1, D-28359 Bremen, Germany, www.marcam.de). This proved successful and produced a complete stainless-steel RPD framework, as shown in Fig. 5.240.

5.36.2.5 Second experiment

The same RPD framework design was manufactured using cobalt—chrome alloy using a layer thickness of 0.075 mm (Sandvik Osprey Ltd.). The principal reason for attempting the design in cobalt—chrome was for direct comparison with traditionally made RPD frameworks, which are typically cast from the same material. Like the previous material, the SLM machine manufacturers have shown cobalt—chrome to be suitable for processing by SLM. As before, the laser had a maximum scan speed of 300 mm/s with a beam diameter 0.150—0.200 mm. The material used was cobalt—chrome spherical powder with a maximum particle size of 0.045 mm (particle size range 0.005—0.045 mm) and a mean particle size of approximately 0.030 mm. The part proved successful and produced a complete cobalt—chrome RPD framework, as shown in Fig. 5.241.

5.36.2.6 Step 4: Finishing

Supporting structures were removed using a Dremel handheld power tool using a reinforced cutting wheel (Dremel, Reinforced Cutting Disc,

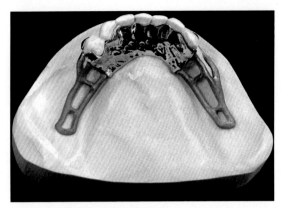

Figure 5.241 Cobalt—chrome RPD framework fitted to patient cast.

Ref. Number 426). The frameworks in their initial form were well formed but showed a fine surface roughness. This roughness was easily removed by bead blasting. This resulted in a framework that showed similar physical appearance and surface qualities as the investment cast items typically used in dental technology. Therefore, the treatment and finishing of the framework from that point onward was conducted in the same manner as any other RPD framework, using normal dental laboratory techniques and equipment.

5.36.3 Results

The successful 316L stainless-steel RPD framework was assessed for the quality of fit by fitting it to the plaster cast of the patient's oral anatomy. The quality of the fit was assessed according to normal dental practice by an experienced dental technician and found to show excellent fit. The frameworks showed a quality of fit that was comparable with investment cast frameworks. However, repeated insertion and removal from the patient cast resulted in small but permanent deformation of the clasp components. The clasp components are the functional parts of the framework and are designed to grip the teeth to provide a firm location of the denture (the clasps are the elements shown in the close-up photographs in Fig. 5.242). Therefore, the permanent deformation reduces the ability of the framework to grip the teeth, and the denture becomes loose. This meant that after several operations, the clasps no longer held the framework as securely to the existing teeth as deemed necessary by the dental technician.

Figure 5.242 Close-up views of the 316L stainless-steel RPD framework fitted to patient cast.

The cobalt—chrome RPD framework was complete, polished, and finished well with the normal dental technology procedures. The framework proved to be an excellent fit, possessing good clasping when test fitted to the patient's cast (see Fig. 5.242). The framework was test fitted to the patient in the clinic and found to be a precise and comfortable fit with good retention, as shown in Fig. 5.243. The framework will therefore be fitted with the artificial teeth and given to the patient to use in exactly the same manner as a traditionally manufactured item. Unlike the previous stainless-steel framework, the clasping forces did not result in permanent

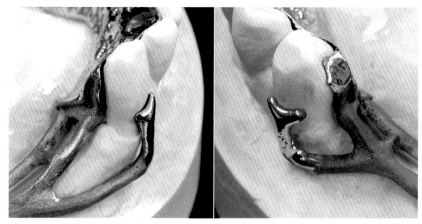

Figure 5.243 Close-up views of cobalt—chrome RPD framework fitted to patient cast.

deformation of the clasps, and the framework withstood repeated insertion and removal cycles.

5.36.4 Discussion

5.36.4.1 Sources of error

Various studies have aimed to assess error in cobalt—chrome partial denture frameworks made using traditional investment-casting techniques [10—12]. However, in the absence of an appropriate intraoral scanning technology, the application of CAD/CAM in dental technology depends on the dental model, which is a plaster cast taken from an impression of the dental anatomy taken by a dentist. Clearly, this paper cannot address issues relating to the quality of the original dental impression or the casting of the dental model from this impression. In addition, human error in the interpretation of the dentist's instructions or in the dental technician's chosen design for the framework is not addressed in this paper. However, the adoption of CAD/CAM/RP technologies may incur several process steps that may contribute to error between the theoretical designs produced using CAD and the final manufactured item. The effect of these processes will be an accumulation of tolerances at each technology stage. However, certain levels of care and skill may still affect the accuracy of these computer-controlled techniques.

Table 5.18 shows the steps in the process investigated here and indicates nominal tolerances associated with the technologies used. The accumulation of tolerances leads to the maximum error that could be expected to result from the technologies alone, assuming no human error is encountered. As human skill level and error cannot be attributed a numeric value and might range from zero to complete failure, discussion is not included here. However, as this study aims to investigate the implications of adopting CAD/CAM/RP technologies in dental technology, it is appropriate to attempt to illustrate their potential contribution to error in the final RPD framework. The tolerances used in this table indicate typical or nominal figures, which are quoted by manufacturers or set as parameters in software.

From the processes stated in Table 5.18 and, it is reasonable to expect a tolerance of approximately 0.2 mm for these parts. It should be noted that cumulative negative and positive tolerances from the various steps might also partially cancel each other out, resulting in a lower overall tolerance. The contribution of each individual step would be difficult to demonstrate without a statistically significant number of cases. The closeness of the fit

Table 5.18 Process steps and associated tolerances.

Process step	Source of error	Tolerance
Impression taking	Human/skill level	No value
Casting study model	Human/skill level	No value
Optical scanning of study model	Scanner	±0.050 mm
Creating polygon computer model from point cloud data	Software setting	±0.050 mm
Import into CAD software	Software	0.000 mm
Design in CAD software	Software setting	±0.001 mm
Export of CAD data in STL file format	Software setting	±0.010 mm
Physical manufacture using SLM	RP machine	±0.100 mm
Removal of RP pattern from machine, cleaning, and support removal	Human/skill level	No value
Surface preparation and polishing	Human/skill level	No value
Total		**±0.211 mm**

and effective clasping observed when fitting the frameworks to the patient cast, as shown in Figs. 5.242 and 5.243, suggest that SLM RPD frameworks are in fact within this tolerance.

5.36.4.2 Error analysis

RPD frameworks are by definition one-off custom-made appliances specifically designed and made to fit a single patient. In addition, the anatomically fitting nature of RPD frameworks means that they are complex in form and do not provide convenient datum or reference surfaces. This makes it difficult to achieve an investigation that provides detailed quantitative analysis of error. Therefore, it is not practical to perform the type of repeated statistical analysis that would be commonly encountered in series production or mass manufacture. Instead, it is normal dental practice to assess the accuracy of an RPD by test fitting the device to the patient cast and subsequently, in clinic, to the patient. In this study, the RPD frameworks created were deemed by a qualified and experienced dental technician to be a satisfactory fit and comparable to those produced by expert technicians; see Fig. 5.244. This suggests that the approach and technologies used are fit for purpose in this application although further experiments with a range of patients with differing RPD designs will be required to ensure that this is in fact the general case.

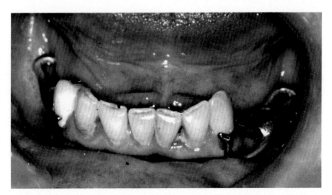

Figure 5.244 The framework fitted to the patient in clinic.

5.36.5 Conclusions

SLM has been shown to be a viable AM method for the direct manufacture of RPD metal alloy frameworks. Parts produced using the SLM process in conjunction with cobalt—chrome alloy result in RPD frameworks that are comparable in terms of accuracy, quality of fit, and function to the existing methods typically used in the dental technology laboratory. The CAD/CAM approach offers potential advantages in terms of reduced interoperator variability, repeatability, speed, and economy over traditional hand-crafting and investment-casting techniques.

References

[1] Willer J, Rossbach A, Weber HP. Computer-assisted milling of dental restorations using a new CAD/CAM data acquisition system. The Journal of Prosthetic Dentistry 1998;80(3):346—53.
[2] Van der Zel J, Vlaar S, de Ruiter W, Davidson C. The CICERO system for CAD/CAM fabrication of full ceramic crowns. The Journal of Prosthetic Dentistry 2001;85(3):261—7.
[3] Duret F, Preston J, Duret B. Performance of CAD/CAM crown restorations. Journal of the California Dental Association 1996;9(9):64—71.
[4] Williams R, Bibb R, Rafik T. A technique for fabricating patterns for removable partial denture frameworks using digitized casts and electronic surveying. The Journal of Prosthetic Dentistry 2004;91(1):85—8.
[5] Williams R, Eggbeer D, Bibb R. CAD/CAM in the fabrication of removable partial denture frameworks: a virtual method of surveying 3-dimensionally scanned dental casts. Quintessence Journal of Dental Technology 2004;2(3):268—76.
[6] Eggbeer D, Williams RJ, Bibb R. A digital method of design and manufacture of sacrificial patterns for removable partial denture metal frameworks. Quintessence Journal of Dental Technology 2004;2(6):490—9.

[7] Eggbeer D, Bibb R, Williams R. The computer aided design and rapid prototyping of removable partial denture frameworks. Journal of Engineering in Medicine 2005;219:195–202.

[8] Manners CR. "STL file format". CA, USA: 3D Systems Inc; 1993. www.3dsystems.com.

[9] Budtz-Jorgensen E, Bochet G. Alternate framework designs for removable partial dentures. The Journal of Prosthetic Dentistry 1998;80(1):58–66.

[10] Stern MA, Brudvik JS, Frank RP. Clinical evaluation of removable partial denture rest seat adaptation. The Journal of Prosthetic Dentistry 1985;53(5):658–62.

[11] Murray MD, Dyson JE. A study of the clinical fit of cast cobalt-chromium clasps. Journal of Dentistry 1988;16(3):135–9.

[12] Barsby MJ, Schwarz WD. The qualitative assessment of cobalt-chromium casting for partial dentures. British Dental Journal 1989;166(6):211–6.

CHAPTER 5.37

Dental applications case study 4—A comparison of plaster, digital, and reconstructed study model accuracy[a]

5.37.1 Introduction

Orthodontic treatment outcome and treatment change have traditionally been recorded with gypsum-based study models that are heavy and bulky, pose storage and retrieval problems, are liable to damage, and can be difficult and time-consuming to measure [1–5]. Legislation relating to the retention of patient records after the completion of treatment [6] has led to huge demands on space for storage that has prompted the development of alternative methods of recording occlusal relationships (Table 5.19) and electronic storage of records [7–13].

The replacement of plaster study models with virtual images has several advantages including ease of access, storage, and transfer [83], and the accuracy of image capture techniques has been reported [3,63,67,75,77,84,85]. However, if the physical restoration of a digital occlusal record is needed, possibly for medicolegal reasons, an accurate method of three-dimensional reconstruction is required.

Rapid prototyping (RP) systems, such as stereolithography, generate 3D models from a digital file through incremental layering of photo-curable

---

[a] The work described in this chapter was first reported in the reference below and is reproduced here in part or in full with the permission of the copyright holders.
Keating AP, Knox J, Bibb R, Zhurov A, "A comparison of plaster, digital and reconstructed study model accuracy", Journal of Orthodontics 2008;35:191–201, http://doi.org/10.1179/146531207225022626.

Medical Modeling
ISBN 978-0-323-95733-5
https://doi.org/10.1016/B978-0-323-95733-5.00042-9

Table 5.19 Alternative methods of recording occlusal relationships.

Two-dimensional techniques	
Conventional photography	[14−17]
Photocopying	[18−24]
Flatbed scanner	[25]

Three-dimensional techniques	
Optocom	[26]
Reflex plotters	[9,27]
Reflex metrograph	[28−31]
Reflex microscope	[28,31−33]
Traveling microscope	[34]
Moire' topography	[35−37]
Stereophotogrammetry	[38−42]
Telecentric lens photography	[43]
Holography	[44−51]
Optical profilometer	[52]
Image analysis system	[53,54]
Three-dimensional computerized tomography (3D-CT)	[55−58]
Structured light scanning methods	[7,9−11,13], [59−62, 64−73]
Intraoral scanning devices	[13,74]
Commercially available 3D digital study models	[1,58,75−82]

Adapted from [13a], Copyright Wiley-VCH GmbH. Reproduced with permission.

polymers [86]. The dimensional accuracy of physical replicas reproduced using the stereolithography technique has been evaluated by a number of authors. Barker et al. [87] found a mean difference of 0.85 mm between measurements made on actual dry bone skulls and physical replicas of the skulls produced by stereolithography from three-dimensional computed tomography (3D-CT) scans of the original dry bone skulls. They concluded that RP models could be confidently used as accurate three-dimensional replicas of complex anatomic structures. Using similar techniques, Kragskov et al. [88] and Bill et al. [89] found mean differences of −0.3−0.8 mm and ±0.5 mm between measurements on 3D-CT images and stereolithography models.

The objectives of this study were the following:

• to assess the reproducibility of a conventional method of using a handheld Vernier caliper to measure plaster study models;

- to develop an efficient and reproducible method of capturing a three-dimensional study model image, in a digital format, using the Minolta VIVID 900 noncontact surface laser-scanning device (Konica Minolta Inc., Tokyo, Japan);
- to assess the reproducibility of measurements made on the on-screen three-dimensional digital surface models captured using the scanning system setup developed;
- to compare the accuracy of measurements made on the three-dimensional digital surface models and plaster models of the same dentitions;
- to evaluate the feasibility of fabricating accurate three-dimensional hard copies of dental models from the laser scan data by an RP (stereolithography).

5.37.1.1 Null hypotheses

- There is no difference in the dimensional accuracy of three-dimensional digital surface models, captured with the surface laser-scanning technique described, and plaster study models.
- There is no difference in the dimensional accuracy of physical model replicas fabricated from the laser scan data by RP and plaster study models.

5.37.2 Materials and methods

5.37.2.1 Manual measurements

The local Research Ethics Committee chair confirmed that no ethical approval was required for this study. A minimum of 7—10 models per group were calculated to be required to allow a 90% chance of detecting a 0.3-mm difference in related sample means (SD = 0.2) at the 5% level of significance (alpha = 0.05 and power = 0.90) [90]. Thirty randomly selected plaster study models, held in the Orthodontic Unit of University Dental Hospital, Cardiff, were used in the study. Each study model was cast in matt white Crystal R plaster (South Western Industrial Plasters, Wilts, England) and conventionally trimmed [91]. To be included in the study, the plaster study models had to completely reproduce the arch, show no surface marks, loss of tooth material, voids, or fractures, and demonstrate varying degrees of contact point and buccolingual tooth displacements.

A handheld digital caliper (Series 500 Digimatic ABSolute caliper, Mitutoyo Corporation, Japan) was used to manually measure the plaster models. This caliper had a measurement resolution of 0.01 mm, was accurate to ±0.02 mm in the 0–200 mm range, and automatically downloaded data, eliminating measurement transfer and calculation errors.

All plaster models were measured in a bright room without magnification. The plaster models were not prepared in any way before measuring, and the anatomic dental landmarks used in the measurements were not premarked. A single examiner conducted all the measurements after an initial training period.

Twenty linear dimensions were measured, on each model, in each of the three planes (x, y, and z) with all measurements being recorded to the nearest 0.01 mm. The following dimensions were selected for measurement.

- **x plane**
  1. Intercanine distance — measured as the distance between the following:
     - **(a)** the occlusal tips of the upper canines
     - **(b)** the occlusal tips of the lower canines
  2. Interpremolar distances — measured as the distance between the following:
     - **(a)** the buccal cusp tips of the upper and lower first and second premolars
     - **(b)** the palatal cusp tips of the upper first and second premolars
     - **(c)** the lingual cusp tips of the lower first premolars
     - **(d)** the mesiolingual cusp tips of the lower second premolars
  3. Intermolar distances — measured as the distance between the following:
     - **(a)** the mesiopalatal cusp tips of upper first and second molars
     - **(b)** the mesiobuccal cusp tips of the upper and lower first and second molars
     - **(c)** the mesiolingual cusp tips of lower first and second molars
     - **(e)** the distobuccal cusp tips of the upper and lower first molars
- **y plane**
  1. In the upper arch, the distance from the mesiopalatal cusp tip of the upper second molar to the following:
     - **(a)** the mesiopalatal cusp tip of the upper first molar
     - **(b)** the palatal cusp tip of the upper first and second premolar
     - **(c)** the cusp tip of the upper canine
     - **(d)** the mesioincisal corner of the upper lateral incisor

2. In the lower arch, the distance from the mesiolingual cusp tip of the lower second molar to the following:

   (a) the mesiolingual cusp tip of the lower first molar and second premolar

   (b) the lingual cusp tip of the lower first premolar

   (c) the cusp tip of the lower canine

   (d) the mesioincisal corner of the lower lateral incisor

- **z plane**

   1. The clinical crown height of all the teeth, in both upper and lower arches, from the second premolar to second premolar inclusive, measured as the distance between the cusp tip and the maximum point of concavity of the gingival margin on the labial surface.

Measurements were made on two occasions separated by at least 1 week.

5.37.2.2 Virtual measurements

A noncontact laser-scanning device (Minolta VIVID 900, Konica Minolta Inc., Japan) was used to record the surface detail of each of the 30 study models using a telescopic light-receiving lens (focal distance $f = 25$ mm) and rotary stage (ISEL–RF1, Konica Minolta Inc., Japan). The rotary stage facilitated the acquisition of multiple range maps by moving the plaster study models, in sequence by a controlled rotation as they were being scanned, thus ensuring the entire visible surface of each plaster model was captured. The stage was controlled by a computer software program (Easy3DScan, Tower Graphics, Italy) and integrated controller box (IT116G, Minolta Inc., Osaka, Japan).

Easy3DScan was used to align, merge, and simplify the range maps acquired at different angles to produce a composite surface dataset that was then imported into the RapidForm 2004 software program (INUS Technology Inc., Seoul, Korea) as a triangulated 3D mesh (Fig. 5.245). An automated measuring tool was used to record the same measurements that had been conducted manually on the plaster study models. The 3D digital surface models were magnified and rotated on-screen to aid identification of the anatomic landmarks as necessary. Linear distances between landmarks were calculated automatically to five decimal places (Fig. 5.246). Replicate measurements were made on all digital model images with a time interval of at least 1 week.

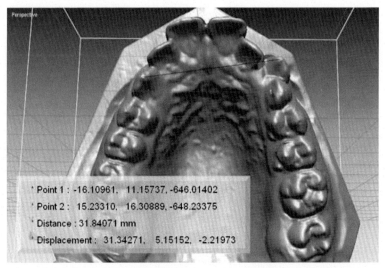

Figure 5.245 On-screen 3D virtual model image — intercanine width measured.

5.37.2.3 Measurement of reconstructed models

One pair of upper and lower plaster models were scanned individually using an identical protocol, adhering to the inclusion criteria listed previously. Only one set of models was evaluated due to the current cost of stereolithography. The scanned data for both upper and lower plaster models were saved as binary STL files and imported into the Magics RP software (Materialise Inc., Technologielaan 15, 3001 Leuven, Belgium).

A 3D Systems (SLA-250/40) stereolithography machine (3D Systems Inc., 26081 Avenue Hall Valencia, CA 91355, USA) containing a hybrid-epoxy based resin (10110 Waterclear — DSM Somos, 2 Penn's Way, Suite 401, New Castle, DE 19720, USA) was used to construct replica (3D Printed) models from the digital files using a build layer thickness of 0.15 mm (Figs. 5.246 and 5.247).

Identical measurements, in x, y, and z planes, were made on the reconstructed stereolithography models to those recorded on the original plaster study models and virtual models. Replicate measurements were made 1 week later.

5.37.2.4 Statistical analysis

Bland Altman analysis [92] was undertaken to determine agreement between repeat model measurements. Intrarater reliability was assessed by

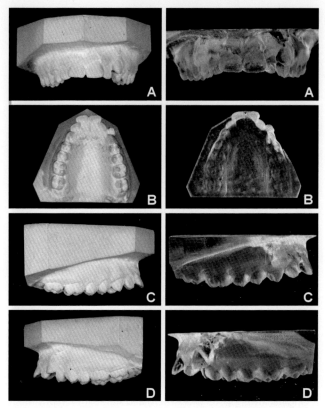

Figure 5.246 Original upper plaster model (left) and stereolithography model (right) as seen from (a) frontal, (b) occlusal, (c) right buccal, and (d) left buccal directions.

visually comparing the difference in repeat measurements and performing nonparametric, Wilcoxon signed rank hypothesis tests. This is described in the Results section.

5.37.3 Results

Data analysis demonstrated a nonnormal distribution of results, and nonparametric tests (Wilcoxon signed rank test) were, therefore, employed in the statistical analysis.

No significant difference ($P > .2$) was demonstrated in measurements at initial time (T1) and 1 week later (T2) for the manual measurement of plaster study models (Table 5.20, Fig. 5.248), 3D digital surface model measurement (Table 5.21, Fig. 5.249), or manual measurement of the

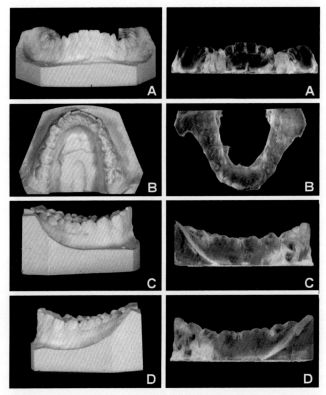

Figure 5.247 Original lower plaster model (left) and stereolithography model (right) as seen from (a) frontal, (b) occlusal, (c) right buccal, and (d) left buccal directions.

Table 5.20 Variation in repeat measurements of plaster models — 20 measurements in each plane were repeated on 30 models.

Plane	N	Mean difference (mm)	Standard deviation (mm)	P value
x plane	20	0.15	0.09	0.601
y plane	20	0.16	0.09	0.313
z plane	20	0.11	0.07	0.489
x, y, z planes	60	0.14	0.09	0.558

stereolithography, reconstructed models (Table 5.22, Fig. 5.250). Almost all points were clustered around the mean difference of zero, within two standard deviations of the mean difference (Figs. 5.248—5.250) indicating good intrarater reliability.

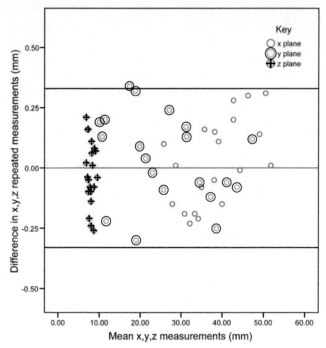

Figure 5.248 Bland altman plot for repeat measurements plaster model. Twenty measurements in each plane were repeated on 30 models (reference lines showing 2 SD).

Table 5.21 Variation in repeat measurements of virtual models — 20 measurements in each plane were repeated on 30 models.

Plane	N	Mean difference (mm)	Standard deviation (mm)	P value
x plane	20	0.15	0.93	0.823
y plane	20	0.12	0.75	0.549
z plane	20	0.14	0.11	0.501
x, y, z planes	60	0.14	0.92	0.965

A comparison of linear measurements made on the plaster study models and 3D digital surface (virtual) models is presented in Table 5.23 and Fig. 5.251. The mean difference in all planes was 0.14 mm (standard deviation 0.10 mm) and was not statistically significant ($P > .2$).

Measurements made in "x" and "y" planes were not significantly different for reconstructed models and plaster models ($P > .3$) or 3D digital

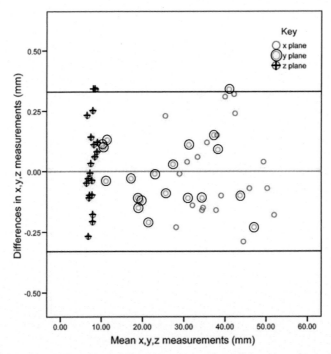

Figure 5.249 Bland altman plot for repeat virtual model measurements. Twenty measurements in each plane were repeated on 30 models (reference lines showing 2 SD).

Table 5.22 Variation in repeat measurements of the reconstructed model — 20 measurements in each plane were repeated on one model.

Plane	N	Mean difference (mm)	Standard deviation (mm)	P value
x plane	20	0.12	0.06	0.985
y plane	20	0.13	0.11	0.985
z plane	20	0.14	0.09	0.550
x, y, z planes	60	0.13	0.09	0.938

surface models $(P > .5)$. However, in the "z" plane, measurement differences were significantly different $(P < .001$, Tables 5.24 and 5.25; Figs. 5.252 and 5.253). All "z" plane reconstructed model measurements were significantly smaller than the corresponding plaster and 3D digital surface model measurements.

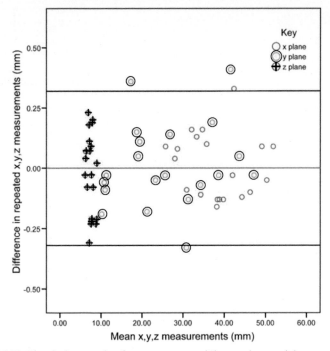

Figure 5.250 Bland altman plot for repeat stereolithography model measurements. Twenty measurements in each plane were repeated on one pair of models (reference lines showing 2 SD).

Table 5.23 Difference between plaster and virtual model measurements (means of 20 measurements in each plane compared).

Plane	N	Mean difference (mm)	Standard deviation (mm)	P value
x plane	20	0.19	0.12	0.765
y plane	20	0.14	0.09	0.501
z plane	20	0.10	0.07	0.218
x, y, z planes	60	0.14	0.10	0.237

5.37.4 Discussion

This study has demonstrated a simple and reproducible method of study model measurement. The excellent reproducibility of plaster, digital, and reconstructed model measurements reported compares favorably with Zilberman et al. [75] and Bell et al. [42] who reported mean intraoperator

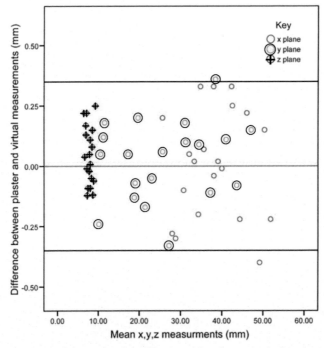

Figure 5.251 Bland altman plot for differences in plaster and virtual model measurements (reference lines showing 2 SD).

Table 5.24 Difference between plaster and reconstructed model measurements (means of 20 measurements in each plane compared).

Plane	N	Mean difference (mm)	Standard deviation (mm)	P value
x plane	20	0.15	0.16	0.645
y plane	20	0.19	0.15	0.360
z plane	20	0.42	0.23	>0.000**
x, y, z planes	60	0.26	0.22	>0.000**

errors of 0.18 and 0.17 mm, respectively, when the same points were measured by the same operator at different times on plaster study models, and Stevens et al. [83] who reported a concordance correlation coefficient of 0.88 for the measurement of digital models using emodel software (GeoDigm, Chanhassen, MN, USA) [38,75,83]. The reconstructed,

Table 5.25 Difference between virtual and reconstructed model measurements (means of 20 measurements in each plane compared).

Plane	N	Mean difference (mm)	Standard deviation (mm)	P value
x plane	20	0.18	0.12	0.550
y plane	20	0.22	0.16	0.513
z plane	20	0.38	0.21	>0.000**
x, y, z planes	60	0.25	0.21	>0.000**

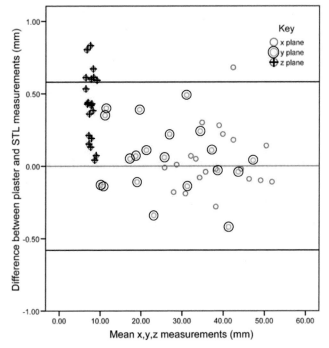

Figure 5.252 Bland altman plot for differences in plaster and stereolithography model measurements (reference lines showing 2 SD).

stereolithography model measurements in this study demonstrated a greater range in repeated absolute measurement differences (0.02–0.41 mm) compared with those for the other two methods (0.01–0.34 mm), reflecting the greater difficulty in measuring these models.

This study has also demonstrated the validity of digital (virtual) models derived from the laser-scanning process described. The problems of trying

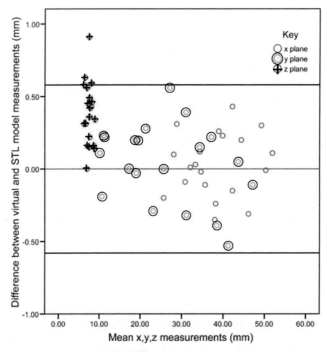

Figure 5.253 Bland altman plot for differences in virtual and stereolithography model measurements (reference lines showing 2 SD).

to acquire dimensionally accurate images using structured light scanning methods have been reported by Bibb et al. [93] and Mah and Hatcher [94]. The light beam from structured light scanners travels in straight lines, so any object surfaces that are obscured or are at too great an angle to the line of sight of the light source will not be scanned. This results in "voids" or "holes" in the scanned surface data. To overcome this problem, the object or the scanner needs to be moved to different angulations and the scan process repeated at each angle. For irregular objects, multiple scans of the same object from different angles may need to be acquired. The data from each of these scans can then be "stitched" (registered and merged) together using special software programs to produce a single composite surface model of the object [38,56,93]. Compounding these difficulties are the errors introduced during computer processing of the acquired data that is necessary to reduce artifacts and yet retain detail, while errors can also be introduced during the merging of the multiple perspectives to form the single composite surface model of the object being scanned [94].

A number of authors, who have evaluated alternative ways of measuring study models, have suggested what they consider a clinically significant measurement difference. Schirmer and Wiltshire [22] regarded a measurement difference between alternative measurement methods of less than 0.20 mm as clinically acceptable. Hirogaki et al. [11] suggested the accuracy required with orthodontic study models to be about 0.30 mm, while Halazonetis [38] reported that an accuracy of 0.50 mm was sufficient for head and face laser scanning but would be inadequate for scanning study models. Bell et al. [42] investigating the accuracy of the stereo-photogrammetry technique for archiving study models decided a mean difference of 0.27 mm (SD = 0.06 mm) between this technique and measurements made by hand on plaster models was unlikely to have a significant clinical impact.

The accuracy of the on-screen virtual models as reported in this study compares favorably with some studies but less favorably than with others. These studies varied greatly in their three-dimensional capture techniques and software analysis systems (Table 5.26).

The statistically significant difference between measurements made directly on the plaster models and those made on the reconstructed models was largely due to errors in the "z" plane. The stereolithography models were built in 0.15-mm layers from a clear resin. Model translucency made landmark identification difficult, and layering resulted in some loss of surface detail particularly at the cervical margin (Figs. 5.246 and 5.247). In addition, errors in data conversion and data manipulation generated while converting digital surface models to stereolithography file format can result in some distortion [94,95], and the RP technique can also introduce errors due to model shrinkage during building and postcuring [87]. However, clinical significance of these errors will depend on the intended purpose of the reconstructed model. The models may not be sufficiently accurate for appliance construction but may be sufficient to demonstrate pre- or post-treatment occlusal relationships. Unfortunately, the current prohibitive cost of stereolithography limited this study to the evaluation of only one pair of reconstructed models.

This study has presented a novel method of digitally recording study model data, offering the profession a valid alternative to the use of conventional plaster models and the potential to significantly reduce the burden of model storage. In addition, the potential for physical reconstruction of a model from the digital archive has been demonstrated, which may go toward addressing medicolegal concerns.

Table 5.26 Accuracy of various 3D capture techniques (average measurement error is the mean difference between measurements made on virtual models and the original plaster models).

Author	Type	Device	Average measurement	Landmark/identification
[61]	Point laser	Mitsubishi, MD1211-40	<0.072 mm	Landmarks not identified
[62]	Line laser	Hitachi, HL6712G laser	0.03 mm	Not specified
[84]	Line laser	UNISN, 3D VMS-250R	0.00–0.20 mm	Not specified
[66]	Laser	Not indicated	<0.10 mm	Not specified
[74]	Line laser	POS-PLD-50, laser 2000	Not more than +0.2 mm	Landmarks identified by marking
[63]	Line laser	Minolta VIVID700	0.08–0.35 mm	Landmarks not identified
[11]	Line laser	Cubesper laser	<0.30 mm	Landmarks not identified
[77]	Destructive laser scanner	OrthoCAD	1.2 mm	Overall bolton discrepancy
[12]	Line laser	Minolta VIVID700	x = 0.20 mm z = 0.70 mm	Landmarks identified by marking
[1]	Destructive laser scanner	OrthoCAD	0.16–0.49 mm	Landmarks not identified
[42]	Stereophotogrammetry	Not indicated	0.27 mm	Landmarks identified by marking
[3]	Destructive laser scanner	OrthoCAD	0.15–0.66 mm	Landmarks not identified

5.37.5 Conclusions

- The use of using a handheld Vernier caliper to measure plaster study models was reliable and reproducible.
- The Minolta VIVID 900 is a reliable device for capturing the surface detail of plaster study models three-dimensionally in a digital format using the protocol described.
- The measurement of the "on-screen" three-dimensional digital surface models captured was reproducible.
- The measurement of three-dimensional digital surface models and plaster models of the same dentitions showed good agreement.
- The detail and accuracy of physical models, reconstructed from digital data, may not be sufficient for certain applications, using the standard stereolithography techniques described
- Improved RP techniques may offer a more accurate method of model reconstruction from digital archives.

5.37.6 Future work

A stereolithography process employing thinner layers or other RP technologies that use significantly thinner build layers may address the deficiencies of the reconstructed models used in this study. Techniques that should be investigated include digital light processing—based machines (EnvisionTEC GmbH, Brüsseler Straße 51, 45968 Gladbeck, Germany) and the various printing based processes such as poly-jet modeling (Objet Geometries Ltd. Headquarters, 2 Holzman Street, Science Park, PO Box 2496, Rehovot 76124, Israel), multijet modeling (3D Systems Inc. Headquarters, 333 Three D Systems Circle, Rock Hill, SC 29730, USA), and single-head jetting (Solidscape Inc., 316 Daniel Webster Highway, Merrimack, NH 03054—4115, USA). These processes are all capable of producing physical models with a layer thickness of up to 10 times thinner than the stereolithography models described in this paper (layer thicknesses range from 0.013 to 0.150 mm).

Contributors

Andrew Keating, Richard Bibb, and Jeremy Knox were responsible for study design. Richard Bibb contributed to this work while at the National Center for Product Design and Development Research, University of

Wales Institute Cardiff, which supplied the stereolithography models. Andrew Keating was responsible for data collection. Andrew Keating and Alexei Zhurov were responsible for data manipulation and processing. All authors were responsible for data analysis and preparation of the manuscript. Jeremy Knox is the guarantor.

References

[1] Santoro M, Galkin S, Teredesai M, Nicolay OF, Cangialosi TJ. Comparison of measurements made on digital and plaster models. American Journal of Orthodontics and Dentofacial Orthopedics 2003;124:101–5.
[2] Hunter WS, Priest WR. Errors and discrepancies in measurement of tooth size. Journal of Dental Research 1960;39:405–14.
[3] Quimby ML, Vig KWL, Rashid RG, Firestone AR. The accuracy and reliability of measurements made on computer based digital models. The Angle Orthodontist 2004;74:298–303.
[4] Ayoub AF, Wray D, Moos KF, et al. A three-dimensional imaging system for archiving dental study casts: a preliminary report. The International Journal of Adult Orthodontics and Orthognathic Surgery 1997;12:79–84.
[5] McGuinness NJ, Stephens CD. Storage of orthodontic study models in hospital units in the UK. British Journal of Orthodontics 1992;19:227–32.
[6] McGuinness NJ, Stephens CD. Holograms and study models assessed by the PAR (peer assessment rating) index of malocclusion - a pilot study. British Journal of Orthodontics 1993;20:123–9.
[7] Yamamoto K, Toshimitsu A, Mikami T, Hayashi S, Harada R, Nakamura S. Optical measurement of dental cast profile and application to analysis of three-dimensional tooth movement in orthodontics. Frontiers of Medical and Biological Engineering 1988;1:119–30.
[8] Kuroda T, Motohashi N, Tominaga R, Iwata K. Three-dimensional dental cast analyzing system using laser-scanning. American Journal of Orthodontics and Dentofacial Orthopedics 1996;110:365–9.
[9] Foong KWC, Sandham A, Ong SH, Wong CW, Wang Y, Kassim A. Surface laser-scanning of the cleft palate deformity- Validation of the method. Annals Academy of Medical Singapore 1999;28:642–9.
[10] Hirogaki Y, Sohmura T, Takahashi J, Noro T, Takada K. Construction of 3-D shape of orthodontic dental casts measured from two directions. Dental Materials Journal 1998;17:115–24.
[11] Hirogaki Y, Sohmura T, Satoh H, Takahashi J, Takada K. Complete 3-D reconstruction of dental cast shape using perceptual grouping. IEEE Transactions on Medical Imaging 2001;20:1093–101.
[12] Kusnoto B, Evans CA. Reliability of a 3D surface laser scanner for orthodontic applications. American Journal of Orthodontics and Dentofacial Orthopedics 2002;122:342–8.
[13] Delong R, Heinzen M, Hodges JS, Ko CC, Douglas WH. Accuracy of a system for creating 3D computer models of dental arches. Journal of Dental Research 2003;82:438–42.
[13a] Kalender WA. Computed Tomography. 2nd ed. 2000. p. 101.
[14] Cookson AM. Space closure following loss of lower first premolars. Dental Practitioner 1970;21:411–6.

[15] Burstone CJ. Uses of the computer in orthodontic practice. Journal of Clinical Orthodontics 1979;13(442—453):539—51.

[16] Dervin E, Gore R, Kilshaw J. The photographic measurement of dental models. Medical and Biological Illustration 1976;26:219—22.

[17] McKeown HF, Robinson DL, Elcock C, Al-Sharood M, Brook AH. Tooth dimensions in hypodontia patients, their unaffected relatives and a control group measured by a new image analysis system. European Journal of Orthodontics 2002;24:131—41.

[18] Singh IJ, Savara BS. A method for making tooth and dental arch measurements. The Journal of the American Dental Association 1964;69:719—21.

[19] Huddart AG, Clarke J, Thacker T. The application of computers to the study of maxillary arch dimensions. British Dental Journal 1971;130(9):397—404.

[20] Mazaheri M, Harding RL, Cooper JA, Meier JA, Jones TS. Changes in arch form and dimensions of cleft patients. American Journal of Orthodontics 1971;60:19—32.

[21] Champagne M. Reliability of measurements from photocopies of study models. Journal of Clinical Orthodontics 1992;26:648—50.

[22] Schirmer UR, Wiltshire WA. Manual and computer-aided space analysis: a comparative study. American Journal of Orthodontics and Dentofacial Orthopedics 1997;112:676—80.

[23] McCance A, Perera S, Woods SJW. Trimming study models for photocopying. Journal of Clinical Orthodontics 1991;25:445—7.

[24] Yen CH. Computer-aided space analysis. Journal of Clinical Orthodontics 1991;25:236—8.

[25] Tran AM, Rugh JD, Chacon JA, Hatch JP. Reliability and validity of a computer-based Little irregularity index. American Journal of Orthodontics and Dentofacial Orthopaedics 2003;123:349—51.

[26] Van der Linden FP, Boersma H, Zelders T, Peters KA, Raaben JH. Three-dimensional analysis of dental casts by means of the Optocom. Journal of Dental Research 1972;51:1100.

[27] Suzuki S. An application of the computer system for three-dimensional cast analysis. Journal of the Japanese Orthodontic Society 1980;39:208—28.

[28] Scott PJ. The reflex plotters: measurements without photographs. Photogrammetric Record 1981;10:435—46.

[29] Scott PJ. Pepper's ghosts observed in Cape Town. South African Journal of Photogrammetry. Remote Sensing and Cartography 1984;40:89—95.

[30] Takada K, Lowe AA, DeCou R. Operational performance of the Reflex Metrograph and its applicability to the three-dimensional analysis of dental casts. American Journal of Orthodontics 1983;83:195—9.

[31] Speculand B, Butcher GW, Stephens CD. Three-dimensional measurement: the accuracy and precision of the reflex metrograph. British Journal of Oral and Maxillofacial Surgery 1988b;26:276—83.

[32] Speculand B, Butcher GW, Stephens CD. Three-dimensional measurement: the accuracy and precision of the reflex metrograph. British Journal of Oral and Maxillofacial Surgery 1988a;26:265—75.

[33] Johal AS, Battagel JM. Dental crowding: a comparison of three methods of assessment. European Journal of Orthodontics 1997;19:543—51.

[34] Bhatia SN, Harrison VE. Operational performance of the travelling microscope in the measurement of dental casts. British Journal of Orthodontics 1987;14:147—53.

[35] Takasaki H. Moire topography. Applied Optics 1970;9(6):1467—72.

[36] Kanazawa E, Sekikawa M, Ozaki T. Three-dimensional measurements of the occlusal surfaces of upper molars in a Dutch population. Journal of Dental Research 1984;63:1298—301.

[37] Mayhall JT, Kageyama I. A new three-dimensional method for determining tooth wear. American Journal of Physical Anthropology 1997;103:463—9.

[38] Halazonetis DJ. Acquisition of 3-dimensional shapes from images. American Journal of Orthodontics and Dentofacial Orthopedics 2001;119:556—60.

[39] Jones ML. An investigation into stereophotogrammetric measurement of routine study casts and its use in relating palatal cortica adaption to incisor movement. University of London; 1979. MScD dissertation.

[40] Richmond S. The feasibility of categorising orthodontic treatment difficulty: the use of three-dimensional plotting. University of Wales; 1984. MScD dissertation.

[41] Ayoub AF, Wray D, Moos KF, et al. A three-dimensional imaging system for archiving dental study casts: A preliminary report. International Journal of Adult Orthognathic Surgery and Osteotomy 1997;12:79—84.

[42] Bell A, Ayoub AF, Siebert P. Assessment of the accuracy of a three-dimensional imaging system for archiving dental study models. Journal of Orthodontics 2003;30:219—23.

[43] Kennedy D. The photography of orthodontic study casts. MScD dissertation. Cardiff: University of Wales; 1979.

[44] Schwaninger B, Schmidt RL, Hurst RV. Holography in dentistry. Journal of American Dental Association 1977;95:814—7.

[45] Burstone CJ, Pryputniewicz RJ, Bowley WW. Holographic measurement of tooth mobility in three dimensions. Journal of Periodontal Research 1978;13:283—94.

[46] Ryden H, Bjelkhagen H, Martensson B. Tooth position measurements on dental casts using holographic images. American Journal of Orthodontics 1982;81:310—3.

[47] Keating PJ, Parker RA, Keane D, Wright L. The holographic storage of study models. British Journal of Orthodontics 1984;11:119—25.

[48] Harradine N, Suominen R, Stephens C, Hathorn I, Brown I. Holograms as substitutes for orthodontic study casts: a pilot clinical trial. American Journal of Orthodontics and Dentofacial Orthopaedics 1990;98:110—6.

[49] Buschang PH, Ceen RF, Schroeder JN. Holographic storage of dental casts. Journal of Clinical Orthodontics 1990;24(5):308—11.

[50] Martensson B, Ryden H. The Holodent system, a new technique for measurement and storage of dental casts. American Journal of Orthodontics and Dentofacial Orthopaedics 1992;102:113—9.

[51] Romeo A. Holograms in orthodontics: a universal system for the production, development, and illumination of holograms for the storage and analysis of dental casts. American Journal of Orthodontics and Dentofacial Orthopaedics 1995;108:443—7.

[52] Berkowitz S, Gonzalez G, Nghiem-Phu L. An optical profilometer — a new instrument for the three dimensional measurement of cleft palate casts. Cleft Palate Journal 1982;19:129—38.

[53] Brook AH, Pitts NB, Renson CE. Determination of tooth dimensions from study casts using an Image Analysis System. Journal of the International Association of Dentistry for Children 1983;14:55—60.

[54] Brook AH, Pitts NB, Yau FS, Sandar PK. An image analysis system for the determination of tooth dimensions from study casts: comparison with manual measurements of mesio-distal diameter. Journal of Dental Research 1986;65:428—31.

[55] Quintero JC, Trosien A, Hatcher D, Kapila S. Craniofacial imaging in orthodontics: historical perspective, current status, and future developments. Angle Orthodontist 1999;69:491—506.

[56] Mah J, Baumann A. Technology to create the three-dimensional patient record. Seminars in Orthodontics 2001;7:251—7.

[57] Darvann TA, Hermann NV, Huebner DV, Nissen RJ, Kane AA, Schlesinger JK, et al. The CT-scan method of 3D form description of the maxillary arch. Validation and an

application. Transactions of the 9th International Con on Cleft Palate and Related Craniofac Anomalies. Goteborg: Erlanders Novum; 2001. p. 223—33.

[58] Kuo E, Miller RJ. Automated custom-manufacturing technology in orthodontics. American Journal of Orthodontics and Dentofacial Orthopedics 2003;123:578—81.

[59] Harada R, Yamamoto K, Ohnuma H, Mikami T, Nakamura S. Three-dimensional measurement of dental cast using laser and image sensor. Japanese Journal of Medical, Electronic and Biological Engineering 1985;23:166—71.

[60] Laurendeau D, Guimond L, Poussart D. A computer-vision technique for the acquisition and processing of 3-D profiles of dental imprints: an application in orthodontics. IEEE Transactions on Medical Imaging 1991;10:453—61.

[61] Wakabayashi K, Sohmura T, Takahashi J, Kojima T, Akao T, Nakamura T, et al. Development of the computerised dental cast form analyzing system — three dimensional diagnosis of dental arch form and the investigation of measuring condition. Dental Materials Journal 1997;16:180—90.

[62] Kojima T, Sohmura T, Wakabayashi K, Nagao M, Nakamura T, Takasima F, et al. Development of a new high-speed measuring system to analyze the dental cast form. Dental Materials Journal 1999;18:354—65.

[63] Sohmura T, Kojima T, Wakabayashi K, Takahashi J. Use of an ultrahigh speed laser scanner for constructing three-dimensional shapes of dentition and occlusion. Journal of Prosthetic Dentistry 2000;84:345—52.

[64] Nagao M, Sohmura T, Kinuta S, Wakabayashi K, Nakamura T, Takahashi J. Integration of 3-D shapes of dentition and facial morphology using a high-speed laser scanner. The International Journal of Prosthodontics 2001;14:497—503.

[65] Hayashi K, Araki Y, Uechi J, Ohno H, Mizoguchi I. A novel method for the three-dimensional (3-D) analysis of orthodontic tooth movement — calculation of rotation about and translation along the finite helical axis. Journal of Biomechanics 2002;35:45—51.

[66] Lu P, Li Z, Wang Y, Chen J, Zhao J. The research and development of a non-contact 3-D laser dental model measuring and analysing system. The Chinese Journal of Dental Research 2000;3:386—7.

[67] Alcaniz M, Montserrat C, Grau V, Chinesta F, Ramon A, Albalat S. An advanced system for the simulation and planning of orthodontic treatment. Medical Image Analysis 1998;2:61—77.

[68] Alcaniz M, Grau V, Monserrat C, Juan C, Albalat S. A system for the simulation and planning of orthodontic treatment using a low cost 3D laser scanner for dental anatomy capturing. Studies in Health Technology and Information 1999;62:8—14.

[69] Brosky ME, Pesun IJ, Lowder PD, Delong R, Hodges JS. Laser digitization of casts to determine the effect of tray selection and cast formation technique on accuracy. Journal of Prosthetic Dentistry 2002;87:204—9.

[70] Brosky ME, Major RJ, Delong R, Hodges JS. Evaluation of dental arch reproduction using three-dimensional optical digitisation. The International Journal of Prosthetic Dentistry 2003;90:434—40.

[71] Delong R, Ko C, Anderson GC, Hodges JS, Douglas WH. Comparing maximum intercuspation contacts of virtual dental patients and mounted dental casts. Journal of Prosthetic Dentistry 2002;88:622—30.

[72] Schelb E, Kaiser DA, Brukl CE. Thickness and marking characteristics of occlusal registration strips. Journal of Prosthetic Dentistry 1985;54:122—6.

[73] Braumann B, Keilig L, Bourauel C, Niederhagen B, Jager A. Three-dimensional analysis of cleft palate casts. Annals of Anatomy 1999;181:95—8.

[74] Commer P, Bourauel C, Maier K, Jager A. Construction and testing of a computer-based intraoral laser scanner for determining tooth positions. Medical Engineering and Physics 2000;22:625—35.

[75] Zilberman O, Huggare JAV, Parikakis KA. Evaluation of the validity of tooth size and arch width measurements using conventional and three-dimensional virtual orthodontic models. The Angle Orthodontist 2003;73:301—6.

[76] Redmond RW. The digital orthodontic office: 2001. Seminars in Orthodontics 2001;7:266—73.

[77] Tomassetti JJ, Taloumis LJ, Denny JM, Fischer JR. A comparison of 3 computerized Bolton tooth-size analyses with a commonly used method. The Angle Orthodontist 2001;71:351—7.

[78] Baumrind S. Integrated three-dimensional craniofacial mapping: background, principles, and perspectives. Seminars in Orthodontics 2001;7:223—32.

[79] Freshwater M, Mah J. The cutting edge. Journal of Clinical Orthodontics 2003;37:101—3.

[80] Hans MG, Palomo JM, Dean D, Cakirer B, Min KJ, Han S, et al. Three-dimensional imaging: the Case Western Reserve University method. Seminars in Orthodontics 2001;7:233—43.

[81] Baumrind S, Carlson S, Beers A, Curry S, Norris K, Boyd RL. Using three-dimensional imaging to assess treatment outcomes in orthodontics: a progress report from the University of the Pacific. Orthodontics and Craniofacial Research 2003;6(1):132—42.

[82] Mah J, Sachdeva R. Computer-assisted orthodontic treatment: the SureSmile process. American Journal of Orthodontics and Dentofacial Orthopaedics 2001;120:85—7.

[83] Stevens DR, Flores-Mir C, Nebbe B, Raboud DW, Heo G, Major PW. Validity, reliability and reproducibility of plaster v's digital study models: comparison of PAR and Bolton analysis and their constituent measurements. American Journal of Orthodontics and Dentofacial Orthopedics 2006;129:794—803.

[84] Motohashi N, Kuroda T. A 3D computer-aided design system applied to diagnosis and treatment planning in orthodontics and orthognathic surgery. The European Journal of Orthodontics 1999;21:263—74.

[85] Garino F, Garino GB. Comparison of dental arch measurements between stone and digital casts. World Journal of Orthodontics 2002;3:250—4.

[86] Chua CK, Chou SM, Lin SC, Lee ST, Saw CA. Facial prosthetic model fabrication using rapid prototyping tools. International Journal of Manufacturing and Technology Management 2000;11:42—53.

[87] Barker TM, Earwaker WJS, Lisle DA. Accuracy of stereolithographic models of human anatomy. Australian Radation Journal 1994;38:106—11.

[88] Kragskov J, Sindet-Pedersen S, Gyldensted C, Jensen KL. A comparison of three-dimensional computed tomography scans and stereolithographic models for evaluation of craniofacial anomalies. Journal of Oral and Maxillofacial Surgery 1996;54:402—11.

[89] Bill JS, Reuther JF, Dittmann W, et al. Stereolithography in oral and maxillofacial operation planning. International Journal of Oral and Maxillofacial Surgery 1995;24:98—103.

[90] Altman DG. Practical statistics for medical research. London: Chapman and Hall; 1991. p. 455—8.

[91] Adams PC. The design and construction of removable orthodontic appliances. 4th ed. Bristol: John Wright and Sons Ltd.; 1976.

[92] Bland JM, Altman DG. Measuring agreement in method comparison studies. Statics Method of Medical Research 1999;8:135—60.

[93] Bibb R, Freeman P, Brown R, Sugar A, Evans P, Bocca A. An investigation of three-dimensional scanning of human body surfaces and its use in the design and manufacture of prostheses. Proceedings - Institution of Mechanical Engineers Part H — Journal of Engineering Medicine 2000;214:589—94.

[94] Mah J, Hatcher D. Current status and future needs in craniofacial imaging. Orthodontics and Craniofacial Research 2003;6(Suppl. 1):10—6.

[95] Cheah CM, Chua CK, Tan KH, Teo CK. Integration of laser surface digitizing with CAD/CAM techniques for developing facial prostheses. Part 1: design and fabrication of prosthesis replicas. International Journal of Prosthodontics 2003;16:435—41.

CHAPTER 5.38

Dental applications case study 5—Design and fabrication of a sleep apnea device using CAD/ AM technologies*

5.38.1 Introduction

In dentistry, a three-stage process is commonly used to produce patient-specific devices: a scanner/digitization tool for data acquisition (either intraoral scanning, scanning of a poured cast obtained through a traditional impression, or direct scan of a conventional impression), computer-aided design (CAD) software for data processing, and a subtractive computer-aided manufacture (CAM) production technology for manufacturing [1,2]. Systems can be categorized into in-office (chair-side) systems, laboratory-based systems, and milling (production) center systems [3] depending on the locations of their components [4]. Additive manufacture (AM) building centers should also now be included in this list. The term appropriate for this study is CAD/AM.

The main principle of AM is that the final model consists of processed layers joined together. The layers are organized and bonded together beginning from the base to the uppermost aspect of the build. AM has previously been employed to build complex 3D models in medicine, but its application to orthodontics is comparatively limited. The technique is adept

---

* The work described here was first reported in the reference below and is reproduced here with kind permission of Sage Publishing.
Al Mortadi N, Eggbeer D, Lewis J, Williams RJ. Design and fabrication of a sleep apnea device using CAD/AM technologies. Proceedings of the Institution of Mechanical Engineers, Part H, Journal of Engineering in Medicine 2013;227(4):350—5.

Medical Modeling
ISBN 978-0-323-95733-5
https://doi.org/10.1016/B978-0-323-95733-5.00043-0

at producing complex medical and/or dental models that have fine details such as cavities, sinuses, undercuts, and thin bars [2,5,6]. AM technology is also suitable for building thin parts because it is difficult or impossible for these parts to be created by milling due to the potential flexure of the parts during production [7]. The technique also has the advantage of potentially allowing translucent structures and internal anatomy to be easily seen [8]. AM has been used in areas such as in the production of occlusal splints [9] and maxillofacial prosthetics [10−12]. The technology has also been used in orthodontic treatments [13] for the invisible orthodontic appliance fabricated without metal wire or brackets (Align Technology, Inc., Santa Clara, CA) and in producing customized lingual brackets [14]. The present report highlights a new application of the fabrication of a sleep apnea device with rotating parts produced in a single AM build.

Snoring and obstructive sleep apnea is a medical condition, which can result in abnormal pauses in breathing due to the soft palate obstructing the oropharynx. Disturbances in normal sleep patterns can leave individuals with daytime fatigue, slower reaction times, difficulties with vision, and moodiness, among other major problems [15,16]. It is estimated that about 80,000 people in Britain suffer from obstructive sleep apnea. They are mostly (but not all) men, mostly (but not all) overweight, especially around the neck, and they all snore. These conditions are treatable with dental sleep apnea appliances that are often designed to posture the mandible forward. Current sleep apnea devices are designed using laboratory-based techniques and require component assembly [17,18].

The problem this study addressed was to apply and evaluate novel CAD/AM design solutions that take advantage of AM to produce a hinged, patient-specific, sleep apnea device that did not require component assembly.

5.38.2 Methods and materials

Opposing dental casts used for technique development available at a university were scanned by a handheld laser sensor (Creaform, HandyScan 3D, 5825 rue St-Georges, Lévis, Québec) (Fig. 5.254). The manufacturer had calibrated this equipment to ±0.075 mm. Laser lines were projected from the scanner onto the surface of each cast to produce a three-dimensional (3D) image that appeared on the computer screen as a faceted surface (Fig. 5.255).

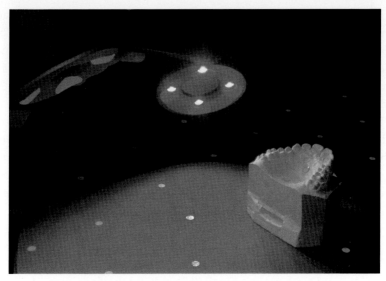

Figure 5.254 Scanning the cast with the handheld scanner.

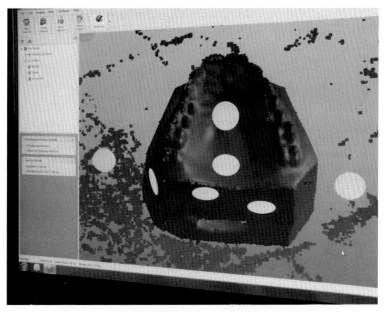

Figure 5.255 Point cloud data.

The scanner was moved around the casts by hand to produce a clear and complete 3D surface in the stereolithography (STL) file format (Fig. 5.256) with minimal missing data. Some manufacturers use a specific data format, which is not compatible with other construction programs [4]. STL file format is defined as a network of triangular facets that describes the 3D model design mathematically [19].

The obscured areas appeared on the screen as "holes." These mesh holes could be filled by software (Creaform, VX elements version 2.1) together with a specific tool. Data were processed and STL files generated using the proprietary scanner software (VX elements version 2.1). Computer software (FreeForm Modeling Plus, version 11, Geo Magics SensAble Group, Wilmington, MA 01887, USA) that enables the design of complex, free-form shaped objects and is ideal for designing anatomically based devices [5,7,11,12] was used and is the one developed in this study. This specialist CAD software was used in conjunction with a Phantom Desktop (Geo-magics Haptics Division) arm (Fig. 5.257) to build the sleep apnea device in a virtual environment.

The software has different tools to accomplish different tasks and ranges from sculpting to trimming and from contraction to stretching of the "clays" applied to the virtual cast. The virtual material, in this case the sleep apnea device, so formed is termed "clay," and it is manipulated and shaped in a similar way to wax in conventional methods. An experienced CAD/CAM dental technician can choose the appropriate tool for the required

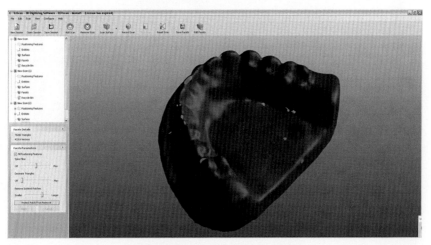

Figure 5.256 Completed 3D model.

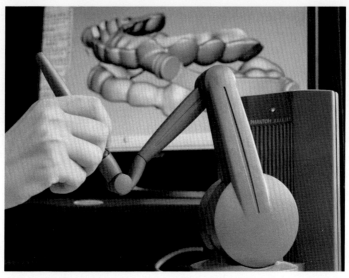

Figure 5.257 Phantom (haptic) arm and appliance being designed using "FreeForm."

steps easily and quickly. The Phantom device works as a 3D mouse mimicking the hand of a dental technician, enabling the creation of appropriate appliances in a similar way to the conventional laboratory method.

Prior to the virtual build, it was necessary to save the scanned dental casts as "buck" files by altering the settings appearing at the foot of the screen. This ensures the models are protected since any change to this saved form is not permitted.

The sleep apnea device was developed in the virtual environment by initially orienting the scanned casts and creating a normal (class I) molar relationship or as close to it as possible. This was achieved within the virtual environment by dragging the mandible forward and by opening the vertical dimension between the upper and lower casts from a closed position until there was a space of 4—5 mm between the opposing premolars, which is considered enough to open the palatal blockage and therefore to help prevent snoring. Spaces were measured using the "ruler" function of the software.

The inclination of upper central incisors toward the lip was selected as a reference for the path of insertion of the upper cast, and the lower cast followed the upper according to the corrected occlusal relationship. The midlines between upper and lower incisors were aligned until coincident.

Reentry points or undercuts on both virtual models that would prevent the seating of the appliance were detected automatically using the "set pull direction" option from the "view" tool. These areas were blocked out using the "layer" tool to add clay with the thickness depending on the size of the undercut. The blocked-out layers were smoothed using the "smooth" tool, and the files were then saved as "buck" files, so that the models were protected and could not be affected by the software tools. It is essential that undercut areas are identified and eliminated to allow the creation of parallel guide planes so that the appliance will fit onto a physical model and ultimately the patient.

In the virtual environment, a 2-mm layer was applied to cover the tooth surfaces, as shown in Fig. 5.258, using the "layer" tool with an offset thickness of 2 mm. The layers covered the lingual tooth surfaces and further extended to 3 mm below the gingival margins. The layers were kept 1 mm above the gingival margins of the molars and premolars buccally to avoid soft tissue irritation. One-millimeter capping was applied on the four incisors of opposing casts to minimize the inclination toward the lips or to prevent overeruption.

The hinges on both sides connected the upper molars to lower canines (Fig. 5.258). The role of these hinges is to move the mandible forward into the corrected position. Four hinges were built for each single appliance: two on the lower canines and two on the upper first molars. A bar on either side connected the hinges. The four hinges were positioned parallel to each other to allow free rotation.

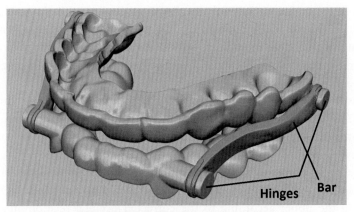

Figure 5.258 Sleep apnea device design.

Each hinge consisted of two parts (Fig. 5.259). The first part was a circular rod (2.5 mm length, 3 mm diameter), with two wide circular "stops" (6 mm diameter) at the ends (Fig. 5.259a). The second part was a hollow cylinder with an internal diameter of 3.5 mm and external diameter of 6 mm (Fig. 5.259b). The two parts were separated in all areas to permit rotational freedom of the cylinder around the rod. There was a gap clearance between the sides of the cylinder and the first part. The inner gap was 0.2 mm, and the outer gap was 0.1 mm. The gap between the inner surface of the cylinder and the rod was 1 mm. The internal stop connected the teeth continuously.

All hinges were built in the same way by use of the "wire cut" tool. This tool can build a 3D model from a 2D sketch by producing an additional prismatic plane, which can either "cut" or "fill" the profile area between it and the original plane. It is critical to position the original plane in the correct place and to determine how far from the additional prismatic plane by "dragging."

For building a hinge virtually, a new plane was created on the frontal view of the models. Three combined circles to make up the hinges were drawn on the plane. The diameter of the outer circle was 6 mm, the middle circle 3 mm, and the inner was 2.5 mm. The outer and inner circles were used to produce the stops labeled as "B" in Fig. 5.259.

All the hinges were drawn on the same plane. It was only possible to modify the plane by moving it using "snap axis," which moves the plane in and out without any rotation in any axis. This ensures the parts of each hinge are parallel and guarantees the four hinges are parallel to each other. A profile of the bar to connect the two hinges was drawn on the same plane too. The thickness of the bar was 2.9 mm, similar to the cylinder's thickness. However, the transverse dimension of the bar was 4 mm.

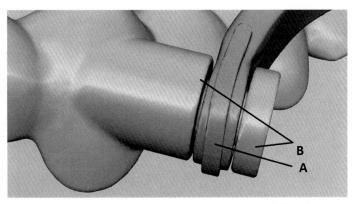

Figure 5.259 The hinge: (a) a rod with its stops and (b) cylinder.

The bars were constructed with a bend to avoid interference with the teeth and device during movement, but at the same time allowing positioning to be as close as possible to them (see Fig. 5.258).

Finally, the design was exported as an STL file. The ProJet 3000 Plus 3D printing system (3D-Systems, Rock Hill, South Carolina) was chosen for fabrication.

The process is capable of building in extremely fine detail ($656 \times 656 \times 1600$ dots per inch in x, y, and z axes) necessary to define the intricate features. It uses an acrylate-based polymer (EX200, 3D-Systems) for the component and, crucially, uses a wax support structure. This enables the production of high-definition hinge mechanisms that can be built without assembly. Once the print is complete, the wax that supports and separates can be melted away to free the hinge mechanisms. This also provides the freedom to incorporate other features such as slides and screws that do not require assembly to be fabricated in a single build. Even with the ability to melt the support structure, it is necessary to ensure a sufficient gap between hinge components to enable trapped wax to be removed. Parallelism of hinges is also very important for hinge function.

The device was transferred into proprietary software (Client Manager) and orientated in the same plane in which it would be worn (i.e., the lowest z height) to obtain the optimum build speed. The building took 8 h to complete. The completed build was postprocessed by using an oven set at 72°C to melt the wax support structure and an oil bath to remove residual wax followed by degreasing to clean the device. This was then trial fitted to the stone casts (Fig. 5.260).

5.38.3 Results

The device was fitted on the physical casts and its features checked and compared with the expectations of conventionally formed appliances. The device had smooth surfaces with no sharp edges that could not be easily dealt with by normal finishing procedures.

Due to the virtual preparation of the models, no areas were undercut. The device was inserted and removed to and from the casts easily without any interference. The sleep apnea device displayed an excellent fit on the teeth and palatal and gingival tissues.

The appliance was considered functional, and the hinges worked well to posture the mandible forward. The hinges allowed free forward and reverse movements. The hinges also permitted some lateral movement. Further

Figure 5.260 Sleep apnea prototype fitted to cast.

experimentation could reduce this, but there may be an argument in favor of maintaining such movement to allow the device to be myodynamic, that is, allowing muscular exercise.

5.38.4 Discussion

Currently, the majority of sleep apnea devices are built relying on fairly intensive handcrafting methods. The design described in this research for a sleep apnea appliance highlighted the technical achievements necessary to use CAD/AM technologies for its fabrication. With the AM technique, it is possible to build all parts of a sleep apnea device in one build, layer by layer, and parts of hinges can be manufactured without assembly. In the future, there may also be opportunities for greater automation.

Another benefit of using CAD/AM technologies as described above is the possibility of linking it to intraoral scanning. This may allow an appliance to be made without the use of a dental cast. Hence the difficulties of tray selection, impression material selection, and inadequate impression-tray adhesive, dimensional changes in impression material, and bubble entrapment during impression pouring could be avoided. There is evidence that misunderstanding between laboratory technicians and dentists [20] and patient discomfort [21] may also be avoided. The technique offers the negation of repeated disinfection necessary as items passing between the laboratory and clinic are found to adversely affect the dimensional accuracy of impression materials [22]. Other advantages include ease of identification of undercuts and the parallel positioning of hinges, which is much less precise when accomplished by hand—eye coordination.

It is suggested that AM techniques are advantageous as there is little waste of material with this technology, and production is potentially fast as any number of objects can be produced at once as long as they fit within the build envelop of the machine. Examples of another interesting application is that there are possibilities for AM use in the fabrication of removable partial dentures [2,7].

Although the techniques described demonstrate the potential benefits in workflow and design freedom, technical and material challenges currently prevent the chosen technologies being suitable for clinical application. The ProJet EX200 material used in this study has not undergone toxicity testing to the necessary standards to enable it to be used in intraoral device production. There are currently no AM polymers approved for long-term intraoral use, which remains a challenge for material developers. However, there is significant pace in the development of AM processes and materials, particularly in the medical device sector, which may shortly overcome these issues.

It is the considered opinion of the authors that the appliance described above is worthy of presentation to a dental clinic.

5.38.5 Conclusion

A method of applying CAD/AM to patient-specific devices requiring rotating parts has been described through employing handheld scanners, a virtual environment, and AM technologies. The benefits, difficulties, and barriers have also been discussed. However, the built example shown above demonstrates the enormous potential for the application of CAD/AM technologies to sleep apnea treatment and the benefits of accuracy and control that become available with these technologies.

References

[1] Bever P, Brown C. The CAD in CAD/CAM: CAD design software's powerful tools continue to expand. Inside Dental Technology: Published by AEGIS Communications 2011;2(9):52.
[2] Van Noort R. The future of dental devices is digital. Dental Materials 2012;28(1):3—12.
[3] Liu P-R, Essig ME. A panorama of dental CAD/CAM restorative systems. CAD/CAM Systems Update 2008;29(8):482—93.

[4] Beuer F, Schweiger J, Edelhoff D. Digital dentistry: an overview of recent developments for CAD/CAM generated restorations. British Dental Journal 2008;204(9):505—11.

[5] Bibb R, Eggbeer D, Williams R. Rapid manufacture of removable partial denture frameworks. Rapid Prototyping Journal 2006;12(2):95—9.

[6] Azari A, Nikzad S. The evolution of rapid prototyping in dentistry: a review. Rapid Prototyping Journal 2009;15(3):216—22.

[7] Williams RJ, Bibb R, Eggbeer D. CAD/CAM-fabricated removable partial denture alloy frameworks. Practical Procedures & Aesthetic Dentistry (PPAD) 2008;20(6):349—51.

[8] Liu Q, Leu MC, Schmitt SM. Rapid prototyping in dentistry: technology and application. The International Journal of Advanced Manufacturing Technology 2006;29:317—35.

[9] Lauren M, McLntyre F. A New computer-assisted method for design and fabrication of occlusal splints. American Journal of Orthodontics and Dentofacial Orthopedics 2008;133:S130—5.

[10] Hughes CW, Page K, Bibb R, Taylor J, Revington P. The custom-made titanium orbital floor prosthesis is reconstruction for orbital floor fractures. The British Journal of Oral & Maxillofacial Surgery 2003;41(1):50—3.

[11] Eggbeer D, Bibb R, Evans P. Digital technologies in extra-oral, soft tissue facial prosthetics: current state of the art. The Institute of Maxillofacial Prosthetists and Technologists (JMPT) — Winter 2007;10:9.

[12] Bibb R, Eggbeer D, Evans P. Rapid prototyping technologies in soft tissue facial prosthetics: current state of the art. Rapid Prototyping Journal 2010;16(2):130—7.

[13] Abe Y, Maki K. Possibilities and limitation of CAD/CAM based orthodontic treatment. International Congress Series 2005;1281:1416.

[14] Wiechmann D, Rummel V, Thalheim A, Simon J-S, Wiechmann L. Customized brackets and archwires for lingual orthodontic treatment. American Journal of Orthodontics and Dentofacial Orthopedics 2003;124(5):593—9.

[15] Dieltjens M, Vanderveken OM, van de Heyning PH, Braem MJ. Current opinions and clinical practice in the titration of oral appliances in the treatment of sleep-disordered breathing. Sleep Medicine Reviews 2012;16(2):177—85.

[16] Malhotra A, White DP. Obstructive sleep apnoea. Lancet 2002;360:237—45.

[17] Stradling J, Dookun R. The role of the dentist in sleep disorders. Dent Update March 2011;38(2):136.

[18] Cistulli PA, Gotsopoulos H, Marklund M, Lowe AA. Treatment of snoring and obstructive sleep apnea with mandibular repositioning appliances. Sleep Medicine Reviews 2004;8:443—57.

[19] Gronet PM, Waskewicz GA, Richardson C. Preformed acrylic cranial implants using fused deposition modeling: a clinical report. The Journal of Prosthetic Dentistry 2003;90(5):429—33.

[20] Polido WD. Digital impressions and handling of digital models: the future of dentistry. Dental Press Journal of Orthodontics 2010;15(5):18—22.

[21] Christensen GJ. The challenge to conventional impressions. The Journal of the American Dental Association 2008;139(3):347—8.

[22] Al Mortadi N, Chadwick RG. Disinfection of dental impressions — compliance to accepted standards. British Dental Journal 2010;209(12):607—11.

CHAPTER 5.39

Dental applications case study 6—CAD/CAM/AM applications in the manufacture of dental appliances*

5.39.1 Introduction

Computer-aided design, computer-aided manufacturing, additive manufacturing (CAD/CAM/AM), and scanning systems have been successfully introduced in dentistry [1–3]. Researchers have used these technologies as a tool for orthodontic diagnosis and treatment planning for determining the position of impacted maxillary canines [4] and for the fabrication of occlusal splints [5]. However, CAD/CAM/AM innovations have not been used successfully on a wide scale for removable orthodontic appliances.

Sassani and Roberts [6] reported that it was possible to use CAD/CAM/AM techniques to build base plates of orthodontic appliances, but they stated that it was not possible to incorporate wires into the built parts. However, this study discusses a method to accomplish such incorporation. The technology mentioned above has been used in orthodontic treatments such as the invisible orthodontic appliance fabricated without metal wire or brackets (Align Technology, Inc., Santa Clara, CA) [7] and in producing customized lingual brackets [8]. Recently, Wiechmann et al. [9] used this

---

* The work described here was first reported in the reference below and is reproduced here with kind permission of the copyright holders.

Al Mortadi N, Eggbeer D, Lewis J, Williams RJ. CAD/CAM/AM applications in the manufacture of dental appliances. American Journal of Orthodontics & Dentofacial Orthopedics 2012;142(5):727–33.

Medical Modeling
ISBN 978-0-323-95733-5
https://doi.org/10.1016/B978-0-323-95733-5.00044-2

computer-based technology to make connectors for Herbst appliance hinges to custom lingual brackets and a more recent study used this technique for a digital titanium Herbst appliance [10].

Given the above lack of application in general to removable appliances, the aim of this study was to apply new methods of producing dental appliances with wires and hinges using CAD/CAM/AM-based techniques and to evaluate these. Andresen and sleep apnea appliances were chosen. Although an Andresen appliance is rarely used, it provides a clear illustration for the inclusion of wire into a build, which is straightforward. A sleep apnea device provides a case where four coordinating hinges can be studied. Both devices were produced in a single build.

5.39.2 Material and methods

Opposing class II case models suitable to be treated with Andresen and sleep apnea removable devices were scanned by a handheld laser sensor (Creaform, HandyScan 3D, 5825 rue St-Georges, Lévis, Québec) (Fig. 5.254 in Chapter 5.38). Laser lines were projected from the scanner onto the surface of each model to produce a 3D image from a "point cloud." The software automatically transformed the point cloud into a stereolithography (STL) facet file, which is shown as it appears on a screen (Fig. 5.255 in Chapter 5.38). The scanner was moved around the casts by hand to produce a clear and complete 3D surface in the STL file format (Fig. 5.256 in Chapter 5.38) with minimal missing data.

Data were processed and STL files were generated using a proprietary scanner software (Creaform, VX elements version 2.1). The appliances were designed using specialist CAD software (FreeForm Modeling Plus, version 11, SensAble Technologies, Wilmington, Massachusetts 01887), in conjunction with a Phantom Desktop (SensAble Technologies) arm (Fig. 5.257 in Chapter 5.38). This apparatus allowed appliances to be "built" on scans of a patient's cast in a virtual environment.

The software enables the design of complex, nongeometric forms and is ideal for designing anatomic-based devices [11]. It has different tools to accomplish specific functions. Some tools are used for building, while others are used for trimming and shaping the constructed parts. The Phantom device works as a 3D mouse as it mimics the work of a dental technician, enabling the design and construction of the appropriate appliances in a similar way to the conventional method. The operator is able to feel force-feedback sensations during the process.

Prior to the virtual build, it was necessary to create a class I molar relationship or achieve this as closely as possible. This relationship was accomplished by posturing the mandible forward, until a class 1 M relationship existed and the vertical opening between opposing premolars was 4–5 mm at the lingual cusps.

The labial inclination of upper central incisors was selected as a reference for the "path of insertion" of the appliance onto the upper model. Undercuts on both virtual models were detected and eliminated virtually automatically using the "extrude-to-plane" tool. The files were then saved as "buck" files, which meant that the models were protected and could not be changed.

The following is a brief description of the design process for each appliance.

5.39.2.1 Andresen

A 2-mm layer was applied to the upper and lower models with capping on the lower incisors. Palatal coverage was made by selecting the required area using a "cursor" and offsetting a 2-mm thickness to form the plate. Upper and lower plates extended to the first molars. The capping was 1 mm thick on the anterior lower teeth to prevent proclination and overeruption. The thickness of the appliance was set at 2 mm to withstand occlusal forces and patient handling. The virtual material so formed is termed "clay," and it is added and shaped in a similar way to physical wax. It was necessary to add clay in some areas of the appliance to make the surfaces smooth and to form junctions of two surfaces continuous to each other. Borders were smoothed.

Finally, the occlusal surfaces of the opposing offset clay areas were connected to produce a one piece "monoblock" at the lingual surfaces of the teeth extending to the middle of the occlusal surfaces (Fig. 5.261).

These blocks were attached to the plates, and the junctions were smoothed using the "sculpting" and "smooth" tools. The mid distances of the blocks were stretched buccally using the "smudge" tool up to the outline of the buccal surfaces to form a sharp "intermaxillary edge." The blocks were cleared from the mandibular teeth to permit upward and forward eruption (Fig. 5.262).

To allow wire to be inserted within the physical build, a square section tube (1.25 × 1.25 mm) on either side of the appliance was created. Each tube consisted of two parts: The first part was horizontal, 15 mm in length, and 1 mm from the lingual walls of the intermaxillary blocks and was

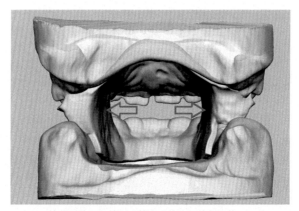

Figure 5.261 Andresen monoblock.

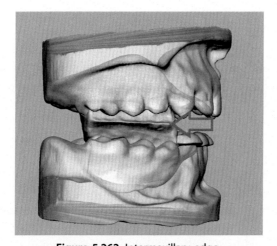

Figure 5.262 Intermaxillary edge.

located in the maxillary part with its lower border at the level of the intermaxillary edge. The second part was vertical projecting into the maxillary part and 5 mm in length. The first part was at right angles to the second part, and these were connected at a point between the second premolar and first molar. These tubes were made so that the retentive part of a 0.9-mm hard stainless-steel wire labial bow could be inserted, as shown in Fig. 5.263.

For the wire to be inserted in the tubes within the appliance, it was necessary to build "guiding jigs." Fitting the wire in the guiding jigs prior to the build ensured that the wire tagging would fit into the appliance during the build.

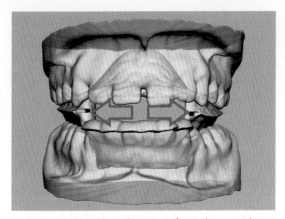

Figure 5.263 The tubes seen from the outside.

5.39.2.2 The guiding jigs

The guiding jigs (i.e., small plastic components around which wire could be bent prior to the main build) were built from rapid prototype material for the 3D Systems Project 3000 Plus Machine, which was acrylate based. The jigs were produced using the "copy" tool to duplicate the upper virtual model in conjunction with the outline of the tubes. Clays were constructed in the space between the model and the level of the tubes (Fig. 5.264a). Also, a small amount of clay was added under the tubes to make sure the wire would not be in contact with the model (Fig. 5.264b). This was accomplished by offsetting a 1-mm layer underneath the tubes using a "sculpting clay" tool.

The tubes formed spaces along which wire would be bent prior to insertion in the final device during the build process. Fig. 5.265 shows 1.25-

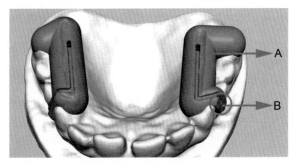

Figure 5.264 Guiding jigs.

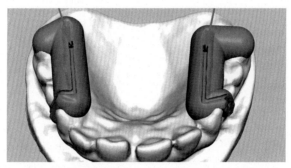

Figure 5.265 Outline of the tubes in guiding jigs.

mm–deep grooves. Sharp angles were smoothed to facilitate positioning of the wire in the tubes.

Next, the guiding jigs were exported as an STL file to an AM machine (3D Systems Project 3000 Plus Machine) for building. VisiJet EX200 (DSM Somos, Elgin, Illinois, USA), a 3DP build material, which is acrylate based and near colorless, was used. The wax supporting material was removed by melting using an oven set at 72°C, and an oil bath was used to remove residual wax. A citrus-based degreaser was then used to clean the jigs before they were fitted to the stone models.

The labial bow was bent in the conventional way and the tagging fitted to the jigs positioned on the gypsum cast (Fig. 5.266). The labial bow was constructed from 0.9-mm hard stainless-steel wire. The bow extended from canine to canine touching the labial surfaces of the upper incisors and the

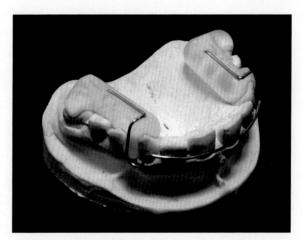

Figure 5.266 Labial bow fitted on the guiding jigs.

canines at the mesial third area. From that point on each side, U loops were formed 1–2 mm below the gingiva on both sides. The wire was then passed between the canines and first premolars to enter the plastic between the occlusal surfaces. The wire tagging followed the tubes formed in the jigs. The horizontal parts of the wire tags were made straight and passively fitted on the jigs.

The next stage was to export the Andresen design to an AM machine as an STL file ready for fabrication in WaterShed XC 11122 (DSM Somos). Stereolithography (SLA) was chosen for fabrication since it enables builds to be paused and prefabricated pieces to be inserted and built around. This was necessary for embedding wires.

Preparation software (Magics, Materialise, Leuven, Belgium) was used to orientate the Andresen design with the recessed section to accommodate the wire set horizontally. The same software was also used to generate support structures that are necessary to attach the part to the build platform. Lightyear (version 1.1, 3D-Systems) was used to prepare and slice the build in 0.1-mm layers before transferring the build to the SLA machine (SLA 250-50, 3D-Systems, Rock Hill, South Carolina). The machine was paused before it reached the top of the tubes to allow insertion of the wire in the correct place and then restarted again. The "zephyr blade" recoating mechanism, which normally sweeps unprocessed liquid from the build surface of the machine, was stopped from sweeping for the three layers after insertion of the wire so that the wire was not disturbed during the process. The procedure continued until building was completed (Fig. 5.267).

Support structures were removed from the device, and it was cleaned in fresh isopropanol solvent (99% minimum) for 20 min and then dried using compressed air and ventilated for a further 6 h. Next the device was postcured for 30 min in UV light to ensure full polymerization. The final design is shown in Fig. 5.268 and the built Andresen fitted to the master model in Fig. 5.269.

5.39.2.3 Sleep apnea device

The material used was VisiJet EX200. In the virtual environment, a 2-mm layer was applied to the upper and lower models and extended along the occlusal and buccal surfaces to approximately 1 mm short of the gingival margins facially to avoid irritation to the gingivae. 1-mm capping was applied on the four incisors to prevent labial inclination and extended

Figure 5.267 The final prototype before cleaning.

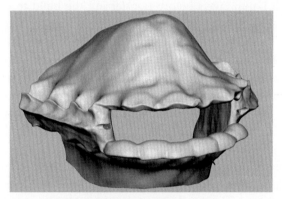

Figure 5.268 Final Andresen in CAD.

2–3 mm below the palatal and lingual gingival margins. The hinges on both sides connected the upper molars to lower canines and were parallel to allow free rotation. The role of these hinges is to move the mandible forward into the corrected position. See the previous case study 5.38 for more detail and figures.

Each bar connected two hinges located on the opposing arches. One bar was built on each side to connect the upper first molar hinge to the lower canine hinge. The four hinges were parallel to each other. Each hinge consisted of two parts. The first part was a circular rod (2.4 mm length,

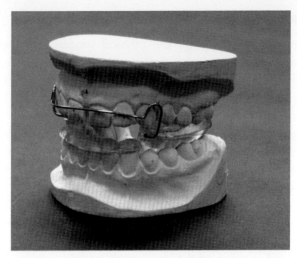

Figure 5.269 Andresen prototype fitted on model.

3 mm diameter) a wide circular "stop" (6 mm diameter). The second part was a hollow cylinder with an internal diameter of 3.5 mm and external diameter of 6 mm. The two parts were separated in all areas to permit rotational freedom of the cylinder around the rod. There was a gap clearance between sides of the cylinder and the first part. The inner gap was 0.2 mm, and the outer gap was 0.1 mm. The gap between the inner surface of the cylinder and the rod was 0.5 mm.

Finally, the design was exported as an STL file. A ProJet 3000 Plus 3D printing machine (3D-Systems) was chosen to manufacture the sleep apnea device. The process fabricates in an acrylate-based polymer with a wax support structure, which effectively allows parts to be separated after building. Fundamentally, the process allows features such as hinges to be fabricated as one piece, with the wax melted away to free the mechanism once produced. The ProJet process is also capable of building extremely fine detail ($656 \times 656 \times 1600$ dots per inch in x, y, and z).

The device was transferred into proprietary software and oriented in the same plane in which it would be worn (i.e., the lowest z height) to obtain the optimum build speed. The printing took 8 h to complete. The completed build was postprocessed by using an oven set at $72°C$ to melt the wax support structure and an oil bath to remove residual wax followed by degreasing to clean the device. This was then trial fitted to the stone models.

5.39.3 Results

The appliances were fitted on the physical models and their features checked and compared with the expectations of conventional appliances. Both appliances built by the AM method had smooth surfaces with no sharp edges that could not be easily dealt with by normal finishing procedures.

With regard to the Andresen, the fit palatally and lingually was very satisfactory. Due to the virtual preparation of the models, no areas were positioned in undercuts. The labial bow fitted firmly in the tubes, was quite well connected to the appliance, and was deemed functional. The acrylic guiding channels could be trimmed according to conventional practice to enhance the anterior movement of the lower teeth and the posterior movement of the upper teeth and to encourage expansion of the upper arch and upward eruption of the lower teeth into the corrected occlusion.

The sleep apnea device displayed an excellent fit on the teeth and palatal and gingival tissues. The hinges allowed free anterior and posterior movements. Some sideways movement was also permitted by the hinges. Further experimentation could reduce this, but there may be an argument in favor of maintaining such movement to allow the device to be myodynamic.

5.39.4 Discussion

The appliances described above could be presented to a clinic for fitting to a patient. The two appliances illustrated in this research highlight the technical possibilities of using CAD/CAM/AM technologies in dental device design. Although the techniques are currently slow and expensive, the benefits of introducing digital technologies into other areas of dentistry are well known. The ability to fabricate layer by layer enables features such as inserting wire, producing hinges, and building threads to be made in one build without assembly. In orthodontics, this has potential application in fabricating devices incorporating built expansion screws and twin blocks, which is part of ongoing research. Other benefits of using CAD are accurate control of the thickness of the material producing an even shape when required and exact positioning of the periphery of the design. Precise measurements can be easily taken as the virtual build progresses or programmed in as the software produces layers. A huge potential advantage is that the techniques could be linked to intraoral scanning, which would help to streamline the process of producing appliances by removing the impression and dental cast production stages.

5.39.5 Conclusion

The above indicates that digital technologies may be applied to the fabrication of appliance types not previously considered to be possible.

References

[1] Stevens B, Yang Y, Mohandas A, Stucker B, Nguyen KT. A review of materials, fabrication methods, and strategies used to enhance bone regeneration in engineered bone tissues. Journal of Biomedical Materials Research Part B: Applied Biomaterials 2007;85:573—82.

[2] Sun Y, Lü P, Wang Y. Study on CAD & RP for removable complete denture. Computers Methods and Programs in Biomedicine 2009;93:266—71.

[3] Williams RJ, Bibb R, Eggbeer D. CAD/CAM-fabricated removable partial denture alloy frameworks. Practical Procedures and Aesthetic Dentistry 2008;20:349—51.

[4] Faber J, Berto PM, Quaresma M. Rapid prototyping as a tool for diagnosis and treatment planning for maxillary canine impaction. American Journal of Orthodontics and Dentofacial Orthopedics 2006;129:583—9.

[5] Lauren M, McLntyre F. A New computer-assisted method for design and fabrication of occlusal splints. American Journal of Orthodontics and Dentofacial Orthopedics 2008;133:S130—5.

[6] Sassani F, Roberts S. Computer-assisted fabrication of removable appliances. Computers in Industry 1996;29:179—95.

[7] Abe Y, Maki K. Possibilities and limitation of CAD/CAM based orthodontic treatment. International Congress Series 2005;1281:1416.

[8] Wiechmann D, Rummel V, Thalheim A, Simon J-S, Wiechmann L. Customized brackets and archwires for lingual orthodontic treatment. American Journal of Orthodontics and Dentofacial Orthopedics 2003;124:593—9.

[9] Wiechmann D, Schwestka-Polly R, Hohoff A. Herbst appliance in lingual orthodontics. American Journal of Orthodontics and Dentofacial Orthopedics 2008;134:429—46.

[10] Farronato G, Santamaria G, Cressoni P, Falzone D, Colombo M. The cutting edge. Journal of Clinical Orthodontics 2011;XLV(5):263—7.

[11] Bibb R, Eggbeer D, Williams R. Rapid manufacture of removable partial denture frameworks. Rapid Prototyping Journal 2006;12:95—9.

PART 5.6

Research applications

PART III

Research applications

CHAPTER 5.40

Research applications case study 1—Bone structure models using stereolithography*

5.40.1 Introduction

To further the understanding of osteoporosis and its dependence upon the material and structural properties of cancellous bone, many experimental studies continue to be undertaken on natural tissue samples obtained from human subjects and animals. A significant difficulty however in analyzing in vitro bone samples is that the structural parameters have to be elucidated by destructive means. In addition, the human in vitro samples most readily available tend to be from an elderly population and therefore may be of limited structural variation compared with the full population age range. The development of a physical model of cancellous bone whose structure could be controlled would provide significant advantages over the study of in vitro samples. This would also enable the relationship between the mechanical integrity and hence fracture risk of cancellous bone and its structural properties to be more exactly defined.

This chapter describes how these complex three–dimensional structures can be physically reproduced using rapid prototyping techniques. As the chapter will show, although use of rapid prototyping techniques allows the generation of physical objects that would have previously been impossible to manufacture, problems are still encountered. The data, generated from microcomputed tomography (μCT), were used to perform finite element

---

* The work described in this chapter was first reported in the reference below and is reproduced here in part or in full with the permission of MCP UP Ltd.
Bibb R, Sisias G. Bone structure models using stereolithography: a technical note. Rapid Prototyping Journal 2002;8(1):25−9.

Medical Modeling
ISBN 978-0-323-95733-5
https://doi.org/10.1016/B978-0-323-95733-5.00045-4

analysis (FEA) on the structure of various human and animal bones. The physical models were required to validate the results of the FEA.

The difficulties encountered in creating physical models of these structures arose from the nature of the structure. Not only does the highly complex porous structure result in extremely large computer files, but it also presents problems of support during the build process.

5.40.2 Human sample data

The techniques of serial sectioning and μCT reconstruction were used to obtain 3D reconstructions of in vitro samples of natural cancellous bone tissue. The physical size of each sample is approximately $4 \times 4 \times 4$ mm, and their relative densities are 9%–25%. The models were to be scaled up by an approximate factor of 10 and physically produced using additive manufacture (AM). The considerable effort that went in to generating the data from which these models that would be made is described elsewhere [1].

5.40.3 The use of stereolithography in the study of cancellous bone

For this project, stereolithography (SL) was the preferred method for several essential reasons. Firstly, it is capable of building models at an exact layer thickness of 0.1 mm. This was desirable as the FEA was generated from voxel data with a voxel size of 0.1 mm. Therefore, the SL model would replicate the FEA mesh exactly, i.e., with no smoothing between the layers or within the plane of each slice. Secondly, it is the most accurate method available (except Solidscape machines, but this would have been extremely slow, and the models would have proved far too delicate to mechanically test). Thirdly, although selective laser sintering and 3D printing had the advantage of not requiring support structures, the finished models are not completely dense or sufficiently accurate. Fused deposition modeling (FDM) parts can also show a small degree of porosity and are unable to match the accuracy desired in this case. However, since these models were built, the water-soluble supports that are now available for FDM would prove extremely useful for structures such as these. Finally, and most importantly, stereolithography could be used to generate models from slice data rather than triangular facetted data. The intention was to use the

SLC file format as it resulted in dramatically smaller files than the same data generated in the STL file format.

A general description of all of these AM technologies can be found in Chapter 4. For a general description of the SLC, STL, and SLI data formats, see Chapter 3.

5.40.4 Single human bone sample (approximately 45-mm cube)

Due to initial problems with the SLC data files, the first single model was built from an STL file; see Fig. 5.270. This was attempted to test the general capability of stereolithography to manufacture these forms. Although the STL file was much larger than the equivalent SLC file, as the model was small, it was still a feasible option. However, the standard procedures for generating supports were simply not suitable as the software automatically attempts to support all of the down-facing areas of the model. This resulted in overly long processing times and a vast number of supports with a correspondingly large support STL file. In addition, the nature of the structure would make it extremely difficult to remove supports from the innermost areas of the model. (Note: this model was constructed before fine-point supports became available. Although they would have been an

Figure 5.270 STL file of the first sample.

improvement in this respect, they would have still resulted in an excessive number of supports with an even larger support STL file.)

To avoid these problems, a novel strategy for producing the necessary support was attempted. The approach was based on two fundamental assumptions. Firstly, as bone is a naturally occurring, load-bearing structure, it is made up of self-supporting arches. Therefore, the structure should in theory, support itself except for the sides of the model where the structure has been sliced through. The second assumption was that the open spaces would all be intercommunicating, and therefore, there would be no "trapped volumes" (a recognized problem in stereolithography).

As automatically generated supports were impractical, to support the base and sides of the model, a very thin crate was designed using CAD that would support the sides and base yet still allow good draining; see Fig. 5.271. This is necessary to avoid the "trapped volume" effect that adversely affects the stereolithography process. The crate was then exported as an STL file. Curtain support structures were then automatically generated (using 3D Lightyear) for the crate only to separate the whole build from the platform. This combination can be seen in Fig. 5.272. The objects were then prepared for stereolithography in the usual manner and built on an SLA-250 using SL5220 resin.

Figure 5.271 The crate structure.

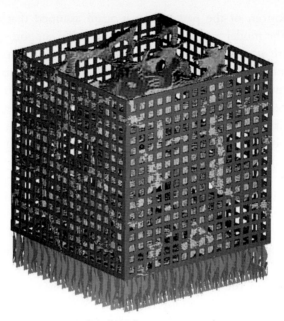

Figure 5.272 Sample, supports, and crate.

The model was built successfully, suggesting that the fundamental assumptions regarding the self-supporting and self-draining nature of the object were indeed sound. The crate structure was carefully broken away along with the supports using a scalpel.

5.40.5 Multiple human samples (approximately 50-mm cube)

For the second batch, five copies each of five types of model were required. The STL approach was not feasible due to the incredibly excessive size of the files. Instead, contour files were generated from the original data in the SLC file format.

However, the problem of how to support the sides and base was encountered again. Although automatic support software is available to generate supports for SLC files, it presented exactly the same problems as the previous attempt. To create minimal supports, a combination of two approaches was used. The first involved the use of C-Sup to generate supports automatically that would separate the bottom of the model from the build platform. These supports were automatically generated to end just above the bottom of the part as it was again assumed that the internal structure of the model would be self-supporting. To support the sides, the crate structure was used again (minus the bottom). Automatic support-generation software was then used to create supports that would separate the crate from the build platform. The STL files of these structures were then converted into the SLC format (using Magics). This resulted in four SLC files that could be prepared for the stereolithography process. These are not shown here, as the files are only contours and therefore cannot be rendered and viewed from an angle.

However, this was not a simple task to achieve. The first problem was encountered when arranging the items in the correct positions for part building due to the use of the SLC format. The SLC file is essentially the contours of the model at the layer thickness intended to build the model. Therefore, it has to be generated in the correct position and orientation relative to the z-axis. Once generated, the files cannot be repositioned or rotated relative to the z-axis. Initially, this was overlooked, resulting in corrupt SLC files after repositioning. The SLC generation code was therefore rewritten to create the first contour and 8 mm in z height to allow room for support structures.

The second more fundamental and difficult problem was discovered with the software that generated the bone structure SLC files. The SLC files created were invalid and were not recognized by current stereolithography software (3D Lightyear). However, no error files are generated by 3D Lightyear, and therefore, it was impossible to ascertain the nature of the problem. However, when using the obsolete Maestro software, the SLC files were again found invalid, but error message files were generated. Reading these error messages showed us that the orientation of the contours was incorrect. Crucially, when attempting to prepare a build using these SLC files, Maestro was able to reorient them, resulting in valid slice files in the SLI format. These error messages also highlighted the fact that the final layer had zero thickness. Although Maestro was unable to correct this error automatically, once it had been discovered, it was a relatively simple matter to correct the code.

The reason that the SLC files were invalid in 3D Lightyear is that they were generated according to an obsolete specification from 3D Systems. This means that they are not recognized by current 3D Systems software releases. A similar issue can be found when attempting to use SLI files generated by CTM (a module of Mimics), which are also not recognized by Lightyear. This effectively renders the SLI export option of CTM redundant unless the user is prepared to maintain and use Maestro. This is an important point as the ability to move from CT or Magnetic Resonance imaging (MR) data directly to the SLI format represents by far the most efficient route to a stereolithography build.

Given the fact that Maestro is old software, running on obsolete UNIX hardware, the SLC files took an incredibly long time to prepare (approximately 3 days). To reduce the file size requirement, four types of models were prepared in a line that would fit across the width of the SLA-250 build platform. Once generated, this vector file was then copied four times at the SLA machine. This method does not increase the size of the vector file but offsets the whole set and repeats it. This enabled us to complete 16 models without creating an unnecessarily large vector file. Even after taking these steps, the vector file was large, and the build took approximately 65 h to build using SL5220 resin. A second build was implemented along the same lines to produce the remaining models of the 25 required.

An example of one of the models with its supporting crate is shown in Fig. 5.273. Due to the extremely delicate nature of the models, they were painstakingly hand finished using scalpels to remove the supporting structures without causing damage. This was complicated by the inability to

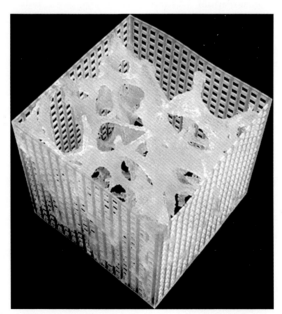

Figure 5.273 Model of one of the samples supported by the crate.

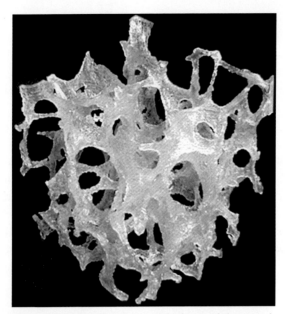

Figure 5.274 Final cleaned model of one of the samples.

familiarize the technicians with the models because it is not possible to generate three-dimensional rendered views of SLC data. One of the finished models is shown in Fig. 5.274.

5.40.6 Conclusion

Issues caused by rapid prototyping software made these items a particular challenge to the application of stereolithography. The machine operators' complete dependence on the preparation software that is supplied with the machine presents two main reasons behind the difficulties.

Firstly, as rapid prototyping develops and improves, the software is increasingly designed to automate as many functions as possible. This is intended to improve the ease and speed of use in the commercial environment. However, this increasing level of automation reduces accessibility to variables and parameters, removing many options from the expert user, particularly those in the research community. In this case, typical semi-automated methods for support generation would have proved utterly impractical for the nature of these objects.

The second issue is the industry's concentration on the STL file format and complete neglect of $2^1/_2$ D alternatives. As this case in particular shows, contour-based formats can be dramatically more efficient. However, current stereolithography users have no functions within the preparation software (3D Lightyear) for the verification, translation, and orientation or crucially support generation of contour formats, specifically the SLC. This is despite the fact that the necessary code for each of these functions exists elsewhere (C-Sup, VerSLC).

These problems meant that to successfully build these models necessitated the utilization and combination of several different pieces of software where appropriate and a thorough understanding of the stereolithography process. Objects that appear unfeasible for the standard practises of STL file and automatic support generation may in fact be perfectly possible if they are considered carefully (there is always more than one way around a problem). For example, these models were made possible because (a) the nature of the objects suggested that they would form self-supporting and self-draining structures in the inner volume, and (b) the SLC file format was used.

As building these models has shown, seemingly obsolete software may still posses a level of accessibility that is extremely useful to the expert user, especially in research. Consequently, for many researchers in this field, it is advisable to maintain copies of superseded rapid prototyping software.

Reference

Sisias G, Phillips R, Dobson CA, Fagan MJ, Langton CM. Algorithms for accurate rapid prototyping replication of cancellous bone voxel maps. Rapid Prototyping Journal 2002;8(1):6—24.

Software

C-Sup - a module of Mimics Version 7.0, Materialise NV, Technologielaan 15, 3001 Leuven. Belgium; 2001.

3 D Lightyear™ version 1.1 - Advanced User License, 3D Systems Inc., 26081. Valencia, CA 91355, USA: Avenue Hall; 2000.

VerSLC, 3D Systems Inc., 26081. Valencia, CA 91355, USA: Avenue Hall; 1993.

Maestro version 1.9.1, 3D Systems Inc., 26081. Valencia, CA 91355, USA: Avenue Hall; 1996.

Slicer — a module of Magics version 6.3. 3001 Leuven, Belgium: Materialise NV, Technologielaan 15; 2000.

CHAPTER 5.41

Research applications case study 2—Recreating skin texture relief using CAD/AM*

5.41.1 Introduction

Maxillofacial prosthetics and technologists (MPTs) seek to meet the needs of patients with various degrees of facial deformity by restoring aesthetic and functional portions of missing tissue using artificial materials. Prosthetic restoration of lost tissue precedes surgical reconstruction, and despite recent advances in surgery, many cases remain where prosthetic rehabilitation is a more appropriate treatment [1]. Patients typically suffer from conditions resulting from traumatic injury (such as road traffic accidents), congenital deformity, or diseases that cause significant tissue damage such as cancer.

Factors that contribute to the aesthetic success of prostheses include skin color match, appropriate contours, and realistic texture [2]. The MPTs who create the prostheses attempt to address these factors, which are conventionally assessed by eye and carved by hand in wax on a plaster replica of the patient's defect. Firstly, the external and fitting surfaces are shaped so that the contours of the prosthesis are established. This is often done with the patient present for test fittings. The detail is gradually refined using sculpting tools to define features, creases, folds, and smaller skin details that

* The work described in this chapter was first reported in the reference below and reproduced here with kind permission of Sage Publishing.

Eggbeer D, Evans P, Bibb R. A pilot study in the application of texture relief for digitally designed facial prostheses. Proceedings of the Institute of Mechanical Engineers Part H: Journal of Engineering in Medicine 2006;220(6):705–14, ISSN: 0954-4119. https://doi.org/10.1243/09544119JEIM38

Medical Modeling
ISBN 978-0-323-95733-5
https://doi.org/10.1016/B978-0-323-95733-5.00046-6

recreate missing anatomy (perhaps using old photographs as a guide) and that match the topography of the surrounding anatomy.

To create a more realistic appearance for the prosthesis, skin texture may be added. This can be achieved in a variety of techniques, such stippling with a stiff brush or taking an impression from orange peel or gauze. Conversely, a flame torch may be used to locally melt or soften the wax to selectively smooth areas or decrease the prominence of textures and bumps. When the wax sculpting is complete, a plaster mold is made from it. The mold surface picks up the texture and detail of the wax carving. Once the plaster mold is set, the wax is melted out, and the mold is packed with silicone elastomer that has been matched to the patient's skin color. Other details may be added at this stage such as the use of red rayon fibers that replicate superficial capillaries and veins. The silicone is heated under pressure to produce the final solid but flexible prosthesis. Depending on the complexity and size of the prosthesis, this process may take 2–3 days.

Improved surgical and medical techniques have led to improved survival rates from accidents and cancer treatments, which has in turn led to increased workload on the MPTs. This has driven growing interest in the application of advanced computer-aided design (CAD) and manufacturing technologies. Technologies such as CAD and rapid prototyping and additive manufacturing (RP/AM) have shown benefits in reducing time and labor in product design and development, and initial research suggests that similar benefits may be realized in the production of facial prosthetics. However, although some RP/AM technologies have been successfully exploited in maxillofacial surgery for many years, their application in facial prosthetics remains relatively unexplored [3]. Recent technologic advances have increased opportunities for MPTs to benefit from these technologies, and this can be seen in the recent research [4–12]. Despite this interest and some promising results, most of this research has focused on the creation of the overall shape of the prosthesis and has not considered the importance of the smaller details that make a prosthesis visually convincing. Given the importance of details such as texture and wrinkles in achieving a natural and realistic result, it was important to explore the problem through the study described here.

This research aimed to identify and assess suitable technologies that may be used to create and produce fine textures and wrinkles that may be conveniently incorporated into prosthesis design and production techniques.

5.41.2 A definition of skin texture

Visible skin texture may be classified according to the orientation and depth of the lines [13]. Primary and secondary lines form a pattern on the skin surface and are only noticeable on close observation. They typically form a polygon pattern ranging from 20 to 200 μm in depth [14]. The back of the hand often shows a good example. It has been suggested that the term "wrinkle" should apply when an extension of the skin perpendicular to the axis of the skin surface change leaves a marked line representing the bottom of the wrinkle [13]. Further, an assessment scale that was subsequently used to assess and quantify deep facial wrinkles has been developed by Lemperle et al. [15]. Wrinkles from various facial locations were subjectively graded from 0 to 5 by dermatologists, where 0 was described as no wrinkles and 5 very deep wrinkles and redundant folds. Following the visual grading, the wrinkles were then measured using profilometry and the results correlated. This produced a graded wrinkle scale table with associated depth of wrinkle values for the various facial locations. Using this scale, a nasolabial wrinkle (side of the nose) with a grading of 1 would correlate to a wrinkle depth of less than 0.2 mm and a grading 5 would be greater than 0.81 mm in depth. This varied with other facial locations with the minimum measured depth being 0.06 mm and maximum of 0.94 mm. The proposed margins on the scale ranged from >0.1 mm to <0.81 mm.

5.41.3 Identification of suitable technologies

5.41.3.1 Specification requirements

Based upon the rating scale developed by Lemperle et al. [15], potential digital technologies must be capable of creating and reproducing wrinkle and texture details with a minimum depth of 0.1 mm. The CAD software used must be capable of creating and manipulating complex anatomic forms and the RP/AM process capable of producing parts or patterns to this resolution in a material compatible with current prosthetic methods [8,11,12].

5.41.3.2 Computer representation and manipulation of skin textures

Three-dimensional CAD packages have traditionally been developed for two main markets, engineering design and computer gaming/animation. Engineering CAD has been developed to define exact shapes using

mathematical geometry (lines, arcs, circles, squares, etc.). Relatively complex surfaces may be generated, for example the surfaces of cars, but the
mathematical geometry used is aimed toward smooth flowing surfaces and
limits the ability to define the levels of contouring, such as creases, folds,
and sharp radii required to represent anatomic forms and textures. In fact, in
applications such as automotive and aerospace design, it is highly desirable
to avoid unwanted creases in the surfaces being created. Software aimed
toward three-dimensional computer gaming and animation, such as 3D
Studio (Discreet, Autodesk Inc., 10 Duke Street, Montreal, Quebec H3C
2L7, Canada), exhibits many of the same limitations as engineering CAD,
but it typically allows a greater freedom for surface manipulation. As these
objects are not actually physically produced, as long as the visual effect on
the screen is convincing, there is no need to go further in terms of detail.
The textures that appear on these animations and games are normally
represented by two-dimensional images that are "wrapped" around the
object. This creates an illusion of texture rather than true three-dimensional
relief. Therefore, for the purposes of prosthesis manufacture, this wrapped
texture cannot be used, because it cannot be physically reproduced using
RP/AM techniques.

Recently, methods of true three-dimensional texture creation have
been explored [16]. The application of textures has been applied in the
jewelry industry, and software such as ArtCAM (Delcam plc, Small Heath
Business Park, Birmingham B10 OHJ, UK) incorporate tools to map three-
dimensional textures around a CAD model. However, ArtCAM and other
jewelry design software construct their shapes in the same manner as engineering CAD and are therefore not suited to the representation and
manipulation of anatomic forms. Recent developments in CAD software
have led to design packages that offer a more intuitive and freehand
interface to the design process. Software such as ZBrush (Pixologic Inc.,
320 West 31st Street, Los Angeles, CA 90007, USA) and Geomagic
FreeForm Plus (3D Systems Inc., 333 Three D Systems Circle, Rock Hill,
SC 29730, USA) may provide a CAD environment that is more analogous
to sculpting by hand, which clearly is more appropriate to the design and
manufacture of a facial prosthesis. Both FreeForm and ZBrush allow
complex three-dimensional forms to be manipulated and given high-
resolution textures in a freehand manner. FreeForm has been shown to
be suitable for facial prosthesis design in previous research [8,11,12].
In addition, the haptic interface between the user and FreeForm software
allows shapes to be manipulated in ways that more closely mimic the

hand-carving techniques used in conventional prosthesis sculpting. In addition, FreeForm has a number of functions that can create relief on a model surface derived from a two-dimensional image.

5.41.3.3 RP/AM reproduction of skin textures

RP/AM offers the most suitable solution to the production of a prosthesis or pattern from CAD data [11,12]. Computer numerically controlled machining (CNC) has also been used to create textures [16] but is not as well adapted to create fitting and undercut surfaces and is also limited by suitable material choice (machining of soft flexible materials is difficult). CNC becomes very slow when creating intricate or small-scale detail such as textures and requires a cutting tool with a very small diameter. A review of the currently available RP/AM technologies highlights a number of technologies that are capable of creating the level of detail required to reproduce realistic skin textures. A critical parameter to achieve the level of detail required is the layer thickness that the RP/AM system uses. To achieve the level of detail identified above, a layer thickness of below 0.1 mm is necessary. Currently available RP/AM technologies that can achieve a layer thickness of below 0.1 mm include the following.

- ThermoJet wax printing (3D Systems Inc., 333 Three D Systems Circle, Rock Hill, SC 29730, USA)
- Perfactory digital light processing (EnvisionTEC GmbH, Elbestrasse 10, D-45768 Marl, Germany)
- Solidscape wax printing (Solidscape Inc., 316 Daniel Webster Highway, Merrimack, NH 03054-4115, USA)
- Objet poly-jet modeling (Objet Geometries Ltd, 2 Holzman St., Science Park, P.O. Box 2496, Rehovot 76124, Israel)
- Stereolithography (3D Systems Inc.)

Of these, only the ThermoJet and Solidscape printing technologies are capable of producing parts in a material directly compatible with current prosthetic construction techniques. Therefore, it was decided that these would provide the focus of the study. The Solidscape process utilizes a single jetting head to deposit a wax material and another one to deposit a supporting material, which can be dissolved from the finished model using a solvent. This process produces very accurate, high-resolution parts but due to its single jetting head is extremely slow. The process is therefore highly appropriate for small, intricate items such as jewelry or dentures but proves unnecessarily slow for facial prosthetic work. Like the Solidscape process, the ThermoJet process deposits a wax material through inkjet-style printing heads, building a solid part layer by layer. The object being built requires supports, which are built concurrently as a lattice, which can be manually removed when the part is completed. The ThermoJet process uses an array of jetting heads to deposit the material and is therefore much faster. The material is also softer than that used by Solidscape, making it more akin to the wax already used by MPTs and is therefore more appropriate for manipulation using conventional sculpting techniques. Although no accuracy specifications are given for ThermoJet, it is advertised as having a very high resolution (300 × 400 × 600 dots per inch in x, y, and z axes) and aimed at producing finely detailed parts (a drop of wax approximately every 0.085 mm by 0.064 mm in layers 0.042 mm thick).

5.41.3.4 Suitable technologies identified

From the review of currently available technologies and the required parameters stated above, the following technologies were selected. Prosthesis design was to be performed using FreeForm CAD software, including the addition of texture. The completed prosthesis design would be produced from this design file using the ThermoJet wax printing process.

5.41.4 Methods

Case studies were conducted that explored utilizing the technologies identified. Firstly, the application of two-dimensional texture maps to create three-dimensional relief on anatomic models was explored using FreeForm CAD and ThermoJet RP/AM. This was followed by a series of test pieces designed and manufactured to test the ability of the selected CAD software to create and export texture relief that would provide a range of realistic skin textures and the capability of the RP/AM machine to reproduce these textures to the required accuracy.

5.41.4.1 Assessment

Subjective analysis of the results was used in this study since the nature of prosthesis production and texture detailing results in complex forms that do not provide convenient datum or reference surfaces from which to measure accuracy. By definition, prostheses are one-off custom-made appliances made to fit individual patients. Therefore, it is not practical to perform the type of repeated statistical analysis that would be commonly encountered in series production or mass manufacture. The aesthetic outcome and accuracy of prostheses is subjectively assessed by the prosthetist and subsequently the patient.

The RP/AM-produced patterns were firstly assessed against the CAD models, and secondly, the level of detail in each of the model sets was assessed by a qualified and experienced prosthetist, and the ability of the technologies to create and manufacture realistic and convincing skin textures was commented on.

5.41.5 Case studies

5.41.5.1 Pilot study one

An initial pilot study was used to assess the capability of FreeForm in creating texture relief and ThermoJet in reproducing the detail. A series of texture images were converted into high-contrast black and white images using photographic manipulation software. These images were used to create relief textures on small rectangular samples (20 × 10 mm by 5 mm high). A four-sided patch was drawn on the top surface of the blocks, and the "emboss with wrapped image" function was used to overlay the texture maps. The scaling tools were used to gauge the approximate depth and density of the required texture relief. The method of creating the textures is described in detail in the experimental section below.

ThermoJet was used to produce the physical models of the test pieces shown in Fig. 5.275. The details were produced with good visual effect. However, the texture was not sufficiently detailed or deep enough to prove conclusive.

5.41.5.2 Pilot study two

Another slightly larger test piece was created with an image that more closely resembled a detailed skin texture. The texture was created on a segment of anatomic data to assess the effect on a contoured piece rather than the simple flat area of the previous test pieces. The test contoured test piece was approximately 27 mm by 21 mm by 10 mm at its highest point. The model was built using the ThermoJet RP/AM machine, shown in Fig. 5.276. Again, the depth was judged by eye, and the resulting pattern suggested that a realistic skin texture could be produced over a contoured surface.

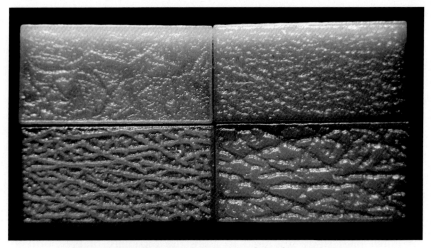

Figure 5.275 Close-up photograph of the four trial pieces.

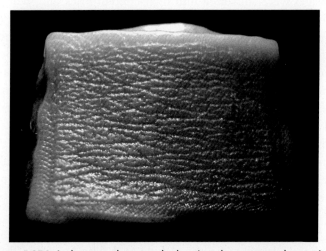

Figure 5.276 A close-up photograph showing the contoured test piece.

5.41.5.3 Experimental series

5.41.5.3.1 Step 1—Producing the texture image

A sample skin texture image was located in the form of a two-dimensional grayscale bitmap, shown in Fig. 5.277. This image was then manipulated to produce a high-contrast, black and white image shown in Fig. 5.278 using Photoshop software (Adobe Systems Inc., 345 Park Avenue, San Jose, California 95110-2704, USA). Suitable texture images may be obtained

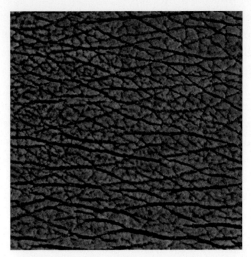

Figure 5.277 The skin texture image used in this study.

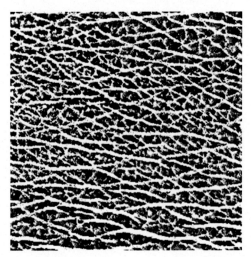

Figure 5.278 The manipulated, high-contrast version used to create the relief.

from databases, digital macro-photographs, or a pad print of skin produced from an impression.

5.41.5.3.2 Step 2—Application of textures in CAD

To assess the effect of creating textures on anatomic shapes associated with maxillofacial prosthetics, a small section of data was taken from a three-dimensional CAD model of a human face derived from a 3D CT scan.

The area was selected to display a variety of compound curved surfaces, although the size of the selected area was kept small to minimize build time and cost. The selected region is shown in Fig. 5.279. A series of test pieces based on the selected data were created in FreeForm with a 0.1-mm edge definition.

A rectangular box was drawn on the contoured surface, and the "emboss with wrapped image" function was used to overlay the sample two-dimensional texture image in the box. The "emboss" function was used to set varying depths of texture relief, at 0.1, 0.15, 0.25, 0.35, 0.5, and 0.8 mm. These depths corresponded to the wrinkle-depth scale and associated measurements in various facial areas developed by Lemperle et al. [15]. The actual emboss depth produced in the model is also influenced by the by the grayscale values in the image. Black areas are embossed to the full depth selected, and gray areas will be proportionally embossed to a lesser degree. The embossing effect may be previewed as an image (Fig. 5.280) or as the resultant texture relief (Fig. 5.281).

The "ruler" function in FreeForm was used to ascertain the depth of the textures by measuring the distance from an original smooth copy of the part to the deepest part of the grooves in the textured part.

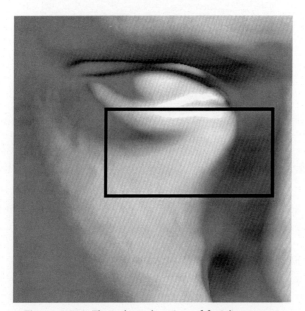

Figure 5.279 The selected region of facial anatomy.

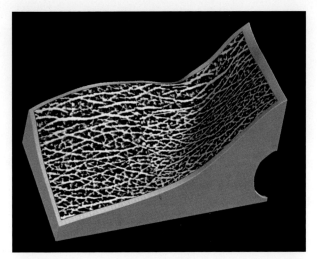

Figure 5.280 Preview of the image for the embossing function.

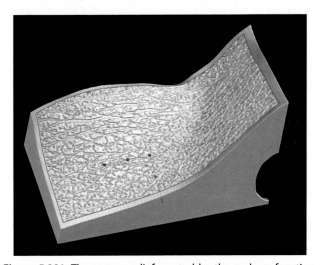

Figure 5.281 The texture relief created by the emboss function.

5.41.5.3.3 Step 3—Additive manufacture

The blocks were manufactured using ThermoJet printing with the contour surface set facing upward. The test pieces were 18.5 mm by 35 mm by approximately 27 mm high at the tallest point. All six patterns were built in less than 3 hours. The results can be seen in Fig. 5.282, with the shallowest relief (0.1 mm) on the left ranging to the deepest relief (0.8 mm) on the right.

Figure 5.282 The 6 ThermoJet-produced skin texture sample patterns.

5.41.6 Results

All of the texture depths were visible on the ThermoJet patterns. This indicated that the process was capable of producing patterns with sufficient detail to describe realistic skin textures. The layer-stepping effect commonly exhibited by layer additive manufacture processes was not visible on the surface and did not interfere with the texture pattern. Textures would not however be well defined on down-facing surfaces due to the dense support structure that when removed, left a rough finish. This may be a problem if the technique was used in the production of complex prosthetic forms where all surfaces are on show such as hands.

5.41.7 Discussion

The tools available in FreeForm were well adapted to creating accurate texture relief from two-dimensional images. One possible limitation is file size. To represent texture faithfully and wrinkle detail at each stage of both methods, large amounts of data are required. A high-resolution model setting that demands a lot of computer processing power must be used in the CAD stage. The output STL (Standard Triangulation Language) file required for the subsequent ThermoJet production stage must also have good detail definition that translates to a large file size. Although more modern, high-specification computers may be able to handle the large file sizes, it may make the process unmanageably slow for others. Further research will be undertaken to optimize the file size versus quality settings.

The ThermoJet process was capable of producing all of the texture samples faithfully and did not exhibit the stair-stepping effect that some other AM processes display. This ability combined with the suitable material properties demonstrates how the process may be integrated into digital prosthesis design and production techniques that are compatible with conventional handcrafting techniques.

5.41.7.1 Limitations

Although the visual effect is ultimately more important to be measured subjectively by a qualified and experienced prosthetist, this study was unable to quantify the accuracy and resolution of the ThermoJet-produced patterns. This study has shown that CAD and AM processes may be used to generate fine texture detailing, but further research is required to quantify the resolution and accuracy requirements of these processes. The authors intend to apply profilometry to assess the surface of the AM-produced patterns and compare these with the CAD models.

5.41.8 Conclusions

This research has identified methods of capturing, creating, and reproducing three-dimensional, skin texture-like relief using CAD and AM technologies. Furthermore, it has highlighted how these techniques may be integrated into digital prosthesis design and construction processes. Limitations of current technologies have also been highlighted, and the authors intend to refine the techniques and evaluate their effectiveness in suitable patient case studies.

References

[1] Thomas K. Maxillofacial prosthetics. London: Quintessence Publishing; 1994. p. 13.
[2] Ansgar C, Cheng AC, Wee AG, Li JTK, Archibald D. A new prosthodontic approach for craniofacial implant-retained maxillofacial prostheses. The Journal of Prosthetic Dentistry 2002;88(2):224–8.
[3] Wolfaardt J, Sugar A, Wilkes G. Advanced technology and the future of facial prosthetics in head and neck reconstruction. International Journal of Oral and Maxillofacial Surgery 2003;32(2):121–3.
[4] Coward TJ, Watson RM, Wilkinson IC. Fabrication of a wax ear by rapid-process modelling using stereolithography. International Journal of Prosthodontics 1999;12(1):20–7.
[5] Bibb R, Freeman P, Brown R, Sugar A, Evans P, Bocca A. An investigation of three-dimensional scanning of human body surfaces and its use in the design and manufacture of prostheses. Proceedings of the Institute of Mechanical Engineers Part H: The Journal of Engineering in Medicine 2000;214(6):589–94.
[6] Cheah CM, Chua CK, Tan KH, Teo CK. Integration of laser surface digitizing with CAD/CAM techniques for developing facial prostheses part 1: design and fabrication of prosthesis replicas. International Journal of Prosthodontics 2003;16(4):435–41.

[7] Cheah CM, Chua CK, Tan KH. Integration of laser surface digitizing with CAD/CAM techniques for developing facial prostheses part 2: development of molding techniques for casting prosthetic parts. International Journal of Prosthodontics 2003;16(5):543—8.

[8] Verdonck HWD, Poukens J, Overveld HV, Riediger D. Computer-assisted maxillofacial prosthodontics: a new treatment protocol. International Journal of Prosthodontics 2003;16(3):326—8.

[9] Reitemeier B, Notni G, Heinze M, Schone C, Schmidt A, Fichtner D. Optical modeling of extraoral defects. The Journal of Prosthetic Dentistry 2004;91(1):80—4.

[10] Tsuji M, Noguchi N, Ihara K, Yamashita Y, Shikimori M, Goto M. Fabrication of a maxillofacial prosthesis using a computer-aided design and manufacturing system. Journal of Prosthodontics 2004;13(3):179—83.

[11] Eggbeer D, Evans P, Bibb R. The application of computer aided techniques in facial prosthetics. In: Abstract in the proceedings of the 6th international congress on maxillofacial rehabilitation, Maastricht, The Netherlands; 2004. p. 55.

[12] Evans P, Eggbeer D, Bibb R. Orbital prosthesis wax pattern production using computer aided design and rapid prototyping techniques. The Journal of Maxillofacial Prosthetics and Technology 2004;7:11—5.

[13] Piérard GE, Uhoda I, Piérard-Franchimont C. From skin micro relief to wrinkles: an area ripe for investigation. Journal of Cosmetic Dermatology 2003;2(1):21—8.

[14] Hashimoto K. New methods for surface ultra structure: comparative studies scanning electron microscopy and replica method. International Journal of Dermatology 1974;13:357—81.

[15] Lemperle G, Holmes RE, Cohen SR, Lemperle SM. A Classification of facial wrinkles. Plastic and Reconstructive Surgery 2001;108:1735—50.

[16] Yean CK, Kai CC, Ong T, Feng L. Creating machinable textures for CAD/CAM systems. International Journal of Advanced Manufacturing Technologies 1998;14:269—79.

CHAPTER 5.42

Research applications case study 3—Comparison of additive manufacturing materials and human tissues in computed tomography scanning*

5.42.1 Introduction

Additive manufacturing (AM) technologies have rapidly developed and can now produce objects in a wide variety of materials ranging from soft, flexible polymers to high-performance metal alloys [1–4], and they have been successfully employed in medicine since the early 1990s [5]. Initially, AM processes, such as stereolithography, were used to make highly accurate models of skeletal anatomy directly from three-dimensional computed tomography (CT) data. Typically referred to as medical modeling or bio-modeling, this has now become widely accepted as good practice with much literature reporting cases and the benefits achieved, particularly in craniomaxillofacial surgery. Medical models have been typically used to plan and rehearse surgery and in the design and manufacture of custom-

---

* The work described in this chapter was first reported in the references below and is reproduced here with the permission of Elsevier publishing and CRDM/Lancaster University respectively.

Bibb R, Thompson D, Winder J. Computed tomography characterisation of additive manufacturing materials. Medical Engineering & Physics 2011;33(5):590–6, ISSN: 1350-4533. https://doi.org/10.1016/j.medengphy.2010.12.015

Winder RJ, Thompson D, Bibb RJ. Comparison of additive manufacturing materials and human tissues in computed tomography scanning. In: Bocking CE, Rennie AEW, editors. 12th national conference on Rapid Design, Prototyping & Manufacture. CRDM Ltd., High Wycombe; 2011, ISBN: 978-0-9566643-1-0, p. 79–86.

Medical Modeling
ISBN 978-0-323-95733-5
https://doi.org/10.1016/B978-0-323-95733-5.00047-8

fitting prostheses. The use of medical models has become routine, and it is not necessary to discuss these applications in detail here. A number of texts and review papers are available that describe a wide range of medical applications and their principal advantages [6—11].

Recently, AM technologies have been used to manufacture custom-fitting medical devices, for example facial prosthetics, removable partial denture frameworks, surgical guides, and even implants directly from 3D computer-aided design (CAD) data [12—14]. AM principles are also being exploited in tissue engineering where the advantages of layer-additive manufacture are being used to build highly complex porous scaffolds that can support the growth of living cells [15,16]. Polymer-based AM materials have not been approved for implantation, and most of the materials tested in this research are not considered biocompatible. However, these materials may be used externally or to produce working templates from which biocompatible devices may be developed.

Some assessment of the physical properties of AM materials has been carried out. For example, dimensional accuracy, roughness of surface, and mechanical properties have been established for ZPrinter 310 Plus and the Objet Eden 330 [17]. Also, much research has been conducted on the utilization of CT data in building objects using AM technologies, and some of the materials have been characterized using CT scanning [18]. A CT scanner measures the spatial distribution of the linear attenuation coefficient or amount of absorption of X-rays. To enable this measure to be compared between scanners, the CT number range (also known as the Hounsfield unit) was developed, which is based on the linear attenuation to X-rays of water. The CT number range is typically from -1024 to 3072. Table 5.27 shows typical CT number ranges of selected human tissues [19]. The characteristics of AM materials under radiological conditions will become important in the future as a variety of medical devices and custom-fitting patient products may be manufactured using AM and subsequently scanned using CT for either design, testing, or treatment purposes. Therefore, the aim of this work was to determine the CT number of a wide selection of AM materials and establish their appearance in CT images.

It should be noted that since this work was carried out, there have been consolidation through mergers and acquisitions in the AM industry. Z Corp was acquired by 3D Systems, and the technology is now marketed under their ProJet label. Stratasys merged with Objet, but the technology remains labeled Objet and Connex. The Huntsman materials business has been acquired by 3D Systems. These developments have not altered the basic

Table 5.27 CT number ranges of selected human tissues.

Tissue	CT number
Air	−1005 to −995
Lungs	−950 to −550
Fat	−100 to −80
Water	−4 to 4
Kidney	20 to 40
Pancreas	30 to 50
Blood	50 to 60
Liver	50 to 70
Spongious bone	50 to 300
Cortical bone	300+

Adapted from Kalender [19] p. 30, Fig. 1.9.

materials or process capabilities of the machines used, and these results remain valid.

5.42.2 Materials and methods

A total of 29 AM samples were constructed from a CAD-generated STL file defining a rectangular block of material with dimensions 40 × 20 × 10 mm. The samples represented a variety of commonly used materials from the most popular AM processes. However, the sample set is not intended to be comprehensive as there are potentially hundreds of processes and material combinations that could have been used. Twenty-five of the blocks were solid, and five were "sparse" or quasihollow. Quasihollow parts are created to reduce material consumption, build time, and therefore cost in some AM processes. In this study, for each quasi-hollow sample, there was an equivalent solid sample. The CT number ranges were determined only for the solid samples, and the quasihollow samples were included only to investigate their appearance in CT images.

Objet "digital materials (DMs)" are made from combinations of the rigid VeroWhite and soft, flexible Tango+ materials. For the range DM9740—DM9795, the final two digits relate to Shore hardness (i.e., from 40 to 95 Shore). DM8410 and DM8430 approach the rigidity of some common thermoplastics.

The samples underwent a CT scan in contact with a tissue-equivalent head phantom, as shown in Fig. 5.283 (phantom supplied by Imaging Equipment Ltd., Bristol, UK) to mimic the situation of the materials

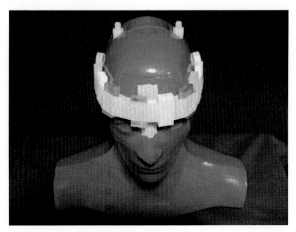

Figure 5.283 AM samples attached to the tissue-equivalent head phantom.

adjacent to the body, as they might be in the case of an AM prosthesis or wearable medical device. CT scanning was performed using a Philips Brilliance 10 multislice system (www.medical.philips.com) using a sinus/facial/head CT protocol (exposure of 67 mAs, peak voltage 120 kV, slice thickness 2 mm, rotation time 1 s, and convolution kernel type "D," software version 1.2.0). CT images were stored in DICOM format and imported into image analysis software, AnalyzeAVW Version 9.0 (Lenexa, Kansas, US), for visualization and CT number measurement. Visual inspection and analysis of the images was also performed using Mimics version 13 (Materialise NV, Leuven, Belgium). The density of each sample (excluding the quasihollow samples) was calculated by measuring the sample weight in grams using a Sartorius precision balance and the sample volume in cubic centimeters using a digital Vernier caliper (g/cm^3).

5.42.3 Results

Table 5.28 shows the sample name, material, mean CT number, the standard deviation of pixel values, and the density of each solid sample. ABS is acrylonitrile butadiene styrene, PPSF is polyphenylsulphone, and PC is polycarbonate. The Objet material VeroWhite is a rigid, white, acrylate-based material, and the Tango+ material is a soft rubber–like material. The Objet DMs are composite materials made from a selective mixture of VeroWhite and Tango+. This produces a range of physical properties that can replicate the stiffness of a variety of thermoplastics. All materials are proprietary and specific to the relevant AM process.

Table 5.28 CT number, standard deviation of CT numbers, and density of AM samples.

No.	Manufacturer and machine	Material	CT number	Standard deviation	Density g/cm³
1	Z Corp 450	Z Bond (cyan-acrylate)	850.17	51.28	1.44
2	Z Corp 450	ZP130 (wax)	1146.41	71.72	1.64
3	EOS P100 Formiga	Nylon 12 (polyamide)	−17.80	29.88	1.00
4	3D Systems 250	ProtoCast AF19120, DSM Somos	168.50	28.57	1.20
5	3D Systems 250	Watershed XC11122, DSM Somos	320.82	27.62	1.17
6	3D Systems 250	9420 EP (white) DSM Somos	251.57	26.35	1.19
7	3D Systems 250	RenShape SL Y-C 9300, Huntsman	142.43	28.67	1.22
8	3D Systems InVision	Visijet SR	126.44	26.05	1.18
9	Dimension 1200 SST	ABS	−115.74	34.43	0.96
10	Fortus 400mc	ABS	−102.86	32.61	0.97
11	Fortus 400mc	ABS+	−358.93	31.08	0.79
12	Fortus 400mc	PPSF	151.60	46.01	1.17

Continued

Table 5.28 CT number, standard deviation of CT numbers, and density of AM samples.—cont'd

No.	Manufacturer and machine	Material	CT number	Standard deviation	Density g/cm³
13	Fortus 400mc	PC	−26.37	29.88	1.11
14	Fortus 400mc	PC/ABS	−30.21	26.34	1.05
15	Fortus 400mc	PC/ABS	110.36	8.86	1.17
16	Objet Connex 500	VeroWhite	99.75	5.06	1.17
17	Objet Connex 500	Tango+	118.28	6.00	1.17
18	Objet Connex 500	DM 9740	111.96	5.99	1.14
19	Objet Connex 500	DM 9750	93.13	5.54	1.13
20	Objet Connex 500	DM 9760	75.09	5.54	1.14
21	Objet Connex 500	DM 9770	72.60	5.70	1.13
22	Objet Connex 500	DM 9785	69.65	6.17	1.14
23	Objet Connex 500	DM 9795	75.96	5.08	1.16
24	Objet Connex 500	DM 8410	71.55	6.16	1.18
25	Objet Connex 500	DM 8430	119.82	6.16	1.17

Data for samples 1−18 have been used with permission from Elsevier [18].

The table presents the average density for each sample, and it should be noted that some samples are not homogeneous, and their density varies considerably across their sections (especially the Z Corp samples). As might be expected, the relationship between CT number and average sample density is essentially linear (Fig. 5.284). It is well known that the CT number of a material is dependent on a range of properties including density, X-ray beam energy, and sample thickness. As X-ray beam energy and section thickness were constant, the large variations in CT number can be attributed to the differences in material and are related to their density. It can be seen that there is a cluster of samples around the density of $1.0-1.2 \text{ g/cm}^3$, which is typical for polymers, and the CT numbers are clustered, suggesting that the CT number for these polymers is also similar. The two denser materials are from the Z Corp process, which are not polymers, but it is interesting to note that their CT number is also proportional to their density.

The mean CT number for the solid samples ranged from a minimum -359 to a maximum of 1146. It is interesting to note that many of the AM sample CT number ranges coincide with or are similar to those of the human tissues, as shown in Table 5.28. For example, samples 1 and 2 have CT number ranges that are similar to cortical bone, which may range from 200 to 1200. Samples 4, 6, 7, and 8 are similar to cancellous bone with a CT number range 50−300. Both samples 9 and 10 have mean CT numbers, which are very similar to the range found for fat tissue in the body at approximately −100. The standard deviation of the sample CT numbers

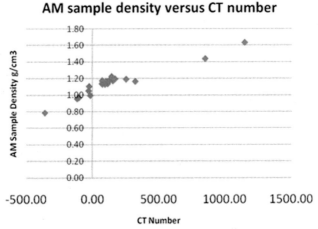

Figure 5.284 Linear relationship between CT number and average sample density.

ranges from approximately 5.0 to over 70.0, the larger deviations measured in samples, which had a higher CT number. This is in keeping with CT scans of human tissue where bone (CT number >300) has the highest standard deviation due to increased noise present in that tissue type, while air (CT number = −1024) had the lowest standard deviation. The standard deviation of the measurements within the AM samples was due to two factors, inherent noise due to the CT imaging system and any material density variation within structure of the AM sample. Samples numbered 21−29 were all constructed using the Objet Connex 500 system, and they demonstrate a limited range of CT numbers, from 70 to 120. This CT number range is similar in CT scanning to contrast-enhanced blood.

Although it is known that certain AM processes produce inherently porous parts, the porosity is not apparent in all of the CT images. This is because the porosity is at a very small scale compared to the resolution of the CT scanner (high-contrast objects less than 0.5 mm can be visualized) and appears uniform throughout the parts. If we consider the example of SLS (selective laser sintering), the process works by sintering together thermoplastic particles with a typical average particle size of around 60 microns (material data sheet for PA2200, EOS GmbH, Munich, Germany). The particles do not fully melt but fuse together to form a sintered, porous structure. Therefore, it is reasonable to assume that this results in a slight lowering of the CT number compared to fully dense nylon produced by injection molding, extrusion, or casting. Further work will be conducted to ascertain whether the difference between SLS nylon and solid nylon can be detected in CT images.

Fig. 5.285 shows the internal structure of two AM samples (3 and 1) using CT scanning. Fig. 5.285a shows an overall uniform internal structure, although there is some noise present within the image. This is typical of CT scanning where the X-ray photons are detected in a random manner creating small variations in pixel values for a constant material, while 5.285(b) demonstrates internal variation in the sample. The variation is caused by the AM method where more material is deposited internally in a structured way to support the sample. This distinct variation of internal

Figure 5.285 Demonstrating uniform CT appearance of (a) (*left*) sample 3, EOS P100 Formiga, and (b) (*right*) nonuniform appearance of sample 1, ZCorp Z450 ZBond.

density was also visible in the images of the 3D printed samples shown in Fig. 5.286b. A variation in the sample density can be seen in the CT images, which show a higher density around the periphery of the sample. This is a result of the 3D printing process, whereby the manufactured part is initially very fragile. The parts are subjected to infiltration of a liquid hardener, typically a cyano-acrylate resin (as in sample 1) or wax (as in sample 2). It is known that the hardeners penetrate into the part through capillary action but that this penetration is limited to a few millimeters. This leads to a higher density "skin" or "shell" that is clearly visible in the CT image. We have observed that the internal structure in Fig. 5.285b simulates variations in human cancellous bone, making this type of modeling ideal for bone simulation. Fig. 5.286a shows a normal CT scan of a human spine vertebra and Fig. 5.286b shows a CT scan of a spine model manufactured using a Z Corp 3D Printing system (the same process as used to produce sample 1). The CT number range for the cortical bone is 1000−1300, and the range within the cancellous bone is 490−815. This mimics the CT number ranges for human cortical and cancellous bone very closely. As described previously, this is due to the AM process, which hardens the outer few millimeters of the model, resulting in an elevated CT number at the periphery. This relatively simple example demonstrates the potential to manufacture anatomically correct, sophisticated test objects with a mixture of hard and soft tissue materials, useful in radiation therapy dosimetry experiments where test objects may be created from a combined approach to model creation.

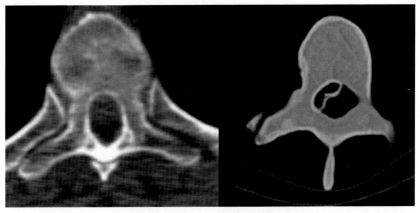

Figure 5.286 (a) (*Left*) showing a CT of human vertebra (used with permission of National Library of Medicine Visible Human Project) and (b) (*right*) a CT scan of a Z Corp lumbar spine model.

We have demonstrated the potential of AM to create a full-size, anatomically accurate test object (phantom) for use in image quality and machine (system) performance testing of a multislice CT scanner. Data from the Digital Human Project was modeled and converted into an STL data file (with permission of National Library of Medicine Visible Human Project). This file was subsequently used to make a life-size physical model of the spine using the ZCorp AM system. Fig. 5.287 shows the ZCorp model viewed from an anterior aspect. To determine how well this model mimicked human bone, the model was scanned in a Siemens Sensation 16 CT scanner (www.medical.siemens.com) with the following spine protocol: 120 kVp, 134 mAs, slice = 1.0 mm, 125.0 mm field of view and 512 × 512 image acquisition matrix.

Fig. 5.288 shows the coronal, sagittal, and transverse reformats through the data. Excellent anatomical detail is demonstrated of the individual vertebrae, spinal canal, nerve foramen, and transverse, spinous, and superior articular processes. Both cortical and cancellous bone simulate the CT image representation of a human scan with cortical bone indicating higher density than the internal cancellous bone. The anatomy is also clearly visualized on surface-rendered images of the posterior and anterior surface of the spine model (Fig. 5.289).

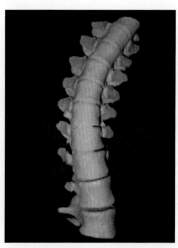

Figure 5.287 Showing the spine model manufactured using the ZCorp rapid proto-typing system. The model represents the thoracolumbar region.

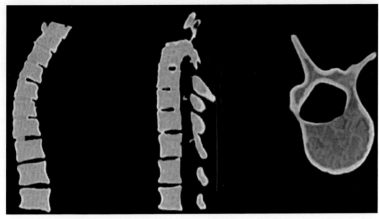

Figure 5.288 Showing coronal, sagittal transverse CT images of the 3D ZCorp spine model. Note the difference in CT number (related to bone density) of cortical bone versus cancellous bone.

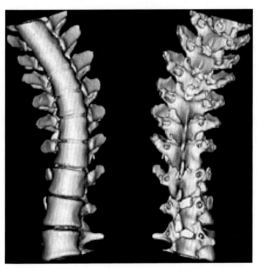

Figure 5.289 Showing surface rendered images of the CT scan; a threshold of CT number = 110 was used to segment the bone anatomy of the model.

5.42.4 Discussion

These CT number and standard deviation findings demonstrate that there is significant potential to use AM materials for sophisticated test objects for use in medical image modality testing, including image quality and radiation dose. Some AM materials have CT numbers very similar to those of human tissues, as summarized in Table 5.28, and therefore, they may be used to develop anatomically accurate phantoms produced from real patient CT scans using AM. Phantoms designed using these materials may have the added advantage of having CT numbers corresponding to real tissues. Anatomically complex, multitissue phantoms could be developed from existing patient CT scan data using well-established image segmentation techniques providing more accurate phantoms for test purposes. Anthropomorphic phantoms have been developed for use in radiation dose studies for diagnostic radiology and therapeutic radiology. Soft tissue, lung, and bone equivalent tissue substitutes (at diagnostic X-ray energy range 80–120 kVp) were created from urethane-based compounds mixed with other materials [20]. This particular phantom suffered from manufacturing difficulties, in that molds would display variation in depth or suffer from physical distortion. AM has the capability to provide accurate anatomic definition, geometric shape, and the appropriate X-ray attenuation.

5.42.5 Conclusions

This study has revealed several interesting facts relating to AM materials in CT images. Firstly, the images provide an indication of material uniformity of density at the macroscale. This analysis can be used to corroborate other observations from visual analysis and mechanical testing. Secondly, the actual CT numbers of a number of commonly used AM materials have been established. This may enable the specification of AM materials for specific medical devices that are required to present a specific CT number or characteristic in CT images. Further work is required to analyze a greater variety of AM materials and in particular samples from AM processes that produce mixed, graded, and multiple material parts.

Acknowledgments

The authors would like to thank the following for the donation of the samples:
Phil Dixon, Loughborough Design School, and Dr. Russ Harris, Wolfson School of Mechanical and Manufacturing Engineering, Loughborough University, UK; Dr. Dominic Eggbeer, National Center for Product Design and Development Research, University of Wales Institute Cardiff; and Jeremy Slater, Technical Sales Engineer, Design Engineering Group, Laser Lines Ltd., Banbury UK.

References

[1] Chua CK, Leong KF, Lim CS. Rapid prototyping: principles and applications. 3rd ed. WSPC; 2010, ISBN 978-9812778987.

[2] Gibson I, Rosen DW, Stucker B. Additive manufacturing technologies: rapid prototyping to direct digital manufacturing. Springer; 2009, ISBN 978-1441911193.

[3] Noorani RI. Rapid prototyping: principles and applications. John Wiley & Sons; 2005, ISBN 978-0471730019.

[4] Hopkinson N, Hague R, Dickens P, editors. Rapid manufacturing: an industrial revolution for a digital age: an industrial revolution for the digital age. Wiley Blackwell; 2005, ISBN 978-0470016138.

[5] Arvier JF, Barker TM, Yau YY, D'Urso PS, Atkinson RL, McDermant GR. British Journal of Oral and Maxillofacial Surgery 1994;32(5):276—83.

[6] Giannatsis J, Dedoussis V. International Journal of Advanced Manufacturing Technology 2009;40(1—2):116—27.

[7] Azari A, Nikzad S. Rapid Prototyping Journal 2009;15(3):216—25.

[8] Bibb R. Medical modelling: the application of advanced design and development technologies in medicine. Woodhead Publishing Ltd.; 2006, ISBN 1-84569-138-5.

[9] Gibson I, editor. Advanced manufacturing technology for medical applications: reverse engineering, software conversion, and rapid prototyping. Wiley Blackwell; 2005, ISBN 978-0470016886.

[10] Petzold R, Zeilhofer H, Kalender W. Computerised Medical Imaging & Graphics 1999;23:277—84.

[11] Webb PA. Journal of Medical Engineering & Technology 2000;24(4):149—53.

[12] Bibb R, Eggbeer D, Evans P. Rapid Prototyping Journal 2010;16(2):130—7. https://doi.org/10.1108/13552541011025852.

[13] Bibb R, Eggbeer D, Evans P, Bocca A, Sugar AW. Rapid Prototyping Journal 2009;15(5):346—54. https://doi.org/10.1108/13552540910993879.

[14] Bibb R, Eggbeer D, Williams R. Rapid Prototyping Journal 2006;12(2):95—9. https://doi.org/10.1108/13552540610652438.

[15] Peltola SM, Melchels FP, Grijpma DW, Kellomäki M. Annals of Medicine 2008;40(4):268—80.

[16] Leong KF, Chua CK, Sudarmadji N, Yeong WY. Journal of the Mechanical Behavior of Biomedical Materials 2008;1(2):140—52.

[17] Pilipović A, Raos P, Šercer M. International Journal of Advanced Manufacturing Technology 2009;40:105—15.

[18] Bibb R, Thompson D, Winder J. Medical Engineering & Physics 2011;33(5):590—6. https://doi.org/10.1016/j.medengphy.2010.12.015.

[19] Kalender WA. Computed tomography. 2nd ed. Wiley VCH; 2000, ISBN 978-3895780813. p. 101.

[20] Winslow JF, Hyer Ryan DE, Fisher F, Tien CJ, Hintenlang DE. Journal of Applied Clinical Medical Physics 2009;10(3):195—204.

CHAPTER 5.43

Research applications case study 4—Producing physical models from CT scans of ancient Egyptian mummies*

5.43.1 Introduction

The development of computed tomography (CT) has allowed archaeologists to gain access to the internal details of mummies without destroying the cartonnage cases or disturbing the wrappings and remains. This nondestructive investigation has proved very successful, and several investigations have been conducted in this way at various locations in the world, improving with advances in CT technology [1–5]. These scans have given archaeologists and forensic experts many insights into the condition of the remains and have provided additional evidence relating to Egyptian funerary practice and the health of the individual. Fig. 5.290 shows an axial CT slice of a mummy.

* This project was conducted in collaboration with Dr. John Taylor, Assistant Keeper at the Department of Ancient Egypt and Sudan. The "Jeni" project was performed on CT data acquired by Clive Baldock, Reg Davies, Ajit Sofat, Stephen Hughes, and John Taylor (British Museum) in 1993. The CT data was gratefully obtained from Stephen Hughes via the Internet. The Nesperennub project was conducted on CT scans acquired at the National Hospital for Neurology and Neurosurgery, London. The facial reconstruction work was undertaken by Dr. Caroline Wilkinson at the Unit of Art in Medicine, The University of Manchester.
Fig. 5.295 is reproduced from Taylor JH, "Mummy: the inside story," 2004, with the permission of the Trustees of the British Museum and Dr. Caroline Wilkinson, University of Manchester.

Medical Modeling
ISBN 978-0-323-95733-5
https://doi.org/10.1016/B978-0-323-95733-5.00048-X

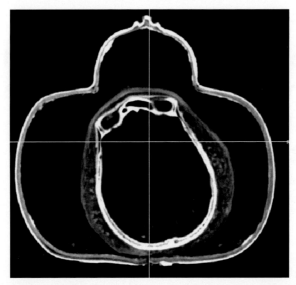

Figure 5.290 Axial CT image of "Jeni." *(Credit: British Museum © 2004 The Trustees of the British Museum.)*

International Centre for Design Research (PDR) at Cardiff Metropolitan University, the Department of Ancient Egypt and Sudan at the British Museum, and the Unit of Art in Medicine at Manchester University have formed a long-term relationship exploring the noninvasive investigation and reconstruction of ancient Egyptian mummies. The first of two mummies investigated, called Tjentmutengebtiu ("Jeni" for short), was the subject of X-ray investigation in the 1960s [6] and subsequent investigation by CT scan more recently in 1993 [7]. The second mummy, called Nesperennub, had also undergone previous X-ray investigation. However, this mummy was recently scanned to capture better data. These CT data, specifically the series of scans of the head, were used in these studies. Both mummies belong to the British Museum.

The aim of this work was to go a step beyond viewing two-dimensional images of the mummies and to use the data to manufacture precise physical replicas of the mummies' skulls. This would allow the skulls to be investigated and handled at will without causing any damage to the original priceless remains. Facial reconstructions could also then be performed on the models. The more recent Nesperennub case was used for investigation into the cause of death, reconstruction of the facial features, and other artifacts of interest for a major new exhibition at the British Museum called "Mummy: the inside story."

5.43.2 Technology

A range of advanced computer software and rapid prototyping hardware was required to achieve accurate digital and physical models of the two mummies. Mimics software was instrumental in importing, segmenting, cleaning, and outputting the required files to produce the 3D model files for visualization and manufacture. The physical models of the skulls were produced by stereolithography [8]. Stereolithography is increasingly used to produce models of patients with many medical conditions and has subsequently been applied to other archeologic and palaeontologic remains [9—11]. Stereolithography and laminated object manufacture were used to reproduce physical models of some objects that were wrapped up with the mummy.

5.43.3 Case studies

Mimics software is typically used in medicine to segment CT data to isolate the tissue of interest. Frequently the tissue of interest is bone. This is accomplished by using the "threshold" functions to select the appropriate limits of density and the numerous editing and segmentation tools to make adjustments to ensure the desired bone structures are isolated.

In contrast, when attempting the first case, "Jeni," a number of challenges made accurate reconstruction of the bony anatomy extremely challenging. In living patients, the difference in density between bone and the adjacent soft tissues is quite marked. This allows the segmentation of bone to be performed relatively easily. However, in these ancient Egyptian mummy cases, all of the soft tissues were completely desiccated by the mummification process. This resulted in the soft tissue remains having an artificially high density compared with the remaining bone (see Figs. 5.291 and 5.292).

This effect is confounded by the demineralization of some bone structures also resulting from the mummification process. Therefore, performing the segmentation by density threshold only results in a poor three-dimensional reconstruction. It loses some data from the skull while including unwanted elements of desiccated soft tissue. This effect can be seen in the reconstruction on the left of Fig. 5.293.

In previous image reconstructions, higher thresholds had been used to try to eliminate some of the desiccated soft-tissue remains. Although this improves matters, it results in the loss of low-density bone structures, while

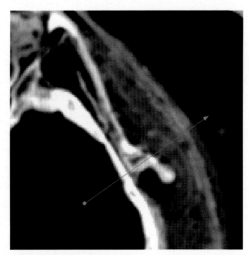

Figure 5.291 Profile through the skull and ear of Jeni. *(Credit: British Museum © 2004 The Trustees of the British Museum.)*

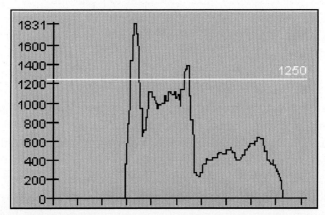

Figure 5.292 Graph of density through the profile. *(Credit: British Museum © 2004 The Trustees of the British Museum.)*

some soft-tissue structures remain. In addition, the high-density artificial objects also remained present in the eyes, mouth, and neck. To produce a model of the skull from these data would have resulted in gaps in the surface of the skull and the absence of some of the more delicate bone structures. For example, gaps could be seen in the temporal bone and zygoma (cheekbones), and the soft tissues of the ears are still present. As facial reconstruction depends on imposing known depths of soft tissue onto the facial bones, these defects would make the whole process more difficult and less reliable.

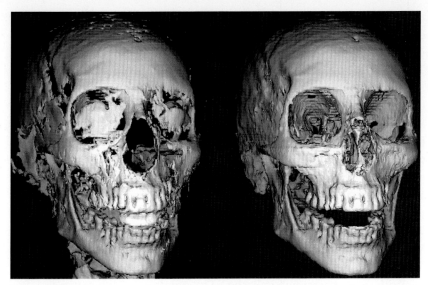

Figure 5.293 3D reconstruction of data segmented by the standard threshold for bone (left) and the manually edited data (right). *(Credit: British Museum © 2004 The Trustees of the British Museum.)*

To improve the data describing the surface of the facial bones required the extensive use of the manual editing tools in Mimics. Some areas were improved by using the local thresholding tools within the edit tools in Mimics. This allowed small areas to be selected according to higher or lower densities without affecting the overall segmentation. However, some areas required specific areas to be edited manually using the edit tools to delete data relating to soft tissue and to draw in data relating to bone. The flexibility of these editing tools combined with good anatomic knowledge resulted in accurate segmentation of the skull. Although this was quite time-consuming, the improved results are well worth the effort and would ensure that the subsequent facial reconstruction was carried out on the best model possible.

The inner surfaces of the skulls were more easily identified as the brains had been removed as part of the mummification process. However, other areas proved much more difficult to pick out from the surrounding tissue. This was especially the case in the mouth, palate, and nasal cavities where desiccated soft tissue remained in place. The nose also posed problems due to the presence of cartilage and the fact that the nasal bones were broken and displaced. Again, this damage resulted from the mummification process.

The 3D reconstruction functions of Mimics were used to assess the quality of the segmentation on screen. Finally, the highest quality 3D reconstructions were created to check the data before building the models. The final reconstruction can be seen on the right of Fig. 5.293.

Once we were satisfied that we had a good segmentation of the skull, the data were prepared for model building using stereolithography. The RP Slice module of Mimics was used to generate SLC files of the skulls. The RP Slice module was then also used to create the necessary supports, also in the SLC format. This direct interface to a layer format suitable for stereolithography results in smaller file sizes while retaining excellent detail and accuracy. In addition, the single support file generated proves simpler to process and subsequently easier to remove from the model when compared with alternative formats. The skull models produced provided Caroline Wilkinson at the Unit of Art in Medicine at Manchester with a sound basis for the facial reconstructions, as shown in Figs. 5.294 and 5.295 [12].

With the Nesperennub case, the STL+ module of Mimics was also used to export the skull data as an STL file. The availability of a high-quality STL file enabled Caroline to attempt a digital facial reconstruction using a sophisticated virtual sculpture system [13].

The Nesperennub CT data also showed a number of other articles of interest that had been wrapped up with the mummy. These included objects found in some other mummies of the period, a snake amulet on the forehead

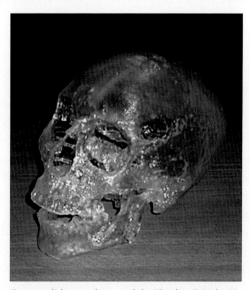

Figure 5.294 "Jeni" stereolithography model. *(Credit: British Museum © 2004 The Trustees of the British Museum.)*

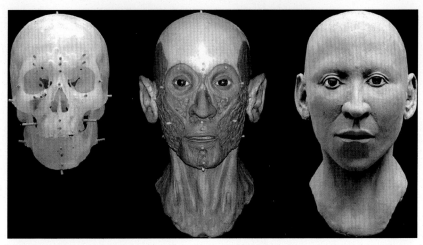

Figure 5.295 The stages of facial reconstruction on the Nesperennub stereo-lithography model. *(Courtesy of the Trustees of the British Museum and Prof. Caroline Wilkinson, University of Manchester.)*

and artificial eyes in the sockets. Uniquely, Nesperennub also appeared to have a bowl on top of his head (see Fig. 5.296). The British Museum was very keen to have replicas of these objects for use in the exhibition.

Mimics was used to segment the data describing these objects so that they could be made using rapid prototyping techniques. The artificial eyes and snake amulet appeared to have high densities and were therefore

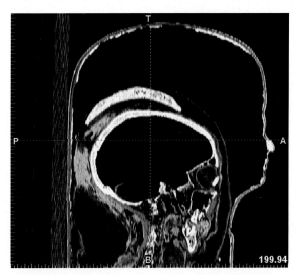

Figure 5.296 A sagittal CT image of Nesperennub. *(Credit: British Museum © 2004 The Trustees of the British Museum.)*

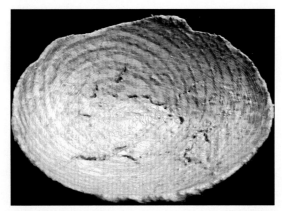

Figure 5.297 The laminated object manufacture model of the bowl. *(Credit: British Museum © 2004 The Trustees of the British Museum.)*

relatively easy to threshold and segment from their surroundings. The bowl, however, proved more difficult as the density was quite high but similar to bone and in close contact with the head in some areas. The bowl was therefore segmented using the same approach used for the skulls. The bowl was of particular interest as it had never been seen in a mummy before, and nobody was quite sure what its function was.

The STL+ module of Mimics was used to generate STL files of the objects that were made by RP. The snake amulet and eyes were made using stereolithography, and the bowl was made using laminated object manufacture (see Fig. 5.297). The model of the bowl was used to help museum staff to identify the object and speculate how it came to be in the mummy's wrappings. Handling and inspecting the bowl led them to the conclusion that it was a simple unfired clay bowl that was probably used to hold the resin used in the mummification process. It had probably become glued to Nesperennub's head during the mummification process by accident and was simply covered up to hide the mistake. The bowl model was also used in a film reconstructing the mummification process, which is shown as part of the exhibition. The amulet, eyes, and bowl are now on display in the exhibition.

5.43.4 Conclusions

"The exhibition at the British Museum has proved a great success, attracting 388,000 visitors. Since the mummy has never been unwrapped, the models of the skull, bowl and amulets are key elements in the display, providing an essential complement to the non-invasive images obtained from CT scans. The models have also been used successfully at the Museum in handling sessions for visually impaired visitors."

Quote by Dr. John Taylor, Assistant Keeper at the Department of Ancient Egypt and Sudan, The British Museum.

References

[1] Harwood-Nash DCF. Computed tomography of ancient Egyptian mummies. Journal of Computer Assisted Tomography 1979;3:768—73.
[2] Marx M, D'Auria HD. Three-dimensional CT reconstruction of an ancient Egyptian mummy. American Journal of Radiology 1998;150:147—9.
[3] Pickering RB, Conces DJ, Braunstein EM, Yurco F. Three-dimensional computed tomography of the mummy Wenuhotep. American Journal of Roentgenology 1986;146:93—6.
[4] Vahey T, Brown D. Comely Wenuhotep: computed tomography of an Egyptian mummy. Journal of Computer Assisted Tomography 1984;8:992—7.
[5] Dawson WR, Gray PHK. Catalogue of Egyptian antiquities in the British Museum: 1 mummies and human remains. London: British Museum Press; 1968. p. 8.
[6] Baldock C, Hughes SW, Whittaker DK, Taylor J, Davis R, Spencer AJ, et al. 3-D Reconstruction of an ancient Egyptian mummy using x-ray computed tomography. Journal of the Royal Society of Medicine 1994;87:806—8.
[7] Foster GS, Connolly JE, Wang JZ, Teeter E, Mengoni PM. Evaluation of an Ancient Egyptian mummy with spiral CT and three-dimensional reconstructions self directed display on the World Wide Web. Radiological Society of North America/Radio-Graphics Archives, Available on the Internet at: http://ej.rsna.org/, accessed June 1999.
[8] Jacobs PF. Stereolithography and other RP&M technologies. Dearborn, MI 48121, USA: Society of Manufacturing Engineering; 1996.
[9] Nedden D, Knapp R, Wicke K, Judmaier W, Murphy WA, Seidler H, et al. Skull of a 5,300 year old mummy: reproduction and investigation with CT guided stereo-lithography. Radiology 1994;193:269—72.
[10] Seidler H, Falk D, Stringer C, Wilfing H, Muller GB, zur Nedden D, et al. A comparative study of stereolithographically modelled skulls of Petralona and Broken Hill: implications for future studies of middle Pleistocene hominid evolution. Journal of Human Evolution 1997;33(6):691—703.
[11] Hjalgrim H, Lynnerup N, Liversage M, Rosenklint A, Stereolithography. Potential applications in anthropological studies. American Journal of Physical Anthropology 1995;97:329—33.
[12] Wilkinson CM. Forensic facial reconstruction. Cambridge, UK: Cambridge University Press; 2004.
[13] Wilkinson CM. Virtual Sculpture as a method of computerised facial reconstruction. In: Proceedings of the 1st International Conference on reconstruction of soft facial parts, Potsdam, Germany; 2003. p. 59—63.

Further reading

[1] Taylor JH. Mummy: the inside story, vol. 46. Bloomsbury Street, London: The British Museum Press; 2004. 0-7141-1962-8.

CHAPTER 5.44

Research applications case study 5—Trauma simulation of massive lower limb/pelvic injury[a]

5.44.1 Introduction

High-fidelity training is a crucial part of skill development for trauma and orthopedic surgeons, yet the tools currently available are frequently poor representations of what is experienced in real-life cases. There are currently four primary training tools available: artificial anatomic models, cadavers, animal models, and clinical cases. Each of these options is flawed.

Artificial mannequin models are extensively used in both undergraduate and postgraduate medical training for fundamental procedures such as intubation, cannulation, and blood gas sampling. These mannequins are characterized by their single function capabilities. The synthetic bones currently available for teaching and training in fracture surgery incorporate very poor quality simulated soft tissue envelopes, which do not reflect human tissue either in terms of anatomic detail or handling characteristics. While some simulation mannequins are anatomically accurate, they are typically limited in terms of the body area reproduced. However, although their tissue-handling properties are far from perfect, they do offer distinct advantages over alternative means for delivering similar trauma training.

---

[a] The work described in this section was undertaken as part of a project led by Professor Ian Pallister (Program Director for the MSc Trauma Surgery (Civilian and Military) Course at the College of Medicine, Swansea University), in collaboration with Professor Mark Waters of MBI (Wales) Ltd and Dr. Dominic Eggbeer at PDR. The project was funded via the Center for Defense Enterprise (project number CDE33698) and was in response to the call "THE MEDIC OF THE FUTURE Challenge 1: SimTraining." Much of the introduction text in this section was drafted by Prof. Pallister as part of the project proposal and has been edited for inclusion in this book.

Medical Modeling
ISBN 978-0-323-95733-5
https://doi.org/10.1016/B978-0-323-95733-5.00049-1

Indeed, the alternative means of simulating such injuries are extremely limited. While animal models may enable simulation in the presence of a live circulation, both the very distinct anatomic differences between animals and humans and the complexities of animal welfare mean that such simulations are both limited in terms of realism and availability. Human cadaveric material, whether soft embalmed or unembalmed tissue is used, may satisfy some of the anatomic shortcomings of animal models, but it is far from perfect. The frail elderly, from whom the material often originates, invariably have suffered major illnesses or have undergone a range of medical procedures, which may then prevent trainees from executing the techniques required. The bone and soft tissues are often involuted and are of very limited value in key procedures such as vascular control maneuvers, especially fasciotomy and damage control external fixation. In both the animal model and cadaver options, it is also very difficult to simulate, control, and repeatedly recreate different permutations of injuries.

This project aimed to create a prototype, realistic, complex trauma simulation system for injuries below the umbilicus reproducing massive lower limb/pelvic injury resulting from blast, gunshot wound, penetrating or blunt injury. It was proposed that the detail of the model and simulation system would be sufficient for complex wound/injury patterns to be reproduced involving major vessels, nerves, muscle groups/compartments, and the skeleton. This would help to reduce the requirements for current animal models and human cadaver materials and create a repeatable and realistic option for surgical training. Specifically, the simulation system was designed to reproduce severe lower limb and pelvic trauma replicating the types of injuries encountered in current conflicts including those typified by improvised explosive devices and gunshot wounds.

5.44.2 Materials and methods

The model at the center of the system was derived from existing computed tomography (CT) scans from a trauma patient who had consented to its use for research and development projects. This was crucial to obtain sufficiently detailed, anatomically correct information on which the simulation system could be based.

5.44.2.1 Step 1: Data segmentation

The CT data (1 mm slice thickness, 0.5 mm increments, and 0.424 mm pixel size on a Toshiba Aquilion Multislice scanner) were imported into

Mimics software (Mimics version 16, Materialise, Belgium). Mimics was used to segment and create separate "masks" of the bones, muscles, urinary tract, bowel, and major blood vessels of the below trunk-level anatomy using a mix of automatic and manual techniques described in Chapter 3. Fig. 5.298 shows 3D renders of the anatomic structures segmented. These structures were exported as high-quality STL files for further modeling. Since the patient's right lower leg anatomy was affected by the surgical

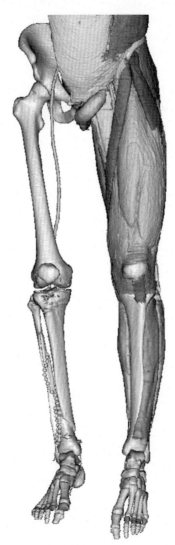

Figure 5.298 The segmented anatomic structures in Mimics.

procedure, it was agreed that their left leg would be mirrored to create the necessary symmetry in subsequent stages.

5.44.2.2 Step 2: Component and mold design

The STL data were imported into FreeForm (3D–Systems) for further modeling, smoothing, and refinement. Each component of the system was individually tailored to fit as an assembly. Lifelike features, such as textures, were added using "emboss area" and "emboss with image" tools to the muscle tissues, bladder, and penis to improve the realism. To avoid duplication of work and avoid the areas affected by the patient's injuries, structures including the muscles and bones of the left leg were mirrored across the midsagittal plane. Remodeling of the skin was required to remove any obvious indications of the mirroring procedure, and a version replicating trauma injuries to the right leg was also created. The completed model schematic is shown in Figs. 5.299a and 5.299b.

FreeForm was also used to undertake mold tool design of components including the outer skin, bladder/penis, and colon/rectum. The outer skin model was segmented axially into sections representing crucial points at which artificial injuries would be reproduced and based on the maximum build volume of an available stereolithography machine (500 × 500 × 500 mm). The outer skin mold was created by offsetting the model to the outside by 4 mm, identifying a split line that avoided significant undercuts and creating NURB surfaces that acted as splitting planes and the basis on which to design flange features. Inserts for the groin and buttock areas were also created to enable the creation of different simulated injuries.

Figure 5.299a Schematic of the model in FreeForm (see color section).

Figure 5.299b FreeForm rendering showing the modeled blast injury (see color sections).

Figure 5.300 The outer skin pelvis region mold tool design in FreeForm.

Flanges for the torso end and mid-thigh were designed to include locating features to hold the internal structures in place, and a separate a version of the right flange that simulated the effects of a blast injury was also created. The completed mold design for the pelvic region is shown in Fig. 5.300.

5.44.2.3 Step 3: Component fabrication

The bone and muscle components were shelled (to reduce the volume of material), split with a dovetail join line, and fabricated using stereo-lithography (SLA 250—50, 3D-Systems). Once completed, the parts were assembled, filled with a pourable, two-part polyurethane, hand finished, and used as master patterns to create flexible silicone tooling for vacuum casting production in tissue-mimicking materials. The circulatory system was split into sections before being fabricated using a ProJet 3000 Plus HD (3D-Systems) machine to create patterns for dip molding.

In parallel to the computer-aided design (CAD) process, materials that mimicked the physical handling and visual characteristics of human tissue were developed in silicone and polyurethane. This included mimics of superficial skin, fat, veins/arteries, muscle, facia, and bone. The system was also designed to incorporate an anatomically correct, pulsatile circulatory system driven using a peristaltic pump.

Final stage component fabrication required a mixture of vacuum casting, bench pouring, dip molding, and low-pressure injection processes depending on the tissue types. The silicone tools created from the bone and muscle master patterns were used to produce vacuum cast parts in the mimic materials. The sections of arteries and veins were used as tools for dip molding in a silicone material. The sections were then assembled as a complete circulatory tree that could be attached to a peristaltic pump. The outer skin mold tool sections were fabricated using stereolithography (SLA 500, 3D-Systems) in a transparent (ClearVue, 3D-Systems) resin to allow observation and evaluation of the molding process. Additional flange strengthening components were fabricated using selective laser sintering (EOS GmbH). Once completed, the mold sections were manually hand finished using grit paper to smooth split lines and improve the ability to assemble the multiple components and flanges.

5.44.2.4 Step 4: Simulation model assembly and molding

The molding process involved numerous stages of tissue layering and material injection to ensure the internal structures remained in the correct anatomic position. Fig. 5.301 shows part of the way through the tissue layering process with the full lower left leg internal structures in place ready for molding the upper portions. The bones and muscles were supported by mounting features integrated into the outer mold sections. Fig. 5.302 shows the closed mold with the internal structures visible through the superior flange.

5.44.3 Discussion and conclusions

This project demonstrated the potential of utilizing patient CT data, advanced CAD techniques, and AM to produce surgical training models

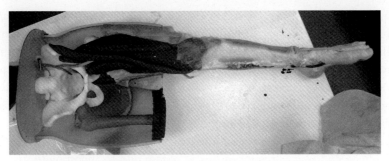

Figure 5.301 Layering the internal structures into the outer mold.

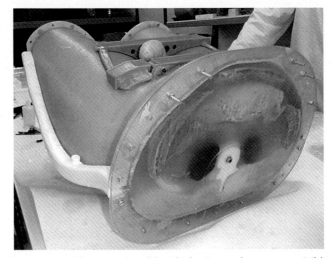

Figure 5.302 The closed mold with the internal structures visible.

with increased fidelity and realism over currently used alternatives. The prototype model developed included rendered and physical details such as specific wound/injury patterns encountered through combat situations. The model also included a dynamic circulation containing a finite volume of synthetic "blood." Active bleeding in the site of the simulated wound/injury allowed for the application of simulated emergency hemostatic techniques (novel dressings, tourniquet, etc.) with subsequent changes in the physiologic measurements being fed back to those participating in the simulation.

The ability to more closely replicate what trauma surgeons encounter is an important step in improving the quality of training. The benefit to the defense and wider medical community from this innovation will be in terms of trauma training and readiness in the event of future conflicts and daily medical emergency scenarios. Free from the constraints of animal welfare, the human tissue act, and the anatomy laws, the prototype systems provides the first platform for a more advanced simulation to be conducted in a range of environments, from simulated contact with the enemy for first responders to specific surgical procedures and, importantly, multiple-stage multidisciplinary resuscitation scenarios.

Given the prototype nature of the first model, further work is necessary to evaluate end user perceptions of the system and to incorporate suggestions to improve the subsequent designs. It is also necessary to refine the fabrication and assembly process to ensure batch/mass production is more economically viable.

CHAPTER 5.45

Research applications case study 6—biomodeling with bio-inspired soft materials[a]

5.45.1 Introduction

Bio-inspired, soft materials may be required for medical modeling for several reasons. Development of compliant prosthetics, surgical task training, and medical device development are some of the most common motives. In this case study, we look at some of the technical aspects behind these applications and reveal how palpable medical models can be created to mimic the human condition. This topic sometimes requires a deep understanding of interdisciplinary practice to be exploited successfully. To help explain the importance of each disciplinary role, some of the methods and materials required for their use are presented. We begin this with a brief look at their origins and historic use. We will explore some of the technologic precursors and related biomechanical dynamics that need to be understood to create synthetic, surrogate soft tissues. Finally, we will take a look at some examples to help illustrate their importance and discuss some of the precursors and barriers to innovation.

5.45.2 A brief history of soft materials in medical modeling

Since the discovery of vulcanization in the middle of the 19th century by tire manufacturers Charles Goodyear and John Dunlop, natural rubbers

---

[a] This case study was kindly provided by Richard Arm, Senior Research Fellow, Nottingham School of Art and Design, Nottingham Trent University, Nottingham, Nottinghamshire, UK.

Medical Modeling
ISBN 978-0-323-95733-5
https://doi.org/10.1016/B978-0-323-95733-5.00050-8

were sometimes used for rehabilitative and restorative maxillofacial and dental prosthesis, sometimes called epitheses. At around the same time, synthetic elastomers like silicones were being developed too, but their potential was not realized until the end of the 20th century. Meanwhile, during the 1940s, basic, vulcanized rubber prostheses were established to help casualties of battlefield trauma recover from the sociopsychological effects of facial disfigurement. By the 1960s, early silicone-based materials were introduced in some medical and cosmetic applications where analogous materials were needed to conform to or simulate the users' physiologies. At the end of the 20th century, silicone-based materials were better understood, more widely available, and became increasingly popular as an alternative to natural rubbers, especially in medical modeling applications. Silicones are used for implantable medical devices and facial prostheses because of their antibacterial properties, chemical stability, and compatibility with soft, human tissues. Within the last 10 years, modified silicone gels molded from additively manufactured (AM)/3D printed molds of human organs and anatomy have been used for teaching too. Most recently, realistic, patient-specific medical models of human organs have been created using silicone gels and fibers for training surgeons. However, the method of their fabrication remains reliant on post-AM production techniques rather than direct AM of organs with soft surrogate materials. Currently, use of soft materials by mainstream AM technologies, frustratingly, remains just out of reach.

5.45.3 Soft materials for medical modeling

The methods, materials, and machines that are adopted for AM medical modeling are often mature technologies borrowed from design and development disciplines like product design. In most product design applications, rigid materials are required to demonstrate an object's appearance or mechanical functionality. But when soft materials are required for medical modeling of soft tissues like organs, even the most advanced AM machines have their limitations.

Considering the huge range of materials that are available for AM manufacture, none are currently capable of recreating the range of soft tissues found in the human body. So, development of materials and machines available for AM of synthetic soft tissues has been the focus for many researchers over the past few decades, mostly without success. This is mainly because of the way most AM machines work. Objects are built layer by layer on a build plate, and this relies on several fundamental factors to achieve the desired outcome. For example, a layer is deposited or cured before another layer is applied, and an object's topology is built by gradually altering the shape of each new layer. Each layer is reliant on the stability of the previous layer to continue the build process, so materials that are rigid with no shear movements are essential, even when supporting scaffolds are used to secure the build during creation. Preferable performance characteristics like extreme softness, viscoelasticity, and slumping are likely to cause deformation in an object during the build process. Moreover, the variable hardness and multilayered nature of most organic soft tissues are simply too complex for existing AM technologies to process. Advances in the diversity of soft materials for AM, as well as the methods in which they are applied, are desperately needed. Most attempts to develop new flexible, elastic materials for AM usually rely on thermoplastic urethane-based elastomers (TPU). TPUs are useful for automotive and footwear product design applications but are not suitable for medical modeling of surrogate soft tissues.

Part of the issue with most extrusion-based AM machines is that the method of material deposition (layer by layer) creates planes of stiffness between each layer. Similar behavior is also seen in other laminated materials like plywood, evident especially when deforming, cutting, or machining. Like most laminated materials, the process of lamination increases rigidity even in flexible base materials. Another part of the problem is that TPUs, in particular, are not known for their native softness or elasticity. They typically have a relatively slow elastic response to deformations like bending, extension, and compression, especially when compared with cast silicones. Silicones generally offer a greater degree of behavioral adjustability with regard to their mechanical performance, which is well within the ranges needed for soft tissue biomimicry. The problem is that current AM technologies are not able to process silicones that are soft enough or with a smooth enough surface to replicate most soft tissues.

In summary, depositing and curing of synthetic elastomers, soft enough to replicate human tissues, are some of the main barriers to innovation of current medical modeling with AM technologies.

5.45.4 Organic additive manufacturing

A new area of medical modeling where soft materials are regularly used is the relatively new discipline of "bioprinting" or "biofabrication." Over the last decade, researchers have successfully created methods that exploit novel AM principles and technologies for the manufacture of soft, organic structures. Living cells suspended in organic materials like alginate and gelatine can be extruded by syringe, directly into a hydrogel or airgel block, which acts as a three-dimensional, soft containment support. However, these experiments are conducted mostly at a very small scale, often at the cellular level using living cells suspended in the deposited extrusion material.

Although still in its infancy, bioprinting offers some potential technologic solutions to the problems associated with AM of soft materials in the future, especially with respect to the use of hydrogels to mitigate problems associated with material rheology. But for now, they remain out of reach for most mass-produced AM machines and users.

5.45.5 Liquid silicone rubber additive manufacturing

Most soft elastomers like silicone require shaping in their liquid state prior to curing. So, they are usually prepared as a liquid and poured into a mold that has the desired shape. Vulcanization (heat curing of a single liquid) or polymerization (mixing two or more liquids together) are the two main processes by which silicone changes its state, from a liquid to a solid. But this can be especially problematic with fused deposition machines, since these rely on melting and cooling a solid filament, which is not possible with cured silicones.

New methods for curing silicone using photochemical reactions with UV (ultraviolet) lasers may offer novel solutions in the near future though. In essence, stereolithography machines use UV lasers to selectively cure layers of silicone in a liquid suspension to create an object with singular material properties. The main issue for developers is that the UV lasers used for curing the silicone weaken the resulting artifacts, robbing the silicone of its desirable mechanical properties.

5.45.6 Craft-enhanced additive manufacturing

Liquid silicone rubbers, or more accurately, polydimethylsiloxanes (PDMSs), are a group of elastomers commonly used in medical applications, like facial prostheses and medical implant fabrication. They are popular because they can mimic the mechanical behavior of soft tissues with some degree of accuracy, and they are safe to use on or inside the human body.

Unfortunately, they currently cannot be used in AM solutions because of their rheological and curing limitations. To get around this problem, craft-based skills can be employed to underpin the process of digital medical modeling.

Rigid, high-resolution AM models of anatomy can be transformed into soft duplicates using craft processes like molding and casting, rather than relying on AM methods to produce artifacts directly. For example, an organ can be scanned using computed tomography (CT) imaging, and the CT data can be segmented and manufactured using almost any AM materials and methods familiar to medical modelers. The next step is to produce molds or tools of the AM organ and cast it in a material that mimics the anthropomorphic qualities of the native tissue. These "qualities" can be classified by coloration, texture, and tactility. For most applications, tactility is one of the most important, and most difficult, qualities to emulate. This is where we must rely largely on the data given in the literature. Most often, crucial literature is written by scientists, well-versed in the principles of biomechanics, physics, and mathematics, which can be difficult to digest for the lay medical modeler. But, before we can unpack this essential information, we must highlight the methods and materials that are required to underpin the literature. Here, we build on current medical modeling practices already discussed in this book, using a step-by-step guide that can be followed to change a rigid model into a soft model.

In Table 5.29, steps 1 and 2 will look familiar to those accustomed with the principles discussed throughout this book. Steps 3 through 6 may be new to most readers, but they provide a useful overview of the transformative processes involved.

5.45.7 Characterization of analogous materials

Defining the behavioral properties of a given soft tissue remains one of the most important challenges when creating realistically palpable medical models from clinical data. Biomechanical studies offer useful insights into the behavior of soft tissues that can be used to help build realistic soft models. But to use information gathered by these scientists, we need to first understand transferability of the data and how they are gathered, in order to recreate the tests for comparable synthetic tissues.

To tease out some useful threads of information from such a huge field of research, consider the variety of contact and noncontact methods used and how they relate to your specific needs, capabilities, and equipment.

Table 5.29 Design and manufacturing steps.

Step 1

Data acquisition

Select appropriate CT or MRI dataset with the smallest possible slice increment.

Check for resolution and image quality.

Complete appropriate local ethics approval forms for anonymized data (this can often take several months, so leave plenty of time for this).

Step 2

Medical modeling

Follow the medical modeling standard procedures already laid out in other chapters of this book.

Create a separate digital model for each soft tissue type required.

Segment data alongside radiologist to verify digital models for accuracy.

AM/3D printing

3D print anatomic components using compatible materials.

Tip: Note that acrylic-based UV-cured polymers should be avoided as they will cause inhibition in the curing of most PDMS polymers.

Step 3

Determine soft material properties

Gather data from the literature to identify mechanical properties of anatomic components and select suitable synthetic surrogates.

Prioritize data sources that use test standards or methods that are repeatable with accessible equipment.

Several standards are useful for soft tissue characterization: ASTM D2240-15(2021), ISO 48−4:2018, BS/ISO 23529:2016, ASTMD412-16(2021), ISO 5893:2019/AMD 1:2020, BS/ISO 37:2017, and BS EN ISO 20932−2:2020. ASTM D2240-15(2021) is the only agreed standard transferable between soft tissues and soft elastomers.

Step 4

Define surrogate materials

Prepare a range of suitable surrogate materials and characterize their properties in alignment with data and test procedures identified in the literature.

Pair the surrogate materials with the data.

Validate surrogate soft materials with surgeons familiar with the defined soft tissues.

(Continued)

Table 5.29 Design and manufacturing steps.—cont'd

Step 5

Mold 3D printed parts
Mold anatomic components using soft tooling materials like firm PDMS.
If PDMS molds are used, release agents like petroleum jelly should be used to prevent bonding of PDMS casting agents.
Latex-based polymers should be avoided as they will cause inhibition in most PDMS polymers.
External details and topology of any organ can be recreated using almost any compatible RP material.
Internal details like blood vessels are best manufactured using water-soluble AM material like PVA filament or gypsum powder (prior to permanent fixing with cyanoacrylate glue).
Tip: When using dissolvable filament like PVA filament, it is preferable to print a hollow object to make removal easier. PVA models can be smoothed to remove surface striations with water and a soft brush. Broken PVA models are also easily repaired with water.

Step 6

Cast surrogate materials
Cast anatomic components using selected surrogate materials. Casting methods include pouring, dipping, brushing, and spraying.
Demold artifacts and assemble final model within 48 h of casting.
Bond separate components with thickened PDMS (added thixotropic agent), pinning components in place until fully cured.
Tip: If multiple colors/materials are used in the final model, it is good practice to coat the entire model in a thin layer of red-tinted PDMS to tie several colors together visually. It also adds strength to the final model before use.

Contact-based mechanical devices like durometers allow measurements to be taken directly from the surface of soft tissues, most useful for ex vivo studies on basic tissue tactility. This method is arguably the easiest and most accessible form of testing and is currently the only agreed method for characterization of both organic and inorganic materials.

Noncontact, acoustic, or optical imaging devices like ultrasound and magnetic elastography enable a deeper look into the soft tissue characteristics, which is useful for gathering internal physiologic data. However, the problem with some noncontact methods (aside from accessibility and cost of apparatus) is that they only work correctly on organic, water-based materials, not oil-based synthetic materials. So direct comparison of data with synthetic surrogate tissue is often impossible.

Table 5.30 Data gathering techniques used to characterize soft tissues.

Contact-based methods	Durometers and indentation, tensile testing, multiaxial/compression testing, swelling, inflating, suction, bending, spinning, torsion, and impact testing
Noncontact-based methods	Magnetic resonance elastography, ultrasound elastography/hardness/vibrometry, sonoelastography, quasistatic compression and strain imaging, resonance frequency shift, dynamic holography, acoustic radiation, aspiration, shear-wave dispersion, and nonempirical hounsfield densitometry
Noncontact, in silico mathematical methods	Finite element analysis, theoretical mathematical modeling, constitutive modeling

A list of techniques used to gather biomechanical data is shown in Table 5.30 to demonstrate the variety of methods available.

Some scientists use novel test devices to gather data, which is useful for their own purposes, but not much good to anyone else, as bespoke equipment is rarely described well enough to recreate the setup. Other investigators adopt devices or test standards borrowed from other disciplines like textile or plastics characterization.

The most useful of these tests is the tensile test (BS/ISO 37:2017). The tensile test standard was specifically designed for the characterization of elastic materials but has been employed successfully for soft tissue characterization since the 1950s. Despite its destructive nature, the data gathered from this test method can deliver directly comparable results not just for modulus but also stress/strain characteristics as well as ultimate tensile strength. Although useful, this test standard is not formally approved for soft tissue characterization though.

Fortunately, there is one tried and tested method that has been internationally agreed in test standard communities: the hardness by durometer test (ASTM D2240). This test can be used for the gathering of data for both organic soft tissues and soft synthetic elastomers alike. The current variant of this standard D2240-15(2021) uses simple, mechanically gathered data to build a profile of hardness, based on the "00" Shore hardness scale. Shore hardness characterization relies on several overlapping scales to measure a materials' resistance to indentation. It can be used to characterize almost any material using various available scales. The "00" scale is used to measure very soft materials and is related to the standard. The "A" scale is used to measure firmer soft materials. Table 5.31 illustrates the relationship between the "00" scale and the "A" scale.

Table 5.31 Relationship between the "00" scale and the "A" scale.

Shore hardness	Unstable gels	Soft gels	Firm gels	Hard elastomers
'000' Shore	5–30	30–80	NA	NA
'00' Shore	5	10–50	50–80	NA
'A' Shore	NA	5–10	10–30	30–70

Both hardness scales span a single order of magnitude from 0 to 100 with 0 being the softest and 100 being the firmest.

Standard tests use a bench-mounted device with an integrated spring-loaded, blunt pin-shaped indenter, a few millimeters in diameter, that measures the force required to penetrate a soft object. The standard sets out clear and repeatable instructions that enable the user to gather data on both soft tissues and soft elastomers using an identical apparatus to compile directly comparable results. Unlike other tests, D2240 has been designed to measure both materials using the same standard, useful for making direct comparisons between not just the materials but between researchers and institutes over time.

5.45.8 PDMS variants

Once a suitable test method has been selected, material selection should be the next focus of any investigation. Collaboration with material specialists and experienced surgeons is the best way to identify and inform selection here. PDMS is a popular choice for reasons already discussed, but due to its almost infinite variability, selecting the correct one for your specific application can be problematic for the inexperienced soft medical modeler. In this section, we define a selection of commonly used PDMS variants and additives used to alter some key mechanical and visual characteristics.

PDMS elastomers can be purchased, commercially, as two main variants. Both versions are two-part liquid materials that require mixing together to form the final elastomer. Condensation polymerization (CP), tin-based systems are used mostly in tooling and product design applications due to its low cost and wide availability. Addition polymerization (AP), platinum-based systems are used in most medical applications, especially when the material comes into direct or prolonged contact with the human body. This is because AP produces a much more stable compound that will not leech chemicals into adjacent surfaces over time. In contrast, the CP process works by ejecting molecules during the curing process, creating a less stable material that breaks down over time as it continues to lose molecules. These molecules can accumulate in organic tissues making them unsuitable for medical use.

Both CP and AP systems have a wide variety of uses and mechanical properties, but in general, all commercial systems are ordered and characterized by their hardness. For example, PlatSil Gel-00—30 has a native hardness of around 30 Shore 00, PlatSil Gel-10 has a native hardness of around 10 shore A, and PlatSil Gel-25 has a native hardness of 25 shore A. Usually, the hardness characteristics are included in the name of the material to make choice easier for the consumer.

Along with these base materials, there are three main additives often supplied by the same distributors, hardener, softener ("deadener"), and oil. Each additive offers the user further opportunity to fine-tune the mechanical behavior of their chosen PDMS. There are also other, less well-known additives that can be used to alter the mechanical and visual properties of PDMS elastomers, such as fiber and powder fillers. As a general rule, most liquid additives (except hardener) will thin the PDMS, lowering the hardness, while solid additives like powder or fiber fillers will thicken and harden it. Almost any dry filler can be mixed with PDMS, so experimentation is king when it comes to finding an additive that works best for your specific application.

It is also worth noting, that when experimenting with PDMS additives, only oil-based liquid additives are compatible; water-based materials are not miscible with PDMS. In contrast, anything with latex, tin, or sulfur will almost certainly inhibit most AP curing systems, preventing it from achieving its intended solid state.

Table 5.32 illustrates some of the most common methods and materials used to alter the mechanical properties of PDMS.

5.45.9 Practical applications

Loose, synthetic, short-strand fibers (polyester or nylon), commonly referred to as "flock," are often mixed into a soft PDMS gel to produce skin-like color/textures and are a popular choice when making clinical maxillofacial and theatrical prosthetics. Their ability to create a sense of depth in surrogate soft tissues is rooted in their intrinsic pigmentation qualities. In other words, the suspension of colored fibers in a colorless gel improves appearance of most synthetic soft tissues. In contrast to liquid pigments that rob the silicone of its translucent qualities, colored fibers can be mixed to achieve almost any pigmentation while retaining translucency and depth associated with organic tissues. But these fibers do more than just pigment the material, they also change the mechanical behavior to make the PDMS perform more like organic membranes such as skin or blood vessels. Like fiber-filled PDMS, skin contains fibers that respond to loading by helping to absorb strain. This effectively makes the PDMS change from being a linear, isotropic elastic material into an anisotropic viscoelastic

Table 5.32 PDMS softeners and hardeners.

Additive	Ratio of mixture	PDMS softeners	
		Effect on mechanical properties	Uses for medical modeling
PDMS oil	1 part PDMS: 1 part oil	• Slightly softens • Increases viscoelasticity • Oily surface • Lowers tear resistance	• Firmer surrogate tissues • Heart • Stomach • Smooth muscle • Subdermal fatty layer under surrogate skin
	1 part PDMS: 3 parts oil	• Softens but holds shape • Increases slumping • Increases viscoelasticity • Greasy, self-lubricating • Tears and stretches easily • Lowers elastic response	• Very soft surrogate organ • Liver • Kidneys • Spleen • Omental membranes • Fatty deposits • Blood clots • Pancreas
	1 part PDMS: 6 parts oil	• Softens and cannot hold shape • Slumps easily • Leaches oils over time • Very high viscoelasticity	
Prosthetic deadener (PD)	1 part PDMS: 1 part PD	• Softens but holds shape • Increased surface tackiness • Improves laminated adhesion • Slows elastic response • Increases viscoelasticity • Increased extensibility	• Firmer surrogate tissues • Heart • Stomach • Smooth muscle • Thin firmer membranes • Epidermis • Organ pleura
	1 part PDMS: 3 parts PD	• Further softens but holds shape • Very tacky surface • Permanent laminated adhesion • Very low elastic response • Greater viscoelasticity • Very extensible	• Softer surrogate tissues • Self-adhesive membranes • Dermis • Subdermal layers • Multilayered membranes • Facial tissues

PDMS hardeners

Short-strand loose fibers (SLF)	1 part PDMS: 0.01–0.04 parts SLF (1%–4% by total volumetric weight)	• Increase hardness (1% = 5 Sh00) • Increases tensile strength • Increases force decay • Increases anisotropic response • Increases viscoelasticity • Decreases elasticity	• Epidermis • Muscle • Heart and lungs • Blood vessels • Membranes • Multilayered membranes
Kaolinite powder ($Al_2Si_2O_5$)	1 part PDMS: 0.01–0.5 parts $Al_2Si_2O_5$	• Increase hardness up to 75 ShA • Increases tensile strength • Reduces surface friction • Decreases elasticity	• Arterial blood vessels • Cartilage • Tendons • Tracheal rings • Vocal cords
PDMS hardener (H)	1 part PDMS: 0.5 parts H	• Increases hardness by 15 ShA • Increases tensile strength • Decreases elasticity	• Arterial blood vessels • Cartilage • Tendons • Tracheal rings • Vocal cords

material, similar in functionality to many organic soft tissues, especially fibrous membranes. By increasing the amount of embedded fibers, users can learn to predict and control the mechanical response with almost any given PDMS gel.

The use of PDMS oil, dispersed in a liquid PDMS gel, is a rarely documented topic in the literature. Oil is especially useful for mimicking very soft organs that have a characteristic slumping behavior, like the spleen, kidneys, and liver. Beginning with a softer PDMS gel, like PlatSil Gel-10, oil can be added to lower the elastic modulus with some degree of accuracy. The more oil is added, the softer and more viscoelastic the material becomes. For example, use of a 1:3 ratio of PDMS gel to oil yields a material similar in hardness and modulus to that of a live human liver, according to data given in the literature [1].

Fig. 5.304 shows a liver biomodel made using a 1:3 ratio (PDMS gel to oil) to saturate a PDMS gel. The resulting mixture was poured into the liver mold with the premade vascular model embedded and allowed to cure overnight. The result was an accurate liver model with realistic hardness and

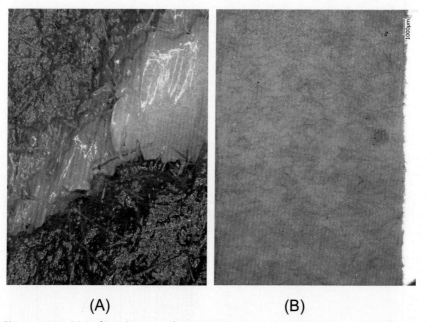

(A)   (B)

Figure 5.303 (a) (Left) A close-up of torn PDMS gel with embedded fibers. Each fiber is approximately 1 mm long. Showing how the addition of fibers lends a texture to PDMS gel. (b) (Right) A detailed view of the effect achieved with mixed pigmented fibers to create skin-like color/textures.

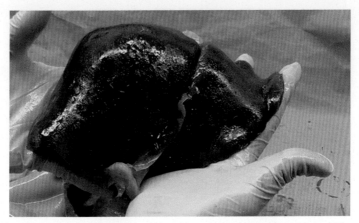

Figure 5.304 Soft biomodel of a cancerous, young human liver. Internal vascular details were produced using fiber-filled, PDMS gel-coated PVA cores. A mold was produced from the rigid AM liver model.

tactility with (synthetic) vascular system and cancerous tumor (Figs. 5.305 and 5.306). Each element had a hardness that had been informed by contact-based mechanical data of human livers from the literature, as well as hardness data gathered from a pig liver. The final model was also validated by a consultant liver surgeon before use.

Table 5.32 highlights some applications for adulterated PDMS, in medical modeling, but in reality, many of the soft tissues and surrogate materials required to replicate them remain elusive. Certainly, the materials and methods that have been discussed here show their potential and occasional employment, but materials that truly emulate soft human tissues still require a great deal of investigation. Work done by small pockets of researchers around the world has helped grow our current understanding of medical modeling with soft materials, but this work has only just begun. Companies like *Syndaver, the Chamberlain Group, Simbodies*, and *Lifecast*, who all make commercial soft biomodels, have helped build a new industry where this knowledge can be applied. But research, development, and mechanical testing of new materials required to feed such businesses is not necessarily within their scope, capacity, or stakeholder interests. So, the onus is largely on research institutes to produce new materials and production techniques to sustain growth in this area.

Since the use of cadavers represents the gold standard in surgical education, some argue that synthetic surrogates that replicate every aspect of the living body are surplus to requirement. Unfortunately for medical

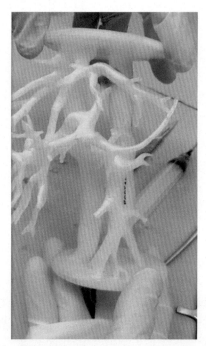

Figure 5.305 PVA, RP model of the livers' vascular system prior to coating with PDMS.

Figure 5.306 Cleaning the PDMS vascular system prior to embedding.

students though, every supplied cadavers' cause of death does not necessarily align with the topic on the curriculum each week. The truth is, that cadavers not only require expensive laboratories, staffing, storage, and ethical use and disposal of donor specimens, but they just are not as accessible as most people think. In the United Kingdom, for example, almost half of all medical schools do not even have a license to use human tissues for teaching on campus. Instead, they rely on plastic models, videos, and textbooks as well as visits to the cadaver lab a handful of times a year. The lucky few may have a simulation lab where they can practice on-screen procedures, but nothing can replace experiential learning. Only accurately constructed biomodels can do this. Medical modeling using soft surrogate tissues that look and feel real offer risk-free, inclusive learning alternatives that can be accessible for all medical students regardless of status or location.

Fig. 5.307 shows a soft biomodel made using the standard medical modeling methods. CT data were segmented and used to produce AM models of each organ, and molds were created of each component. Literature was used to inform the appropriate organ hardness, and samples were created and compared with feedback from surgeons. The final assembly of the thoracic organs was made using multilayered PDMS gels and fibers to pigment.

Fig. 5.308 shows a biomodel of a human trachea, made using multiple molds of CT-derived AM components. Each layer mimicked the related tissues provided in Table 5.32. Softeners in PDMS gel were used for the glands and membranes. Kaolinite was blended with PDMS hardener for the tracheal rings that were in the order of 75 Shore A hardness, in line with

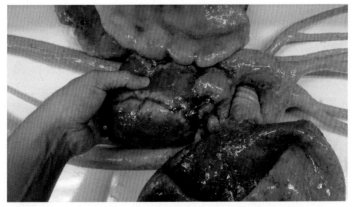

Figure 5.307 Soft biomodel made using the standard medical modeling methods.

Figure 5.308 A biomodel of a human trachea, made using multiple molds of CT-derived AM components.

data in the literature. Other layers were pigmented using only fibers to maintain translucency. Fig. 5.309 shows a view of the internal tracheal detail, including kaolinite-hardened vocal cords and fiber-colored upper throat membrane.

In summary, soft materials like PDMS have been shown to successfully recreate the human condition for a variety of purposes. Molds and cores can be created using CT data-derived rapid prototypes of anatomy. The molds can be used to create any assembly of complex anatomy, from the heart, lungs, and airways to more complex models of solid organs like the liver with internal blood vessels. All the models shown in this chapter were

Figure 5.309 A view of the internal tracheal detail.

created using the methods and materials discussed herein and were used to demonstrate the variety of potential applications soft medical modeling can offer.

Multilayered skin models, for example (Fig. 5.303), were used to train student surgeons how to close wounds, while the liver model (Fig. 5.304) was created to help experienced surgeons prepare for elective, patient-specific tumor removal. The heart and lungs assembly (Fig. 5.307) was integrated into a life-size manikin to train battlefield surgical teams how to treat emergency chest trauma. The trachea model (Fig. 5.308) was used by a medical product development team to help develop better medical devices.

5.45.10 Conclusion

The biggest barrier to innovation in this area is the rarity of interdisciplinary research groups. Academically diverse research groups are required to design, build, and test prototypes that are informed by data, as well as validation by experienced users. More teams of collaborating specialists with combined backgrounds in medical, biomechanical, and material sciences, electronic engineering, manufacturing, craft, and design are essential to encourage growth in this sector.

To explore some of the topics discussed in this chapter in more depth, you may be interested in reading these books and research papers.

References

[1] Estermann S, Pahr D, Reisinger A. Quantifying tactile properties of liver tissue, silicone elastomers and 3D printed polymer for manufacturing realistic organ models. Journal of the Mechanical Behavior of Biomedical Materials 2020;104:103630. https://doi.org/10.1016/J.JMBBM. 103630.

Further reading

Visual effects and prosthetics materials

[1] Debrecini T. Special makeup effects for stage and screen: making and applying prosthetics, vol. 6. Oxon, UK: Taylor and Francis Publishing; 2015. p. 228—9.
[2] Abdullah H, Abdul-Ameer F. Evaluation of some mechanical properties of a new silicone elastomer for maxillofacial prostheses after addition of intrinsic pigments. Saudi Dental Journal 2018;4:330—6.
[3] Montgomery P, Kiat-Amnuay S. Survey of currently used materials for fabrication of extraoral maxillofacial prostheses in North America, Europe, Asia, and Australia. Journal of Prosthodontics and American College of Prosthodont. 2010;19:482—90.
[4] Polytek Dev' Corp®. PlatSilGels technical bulletin polytek- soft, translucent, RTV liquid silicone rubbers for theatrical prosthetics, lifecasting & mold making applications. Technical bulletin. Mouldlife.net. 2017 [cited 4 February 2022]. Available from: https://www.mouldlife.net/ekmps/shops/mouldlife/resources/Other/gel-00-10-data.pdf.

Biomechanics and mechanical characterisation

[1] Fung Y. Biomechanics: mechanical properties of living tissues, vol. 7. New York, NY, USA: Springer; 1993. p. 259—63.
[2] Arm R, Shahidi A, Dias T. Mechanical behaviour of silicone membranes saturated with short strand, loose polyester fibres for prosthetic and rehabilitative surrogate skin applications. Materials 2019;12(22):3647. https://doi.org/10.3390/ma12223647.
[3] Arm R, Shahidi A, Clarke C, Alabraba E. Synthesis and characterisation of a cancerous liver for pre-surgical planning and training applications. Open Gastroenterology. British Medical Journal 2022. https://doi.org/10.1136/bmjgast-2022-000909.
[4] Shergold O, Fleck N, Radford D. The uniaxial stress versus strain response of pig skin and silicone rubber at low and high strain rates. International Journal of Impact Engineering 2006;9:1384—402. https://doi.org/10.1016/j.ijimpeng.2004.11.010.
[5] Jansen L, Rottier P. Some mechanical properties of human abdominal skin measured on excised strips; A study of their dependence on age and how they are Influenced by the Presence of Striae. Dermatologica 1958;117:65—83.

Test standards for the characterisation of soft medical models are:

[1] ASTM D2240-15 standard test method for rubber property- durometer hardness. West Conshohocken, PA: ASTM International; 2022 [cited 4 February 2022]. Available from: Astm.org. https://www.astm.org/d2240-15.html.

[2] ISO 48-4:2018 Rubber, vulcanized or thermoplastic- determination of hardness. Part 4: indentation hardness by durometer method. International Organization for Standardization; 2022 [cited 4 February 2022]. Available from: https://www.iso.org/standard/74969.html.

[3] ISO 23529:2016 General procedures for preparing and conditioning test pieces for physical test methods. International Organization for Standardization; 2016 [cited 4 February 2022]. Available from: https://www.iso.org/standard/70323.html.

[4] ASTM D412-16(2021), standard test methods for vulcanized rubber and thermoplastic elastomers- tension. West Conshohocken, PA: ASTM International; 2021 [cited 4 February 2022]. Available from: Astm.org. https://www.astm.org/d0412-16r21.html.

[5] ISO 5893:2019/AMD 1:2020. Rubber and plastics test equipment- tensile, flexural and compression types (constant rate of traverse). Specification amendment 1. International Organization for Standardization; 2020 [cited 4 February 2022]. Available from: https://www.iso.org/standard/80532.html.

[6] ISO 37:2017 Rubber, vulcanized or thermoplastic- determination of tensile stress–strain properties. International Organization for Standardization; 2017 [cited 4 February 2022]. Available from: https://www.iso.org/standard/68116.html.

[7] ISO 20932-2:2020 Determination of the elasticity of fabrics- part 2: multiaxial tests. International Organization for Standardization; 2020 [cited 4 February 2022]. Available from: https://www.iso.org/standard/69490.html.

CHAPTER 5.46

Research applications case study 7—Three-dimensional bone surrogates for assessing cement injection behavior in cancellous bone*

5.46.1 Introduction

Osteoporosis and other skeletal pathologies, such as spinal metastasis and multiple myeloma, compromise the structural integrity of the vertebra, thus increasing its fragility and susceptibility to fracture [1–3]. During vertebral augmentation procedures, bone cement is injected through a cannula into the cancellous bone of a fractured vertebra with the goal of relieving pain and restoring mechanical stability. However, prophylactic surgical stabilization is often performed to reinforce a structurally compromised vertebra and decrease its susceptibility to fracture [4,5]. The bone cements used are chemically complex, multicomponent, and significantly non-Newtonian with their viscosity having differing degrees of time and shear rate dependency. These cements interact with the porous structures though which they flow and with other fluids present within the porous media. The most widely used cement, polymethyl-methacrylate (PMMA), is generally assumed to be insoluble in any biofluid (bone marrow) it comes into contact with; thus the cement-marrow displacement is characterized as a two-phase immiscible flow in porous media [6,7]. As vertebral cancellous bone has highly complex geometric structures and displays architectural inhomogeneities over a range of length scales, the

* The work described here was undertaken by Antony Bou Francis, Richard M. Hall, and Nikil Kapur, School of Mechanical Engineering, University of Leeds, UK, in collaboration with PDR, Cardiff Metropolitan University, UK.

Medical Modeling
ISBN 978-0-323-95733-5
https://doi.org/10.1016/B978-0-323-95733-5.00051-X

pore-scale cement viscosity varies due to its nonlinear dependency on shear rates, which are affected by variations in the local tissue morphology. Furthermore, the vertebral cancellous bone microarchitecture varies among the patients being treated, thus making the scientific understanding of the cement flow behavior difficult in clinical or cadaveric studies [8,9]. Previous experimental studies on cement flow [8,10−13] have used open-porous aluminum foam to represent osteoporotic bone. Although the porosity was well controlled, the geometric structure of the foams was inherently unique. We propose a novel methodology using reproducible and pathologically representative three-dimensional bone surrogates to help study biomaterial—biofluid interaction, providing a clinical representation of cement flow distribution and a tool for validating computational simulations.

5.46.2 Methods

Three-dimensional bone surrogates were developed to mimic the human vertebral body. Fig. 5.310 shows the boundary of the 3D bone surrogates showing: (a) two identical and symmetrical elliptical openings 2 mm in

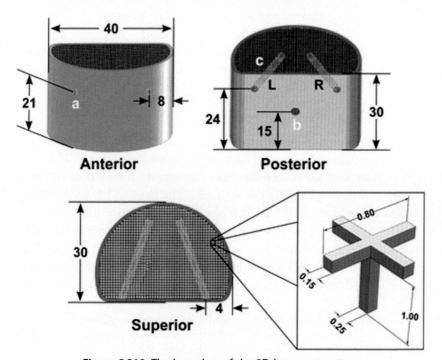

Figure 5.310 The boundary of the 3D bone surrogates.

height and 1 mm in width applied to mimic breaches due to anterior blood vessels, (b) one circular opening 3 mm in diameter applied to mimic posterior breaches due to the basivertebral veins, and (c) the insertion channels that were incorporated to allow consistent needle placement during injection. The superior and inferior surfaces of the surrogates were kept open due to manufacturing restrictions. All dimensions are in millimeters.

The surrogates were designed in SolidWorks (Dassault Systèmes, Vélizy, France) and then manufactured using rapid prototyping (ProJet HD 3000 Plus, 3D Systems, Rock Hill, South Carolina, USA). The structure of the surrogates was tailored to mimic three skeletal pathologies: osteoporosis (Osteo), spinal metastasis (Lesion), and multiple myeloma (MM). Fig. 5.311 illustrates the developed bone surrogates, and Table 5.33 describes the elements incorporated into each surrogate. Once the surrogates were manufactured, microCT (μCT 100, Scanco Medical, Switzerland) was used to assess the variability in their morphology. Eight of the Osteo surrogates were scanned at a spatial resolution of 24.6 μm (isotropic voxel size) with a 300-ms integration time, a 70-kV tube voltage, a 114-μA tube current, and a 0.1-mm aluminum filter. Then, a cylindrical volume of interest 15 mm in diameter and 15 mm in length was consistently defined at the center of each specimen. Within this volume of interest, three-dimensional morphometric indices were determined using the following settings: Sigma 1.2, Support 2.0, Threshold -120 HA mg/ccm (based on Ridler's method [1]), and proprietary software provided by the manufacturer. Only the bone volume fraction, BV/TV (%), trabecular thickness, Tb.Th (mm), and trabecular separation, Tb.Sp (mm), were compared. The porosity of the specimens was obtained from the microCT data (100, BV/TV) and validated using Archimedes' suspension method of measuring volume [2], which was performed using six cubes (2 cm^3 volume) with the same structure as the Osteo surrogates. One of the six cubes was also used to measure the permeability of the Osteo structure using Darcy's law [3,4]. Furthermore, static contact angle analysis (FTA 4000, First Ten Angstroms, Virginia, USA) was performed on the material to compare the surface wettability to that of cortical bone from a dry human femur.

5.46.3 Results

Fig. 5.312 shows an example of the cylindrical volume of interest that was consistently defined at the center of each 3D Osteo surrogate and used to obtain the BV/TV in %, Tb.Th in mm, and Tb.Sp in mm (see color

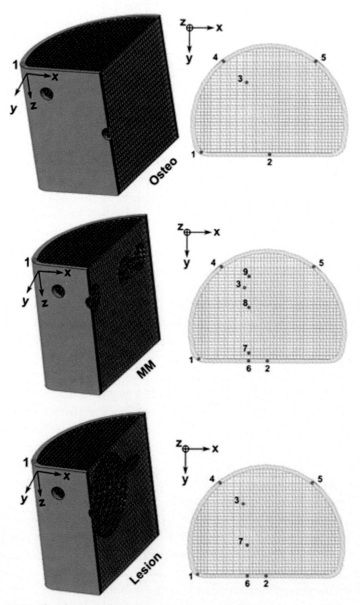

Figure 5.311 The developed 3D bone surrogates. (*Left*) Section view of the Osteo, Lesion, and MM surrogates. (*Right*) Craniocaudal view showing the location of all elements incorporated into each surrogate (refer to Table 5.33).

Table 5.33 The location and size of all elements incorporated into the 3D bone surrogates. All coordinates are measured with respect to the geometric center of each element.

Surrogate	Element	Coordinate			Description
		x	y	z	
Osteo, Lesion and MM	1	0.0	0.0	0.0	Reference point
	2	20.0	0.0	15.3	Outlet, circular Ø 3.0 mm
	3	13.5	−20.6	8.0	Inlet, circular Ø 2.1 mm
	4	8.2	−26.1	9.3	Outlet, elliptical width 1.0 and height 2.0 mm
	5	31.8	−26.1	9.3	Outlet, elliptical width 1.0 and height 2.0 mm
Lesion	6	14.3	0.0	10.0	Outlet, circular Ø 2.7 mm
	7	14.3	−8.9	10.0	Spherical void Ø 19.0 mm
MM	6	14.3	0.0	8.0	Outlet, circular Ø 2.6 mm
	7	14.3	−2.2	8.0	Spherical void Ø 6.0 mm
	8	14.3	−14.9	8.0	Spherical void Ø 6.0 mm
	9	14.3	−23.4	8.0	Spherical void Ø 6.0 mm

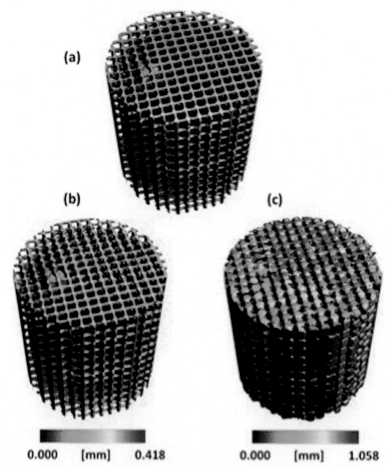

Figure 5.312 (a) Cylindrical volume of interest defined at the center of the osteoporotic type of 3D bone surrogates with its associated thickness map (b) and spacing map (c).

section). Fig. 5.313 shows the histograms associated with the respective thickness and spacing maps (see color section). The thickness map revealed three distinct peaks at approximately 0.18, 0.27, and 0.34 mm. The first two peaks were associated with the horizontal and vertical struts of the 3D bone surrogates, which had a nominal thickness of 0.15 and 0.25 mm, respectively. The third peak could be related to manufacturing artifacts associated with residual support material within the structure. On the other hand, the spacing map revealed one distinct peak at approximately 0.89 mm, which corresponded to the nominal spacings in the horizontal

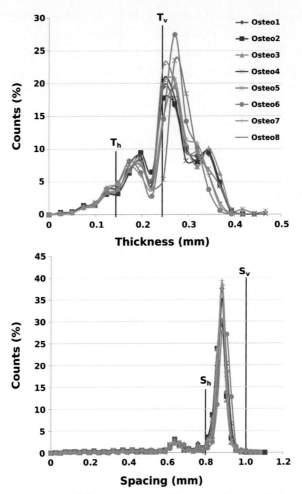

Figure 5.313 Histograms of the trabecular thickness (*top*) and trabecular separation (*bottom*) obtained from the CT data of eight osteoporotic type (Osteo) 3D bone surrogates. The black lines represent the nominal horizontal and vertical thickness (Th and Tv) and spacing (Sh and Sv), respectively.

(i.e., between vertical struts) and vertical (i.e., between horizontal struts) planes of the 3D surrogates, which were 0.8 and 1.0 mm, respectively. The variability in the morphology of the osteoporosis 3D bone surrogates was very low with an overall strut thickness (Tb.Th) of 0.25 ± 0.04 mm and an overall pore spacing (Tb.Sp) of 0.89 ± 0.03 mm. The average porosity that was obtained from the microCT data (100, BV/TV) and validated using Archimedes' principle was $82.6\% \pm 1.1\%$. The measured permeability of the Osteo structure was $57.1 \pm 6.1 \times 10^{-10}$ m^2. The surface wettability was comparable between all materials with contact angles ranging from 49 to 77 degrees for the material used to manufacture the bone surrogates and 60 to 75 degrees for bone to form a dry human femur.

5.46.4 Discussion

It is extremely important to control the surrogate morphology, as bone cement precursors are heterogeneous, especially their powder component, which varies in composition, size, and molecular weight of the prepolymerized polymer beads as well as the morphology of the radiopacifier particles. All these factors have a significant effect on the interaction between the liquid and the powder components during mixing and injection, resulting in different flow behavior for different cement formulations. The developed bone surrogates can be assumed to be constant in geometric structure, as the variability in their morphology was very low. Achieving constant geometric structure is crucial to reduce the variability, render the experiments reproducible, and shift the focus onto understanding the influence of cement properties on the injection behavior. The main disadvantage of the surrogates is that they have a uniform structure and do not simulate the highly complex geometric structures and architectural inhomogeneities of vertebral cancellous bone. However, this was necessary to simplify the representation of the bone morphology, facilitate the manufacturing process, and ensure the reliable reproduction of the surrogates. When complex structures are involved, more support material is required during the manufacturing process. Thus, it is more difficult to remove the support material without damaging the actual structure, especially when regions within the structure (i.e., vertebral cancellous bone) are thin and not well connected.

Although the structure of the surrogates was uniform, the pore spacing of the Osteo surrogates (0.89 ± 0.03 mm) was comparable to that reported in the literature for human vertebral cancellous bone [9–13]. The surrogates also simulate the rheological environment within the vertebral body. The measured permeability of the 3D bone surrogates ($57.1 \pm 6.1 \times 10^{-10}$ m^2) was comparable to that reported by Nauman et al. [14] for human vertebral cancellous bone, which was $80.5 \pm 47.5 \times 10^{-10}$ and $35.9 \pm 19.0 \times 10^{-10}$ m^2 in the longitudinal and transverse directions, respectively. Furthermore, based on contact angle measurements, the surface wettability of the 3D surrogates matches that of bone. Previous experimental studies on cement flow [5–8] have used open-porous aluminum foam to represent osteoporotic bone. Although the porosity of the foam was well controlled, its geometric structure was inherently random. The proposed bone surrogates overcome the limitations of previous materials as their geometric structure is well controlled and can be tailored to mimic the morphology of specific bone conditions at different skeletal sites in the body. Another advantage is that the developed 3D bone surrogates have a boundary to simulate the vertebral shell that confines the flow and controls the intravertebral pressure, significantly affecting the filling pattern [15]. The boundary, including the inlet and the flow exit points, was kept constant for all the surrogates. The openings in the boundary simulate breaches through the cortex due to a fracture, a lesion, and/or a blood vessel exchanging blood in and out of the vertebral body. This is important as such breaches create paths of least resistance, providing means for leakage into the surrounding structures.

References

[1] Ridler TW, Calvard S. Picture thresholding using an iterative selection method. Systems, man and cybernetics. IEEE Transactions on Systems, Man, and Cybernetics 1978;8(8):630–2.
[2] Hughes SW. Archimedes revisited: a faster, better, cheaper method of accurately measuring the volume of small objects. The Physical Educator 2005;40(5).
[3] Baroud G, et al. Experimental and theoretical investigation of directional permeability of human vertebral cancellous bone for cement infiltration. Journal of Biomechanics 2004;37(2):189–96.
[4] Widmer RP, Ferguson SJ. On the interrelationship of permeability and structural parameters of vertebral trabecular bone: a parametric computational study. Computer Methods in Biomechanics and Biomedical Engineering 2012;16(8):908–22.
[5] Baroud G, Crookshank M, Bohner M. High-viscosity cement significantly enhances uniformity of cement filling in vertebroplasty: an experimental model and study on cement leakage. Spine 2006;31(22):2562–8.
[6] Bohner M, et al. Theoretical and experimental model to describe the injection of a polymethylmethacrylate cement into a porous structure. Biomaterials 2003;24(16):2721–30.
[7] Loeffel M, et al. Vertebroplasty: experimental characterization of polymethylmethacrylate bone cement spreading as a function of viscosity, bone porosity, and flow rate. Spine 2008;33(12):1352–9.
[8] Mohamed R, et al. Cement filling control and bone marrow removal in vertebral body augmentation by unipedicular aspiration technique: an experimental study using leakage model. Spine 2010;35(3):353–60.
[9] Chen H, et al. Regional variations of vertebral trabecular bone microstructure with age and gender. Osteoporosis International 2008;19(10):1473–83.
[10] Gong H, et al. Regional variations in microstructural properties of vertebral trabeculae with structural groups. Spine 2006;31(1):24–32.

[11] Hildebrand T, et al. Direct three-dimensional morphometric analysis of human cancellous bone: microstructural data from spine, femur, iliac crest, and calcaneus. Journal of Bone and Mineral Research 1999;14(7):1167−74.

[12] Hulme PA, Boyd SK, Ferguson SJ. Regional variation in vertebral bone morphology and its contribution to vertebral fracture strength. Bone 2007;41(6):946−57.

[13] Lochmuller EM, et al. Does thoracic or lumbar spine bone architecture predict vertebral failure strength more accurately than density? Osteoporosis International 2008;19(4):537−45.

[14] Nauman EA, Fong KE, Keaveny TM. Dependence of intertrabecular permeability on flow direction and anatomic site. Annals of Biomedical Engineering 1999;27(4):517−24.

[15] Baroud G, et al. Effect of vertebral shell on injection pressure and intravertebral pressure in vertebroplasty. Spine 2004;30(1):68−74.

CHAPTER 5.47

Research applications case study 8—Full-color medical models—Worked examples*

5.47.1 Project background

Medical professionals are in constant need of better tools that help them to save lives. 3D printing can be used to improve patient care, reduce costs, and increase the speed of every step in the medical value chain. Positive medical outcomes are usually decided by several factors: well-briefed and prepared surgeons, efficient completion of the surgical procedure within the shortest possible time, and an understanding of patient-specific risks to avoid complications during surgery. All of these factors benefit greatly from using high-fidelity anatomic models made with 3D printing technologies.

As surgeons work in a real 3D world, it makes sense for them to employ patient-specific 3D printed models, particularly when complex pathologies are involved. This allows procedures to be refined during presurgery, meaning fewer complications, shorter procedures, and faster patient recovery times.

From magnetic resonance imaging (MRI) or computer tomography (CT) scan data, it is now possible to recreate highly accurate models of patient anatomy that allow surgeons to interact directly with the patient anatomy/pathology before entering the operating theater—Fig. 5.315. This helps them optimize their surgical plan, predict patient-specific issues, and convey to patients, and other medical professionals, exactly what is involved in a procedure. Detailed models can also be used for teaching anatomy—see Fig. 5.314.

---

* The work described here was undertaken by Olaf Diegel, Professor of Mechanical Engineering, Faculty of Engineering, University of Auckland, New Zealand.

Medical Modeling
ISBN 978-0-323-95733-5
https://doi.org/10.1016/B978-0-323-95733-5.00052-1

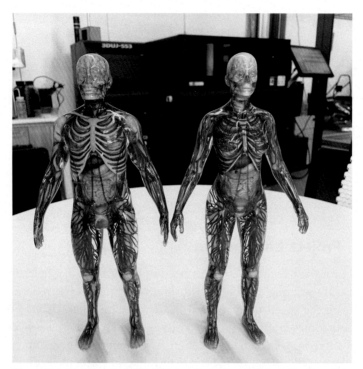

Figure 5.314 Example of highly detailed 3D printed medical models.

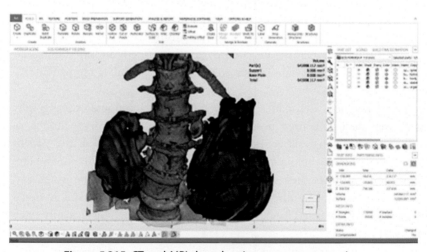

Figure 5.315 CT and MRI data showing tumor next to spine.

From this 3D medical data, it is now possible to 3D print highly accurate, full-color models that can transform these virtual models into real physical models. These 3D printed medical models are reminiscent of the internationally acclaimed work of Dr. Gunther von Hagens who pioneered the plastination of cadavers for educational display. Dr. von Hagens exhibitions have taken place all around the world with his plasticized models being shipped to international destinations.

The use of full-color 3D printing promotes a new way of demonstrating complex ideas and provides a wonderful opportunity to introduce tactile education and promote kinesthetic learning, such as the aorta model in Fig. 5.316. Unlike Dr. von Hagens' work, 3D printing is not limited by supply chain frameworks, and the use of synthetic materials enables students to interact with, and further investigate, the subject matter in new ways. The use of 3D printing for medical models is indicative of the incredible complexity of, and future potential for, 3D printing and its implications on the human experience in the form of bespoke wearable technology, product development, tissue engineering, and organ manufacturing.

5.47.2 How the project started

This project started with the arrival of a new Mimaki 3DUJ-553 color 3D printer at the Creative Design and Additive Manufacturing Lab (CDAM Lab), at the University of Auckland in New Zealand [1]. This 3D printer is

Figure 5.316 3D printed model of aorta with stent for patient communication.

a material jetting printer that inkjet prints a range of photocurable resins (including cyan, magenta, yellow, black, clear, white, and support resins) in a layer-upon-layer process. Lab staff started, as part of their "Creative Fridays" program, to experiment with the machine to test its limits and work out how far they could push the machine. In the CDAM Lab, "Creative Fridays" are used as an opportunity to let all staff minds explore creative ways to find new uses for the additive manufacturing technologies that the lab is all about. This resulted in art works such those shown in Fig. 5.317.

Lab staff started off making several "art" projects to understand how to combine the use of printing in clear material, together with printing in full-color translucent and solid material (cyan, magenta, yellow, black, and white). It was also necessary to understand how fine models could go in the computer-aided design (CAD) design before the printed version started to lose detail, or strength, and how to postprocess the model to achieve an acceptable level of transparency and translucency. In other words, lab staff needed to understand the technical details of what the technology could do and then push it to the very limit and beyond.

The Mimaki clear material, for example, was found to have a yellowish tinge. But, with the addition of 10% cyan and 10% magenta, it was found that the yellowness of the resin was, somewhat, reduced. It was also found

Figure 5.317 Examples of 3D printed trial sculptures to understand the limits of the technology.

that, although geometric features below 0.5 mm in size became hard to recognize by geometry alone, the addition of a colored texture could greatly enhance the perceived resolution of the model. A flat surface could, for example, have a brick texture with color and shadows applied to it and appear to become three-dimensional. However, it was also found that, as geometries became thinner than 0.5 mm, the intended solid colors became translucent. This is because this particular material jetting technology uses opaque white resin on the inside of the part, and only applies color to a thin layer on the outside surface. So, as the geometry becomes too thin, there is no longer enough white resin inside the feature to make it fully opaque.

5.47.3 Project vision

Once CDAM Lab staff understood the technology, a proof-of-concept application beyond art was needed, and because of the lab's extensive work with the medical community of prosthetists and experience with developing other medical products, lab staff consulted with their network of medical professionals and asked them how such a technology might be applied to help them. From these conversations, it was decided to try developing highly accurate medical models for surgical planning and education.

The CDAM Lab wanted these models to be the very best-in-class possible models, which conveyed as much useful detail as possible to the surgeon so that they could plan their surgical approach in the most detailed way possible. The goal was for the models to enable the medical professional to clearly see whatever organs or physiology that they needed to see for the purpose of their work.

5.47.4 Design approach

From the lab's work with understanding the limits of the full-color 3D printing technology, a set of minimum viability print criteria was developed to ensure that the model had recognizable and useful features, for example what the minimum thickness had to be, related to the overall scale of the model, before the detail was lost. The lab then developed the software methodology for rapidly being able to hide or show whatever organs or features were required from the medical CAD data and then saving a fully textured model into a printable file format.

The methodology involved two separate approaches. For educational and generic planning models, a library of 3D models was developed in which organs and bone were alphabetically listed and could be shown or hidden as needed. This was done in 3D Studio Max, a software package by Autodesk [2]. For custom medical models for surgical planning, a workflow was developed that allowed standard DICOM files, the most common file type used in MRI and CT scans, to be converted into 3D printable models. This was done using a software package by Oqton (previously 3D Systems) called D2P that allows different organs in an MRI slice to be identified, and extrapolated through subsequent model slices, to create 3D models of the organs and/or bones [3]. The results can be seen in Fig. 5.318.

As with any other additive manufacturing technology, this full-color 3D printing technology required substantial postprocessing after the print, so the CDAM Lab had to develop the most efficient and cost-effective methodology for achieving this. It involves a combination of manual cleaning, ultrasonic cleaning, media blasting, wet-and-dry sanding, and finally, clear coating. All this knowledge has been well documented within the CDAM Lab and can, of course, be used on all future 3D color printing or material jetting projects. Fig. 5.319 shows the support material and its removal. Fig. 5.320 shows the result of ultrasonic cleaning to remove remainder of support material and after sanding and clear coating.

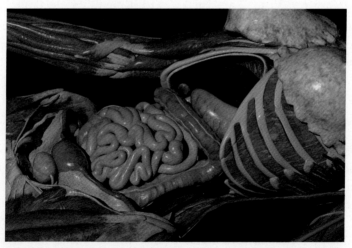

Figure 5.318 Example of CAD high-fidelity anatomic model with certain organs/muscles/systems hidden.

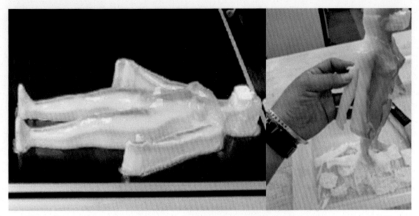

Figure 5.319 Model as printed and with the bulk of the support material being removed by hand.

Figure 5.320 Model after ultrasonic cleaning to remove remainder of support material and after sanding and clear coating.

5.47.5 Impact on the engineering and medical professions

The application and deeper understanding of full–color 3D printing and how to apply it in ways that add true value benefit all of engineering. But this case study of using this technology to produce high-fidelity anatomic models shows how it can also be of enormous benefit to the medical profession. It enables surgeons to use augmented 3D models to determine, and even practice performing, specific surgeries. These 3D printed models can display physical, spatial, and tactile information that computer models simply cannot provide. These can then also be used to assist with the planning of patient-specific molds and surgical guides to allow for more accurate cuts, while

Figure 5.321 Example of 3D printed full-color and clear skull.

implants can also be shaped to an exact replica of the anatomy prior to the surgical procedures. Additionally, these models can be used to aid more effective communication between the surgeon, their colleagues, and the patient. Fig. 5.321 shows a final full-color medical model.

5.47.6 Future work

The CDAM Lab is currently in conversation with the 3D printer manufacturer to assist them in developing flexible/soft clear and colored materials to allow the production of the next generation of medical models. With transparent, soft materials, systems will be able the produce models that will enable the surgeon to not only visualize the anatomy but also to physically cut into the model to practice their surgery before performing it for real.

Another challenge to be resolved, particularly for custom medical models, is in training imaging specialists to produce 3D scan data that is suitable for creating 3D printable models. Spacing between the slices, for example, should be as small as possible, whereas for conventional imaging files, a slightly larger slice spacing may be acceptable. Medical data is also often taken from two different imaging sources. Soft tissue, for example, may be taken from an MRI, whereas bone data may be more accurate from a CT scan. However, different imaging scan types are generally taken in different machines, which means that the patient is usually in a different position in each machine. A great deal of effort is therefore required to realign data from two different data sources. Effective communication with the imaging technician, and some training specific to 3D printable models, goes a long way toward resolving these challenges.

References

[1] Mimaki 3DUJ-553 Colour 3D Printer. Available from: https://mimaki.com/product/3d/3d-inkjet/3duj-553/.
[2] Autodesk 3D Studio Max. Available from: https://www.autodesk.co.uk/products/3ds-max/.
[3] D2P. Available from: https://oqton.com/d2p/.

CHAPTER 6

Future directions in medical modeling

6.1 Background

Reflecting on the myriad of applications detailed in the previous chapters, it is clear that digital fabrication techniques when applied to medical modeling will be a key enabler in the future of healthcare. Medical professionals have historically been accustomed to taking the complex 3D form of patients and to convert these into a series of 2D medical scan slices of the individual, much like how slicing software processes an STL file. It should therefore not come as a surprise that the medical profession became some of the first early adopters of the technology, creating generations of medical models to aid surgical planning in the early 1990s. In those early days, no one could have foreseen the opportunities computer-aided design (CAD) and additive manufacturing (AM) would afford the medical field.

More recently, CAD and AM technologies are being increasingly adopted by hospitals and supporting industries around the world to advance routine medical practises with a growing focus at the point of care (PoC). It was estimated by the company Materialise that in 2019 there were 268 hospitals globally making use of their segmentation software for medical modeling purposes, with North America, the United Kingdom, and Japan being the majority users. This has resulted in the medical AM sector being valued at approximately $1.45 billion in 2021 and with predictions forecasting this to rise to $6.21 billion by 2030, with the largest market being within North America and the fastest growing region as the Asia Pacific [1]. While the United States has been leading the medical modeling revolution, it is evident that the proliferation of activity is increasing becoming a global phenomenon. It is also speculated that as the global population increases and as increased standards of living are increasing life expectancies, such

Medical Modeling
ISBN 978-0-323-95733-5
https://doi.org/10.1016/B978-0-323-95733-5.00053-3

technology will increase in demand as we see a rise in the aging population, who are often the primary recipients of such technology.

Indeed, there is potential now for medical modeling to add clinical value from the point of initial contact with a patient, the development of specific treatment options, the development of the tools to implement these solutions, and in creating systems to aid patient recovery. Equally, due to the increased accessibility and maturity of CAD, AM technology, and processable materials, we have seen an explosion of research activity, which continues to push the boundaries toward the generation of clinical innovations. Traditional single-material medical models, with print resolutions of the order of millimeters, have now been superseded by the routine use of complex multimaterial medical models, printed in an array of colors to allow for enhanced visualization and at print resolutions down to tens of microns, reproducing the internal human form in unprecedented detail and accuracy. Patient-specific implants are now possible in an array of biocompatible metallic, ceramic, and polymer materials, with internal scaffold designs opening possibilities of overcoming stress-shielding issues, encouraging osseointegration, and tailoring of the weight of an implant to match that of human bone tissue. There has also been a recent explosion of upper and lower limb orthic devices, which are now tailored to the external topography of a patient's anatomy, providing a better fit but also integrating intricate design details allowing for lightweighting, moisture release, and even tailoring to the aesthetic preferences of the user, blurring the line between medical device and aesthetic expression.

It has been clearly demonstrated through the previous chapters that CAD and AM applied to medical modeling is an exciting field of research that is constantly evolving as we see increasing innovation and convergence across the core enabling technologies. Computational advances are unlocking greater capacity to increase medical model design details and complexity, advances in computational design allowing greater design possibilities and automation of routine design task, while maturity of AM technology is unlocking a wider array of material combinations and print resolutions allowing for unprecedented levels of anatomic reproduction. All such advances are now enabling an increasing range of potential applications, to a level of what was once seen as the realm of science fiction becoming a distinct reality, particularly in areas of regenerative medicine where the dream of rapidly reproducing fully functional human organs from a person's own cell lines are drawing ever closer to being science fact rather than fiction. It could be argued that the major ambition of medical model—based 3D printing is to be embedded as part of routine practice within a clinical setting, with all complementary technologies, materials, and workflows logically structured, with regulatory barriers resolved and approval procedures to minimize hurdles to use and all the while maximizing therapeutic benefit to patients.

6.2 Technological advances

For some time, the development of imaging, CAD, and AM systems seemed to have reached a plateau, with most development being incremental improvements on existing platforms. However, the past decade has seen a flurry of development across all enabling sectors of software, AM processes, process automation, and materials. There have been too many developments to cover them all here, but we provide the reader with a glimpse of some of the more exciting developments in the field.

Computer tomography (CT) scanning continues to be the imaging modality of choice when creating medical models and patient-specific devices, despite the superior contrast magnetic resonance imaging (MRI) provides for soft tissues. It is speculated that this may be due to most interest being focused on custom implants, for which CT scans provide excellent distinctions between bone and soft tissues. CT technology has seen some impressive recent developments. Researchers at the Mayo Clinic had developed the first photo-counting detector (PCD)-CT system [2], which bypassed the traditional two-step process of converting X-rays into light before a second conversion to the electrical signal that instrumentation converts into the CT image. The novel approach converts X-rays directly into the electrical signal, while they are counted and sorted based on their energy as they enter a detector. This approach allows for the system to differentiate different material types based on the X-ray energy, resulting in a 47% noise reduction and greater spatial resolution compared with conventional CT scanners. Additionally, it was found in early studies that the improved performance resulted in a reduction of 30% in the required dose of contrast agent, which could equally be equated to a reduction in radiation dosage during imaging. The next major advancement beginning to make a major impact introduces the use of artificial intelligence (AI) and machine learning methodologies for image processing. Using these techniques, researchers have demonstrated that superior image quality from a conventional CT scanner is possible at lower radiation doses making use of AI processing. Recent studies have demonstrated radiation dosage can be reduced by 50%—75% compared with standard imaging, while the ability of AI to conduct rapid calculations has revealed no disruptions to clinical workflows. Impressively, the software has received Food and Drug Administration (FDA) 510(k) approval.

6.3 4D printing: The next dimension in the healthcare

The ever-evolving landscape of AM has brought forth a groundbreaking innovation: 4D printing. Building upon the foundation of AM, 4D printing introduces the dimension of time, enabling objects to transform, adapt, and respond to external stimuli over time. This dynamic capability holds immense promise for a wide range of medical applications.

Current state of research: 4D printing in medicine is still in its infancy stages but has already shown remarkable potential [3,4], some of which is listed below.

Biocompatible materials: Researchers are developing biocompatible materials that are printable and respond to environmental factors such as temperature, pH, and moisture. They can be used for 4D printing of implants, drug delivery systems, and tissue scaffolds that adapt to the body's changing needs.

Drug delivery systems: 4D-printed drug delivery devices can release medications at specific times or in response to physiologic signals. This precise control could improve treatment efficacy and reduce side effects.

Patient-specific smart implants: Traditional implants, once implanted, remain static and cannot adapt to changing conditions. Smart implants, created through 4D printing, can change shape and morphology in response to specific conditions. For instance, in cases of severe tracheobronchomalacia disease in children, 4D printing has enabled the creation of implants that grow and transform within the body, providing a customized and dynamic solution.

4D bioprinting: It has enabled the construction of dynamic and responsive biologic structures like living tissues and organs layer by layer. Researchers are exploring the potential of 4D bioprinting for tissue engineering and regeneration. Stem cells and biomaterials are combined to create bio-inks that define the biomechanical properties of fabricated tissues/organs.

Soft robotics: 4D-printed soft robots can mimic the movements of biologic organisms. These robots can be used for minimally invasive surgeries, targeted drug delivery, and even assisting with rehabilitation.

Future potential: The future of 4D printing in medicine is exceptionally promising, with several areas offering tremendous potential for mainstream applications.

Customized orthopedic implants: One of the most exciting prospects is the development of 4D-printed orthopedic implants that can grow and adapt within the body. These implants could revolutionize orthopedic surgery by providing dynamic solutions that respond to the patient's changing needs. Imaging techniques like 4D CT scans and 4D MRI are expected to play a pivotal role in creating intelligent orthopedic implants that mimic natural tissues.

Precision drug delivery: 4D bioprinting is poised to transform drug delivery systems. Researchers are exploring the creation of drug-releasing scaffolds that respond to specific physiologic conditions, allowing for precise and controlled drug administration. Smart dressings with pH-responsive sensors and drug-releasing capabilities are being developed for wound diagnostics and treatment, offering simultaneous monitoring and therapy.

Tissue engineering and regeneration: 4D bioprinting will continue to advance the field of tissue engineering. Dynamic tissue scaffolds with the ability to self-repair and regenerate hold the potential to address various medical conditions and injuries. From skeletal muscles to cardiac and neural tissues, 4D-printed constructs are expected to play a vital role in creating functional and responsive biologic tissues.

Customized organs: Imagine 4D-printed organs and functional human tissues that can adapt to a patient's unique physiology. These organs could replace damaged tissue and offer a revolutionary solution to organ transplantation and regeneration challenges.

Medical devices: 4D printing has the potential to reshape the landscape of medical devices. From absorbable occluders that adapt to heat and magnetic stimuli to programmable culture substrates for neural tissue engineering, the range of applications is vast. Medical tools and devices that respond to changing conditions within the body will enhance patient care and treatment outcomes.

Smart materials and multistimuli responsiveness: The development of new smart materials that can respond to multiple stimuli will be a significant change. Materials capable of reshaping, changing color, and altering functions simultaneously will open new avenues for innovation in medical applications. Researchers are working on materials that mimic the dynamic properties of natural tissues, ensuring seamless integration within the human body.

Remote medical interventions: Telemedicine could benefit from 4D printing. Surgeons might remotely control 4D-printed instruments that adapt to the patient's anatomy, enabling complex procedures from a distance.

Despite the remarkable progress, there are challenges to overcome in the field of 4D printing for medical applications. Technologic limitations, including the need for multiphysics simulation and specialized software, must be addressed to optimize printing methods and structures. Materials research is crucial, with a focus on finding or developing smart biomaterials that can respond to diverse physiologic signals. Design challenges, particularly in controlling the speed and rate of action/reaction in biologic structures, call for innovative methods like the internet of things and AI. We believe that 4D printing will reshape the landscape of medical applications. As 4D printing technology matures, it holds the promise of transforming how we approach medical treatments and improving the quality of life for countless patients.

6.4 Clinical practicality at the point of care

Medical devices have the capacity to realize profound positive benefits to patients but equally can cause significant harm or mortality should their design, fabrication, or materials be compromised, which has resulted in legislation to regulate such devices. However, the greatest advantage of patient-specific devices also poses its greatest conundrum; specifically, how can we apply a classification on a product that will be different for each patient? The bespoke nature of such devices means either long-term studies must be produced for each product, which then negates the benefit to a patient who seeks timely treatment, or that regulations are devised to provided flexibility to create unique medical devices while the manufacturing process and general end device efficacy and risk are scrutinized relative to existing products, such as found with the FDA 510(k) approvals.

In 2016, the US FDA, recognizing both the potential of AM to improve patient outcomes in addition to the risks posed by the technology, issued a general set of guidances to the medical AM community on device design, manufacturing, and testing considerations [5]. The FDA classifies a given device based on the risk to a recipient, alongside the controls to provide assurance in the efficacy and safety of use and extends from class I to class III covering the use of imaging devices, software components, and to some degree materials. Class I devices (e.g., bandages, wheelchairs, etc.) are the lowest risk, and class III (e.g., pacemakers, cochlear implants, etc.) are of the highest risk due to providing life-supporting or sustaining use, so scrutiny increases with an increase in classification. The FDA has also provided an additional pathway for approval that included "custom-made devices" such as personalized implants, which can be made in an emergency or in exceptional use scenarios provided detailed safety requirements are met [6]. This exemption has showed great foresight of the FDA to balance the risks and benefits afforded by medical modeling approaches to device manufacturing. As such, the recent boom in patient-specific devices, facilitated by medical modeling, has only been made possible by such actions.

Looking forward to the future, the debate on regulation has continued, and the FDA has now begun a similar discussion on medical AM at the PoC. Translation of CAD and AM facilities to within a hospital is the logical progression for such technology and could realize a new paradigm of on-demand patient-specific device production. Consolidating all design, fabrication, postprocessing, and delivery in one place could result in increased efficiency and cost benefits to healthcare providers, but importantly, it could provide more timely care to patients, particularly those with immediately life-threatening/debilitating conditions. It also places the control of design, manufacture, and quality within the clinical environment, potentially enabling faster decision-making. Consequently, in 2021, the FDA invited another discussion of this topic with the wider medical AM community [7]. Regulatory discussions hope to now appreciate assuring PoC 3D printed devices are safe and efficacious and that protocols exist so product design and manufacturing approaches at the PoC meet exacting product specifications.

Currently, approaches are well defined for traditional manufacturing process at the PoC but less so when applied to 3D printing, so design and manufacturing for AM workflows need to be developed and validated. If fabrication were to be conducted within a hospital, it becomes unclear then who the "responsible entity" may be for routine tasks, such as the design, manufacturing, and testing; who is responsible for FDA approval submissions, quality control aspects, complaints handling, reporting of adverse events, and maintenance; or who audits and conducts corrective actions. All such tasks would require a team of highly skilled individuals to conduct them, and with healthcare providers often working with financial constraints, does subcontracting such activities become more cost effective and efficient? Equally, it is unclear if there is a one-size-fits-all approach given the wealth of different AM technologies, each of which has particular strengths and weaknesses in a variety of clinical applications, in addition to operational requirements. In particular, powder-based processes require substantial space, postprocessing requirements, and technical requirements, so they may be impractical for clinical use in their current forms.

Many researchers have also been reflecting on the paradox of how AM has been used to enhance patient care, yet little is known about the inherent risks of using the technology itself. Such questions would need to be answered should PoC AM become a reality. Recent research in this space has discovered that material extrusion AM processes, which allow for cost-effective production of medical models, have been found to release volatile organic compounds, nanoparticles, and wider contaminants, which expose users to possible risks of respiratory disorders, irritations, nervous systems alterations, and cardiovascular issues when using polymers such as acrylonitrile butadiene styrene or polycarbonate [8]. Currently, the evidence is inconclusive, and it has been suggested that exposure rates for individual printers are often below acceptable health and safety levels, while modern material extrusion systems are increasingly supplemented with the use of enclosures and HEPA filters to remove airborne contaminants, which further minimizes associated health risks. Despite this, several studies have linked polymer airborne contaminants to increased risks of asthma, COPD, and development of carcinomas. Indeed, these studies have raised particularly important questions about the appropriate operational conditions of 3D printers, both clinically and within shared spaces, alongside what the longer-term affects may be from continual airborne contaminant exposure.

It is clear there are no immediate and easy answer to PoC clinical AM and the regulatory challenges that this may introduce. There is also the possibility that current AM techniques may be eventually superseded by the next generation of bioprinting given the potential to fabricate with organic substrates, which would introduce a new gold standard in treatment. However, we forecast such advances are likely to still be the realm of research for some time to come, while applications such as medical models, resection guides, and orthotics will always mean there is a place for current AM approaches at the PoC.

6.5 Medical modeling for all

This book has detailed how medical modeling can be leveraged at all clinical phases. However, this vision of future healthcare has yet to address the reality that the recipients of both the enabling technology and the end solutions are located in affluent regions of the globe. While this scenario may be typical of "industrial revolutions" and wider political and socio-economic factors may be more largely influential, there has yet to be any meaningful debate on how we ensure equitable healthcare for all given the cost barriers to wider access.

However, it is interesting to see how the mass proliferation and ever decreasing cost for entry to make use of CAD and AM technology alongside a growing global open-source community (free redistribution of designs, software, hardware, and methods of use) have begun to address this shortfall. The most notable example is the activities of the group "Enable" to the wider benefit of all, and many challenges persist, limiting the mass proliferation of such technologies.

Despite these barriers, considerable progress is being made to lay the foundations for future use in clinical settings and to explore the possibilities that technologic advances are allowing, particularly in the area of increased anatomic realism to surpass the current cadaveric benchmarks.

6.6 Summary

We appear to be at an exciting junction in history, where a perfect convergence of key enabling technologies has surpassed the original expectation of what could be possible with medical device technologies, heralding a new paradigm of bespoke, patient-specific solutions. It is an exciting time to be working in the field of CAD and AM for use in medical modeling applications given the wealth of potential that has been realized and has yet to be unleashed. Working in this field can equally be highly rewarding, but also, it does come with its potential risks, which would be expected given the nature of some clinical uses to negatively impact a patient's wellbeing in an attempt to resolve an underlying condition, and we should never lose sight of this delicate balance custodians of this technology tread.

Given the mass proliferation of CAD and AM technologies, there is also a risk that the regulations will not keep pace with the rapid evolution of the software, hardware, materials, and designs and the active solutions that are implemented. Therefore, it is imperative that the community works collaboratively to ensure that we can help regulate the technology as quickly as we unlock the potential of what can be achieved. Equally, we should work collaboratively to address healthcare inequalities between developing and developed nations, to ensure that the remarkable innovations we are witnessing are available to all.

What will the future of medical modeling look like? Given the present advances, the answer may be closer to what we believe to be the realm of science fiction, but only time will tell! What we do know is that the future of healthcare will be very different from healthcare of today, and medical modeling will be a key enabling technology in advancing us into this exciting transition.

References

[1] Available from: https://www.precedenceresearch.com/3d-printing-in-healthcare-market.

[2] Available from: https://pubs.rsna.org/doi/epdf/10.1148/radiol.212579.

[3] Sahafnejad-Mohammadi I, Karamimoghadam M, Zolfagharian A, Akrami M, Bodaghi M. 4D printing technology in medical engineering: a narrative review. Journal of the Brazilian Society of Mechanical Sciences and Engineering 2022;44(6):233.

[4] Naniz MA, Askari M, Zolfagharian A, Naniz MA, Bodaghi M. 4D printing: a cutting-edge platform for biomedical applications. Biomedical Materials 2022;17(6):062001.

[5] Available from: https://www.fda.gov/medical-devices/3d-printing-medical-devices/fdas-role-3d-printing.

[6] Available from: https://www.sciencedirect.com/science/article/pii/S2589750019300676#bib10.

[7] Available from: https://www.fda.gov/media/154729/download.

[8] García-Gonzáleza H, López-Polab T. Health and safety in 3D printing. International Journal of Occupational and Environmental Safety 2022;6(1). https://doi.org/10.24840/2184-0954_006.001_0003.

CHAPTER 7

Glossary and explanatory notes

7.1 Glossary of terms and abbreviations

7.1.1 Technical terms and abbreviations

3D printing	Commonly used term for rapid prototyping/additive manufacturing
3D Systems Inc.	USA-based company producing a wide range of RP machines
AM	Additive manufacturing
Arcam	Electron beam melting machine manufacturer (now part of GE Additive)
Boolean	Mathematical operations such as union, subtraction, intersection, etc.
CAD/CAE/CAM	Computer-aided design/engineering/manufacture
CNC	Computer numeric control, typically of lathes and milling machines
Concept Laser	Manufacturer of laser melting machines (now part of GE Additive)
Connex	Previous name for Objet/Stratasys multiple-material jetting machines
Digital materials	Objet terms for technology-enabled mixed/multiple materials
Digitizing	Capturing data points from an object's surface, 3D scanning
DLP	Digital light processing (a digital micromirror device)
EnvisionTEC	Manufacturer of resin-based machines (now alled ETEC)
EOS GmbH	Electro Optical Systems, manufacturer of laser sintering machines
FDM	Fused deposition modeling, Stratasys term

(Continued)

Medical Modeling
ISBN 978-0-323-95733-5
https://doi.org/10.1016/B978-0-323-95733-5.00054-5

FEA	Finite element analysis, computer testing of strength or stiffness
Invesalius	Free, open-source software to segment structures in biomedical images
ITK–Snap	Free, open-source software to segment structures in biomedical images
LOM	Laminated object manufacture
LS	Laser sintering
Magics	STL manipulation and AM preparation software (Materialise N.V.)
Mimics	Software for processing medical scan data for AM (Materialise N.V.)
MJF	Multi-jet fusion (HP term)
MJM	Multi jet modeling (3D Systems term)
Objet	Machine manufacturer now incorporated into Stratasys
PEEK	Polyether ether ketone, thermoplastic
Photo-polymerizable	A liquid or resin that will solidify (cure) when exposed to certain wavelengths of light
Poly-jet modeling	Stratasys term for multiple jet printing process
QuickCast	SL build style for sacrificial investment casting patterns (3D Systems trademark)
Realizer	German manufacturer of metal-based laser melting machines
Renishaw	UK manufacturer of metal-based laser melting machines
Reverse engineering	To create computer model of an object, by digitizing or scanning it
RP	Rapid prototyping, general term for free-form build technologies
RP&M	Rapid prototyping and manufacturing
RP&T	Rapid prototyping and tooling
Sintering	Fusing a powder, usually by heat
SL	Stereolithography
SLA	Stereolithography apparatus (3D Systems trademark)
SLC	A slice format for three-dimensional computer data
SLI	A slice format for three-dimensional computer data
SLM	Selective laser melting
SLM Solutions	German manufacturer of metal-based laser melting machines (now Nikon SLM)

(Continued)

SLS	Selective laser sintering (3D Systems trademark)
Solidscape	Manufacturer of wax deposition type machines
STL	Derived from stereolithography, de facto standard computer file format
TCT	Time compression techniques (Technologies)
ThermoJet	Obsolete machine utilizing wax material (3D Systems trademark)
UV	Ultraviolet light
Vacuum casting	A method of producing plastic parts from a silicone mold from a master pattern

7.1.2 Medical terms and abbreviations

Alloplastic	Artificial material
Artifact	A distortion that does not reflect normal anatomy or pathology, used in radiology
Arthroplasty	The surgical repair of a joint, usually joint replacement
Autologous	Derived from an organism's own tissues or DNA
Benign	Noncancerous; treatment or removal is curative
Collimation	The operation of controlling a beam of radiation
Craniofacial	Relating to both the face and the head
Cranioplasty	Plastic surgery of the skull; a surgical correction of a skull defect
Hounsfield number	X-ray absorption coefficient of a pixel in CT (after the inventor of CT)
In vitro	In glass, meaning in the laboratory
In vivo	In life, meaning within the body
Malignant	Said of cancerous tumors, tending to become progressively worse
Maxillofacial	Pertaining to the jaws and face, particularly with reference to surgery
Morphology	The configuration or structure of
Occlusion	The relationship between the upper and lower teeth, the bite
Orthosis	Wearable medical device to immobilize or support a limb, hand, or foot
Osteoporosis	Reduction in the amount of bone mass, leading to fractures after minimal trauma

(Continued)

Osteotomy	The surgical cutting of a bone
Pathology	Concerned with disease, especially its structure and its functional effects on the body
Prosthesis	Artificial substitute for a missing body part, for functional or cosmetic reasons or both
Radiography	Making film records (radiographs) of internal body structures by passage of X-rays
Tibia	Bone in the lower leg, the shin bone
Valgus	An abnormal position in which part of a limb is twisted outward away from the midline
Varus	An abnormal position in which part of a limb is twisted inward toward the midline

7.2 Medical explanatory notes

7.2.1 Osseointegrated implants and implant-retained prostheses

Osseointegrated implants are titanium screws that are driven into patient's bones to form rigid and permanent fixation points for prostheses. The screw is driven into the bone in carefully selected positions during an operation. The screw is then typically left in place under the skin until the bone has healed and regrown around the screw. The screw is then exposed through the skin, and an abutment is added. This abutment then forms the anchor point for the rigid and strong fixation of prostheses, i.e., implant-retained prostheses. These types of implants are extensively used in dental restoration and increasingly in facial prosthetics.

For more information see

The Osseointegration Book: From Calvarium to Calcaneus Author(s)/ Editor(s): Brånemark, Per-Ingvar Quintessence Publishing, ISBN 0-86715-347-4.

7.2.2 Hypertrophic scar

A hypertrophic scar is a thick, raised scar resulting from skin injury such as burns. The formation of this kind of scar is not a part of normal wound healing and develops over time. They are more likely to be a problem in patients with a genetic tendency to scarring and in deep wounds that require a long time to heal. In general, they are more likely to form in areas of the body that are subject to significant pressure or movement. Treatment is typically by applied pressure for long periods.

7.2.3 Distraction osteogenesis

Distraction osteogenesis is a method of lengthening bones by cutting through the bone and then controlling a steady gap between the two sections of bone. If the gap is kept small but constant, the bones attempt to grow in order to fill the gap, as they would when healing from a fracture. The gap is typically maintained by mounting the two parts of the bone onto a precision screw mechanism. Adjusting the screw mechanism on a daily basis maintains the required gap. Typically the device is removed after the desired growth has occurred and has been allowed to heal fully. The technique can be used in three dimensions to correct any number of skeletal abnormalities and has been used successfully in orthopedic and maxillofacial surgery.

7.2.4 Benjamin's double osteotomy

The Benjamin double osteotomy was a double transverse cut, one cut through the distal femur and one through the proximal tibia. The rationale was that it would relieve the interosseous pressure underneath the subchondral bone, which was thought to accelerate the degeneration of the cartilage and sever the nerve in the subchondral bone plate, which was also thought to contribute to pain relief. A Charnley compression clamp was used to fix the osteotomy. The procedure developed a poor reputation because the condyles and tibial plateaux blood supplies were compromised, and occasionally bone necrosis led to above-knee amputations in some cases. In addition, a long rehabilitation period was required for a minimal benefit.

7.2.5 Tracheobronchomalacia (TBM)

TBM is a condition in which the airway collapses, caused by deterioration in the cartilaginous walls of the trachea and or bronchi leading to coughing, wheezing. and breathlessness:

https://www.cuh.nhs.uk/patient-information/tracheobronchomalacia-and-physiotherapy/

Index

Note: 'Page numbers followed by *f* indicate figures and *t* indicate tables.'

Printed in the United States
by Baker & Taylor Publisher Services